Short Rotation Woody Crop Production Systems for Ecosystem Services and Phytotechnologies

Short Rotation Woody Crop Production Systems for Ecosystem Services and Phytotechnologies

Special Issue Editors

Ronald S. Zalesny Jr.
William L. Headlee
Raju Y. Soolanayakanahally
Jim Richardson

MDPI • Basel • Beijing • Wuhan • Barcelona • Belgrade

MDPI

Special Issue Editors
Ronald S. Zalesny Jr.
USDA Forest Service
USA

William L. Headlee
Weyerhaeuser Co.
USA

Raju Y. Soolanayakanahally
Agriculture and Agri-Food Canada
Canada

Jim Richardson
Poplar and Willow Council of Canada
Canada

Editorial Office
MDPI
St. Alban-Anlage 66
4052 Basel, Switzerland

This is a reprint of articles from the Special Issue published online in the open access journal *Forests* (ISSN 1999-4907) from 2017 to 2019 (available at: https://www.mdpi.com/journal/forests/special_issues/woody_crop_production)

For citation purposes, cite each article independently as indicated on the article page online and as indicated below:

LastName, A.A.; LastName, B.B.; LastName, C.C. Article Title. *Journal Name* **Year**, *Article Number*, Page Range.

ISBN 978-3-03921-509-6 (Pbk)
ISBN 978-3-03921-510-2 (PDF)

Cover image courtesy of Ronald S. Zalesny Jr.
Industrial plantation of hybrid poplar grown in the Midwestern United States.

Contents

About the Special Issue Editors

Ronald S. Zalesny Jr. is a Supervisory Research Plant Geneticist at the USDA Forest Service, Northern Research Station. He specializes in bioenergy, biomass, ecophysiology, ecosystem services, forest genetics, intensive forestry, phytotechnologies, and short rotation woody crops.

William L. Headlee is a Forest Biometrician at Weyerhaeuser Co. He specializes in the effects of genotype, environment, and management on tree/forest growth, stand dynamics, and developing harvesting schedules for industrial forests.

Raju Y. Soolanayakanahally is a Research Scientist at the Agriculture and Agri-Food Canada, Saskatoon Research and Development Centre. He specializes in poplar and willow genetic improvement, agroforestry, ecophysiology, production of bioenergy crops on marginal lands, and phytoremediation.

Jim Richardson is Technical Director of the Poplar and Willow Council of Canada. He specializes in poplar and willow production, management and genetic improvement, agroforestry, bioenergy crops, ecosystem services, and silviculture.

Preface to "Short Rotation Woody Crop Production Systems for Ecosystem Services and Phytotechnologies"

While international efforts in the development of short rotation woody crops (SRWCs) have historically focused on the production of biomass for bioenergy, biofuels, and bioproducts, research and deployment over the past decade has expanded to include broader objectives of achieving multiple ecosystem services. In particular, silvicultural prescriptions developed for SRWCs have been refined to include woody crop production systems for environmental benefits such as carbon sequestration, water quality and quantity, and soil health. In addition, current systems have been expanded beyond traditional fiber production to other environmental technologies that incorporate SRWCs as vital components for phytotechnologies, urban afforestation, ecological restoration, and mine reclamation. In this *Forests* Special Issue titled *Short Rotation Woody Crop Production Systems for Ecosystem Services and Phytotechnologies* we explore the broad range of international current research dedicated to our topic. The issue can be found online at https://www.mdpi.com/journal/forests/special_issues/woody_crop_production.

There are a total of 18 papers in this book, reprinted from the online Special Issue. Ten of the papers are from the 2018 Woody Crops International Conference that was held from 22–27 July 2018 throughout Minnesota and Wisconsin, USA (these papers are indicated with an asterisk following the lead authors' names below). The remaining 8 manuscripts are open submissions from worldwide SRWC researchers. The first paper of the book is a Conference Report by Gardiner* et al. that provides a compilation of abstracts from each of the 38 oral and poster presentations delivered during the technical program. The conference included technical sessions framed under these topics: *Genetics and Physiology*; *Phytotechnologies*; *Stakeholders, Bioproducts, Harvesting and Logistics*; *Biomass Production*; and *Ecosystem Services*. To be consistent with this structure, the remaining 17 papers are organized according to these topic areas, with conference papers listed first and followed by open submissions.

The book includes one paper for Genetics and Physiology from Hu* and Thomas, who tested the hypothesis that intra-specific breeding of disparate populations of balsam poplar in Alberta and Québec, Canada, would lead to the expression of hybrid vigour. They also determined the role of endogenous hormones linked to ecophysiological and growth performance.

Three *Phytotechnologies* manuscripts are included. Zalesny et al. tested the survival, height growth, and phytoextraction potential of eight hybrid poplar clones (versus Russian olive) grown on soils categorized according to low, medium, and high initial salinity levels in North Dakota, USA. Rogers et al. evaluated how root–shoot allocations related to biomass production and the overall health of poplar and willow clones grown in soil treatments from landfills in southeastern Wisconsin, USA. Finally, Mosseler and Major assessed soil properties in relation to foliar concentration of nutrients and metals in 1-year-old coppice regrowth from willow clones grown at two adjacent coal mine spoil sites in New Brunswick, Canada. They quantified these growth differences for selection of superior clones for phytoremediation, biomass production, and land reclamation purposes.

Furthermore, there are 4 papers related to *Stakeholders, Bioproducts, Harvesting and Logistics*. Hart* et al. (including an erratum) conducted a qualitative case study and presented results of an exploratory feasibility investigation based on conversations with agricultural and natural resources stakeholders in southwestern Washington, USA, which complemented a techno-economic modelling of a hypothetical biorefinery near Centralia, Washington, USA. Rockwood* et al. described eucalyptus' potential for maximizing SRWC productivity through site amendment and genetic improvement in Florida, USA; they also documented eucalyptus' suitability for biochar production and assessed biochar's potential for improving soil properties, tree nutrition, and eucalyptus' growth. Therasme* et al. evaluated changes in the fuel quality (moisture, ash, and heating value) of stored spring harvested shrub willow and hybrid poplar chips with respect to pile protection treatments, location within the storage piles, and length of storage in northern New York, USA.

Biomass Production includes 8 papers. Niemczyk* et al. assessed the potential of tree improvement through hybridization to enhance aspen productivity in northern Poland, and they investigated the effects of hybridization on the growth and cellulosic pulp properties for papermaking purposes. Stanton* et al. evaluated the influence of gains in biomass yield and quality on investment in rapid propagation techniques that speed the time to commercial deployment of hybrid poplar in the Pacific Northwest, USA. Ghezehei* et al. documented biomass, biomass allocation, and wood properties of poplars in relation to growing conditions, physiography, and topography of three physiographic provinces in North Carolina, USA. In Estonia, Heinsoo and Tali evaluated shoot and root production of willow grown in hydroponic conditions to test whether this methodology was valid for predicting early biomass growth. Nissim et al. investigated the effects of genotypes, environments, and their interactions on the growth and biomass yield of willows in short-rotation coppice (SRC) under different harvesting cycles (i.e., 2 vs. 3 year rotations) in Québec, Canada. Truax et al. evaluated planting density and site effects on hybrid poplar productivity and stem dimensions at 8 and 14 years in southern Québec, Canada, as well as how density and site conditions affected biomass accumulation and carbon stocks in different plantation compartments, biomass partitioning at the stand-level, soil carbon stocks, and soil nutrient supply rate (after 14 years). Pray et al. tested whether current inoculation methods with mycorrhizal fungi, coupled with reduced fertilization, can demonstrate a growth benefit in SRC willows on marginal lands in Québec, Canada. Lastly, Tripathi et al. evaluated the potential of radiation use efficiency (RUE) and leaf area index (LAI) as indicators for bioenergy production and responses to changing environmental conditions for hybrid poplar, barley, wheat, maize, and oilseed rape in the Czech Republic.

The book includes one paper for *Ecosystem Services* from McIvor* and Desrochers, who investigated the early root development of willow grown from cuttings in different riverbank sediments (silt, sand, and stones) in New Zealand.

The aforementioned conference was designed as a forum to enhance information exchange while also building a platform for developing future collaboration around SRWC production systems. We believe this book will help move this information exchange forward, using key results to inform the development of SRWC technologies that enhance ecosystem services, regardless of their end use. Enjoy.

Ronald S. Zalesny Jr., William L. Headlee, Raju Y. Soolanayakanahally, Jim Richardson
Special Issue Editors

![forests logo] *forests*   MDPI

Conference Report

The 2018 Woody Crops International Conference, Rhinelander, Wisconsin, USA, 22–27 July 2018

Emile S. Gardiner [1,*], Solomon B. Ghezehei [2], William L. Headlee [3], Jim Richardson [4], Raju Y. Soolanayakanahally [5], Brian J. Stanton [6] and Ronald S. Zalesny Jr. [7]

[1] Center for Bottomland Hardwoods Research, Southern Research Station, USDA Forest Service, Stoneville, MS 38776, USA

[2] Department of Forestry and Environmental Resources, North Carolina State University, Raleigh, NC 27695, USA; sbghezeh@ncsu.edu

[3] Arkansas Forest Resource Center, University of Arkansas, Monticello, AR 71656, USA; headlee@uamont.edu

[4] Poplar and Willow Council of Canada, Ottawa, ON K1G 2C5, Canada; jrichardson@on.aibn.com

[5] Indian Head Research Farm, Agriculture and Agri-Food Canada, Indian Head, SK S0G 2K0, Canada; raju.soolanayakanahally@agr.gc.ca

[6] GreenWood Resources, Inc., Portland, OR 97201, USA; brian.stanton@gwrglobal.com

[7] Institute for Applied Ecosystem Studies, Northern Research Station, USDA Forest Service, Rhinelander, WI 54501, USA; rzalesny@fs.fed.us

* Correspondence: egardiner@fs.fed.us; Tel.: +1-662-336-4820

Received: 30 August 2018; Accepted: 2 November 2018; Published: 7 November 2018

Abstract: The 2018 Woody Crops International Conference was held from 22 to 27 July 2018 throughout Minnesota and Wisconsin, USA to unite world-leading short rotation woody crop (SRWC) organizations at a forum designed to enhance information exchange while also building a platform for developing future collaboration around SRWC production systems. The meeting included pre-conference and post-conference tours in Minnesota and Wisconsin and technical sessions in Rhinelander, Wisconsin. Technical sessions were framed under the topics: Genetics and Physiology, Phytotechnologies, Stakeholders, Bioproducts, Harvesting and Logistics, Biomass Production, and Ecosystem Services. This Conference Report provides a compilation of abstracts from each of the 38 oral and poster presentations delivered during the technical program. It should serve to enhance future discussions among scientists, academicians, regulators, and the general public relative to sustainable application of SRWC technologies for a multitude of current objectives.

Keywords: *Populus* L.; *Salix* L.; biomass production; bioenergy; tree breeding; ecosystem services; environmental technologies; phyto-technologies; ecological restoration; production systems

---

1. Preface

International efforts supporting the development of short rotation woody crops (SRWCs) have historically focused on the production of biomass for bioenergy, biofuels, and bioproducts while research and deployment over the past decade have expanded to include broader objectives of achieving multiple ecosystem services [1]. In particular, silvicultural prescriptions developed for SRWCs have been refined to include woody crop production systems for environmental benefits such as carbon sequestration, water quality and quantity, and soil health [1]. In addition, current systems have been expanded beyond traditional fiber production to other environmental technologies that incorporate SRWCs as vital components for phyto-technologies (e.g., phytoremediation), urban afforestation, ecological restoration, and mine reclamation [2].

The 2018 Woody Crops International Conference was held from 22 to 27 July 2018 throughout Minnesota and Wisconsin, USA, under the sponsorship of the Short Rotation Woody Crops Operations

Working Group, the Poplar and Willow Council of Canada, the International Union of Forest Research Organizations (IUFRO) Working Party 2.08.04 (Physiology and Genetics of Poplars and Willows), the IUFRO Working Party 1.03.00 (Short Rotation Forestry), the International Energy Agency Task 43 (Biomass Feedstocks for Energy Markets), and the International Poplar Commission Environmental and Ecosystem Services Working Party. The goal of the Conference was to unite these world-leading SRWC organizations at a forum designed to enhance information exchange while also building a platform for developing future collaboration around SRWC production systems. In particular, the event consisted of an optional pre-conference tour (22–23 July 2018) that showcased poplar (*Populus* spp.) tree improvement in Minnesota, a technical program (23–25 July 2018) held on the campus of Nicolet College in Rhinelander, Wisconsin, that delivered 38 presentations on current advancements in science and technologies and an optional post-conference tour (25–27 July 2018) that highlighted phyto-technology applications in Wisconsin. The technical program began with plenary presentations from four speakers and continued with five technical sessions and a poster session. Technical sessions were framed under the topics: Genetics and Physiology, Phyto-technologies, Stakeholders, Bioproducts, Harvesting and Logistics, Biomass Production, and Ecosystem Services. The conference attracted accomplished scientists, students, practitioners, business managers, and policy advocates from nine countries. This conference report provides an abstract of each of the 38 oral and poster presentations delivered during the technical program. The reader is referred to Appendix A for an index listing of each presentation title. We hope these efforts will enhance significant future discussions among scientists, academicians, regulators, and the general public.

2. Summary of Scientific Presentations

2.1. Current Trends and Challenges in North American Poplar Breeding

Thomas, B.R.

Background: Poplar (*Populus* spp.) breeding in North America, like everywhere, is dependent on funding and markets. The opportunities afforded by poplars and poplar breeding, however, remain far greater and more diverse than with conifer programs. As the needs have shifted, so have the selection traits, while the basics of any breeding, testing, selection, and production remain the same. Diverse parent populations are costly to maintain and programs are always at the mercy of changing priorities. In Canada, the number of active programs has diminished while the basic resources are still in place. Government programs have become the only programs able to maintain long-term genetic and landrace trials and only such programs have the luxury of conducting the testing that is required to fully determine if new clones are stable in a given set of environments. Examples of current poplar breeding and research are presented, which highlights the risks faced to maintain these programs. In the USA, the number of active poplar and willow (*Salix* spp.) breeding programs has also significantly decreased while those that are active have become integrated across much of the country. The regional nature of programs and poplar adaptability, however, make integrating programs challenging while the installation of plantations on private land ensures new hybrids with potential can be deployed and used to their maximum potential for energy, reclamation/restoration, phytoremediation, or fiber.

Conclusions: Poplar and willow programs are under threat and this primarily means having the funds and local interest to maintain live plant material in a clone bank or in situ genetics trial. Markets driving demand for poplar wood and fiber are also often cyclical. Parent material that is typically of exotic pure species often takes decades to acquire and bring to full flowering production for breeding and this material is always at risk due to reprioritization, diseases and insects, or simply neglect. Despite these challenges, poplars remain one of the most flexible timber species available for the industry (forestry and energy sector), communities, farmers, and individual homeowners. Collaborative efforts across North America and globally continue to allow us to benefit from the exchange of genetic material, institutional knowledge, and continued learning about the potential uses of this fiber including uses as varied as fuelwood from structural building materials to nanocrystals.

2.2. Investigation of Phytoremediation Potential of Poplar and Willow Clones in Serbia: A Review

Pilipović, A.; Orlović, S.; Nikolić, N.; Borišev, M.; Župunski, M.; Arsenov, D. and Kebert, M.

Background: The Institute of Lowland Forestry and Environment (ILFE) at the University of Novi Sad was established in 1958 as the Poplar Research Institute with the objective of implementing and improving forestry practices for the production of poplar (*Populus* spp.) and willow (*Salix* spp.) in the Socialist Federal Republic of Yugoslavia. Decades of research and international and national cooperation resulted in registration and implementation of 16 poplar and 4 willow clones exhibiting nearly 10-fold increases in biomass compared to 1958 productivity levels. Following contemporary worldwide trends in research and practice related to poplars and willows, which are collaborations since the early 2000s with colleagues from the Faculty of Sciences-Department of Biology and Ecology at the University of Novi Sad have focused on the investigation of phytoremediation potential of poplar and willow clones from the Institute's genepool. This resulted in numerous publications and methods. The objective of this presentation is to review 15 years of research results obtained at the University of Novi Sad, which are related to phytoremediation using poplars and willows.

Methods: Phytoremediation research has focused on fast-growing trees from two genera (*Populus* L. and *Salix* L.) and three different groups of contaminants (nitrates, heavy metals, and petroleum hydrocarbons). Poplar clones studied were primarily *Populus* × *canadensis* Moench, *Populus deltoides* Marsh. and intersectional hybrids while willow clones studied were primarily *Salix alba* L., *Salix viminalis* L., *Salix nigra* Marsh., and *Salix matsudana* Koidz. Research was conducted in the greenhouse with hydroponics and pot experiments and, to a lesser extent, in the field. Traditional growth variables, physiological variables, morpho-anatomical variables, biochemical markers, and degradation/extraction efficacy were investigated for various clones.

Results: Research results have shown broad variability in the response of clones relative to their phytoremediation potential. For example, growth responses ranged from high phytotoxicity to a lack of growth impacts in the presence of contaminants. Physiological processes were either suppressed or increased depending upon species and genotype. These responses are often the result of different metabolic and anatomical changes expressed as genotype × environment interactions.

Conclusions: Results obtained have led to the overall conclusion that there is extensive diversity among the clones in the Institute's gene banks. The best performance for phytoremediation of both heavy metals and organic compounds was shown for *P. deltoides* clones. The *P.* × *canadensis* clones most commonly used in Europe showed limitations for phyto-extraction due to their anatomical properties. In contrast, differences among willows were not driven by their species origin. Their potential for heavy metal phytoremediation, as expressed through high phyto-extraction, was even more pronounced than that of poplars.

2.3. Potential for the Agricultural Sector to Produce Poplar Wood as Contribution to the Forestry Wood Industry Chain

Van Acker, J.

Background: The potential of woody biomass production in both volume and quality is key for the future of the forest-based sector. The balance between wood material use and bioenergy use will inevitably lead to a higher competition for the same resource and could evolve into a critical shortage. Vertical integration alongside a better tree and wood quality concept should lead to a more structured approach for dealing with whether some wood products need to be prioritized and how we could deal with the substitution of man-made (building) materials that require more energy. Combining the production of low energy input wood products with the important option to produce green energy based on woody biomass is a major challenge for the future.

The wood resource obtained from the fast growing tree species of the genera *Populus* (poplars) and *Salix* (willows) is considered important in order to enable higher production in the future. Selection and breeding of these deciduous trees has long been a major part of silvicultural and even agricultural frameworks. Furthermore, in many ways, poplar trees can be considered having a potential as a prime alternative for softwood species use in engineered wood products. Applications related to biomass for energy and other end uses that do not require high tree quality should be part of an integrated approach as well.

For traditional products like plywood, but also for constructional timber, poplars are readily available. For example, aspen-OSB (Oriented Strand Board) has been an established product for decades in North America given specific strength and stiffness, which are surely interesting characteristics. However, the ability to select quality trees with a major impact on production yield are also an asset. Today, researchers are reassessing the potential of solid timber products using poplar wood. Dimensional stability and biological durability are improved by using modern wood modification methods besides traditional treatments. In this respect, both glulam and CLT (cross laminated timber) show major potential.

Conclusions: Short rotation woody crops are primarily intended for the production of biomass for energy but can also contribute in the future to an integrated forestry wood industry chain. Both the production of chips and small trees are options and different new production methods are emerging partly based on agroforestry. Among the new methods to produce woody biomass, options to produce chips as well as timber are being assessed in a wide range of agricultural systems. Some are indicated as polycyclic plantations while others are classified under short rotation agroforestry systems. These dual-purpose production systems could be an option to provide, in a sustainable way and based on mid-term strategic decision-making processes, lignocellulosic raw materials for material and energy use. The use of fast growing poplar for high end-products will also profit from developments in using hardwoods for construction and several wood modification options.

2.4. Reaching Economic Feasibility of Short Rotation Coppice (SRC) Plantations by Monetizing Ecosystem Services: Showcasing the Contribution of SRCs to Long Term Ragweed Control in the City of Osijek, Croatia

Kulišić, B.; Fištrek, Ž.; Gantner, R.; Ivezić, V.; Glavaš, H.; Dvoržak, D. and Pohajda, I.

Background: In Europe, biomass for the energy from Short Rotation Coppice (SRC) plantations is seldom feasible except in unique site-specific cases. If society placed priority and value on ecosystem services, then SRC plantations would increase in favorability for providing bioenergy feedstock. Showcasing examples of plantations providing multiple ecosystem services along with monetizing those services could prove important in moving public opinion. We report the case study of Osijek, which is a city of 100,000 inhabitants in Northern Croatia, where an interdisciplinary team of experts investigated planting SRCs for bioenergy as a long-term measure for control and eradication of an allergen and invasive plant, common ragweed (*Ambrosia artemisiifolia* L.). The ultimate goal is to improve living standards of the citizens of Osijek.

Methods: Based on a literature review of ragweed biology and the effectiveness of eradication trials, an assumption was made that SRC plantations in the outer ring of the town could serve as wind/pollen barriers and, thus, a long-term hygiene measure at infestation pathways. Based on local meteorological data, prevailing wind currents for pollen distribution were established and suitable locations for SRC wind barriers were identified. Our analysis considered the attractiveness of planting SRC willow (*Salix* spp.) plantations on agricultural land, public funds spent for short-term interventions against ragweed, and costs (days of sick leave) to the local economy. Economic models used in our analysis were based on: (1) a societal interest in planting SRC plantations for long-term ragweed control that would offset public money invested in the compensation for allergy symptoms and (2) opportunities for farmers to change land use from agricultural crop production to SRC plantations. Net revenues from expected biomass yields for bioenergy markets were assessed to offset investment costs.

Results: Positioning SRC investments as wind barriers together with net revenues from biomass for bioenergy illustrated a new perspective on the economics of SRCs. While production of biomass for bioenergy from SRC plantations was not an attractive option to crop farmers, the addition of ecosystem services from strategically located planting sites changed investment attractiveness. A high replicability potential has been identified based on model results for common ragweed distribution, projected climate change, and expected increased sensitivity of the human population to ragweed pollen.

Conclusions: Biomass for bioenergy is needed to supply the society with renewable carbon. SRC plantations have demonstrated carbon savings, various ecosystem services, and sustainable biomass supply. Yet, the feasibility of SRC plantations remains a challenge. Overlapping ecosystem services with biomass production can yield positive benefits, but interdisciplinary research is needed to identify and monetize ecosystem services that would have positive economics and high reproducibility. This case demonstrates viable methods that can be presented to policy makers and the general public who are not directly familiar with SRC plantings, bioenergy, or ecosystem services.

2.5. Genetic Parameter Estimates for Coppiced Hybrid Poplar Bioenergy Trials

Gantz, C.; Stanton, B.J.; Shuren, R.; Espinoza, J. and Murphy, L.

Background: Selection of highly productive hybrid poplar (*Populus* spp.) varieties is perhaps the most cost-efficient way of increasing biomass plantation yields. Genetic parameters such as clonal repeatability and genetic correlations among different traits obtained from clonal trials are critical to make informed selection decisions. As part of the Advanced Hardwood Biofuels program for sustainable bioenergy production, GreenWood Resources established four hybrid poplar clonal screening trials in Western USA. Trials were established at Hayden, Idaho, and Jefferson, Oregon, in 2012 and at Pilchuck, Washington, and Clarksburg, California in 2013. Our objective was to estimate genetic parameters for coppiced clonal trials to understand the level of genetic control and relationships among traits during the coppice cycle. We discuss our work from Jefferson and Hayden in this report.

Methods. Seventy clones were established in clonal trials at Jefferson, Oregon, and Hayden, Idaho, using a biomass density of 3600 trees per ha. Trials were coppiced two years after planting. Diameter at breast height, diameter of the largest stem of the stump, and number of stems per stump were measured three years after coppice. Mean diameter, basal area, and dry matter per stump were estimated. Specific gravity and moisture content were measured the third year after coppice on the best twenty clones. Statistical analysis was done using mixed linear models with Proc Mixed in Statistical Analysis Software (SAS version 9.4, Cary, North Carolina, USA) to estimate variance components. Clonal repeatability, genetic correlations among traits, and age-age genetic correlations were estimated. Genetic gains in biomass from selecting the best clones are also reported.

Results: Agronomic traits showed a two to three-fold increase between years one and three after coppice and a decrease in the number of stems per stump. Clonal repeatability for agronomic traits were moderate to high at the Jefferson trial. Wood traits repeatability was higher than for agronomic traits. Genetic correlations among traits were high and positive between diameter related traits, but negative between the diameter and the number of stems per stump. Specific gravity had a high negative genetic correlation with wood moisture content. The genetic correlations between age three and age two after coppice were all high. Important genetic gains in dry matter per tree can be obtained by selecting the best 10 clones in the trials.

Conclusions: Clonal repeatability at the Jefferson trial indicated that agronomic traits were under strong genetic control after the second and third coppice seasons. Genetic correlations among agronomic traits suggested that it would be possible to select clones based on the measurement of the main stem of each stool. Age-age genetic correlations indicated that it would be possible to reduce the selection cycle by selecting after the second growing season instead of waiting for one additional year. Important genetic gains in dry matter per stump were obtained by selecting the best five clones at each site.

2.6. Genetic and Environmental Effects on Variability in First-Rotation Shrub Willow Bark and Wood Elemental Composition

Fabio, E.S. and Smart, L.B.

Background: The elemental composition of woody biomass has important consequences for both energy conversion processes and agronomic management of nutrient budgets. Elemental ash content in bark and wood can be problematic for thermal conversion and can decrease pretreatment efficiencies in liquid fuel conversion. It is assumed that a high bark-to-wood ratio in short-rotation crops like shrub willow (*Salix* spp.) will necessarily contribute to high elemental ash content, but little data exists on the genetic and environmental contributions to elemental composition in willow. In support of previous research describing variability in biomass composition of shrub willow crops, we sought to evaluate the contributions of genotype, environment, and their interactions on elemental composition of shrub willow biomass. Specifically, we tested concentrations of macronutrients and micronutrients as well as heavy metals in biomass along with export rates across a range of species and growing environments.

Methods: First rotation biomass samples from six genotypes in four yield trial locations were separated into bark and wood samples, milled and analyzed for C, N, and H by the combustion method, and, for elemental composition, inductively coupled plasma atomic emission spectroscopy (ICP-AES) with acid digestion. An additional set of biomass samples for the same six genotypes from an additional six yield trial locations were milled whole (bark + wood) and analyzed for C, N, H, and elemental composition, which resulted in 252 total independent observations. Factorial analyses of variance were conducted separately for each of the 19 elements in bark and wood to assess the influence of cultivar, the environment, and their interaction. Tukey's honestly significant difference test was used to determine significant differences among factor levels. Multivariate principal components analysis (PCA) bi-plots were used to visualize relationships among elements and growing environments. Lastly, yield values were combined with biomass elemental concentrations to obtain nutrient harvest export rates.

Results: There were significant genotypic differences in elemental concentrations especially for macronutrients and micronutrients. The general trends were that *Salix purpurea* L. genotypes had high N, Mg, and Fe concentrations in bark while *Salix miyabeana* Seemen genotypes had high Ca concentrations in both bark and wood. Hybrids between these two species displayed intermediate values in elemental composition, which suggests an additive mode of inheritance for these traits. Multivariate analyses of the broader suite of biomass samples showed strong environmental effects on elemental composition, but there were a number of significant genotype × environment interactions. Nutrient export rates at harvest were largely a function of yield, but some genotypes showed signs of luxury uptake when N was abundant. Evidence from this study did not support the expectation that greater bark-to-wood ratios increase total ash content, but rather there are strong genotypic effects on element deposition in stems.

Conclusions: These results showed important genotype-specific preferences in nutrient uptake, which will likely need to be addressed through nutrient management plans. Although our test sites did not include heavy metal contaminated soils, significant genotype × environment interactions for a number of elements suggest that there may be capacity to exploit existing genetic resources for improving phytoremediation applications and for the development of effective riparian buffer strips to reduce excess nutrient export.

2.7. Is Hybrid Vigor Possible in Native Balsam Poplar Breeding?

Hu, Y. and Thomas, B.R.

Background: When two or more species are crossed to produce hybrid progeny, some of them can be expected to yield growth performance far superior than either parent (i.e., hybrid vigor/heterosis). To date, there have been few attempts to examine whether hybrid vigor can be achieved by crossing disparate populations of the same species. Balsam poplar (*Populus balsamifera* L.) is a transcontinental tree species in North America ranging from Alaska to Newfoundland. This wide range makes balsam poplar an ideal species to study intraspecies hybrid vigor as a tool for increasing genetic gain in growth through increased genetic diversity (i.e., heterozygosity). We tested the hypothesis that intraspecies breeding of widely spaced populations of balsam poplar would lead to the expression of hybrid vigor through both field and greenhouse assessments. We determined the role of exogenous hormones linked to physiological and growth performance.

Methods: In September 2009, three field trials (two in Alberta (AB), Canada (Field 7 and 23), and one in Quebec (QC), Canada (Field QC)) were established in conjunction with Alberta-Pacific Forest Industries Inc. (Al-Pac) and Mr. Pierre Périnet (Ministry of Forestry, Quebec). Five male parents from each province with five female parents from Quebec and four female parents from Alberta were used for breeding intra-region and inter-region crosses. Based on six-year height and diameter results from the Alberta field trials, the AB × QC cross type was selected for further study. In 2016, clones from within this cross-type were grown in a randomized complete block design under near-optimal greenhouse conditions with families identified as slow and fast growing in cross-type AB × QC to use extremes for examining the relationship of hormone levels and growth performance of the genotype and/or family. Diameter and height growth was measured biweekly and gas exchange data were collected on Day 33 (after transplanting) just prior to harvest. In late June, internode tissue samples were collected from two to three trees from each cross to analyze for the hormones that included gibberellic acids (GAs), indole-3-acetic acid (IAA), and abscisic acid (ABA). These hormones have been shown to play a role in regulating plant growth and will help to elucidate whether hybrid vigor was correlated with hormone levels and/or linked to photosynthetic performance. All growth data (height, diameter, stem volume) were analyzed with ANOVA procedures using Statistical Analysis Software (SAS version 9.4, Cary, North Carolina, USA). Following significant main effects, multiple comparisons among means were completed by using the Student-Newman-Keuls test. Correlation coefficients (r or r^2) and probability (P) were determined with Pearson's correlation analysis using the statistical package in SigmaPlot 13.0 (Systat Software, Inc., Chicago, Illinois, USA). We used $\alpha \leq 0.05$ to indicate the significance of all tests.

Results: Preliminary results indicated that stem volumes calculated from height and diameter of two-month-old rooted cuttings grown under optimal greenhouse conditions were positively and significantly correlated with stem volumes of eight-year old field-grown trees (Field 7, $R^2 = 0.396$, $p = 0.012$; Field 23, $R^2 = 0.384$, $p = 0.014$, Field QC, $R^2 = 0.346$, $p = 0.021$, respectively). Additionally, hybrid vigor is correlated with hormone levels and linked to photosynthetic performance (i.e., $R^2 = 0.679$, $p = 0.001$) for the greenhouse photosynthesis rate (A) vs. GA_{19} content (ng g^{-1} DW).

Conclusions: The project will determine the potential of using disparate, native populations of balsam poplar to produce superior progeny with enhanced stem growth traits.

2.8. Uncovering the Genetic Architecture of Growth-Defense Tradeoffs in a Foundation Forest Tree Species

Riehl, J.F.; Cole, C.; Morrow, C. and Lindroth, R.

Background: Physiological tradeoffs in plant allocations to growth and defense govern the ecology and evolution of tree-insect interactions and influence the sustainable production of forests. Despite the importance of plant defenses to forest health, a fundamental gap remains in our knowledge concerning genetic architecture of growth-defense trade-offs. *Populus* L. provides an ideal model system to study the underlying genetic architecture of different resource allocation strategies since phenolic defense compounds (e.g., phenolic glycosides, condensed tannins) are strongly and negatively correlated with growth. This study aims to identify the genetic architecture underlying variation in key growth and chemical defense traits that determine growth-defense trade-offs in a foundation forest tree species (*Populus tremuloides* Michaux) by using genome-wide association analysis combined with complementary methods (e.g., multiple marker association analysis, extreme phenotype sampling, and transcriptomic data).

Methods: A large association mapping common garden of *P. tremuloides* was established in 2010 with four replicate blocks of genotypes (*n* = 515) collected from a north-south transect throughout Wisconsin, USA (WisAsp population). We evaluated a suite of important tree traits (including tree height and diameter at breast height (DBH), leaf morphology, bud break, and phenolic glycoside concentrations) between 2014 and 2017. Approximately 45,000 probes were used to perform sequence capture genotyping of the full WisAsp population via the Illumina HiSeq2000 platform (San Diego, California, USA), which resulted in the discovery of ~170,000 single nucleotide polymorphisms (SNPs) used in the genome-wide association (GWA) analyses. Leaf material from a subset of genotypes with very high or very low levels of phenolic compounds were collected for RNA (ribonucleic acid) sequencing during the last phenotypic data collection in 2017.

Results: In our preliminary analysis, bud break and phenolic glycosides showed high broad-sense heritability (H^2 = 0.70–0.89) while growth traits (volume, relative growth, leaf morphology) showed low to moderate heritability (H^2 = 0.24–0.49). Traditional single-marker GWA analysis revealed significant SNPs in only two traits known as the specific leaf area (SLA) (uncharacterized gene putatively involved in growth regulation) and the spring bud break (three genes involved in cell wall biogenesis and environmental stress response). Multiple-SNP GWA analysis further elucidated potentially why so few significant SNPs were identified. Traits with both high and low heritability, bud break, and SLA, respectively, exhibited a higher amount of variability explained by our SNP set as compared to other growth and defense traits. This indicated that our SNP set may not have covered portions of the genome controlling the other highly heritable traits (e.g., phenolic glycosides). However, for all traits, no more than 40% of the variation explained by our SNP set was attributed to SNPs with larger effects, which means that these traits were likely controlled by many small-to-moderate effect genes.

Conclusions: From these initial results, we have determined that traditional single-marker GWA alone is unlikely to provide concrete answers as to what genes underlie important tree traits. Thus, future association analyses should begin incorporating complementary methods such as multiple-SNP GWA and expression data (in progress) to help dissect the genetic architecture of important quantitative tree traits to improve tree breeding and conservation.

2.9. The Great Lakes Restoration Initiative: Reducing Runoff from Landfills in the Great Lakes Basin, USA

Zalesny, R.S., Jr.; Burken, J.G.; Hallett, R.A.; Pilipović, A.; Wiese, A.H.; Rogers, E.R.; Bauer, E.O.; Buechel, L.; DeBauche, B.S.; Henderson, D.; Peterson, M. and Seegers, R.

Background: Increasing human population growth and associated industrial development since the 1960s has greatly influenced water quality in the Great Lakes and their watersheds. Closed landfills, dumps, and similar sites contribute to non-point-source pollution of nearshore health especially given the potential impacts of runoff and leakage from such sites. Short rotation woody crops such as poplars (*Populus* spp.) and willows (*Salix* spp.) are ideal for phytoremediation (i.e., the direct use of plants to clean up contaminated soil, sediment, sludge, or groundwater) because they grow quickly, have extensive root systems, and hydraulic control potential. All of these factors serve as biological systems that remediate such pollution. The United States Department of Agriculture Forest Service researchers have developed phyto-recurrent selection, which is a tool for choosing generalist plant varieties that remediate a broad range of contaminants, or specialist plants that are matched to specific pollutants. The ability to select varieties across contaminants allows for broad applicability of these phytoremediation systems. While the science of phytoremediation has undergone rapid growth in the last two decades, there is some uncertainty about the efficacy of using existing forests to remediate liability sites. However, the recent development and patenting of phytoforensic technologies promote the use of plants not only for remediation but also to delineate polluted sites and to monitor remediation progress. Phytoforensics is the use of plant sampling as a way to detect and quantify pollutants in the environment around the plants. Using both phyto-recurrent selection and phytoforensics, the overall objective of this project is to establish 15 phytoremediation buffer systems throughout the Great Lakes Basin to reduce untreated runoff from landfills, which mitigates non-point-source pollution impacts on nearshore health. The objective of this presentation was to describe the landfill runoff reduction project in the context of the overall Great Lakes Restoration Initiative Program, so as to provide context for potential collaborations as well as for abstract 2.10 (Hallett et al.), which is a study within the overall research program. To date, a total of 10 phytoremediation systems have been established, which included 10,000 trees being planted. During summer 2019, we will establish another 5000 trees across five more phytoremediation systems.

2.10. Evaluating Poplar Genotypes for Future Success: Can Phyto-Recurrent Selection Assessment Techniques be Simplified?

Hallett, R.A.; Zalesny, R.S., Jr.; Rogers, E.R.; Wiese, A.H.; DeBauche, B.S.; Bauer, E.O. and Pilipović, A.

Background: Anthropogenic impacts (e.g., runoff from landfills) have been detrimental to water quality and quantity in the Great Lakes. Phyto-technologies could contribute to reducing these ecological impacts. Short rotation woody crops such as poplars (*Populus* spp.) and willows (*Salix* spp.) are ideal for these systems given their elevated biomass relative to slower-growing species and their associated hydraulic control potential. Phyto-recurrent selection is a phyto-technologies method that is used for genotype selection of generalist varieties that are adapted to broad productivity zones or specialists that are matched to specific site conditions such as those at landfills impacted from runoff. We developed phyto-recurrent selection techniques to select poplar and willow genotypes that are best suited to individual field sites by evaluating performance in the greenhouse using weighted rank summation indices consisting of height, root number, root dry mass, leaf number, leaf area, leaf dry mass, and stem dry mass. These variables were selected based on their importance for early establishment. Our overarching objective was to test whether total biomass during evaluation in the greenhouse may be an indicator of future success in the field.

Methods: We used three greenhouse phyto-recurrent selection cycles to reduce a base population of 174 poplar and willow genotypes (cycle 1) to 60 and 24 varieties in cycles 2 and 3, respectively. To make final genotype selections for planting (i.e., cycle 4) at five landfills along the western shore of Lake Michigan, we tested whether genotypes differed for the previously mentioned traits when grown in soil treatments from the five landfills and a commercial potting soil control. For the current presentation, data from cycle 3 ($n = 270$) were compared by using stepwise linear regression techniques to create a simplified model that predicted growth in the greenhouse, which was defined by the total biomass.

Results: We found that the diameter and total leaf number can be used to predict total biomass ($R^2 = 0.82$, $p < 0.001$). In addition, total biomass was significantly correlated with height growth ($r = 0.85$, $p < 0.001$), which is another key factor in rapid establishment.

Conclusions: Diameter and leaf number are easily measured in the greenhouse and may be the only variables needed to predict success in the field. Future studies will need to establish the broad scale applicability of this relationship and include field trials and validation.

2.11. Growth of Poplars in Soils Amended with Fibercake Residuals from Paper and Containerboard Production

Rogers, E.R.; Zalesny, R.S., Jr.; Wiese, A. and Benzel, T.

Background: Phyto-technologies that use poplars (*Populus* spp.) and other woody plants are commonly implemented to remediate polluted soils and groundwater. In addition to environmental clean-up applications, phyto-technologies include the reuse of industrial byproducts. For example, fiber cake residuals are byproducts from paper and containerboard production that have been used as amendments to provide nitrogen and other benefits (e.g., increased organic matter content) to agricultural and forest soils. In this study, we tested early survival and development of poplars grown in fiber cake residual-amended soil treatments to determine the most effective soil-fiber cake combinations for future nursery and field production of selected poplar genotypes.

Methods: Fiber cake residuals were obtained from two different Northern Wisconsin, USA sources. Expera Specialty Solutions (Rhinelander, WI, USA) provided papermaking residuals while Packaging Corporation of America (PCA) (Tomahawk, WI, USA) provided residuals from containerboard production. Both residuals were combined primary (wood fiber) and secondary (bacterial biomass) blended byproducts. The Expera fiber cake was known to release available N while the PCA residual was known to immobilize N when applied to agricultural sites. Three hybrid poplar genotypes (*Populus deltoides* Marsh. × *Populus maximowiczii* Henry 'DM114'; *Populus* × *canadensis* Moench 'DN170'; *Populus nigra* L. × *P. maximowczii* 'NM2') were grown in a greenhouse for 35 days. Cuttings (12.7 cm long) were established in eleven soil treatments that included a commercial potting mix control, a nursery soil collected at the United States Department of Agriculture Forest Service, Rhinelander Experimental Forest, Rhinelander, WI (site of the future out planting), and nine nursery soil-fiber-cake mixes that were blended based on expected N uptake of the trees and the Wisconsin Department of Natural Resources regulations for field N loading. In particular, there were three blends of nursery soil mixed with Expera residuals and three blends with PCA residuals as well as a three-component blend of nursery soil + Expera + PCA. Soil concentrations were determined volumetrically and mixed to achieve application rates of 50, 100, and 200 kg available N ha^{-1}. Analyses followed a split-plot experimental design with three blocks, 11 soil treatments (whole plots), three clones (subplots), and five trees per treatment × clone combination.

Results: Trees grown in the 50 kg N ha^{-1} nursery-Expera blend (A1) had the largest average height (19.54 cm) and diameter (3.02 mm) among all nursery-fiber-cake treatments. Expera soil treatments produced the darkest average leaf color overall (2.92 out of 3) and, thus, the healthiest trees. Trees grown in treatment A2 (100 kg N ha^{-1} nursery-Expera blend) had the lowest average above-ground and below-ground biomass (4.26 g and 0.032 g, respectively) while treatment A1 had the greatest aboveground biomass (4.52 g) and treatment B1 (50 kg N ha^{-1} nursery-PCA blend) had the greatest belowground biomass (0.11 g). Treatments B2 (100 kg N ha^{-1} nursery-PCA blend) and B3 (200 kg N ha^{-1} nursery-PCA blend) produced trees with the smallest average leaf area (48.95 cm^2 and 48.60 cm^2, respectively).

Conclusions: Based on preliminary analyses, there is a wide variability in the measured traits across the differing blends of nursery and fiber cake residuals from both industrial sources. Further data analyses are being conducted to identify the optimal soil treatments for each genotype.

2.12. Growth and Physiological Responses of Three Poplar Clones Grown on Soils Artificially Contaminated with Heavy Metals, Diesel Fuel and Herbicides

Pilipović, A.; Zalesny, R.S., Jr.; Orlović, S.; Drekić, M.; Pekeč, S.; Katanić, M. and Poljaković-Pajnik, L.

Background: Land use associated with agricultural practices is often limited by environmental degradation including soil contamination. Biologically sustainable cleanup methods such as phytoremediation are needed to rehabilitate such soils for agricultural and forestry production. Short rotation woody crops such as poplars, cottonwoods, and aspens (*Populus* spp. and their hybrids) have proven effective for phytoremediation of a broad spectrum of contaminants, which results in economic and ecological benefits. The objective of this study was to test the growth and physiological responses of three different poplar clones grown on soils contaminated with heavy metals, diesel fuel and herbicides.

Methods: Poplar trees belonging to three clones (*Populus deltoides* Marsh. 'Bora' and 'PE 19/66', and *Populus* × *canadensis* Moench 'Pannonia') were grown for three years at the Experimental Estate of the Institute of Lowland Forestry and Environment, University of Novi Sad (former Poplar Research Institute), Serbia. Trees were spaced 2.0×0.5 m apart, which is consistent with biomass production plantings throughout Europe. The field was divided into seven plots containing an uncontaminated control and the following six artificially polluted soil treatments: (1) 10.6 kg Ca ha^{-1}, (2) 183.3 kg Ni ha^{-1}, (3) 247 kg Cu ha^{-1}, (4) 6667 L diesel fuel ha^{-1}, (5) 1320 g Pendimethalin ha^{-1}, and (6) 236 g Oxyfluorfen ha^{-1}. Net photosynthesis, transpiration, stomatal conductance, and chlorophyll fluorescence were assessed during the first two vegetation periods while height and diameter were measured after cessation of growth each year. After the third growing season, plants were harvested and weighed to assess biomass production.

Results: There were significant differences among clones for growth and biomass production ($p < 0.0001$). Treatment with heavy metals affected growth of clones at the beginning of the experiment but the effects were not significant during the third growing season ($p > 0.05$). Net photosynthesis, stomatal conductance, and transpiration were significantly affected by the treatments in the first growing year for all clones ($p < 0.0001$) while chlorophyll fluorescence showed fewer differences between treatments and investigated clones.

Conclusions: Preliminary results showed promise for clones 'Bora' and 'PE 19/66', which had 5.3 and 7.9 times greater biomass than 'Pannonia', respectively, across all soil treatments. Despite non-significant genotype $\times$ treatment interactions at the end of the study, 'Bora' and 'PE 19/66' grown in all treatment soils exhibited greater biomass than 'Pannonia' with trees growing in the control soils exhibiting 13.8 and 19.6 times greater biomass than 'Pannonia', respectively. Unfavorable and favorable genotype $\times$ soil treatment interactions need to be considered along with the applicability of using 'Bora' and 'PE 19/66' in larger-scale systems.

2.13. Mitigating Downstream Effects of Excess Soil Phosphorus through Cultivar Selection and Increased Foliar Resorption

Da Ros, L.M.; Soolanayakanahally, R.Y. and Mansfield, S.D.

Background: Phosphorus is a non-renewable resource pivotal to global food security since it often limits plant growth and development. Concomitantly, due to its diffuse source run-off from agricultural land and role in promoting algal growth, phosphorus is also characterized as a pollutant in aquatic environments. Movement of nutrient contaminants such as phosphorus into riparian ecosystems could be limited via the widespread planting of *Populus* L. (poplar) and *Salix* L. (willow) hybrids on the agricultural landscape. A comparative study using commercially available high biomass-producing tree species was performed to: (1) evaluate the natural variation in phosphate uptake, storage, and resorption in Salicaceae hybrids, (2) recommend promising genotypes for field trials, and (3) identify current barriers to the efficient phytoremediation of phosphorus-rich sites.

Methods: Multiple greenhouse trials involving 15 genotypes were conducted at the University of British Columbia Horticulture Greenhouse, Vancouver, British Columbia, Canada. Trees were grown from cuttings and are exposed to five phosphorus treatments ranging from optimal (0.4 mM P) to 15× optimal levels (6.3 mM P). Half of the trees were harvested at bud-set after 100 days of growth while the remaining trees were harvested after autumnal leaf senescence. Phenotypic differences in phosphorus storage and allocation were analyzed by using inductively coupled plasma atomic emission spectroscopy (ICP-AES) and high-performance liquid chromatography (HPLC). Semi-quantitative polymerase chain reaction (SQ-PCR) was used for gene expression analysis. R version 3.3.2 (R Core Team 2016; https://www.r-project.org/) was used to fit linear mixed-effects models with the packages lmerTest and nlme. Bland-Altman analysis for the method comparison was done using the package MethComp.

Results: Toxicity symptoms were not observed in any genotype across the phosphorus treatments. Species that demonstrated luxury uptake and storage of phosphorus had leaf concentrations >5 mg P g^{-1} as opposed to the 3–4 mg P g^{-1} observed in low or non-accumulating species at optimal levels of external phosphorus. Phosphorus resorption efficiencies ranged from 0% to 32% for hybrid poplar and 35% to 52% for hybrid willow with a high concentration of mobile phosphate remaining in leaves after senescence.

Conclusions: A high degree of intraspecies and interspecies variation in luxury consumption and resorption efficiency exists in poplar and willow genotypes. The poplar hybrid genotype 'Tristis' (*Populus* Tristis) is a promising candidate with luxury consumption of phosphate, large root systems, and the highest resorption efficiency among the poplar. An alternative candidate would be the hybrid willow genotype 'AAFC-5' (*Salix discolor* Muhlenberg × *Salix dasyclados* Wimmer) since it has higher biomass allocation to the stem, which allows for greater phosphorus removal from the site after coppice and higher phosphorus resorption efficiency. For the efficient-use of poplar and willow in agro-ecological and phytoremediation applications, improvements in the translocation of phosphate out of leaves during autumnal senescence would be beneficial.

2.14. Survival and Growth of Poplars and Willows Grown for Phytoremediation of Fertilizer Residues

Zalesny, R.S., Jr. and Bauer, E.O.

Background: Species and hybrids belonging to the genera *Populus* L. (poplar) and *Salix* L. (willow) have been used successfully for phytoremediation of contaminated soils. However, genotypic screening using phyto-recurrent selection is necessary prior to large-scale deployment because of the broad amount of variation among and within poplar and willow clones. To identify promising genotypes for potential use in future phyto-technologies, the objectives of the current study were to: (1) evaluate the genotypic variability in survival, height, and diameter of poplar and willow clones established on soils heavily contaminated with nitrate fertilizer residues and (2) assess the genotypic stability in survival and diameter of selected poplar clones after one and eleven growing seasons.

Methods: We evaluated traits after the first year bud set by testing 27 poplar and 10 willow clones planted as unrooted cuttings along with 15 poplar clones planted as rooted cuttings. The cuttings were planted in randomized complete blocks at an agricultural production facility in the Midwestern United States. After eleven growing seasons and using phyto-recurrent selection, we surveyed survival and a measured diameter of 27 poplar clones (14 unrooted, 13 rooted) that were selected based on superior survival and growth throughout plantation development.

Results: There was a broad amount of genotypic variability in survival, height, and diameter during the establishment. Overall, willow exhibited the greatest survival while poplar had the greatest height and diameter. At eleven years after planting, superior clones were identified that exhibited an above-average diameter growth at establishment and rotation-age in which most had stable genotypic performance over time.

Conclusions: Our results verified that the broad clonal variation among and within poplar and willow genotypes was an important positive indicator for the potential of long-term phytoremediation of $NO_3{}^-$-contaminated soils because the selection is proportional to variation and establishment is the first requirement for any long-term success of the system. Given the broad range in height and diameter among genotypes and the ability to select better clones within this variation, our findings suggest that the selection of specific clones was favorable to genomic groups based on the geographic location and soil conditions of the site.

2.15. Stakeholder Assessment of the Feasibility of Poplar as a Biomass Feedstock and Ecosystem Services Provider in Rural Washington, USA

Hart, N.M.; Townsend, P.A.; Chowyuk, A. and Gustafson, R.

Background: Growing poplar (*Populus* spp.) as an energy crop in southwestern Washington, USA could also provide ecosystem services for wastewater management and flood control. Models of poplar growth and land use show that the region has enough suitable land to support a poplar-based bio-products industry. In addition, the region's economically depressed rural communities could benefit from new cropping and industrial production opportunities. Evaluating the feasibility of a poplar-based industry requires an understanding of the local social-ecological context and stakeholder opinions on converting land to poplar. This study was conducted as part of Advanced Hardwood Biofuels Northwest, a United States Department of Agriculture, National Institute of Food and Agriculture-funded consortium of university and industry partners developing the framework for a poplar-based biofuel and bio-based chemical industry in the Pacific Northwest. We conversed with stakeholders to: (1) capture information about the context of the agricultural landscape in the study region, (2) identify perceived opportunities and challenges to growing poplar for a bio-products industry, (3) explore the validity of techno-economic modeling assumptions and ecosystem services potential, and (4) compile stakeholder questions.

Methods: During the winter of 2018, we conducted 15 in-depth, semi-structured interviews (30–60 min each), a focus group meeting (90 min), and a short discussion (20 min). The stakeholder conversations were recorded and then fully transcribed. The transcripts were coded using descriptive and initial (i.e., open) coding methods to sort the data based on recurrent topics.

Results: Stakeholders expressed a desire for a new economic opportunity that "pencils out" and were generally optimistic about win-win scenarios for floodplain management (when outside of riparian restoration buffers). Stakeholders had questions about the business model for farmers, the inputs and outputs of the biorefinery, and the technical feasibility of growing and harvesting poplar in the region. They also discussed several salient obstacles to growing poplar for bio-products such as past failures to profit from poplar production (for pulp/sawlogs), competing land use practices (for conservation and forage), and limited drivers for alternative wastewater solutions.

Conclusions: A successful poplar-based bio-products industry could meet multiple needs in rural, Southwestern Washington. Given a credible business model, participants did not think it would be difficult to find suitable land and secure willing growers. However, landowners require assurance of turning a profit to convert to poplar and large initial investments (e.g., covering the costs of bio-refinery construction, harvesting equipment, and poplar farm establishment) would be needed to start the industry.

2.16. Barriers and Opportunities for Use of Short Rotation Poplar for the Production of Fuels and Chemicals

Townsend, P.A.; Dao, C.; Bura, R. and Gustafson, R.

Background: Poplar (*Populus* spp.) is widely recognized as excellent feedstock for the production of bio-based fuels and chemicals. The rapid growth of poplar, ease of fractionation, ample sugar content, and potential for year round harvesting make it an attractive candidate as a dedicated feedstock for a bio-refinery. To demonstrate this potential, Advanced Hardwood Biofuels Northwest (AHB) established four demonstration sites with short-rotation poplar in the Pacific Northwest. To produce a reliable biomass supply, we envision that poplar tree farms would be harvested approximately every three years allowing for coppice regeneration. Whole tree harvesting would be enabled by forage harvesters that cut and chip the poplars in a single pass. The short rotation cycle and use of biomass from the entire tree that includes wood chips as well as leaves, stems, and bark results in a poplar feedstock that is quite different from clean poplar chips that have been investigated in past bioconversion research projects. It is important that the bioconversion of poplar biomass be thoroughly investigated before commercial short rotation tree farms are established. The objective of this research is to investigate the bioconversion of whole tree poplar biomass to produce fuels and chemicals. Specific issues investigated were: (1) the impact of leaves on conversion processes, (2) the ease of sugar release from different poplar clones, and (3) the efficacy of preprocessing poplar chips to improve enzymatic hydrolysis and fermentation processes.

Methods: Whole tree poplar biomass harvested at the AHB tree farms were processed at the University of Washington laboratories by acid catalyzed steam explosion, enzymatic hydrolysis, and fermentation to ethanol. Liquid and solid phases from each processing step were analyzed to assess sugar release and fermentation yields.

Results: Our results show that leafy material had a significant impact on enzymatic hydrolysis and that poplar leaves would need to be removed either prior, during, or after harvesting to have reasonable product yields. Differences in sugar release were found between various poplar clones grown at the same demonstration sites. Moreover, consistent differences in sugar release were observed even when the same poplar clones were grown at different tree farm locations. More rapidly growing clones tended to have lower sugar releases. Lastly, we observed whole tree poplar chips with both inorganic and organic components that should be removed for good sugar release and fermentation yields. Use of acidic preprocessing, followed by water washing, was more effective than just water washing of poplar chips to remove detrimental components. Water washing alone, however, resulted in higher sugar and fermentation yields than was observed with poplar chips that were not preprocessed.

Conclusions: Poplar remains an excellent feedstock for the production of fuels and chemicals. Different poplar clones have significant differences in their ease of sugar release and this will need to be investigated further before establishing commercial tree farms. Effective use of whole tree poplar biomass obtained from trees grown on short-rotation cycles may require some additional processing steps to increase sugar and final product yields. The potential to extract high value chemicals from whole tree biomass using preprocessing needs to be investigated further.

2.17. Short Rotation Eucalypts: Opportunities for Bioproducts

Rockwood, D.L.; Ellis, M.F.; He, Z.; Liu, R. and Cave, R.D.

Background: Eucalypts (*Eucalyptus* spp.) are the world's most valuable and widely planted hardwoods (18 million ha in 90 countries) and have numerous potential applications as short rotation woody crops (SRWC) due to many well understood conversion technologies and several promising technologies under development. Using experience in Florida, USA, we describe the potential for eucalypts to maximize SRWC productivity through site amendment and genetic improvement, document their current energy applications, and assess their potential for short-term and likely long-term bio-products.

Methods: An intensively cultured eight plus ha demonstration plot of *Eucalyptus grandis* W. Hill ex Maiden × *Eucalyptus urophylla* S.T. Blake cultivar 'EH1' was planted in May 2011 on former citrus beds near Hobe Sound, FL. Plantings were established on two spacings and they were monitored through harvest in December 2017. In 2012, a *Eucalyptus grandis* cultivar 'G2' was commercially planted near Ft Pierce, FL along with *Corymbia torelliana* (F. Muell.) K.D. Hill & L.A.S. Johnson progenies. Stemwood samples from these and two other species were assessed for biochar characteristics. Biochar and organic fertilizer treatments applied to a windbreak of *E. grandis* W. Hill ex Maid. Cultivars and *C. torelliana* progenies at the Indian River Research and Education Center (IRREC) in February 2018 were periodically evaluated through soil, foliage, and tree growth monitoring.

Results: Through 81 months, the 'EH1' cultivar was more productive at the nominal 3.05 × 1.22 m spacing than the 3.05 × 2.13 m spacing (63.6 vs. 57.6 green MT ha^{-1} year^{-1}). 'EH1', *C. torelliana*, and 'G2', as well as *Eucalyptus amplifolia* Naudin and live oak (*Quercus virginiana* Mill.) are suitable for commercial biochar production in Florida. Soils data from the IRREC biochar test suggest that biochar greatly enhances the nutrient properties of inherently poor Florida soils and four-month foliage and growth results document rapid response to the biochar amendment. We also review other potential *Eucalyptus* bio-products that may be classified as naturally occurring, generated by biochemical processes, or as the result of thermochemical processes.

Conclusions: Eucalypts can be very productive when intensively grown as SRWCs. Biochar produced from *Eucalyptus* spp. and other species may be a useful soil amendment for their intensive culture. Many products currently derived from petrochemicals can be produced from SRWCs and these bio-products have a broad and exciting range of applications for enhancing the value of SRWCs.

2.18. Variability of Harvester Performance Depending on Phenotypic Attributes of Short Rotation Willow Crop in New York, USA

de Souza, D.P.; Volk, T.A. and Eisenbies, M.H.

Background: Fossil fuels remain the dominant energy source despite current environmental, social, and economic concerns. However, renewable energy sources such as short rotation woody crops (SRWC) like shrub willow (*Salix* spp.) have been receiving more attention in recent years. Harvest is the single largest cost in the production of SRWC, but the systems previously used in North America have not been optimized for these crops. Contemporary research has focused on the performance of single pass cut and chip systems and, while there is implicit recognition that crop attributes such as standing biomass, field size and condition, and operator experience could affect the system, there is little concrete information about their effects. The objective of this project is to determine if phenotypic attributes of shrub willow crops in New York State are related to the operating performance of a New Holland FR9090 (Turin, Italy) forage harvester fitted with a 130FB coppice header.

Methods: The study is based on data collected from a 5-ha harvest conducted in Solvay, New York. Prior to harvesting, willow characteristics such as height, stem diameter, number of stems, and plant form were measured. Material capacity (Mg hr^{-1}), field capacity (ha hr^{-1}), standing biomass (Mg ha^{-1}), and other attributes of the harvest operation were determined.

Results: Despite promising results shown by the FR-series forage harvester in past harvests, up to 10 percent of woody biomass is not processed by the harvester and remains on the field. Drop losses (biomass left on the site) were collected after the harvest to determine the amount of harvestable biomass remaining on site and its relationship with the crop's phenotypic attributes. Preliminary results show slight but positive effects of diameter ($p \leq 0.0001$), height ($p = 0.0002$), and stems > 50 mm ha^{-1} ($p \leq 0.0001$) on material capacity and negative (also slight) effects of stems ha^{-1} ($p = 0.0359$) and yield ($p \leq 0.0001$). Additionally, field capacity appears to be slightly and negatively affected by diameter ($p = 0.0119$), stems > 50 mm ha^{-1} ($p = 0.0178$), yield ($p = 0.0009$), and cultivar ($p \leq 0.0001$).

Conclusions: This ongoing study will help illuminate the characteristics of willow crops that limit the harvester's operation and what plant phenotypes optimize harvester performance and minimize operating costs.

2.19. Cover Protection Affects Fuel Quality and Natural Drying of Mixed Leaf-On Willow and Poplar Woodchip Piles

Therasme, O.; Eisenbies, M.H. and Volk, T.A.

Background: Short-rotation woody crops (SRWC), which include shrub willow (*Salix* spp.), have the potential to make substantial contributions to the availability of biomass feedstock for the production of biofuels and bio-products. The objective of this study was to evaluate changes in fuel quality of stored leaf-on willow and poplar (*Populus* spp.) chips with respect to the degree of protection, location within the storage piles, and duration.

Methods: Harvested leaf-on willow and poplar wood chips with moisture content in the range of 42.1% to 49.9% (wet basis) were stored at the State University of New York, College of Environmental Science and Forestry field station in Tully, New York in piles for five months from May to October 2016. Three piles contained 25–30 Mg and three piles contained 35–40 Mg of wood chips. Three protection treatments were randomly assigned within each of the two groups of piles. The unprotected treatment exposed piles to direct solar radiation and rainfall. The second treatment had canopies covering the piles to limit direct rainfall. The final treatment had canopies plus a dome aeration system installed over the piles.

Results: Results indicate that cover protection can play a significant role in reducing and maintaining low moisture content in wood chip piles. Within 30 days of storage, moisture content in the core of covered piles decreased to less than 30% and was maintained between 24% to 26% until the end of the storage period. Conversely, there was an increase of moisture content in unprotected piles during the first two months of storage. For the conditions tested, core material (>45 cm deep) dried faster than shell material (<45 cm deep). The lower heating value, which strongly depends on the moisture content, was the highest for covered piles at the end of the storage.

Conclusions: Leaf-on SRWC biomass stored in piles created in late spring under climatic conditions similar to central and northern New York showed differing moisture contents when stored for more than 60 to 90 days. Overhead protection could be used to preserve or improve the fuel quality in term of moisture content and heating value.

2.20. Historical Perspective and Evolution of the Short Rotation Woody Crops Program at Rhinelander, Wisconsin, USA

Isebrands, J.G.

Background: Tree improvement research began in Rhinelander, Wisconsin, USA in 1931 at the Hugo Sauer Nursery. It was part of the United States Department of Agriculture (USDA) Forest Service, Lake States Forest Experiment Station's effort to develop superior planting stock of northern forest species for reforestation. In 1957, the Northern Institute of Forest Genetics (IFG) was founded at the present location near the Nursery. Over time, the IFG staff expanded with increased political support and funding for USDA Forest Service research priorities. The expansion was largely due to significant accomplishments of IFG scientists. Soon the IFG rose to national and international prominence in disciplines of radiation biology, ecology, physiology, genetics, and phytoremediation.

In 1966, the IFG became part of the North Central Forest Experiment Station and later became the Forestry Sciences Laboratory (FSL). There were several building additions as the research program grew. In 1971, the multidisciplinary Maximum Yield Program was founded under the direction of David H. Dawson with the goal to maximize wood production to meet growing national needs. A worldwide oil embargo and energy shortage in 1973–1974 accelerated the need for increasing wood yields on short rotations for wood and bioenergy. After testing, poplars (*Populus* spp.) and willows (*Salix* spp.) surfaced as promising species because of their rapid growth rate and ease of genetic improvement and culture. The Program soon became known as the Short Rotation Intensive Culture research program. Then the United States Department of Energy became a continuing important funding partner for the research.

There was close collaboration with Canadian scientists from the beginning as well as numerous university scientists. In the 1980s, global climate change became an international issue as the world population grew. Poplars and willows became model forest species for studying bioenergy and global climate change.

Research at the FSL led to the first genetically modified tree in the world, which was a poplar at Rhinelander. It also led to the establishment of the world's largest field experiment on carbon dioxide and tropospheric ozone enrichment of forest trees at the Harshaw Experimental Farm near Rhinelander. Canadian forest scientists were part of the multidisciplinary team along with numerous international university scientists.

Conclusions: More recently, the FSL became part of the reorganized Northern Research Station. Today's emphasis is on landscape ecology and ecosystem services research such as phytoremediation. The FSL has a long history of phytoremediation research since the late 1980s. Archiving important genetic materials for bioenergy and phytoremediation is also a priority. The FSL was renamed the Institute of Applied Ecosystem Studies to reflect these priorities. Numerous examples of historical photos, scientific milestones, international awards received, and international meetings held at Rhinelander are given throughout the presentation.

2.21. Tree Willow Root Growth in Sediments Varying in Texture

McIvor, I. and Desrochers, V.

Background: Willows (*Salix* spp.) are used for riverbank stabilization in New Zealand where they are planted into a range of sediments. River engineers have expressed concern that willow root systems failing under flood conditions may be weakened by the giant willow aphid (*Tuberolachnus salignus* (Gmelin)). However, the river engineers lack sound information on how a 'normal' root system develops in varying river sediments. We investigated the early root development of *Salix nigra* Marsh. in sediments typical of the sediments found in New Zealand riverbanks. The sediment types largely sort on particle size including silt, sand, and stones.

Methods: Cuttings of *S. nigra* were grown for 10 weeks in layered sediment types in five large planter boxes established outside. Each box differed in the proportion of silt, sand, and stones (gravel) with stones layered in the bottom and silt layered on the top in each box. Boxes were kept in ambient conditions and watered by the overhead sprinkler twice a day. Three, 90-cm long cuttings were planted to 5 cm off the floor of each box. At 10 weeks, the boxes were disassembled to extract roots and roots were sorted into diameter classes, according to sediment type and depth. Root length and dry mass were measured and root length density (RLD) and root mass density (RMD) calculated. Shoot number and mass were also recorded.

Results: Root development of *S. nigra* cuttings varied with sediment type, either silt, sand, or stones. Roots were initiated from the entire length of the cutting in the substrate, but root initials were concentrated at the bottom and close to the bottom of the cutting. There was substantial root extension into all three sediments and at all depths. Root length for roots ≥1 mm diameter was greatest for plants grown in approximately even proportions of silt, sand, and stones and least for plants grown in the lowest proportions of silt (30%, 20%) and the highest proportions of stones (50%, 70%). Root mass was greatest for cuttings in the box with the highest proportion of silt. Generally, RMD was highest in the stones because it was influenced by having the bottom of cuttings located in stones for four of the five treatments. RMD was highest for roots <1 mm diameter. However, RLD for roots >0.5 mm diameter was highest in sand while RLD of roots with diameter <0.5 mm was lowest in sand. Roots were extracted easiest from sand and with the most difficulty from silt.

Conclusions: Roots of *S. nigra* were least effective in binding sand primarily because of the low RLD of root <0.5 mm diameter. It is surmised that sand lacks water and nutrients sufficient to sustain growth of fine roots compared with silt and even stones. RLD for roots >0.5 mm diameter was the lowest in silt likely due to the greater resistance of the substrate to root penetration or possibly the greater investment into smaller roots that have stronger absorption capability.

2.22. Environmental Benefits of Shrub Willow as Bioenergy Strips in an Intensively Managed Agricultural System

Cacho, J.F.; Negri, M.C.; Zumpf, C.R.; Campbell, P.; Quinn, J.J. and Ssegane, H.

Background: Integrating perennial biomass crops into predominantly agricultural landscapes is gaining interest because of the potential to address sustainability issues in commodity crop production while providing multiple environmental benefits. This approach takes advantage of intra-field variation in soil productivity and suitability of second-generation biomass crops with their perennial growth characteristics and distinct physiological traits (e.g., deep rooting system, greater resiliency to marginal conditions, etc.). To test the viability of such a system, we are conducting a study in east-central Illinois where shrub willow (*Salix miyabeana* Seemen 'SX61') was introduced in areas of a 6.5-ha intensively managed corn (*Zea mays* L.) field that were considered less productive and/or vulnerable to environmental degradation (marginal areas). Specifically, we were interested in evaluating (1) the ability of shrub willow to intercept and utilize leached nutrients (particularly nitrate) from adjacent corn and (2) relative influence of shrub willow on other environmental quality indicators including water quantity, soil health, and greenhouse gas emissions.

Methods: Our research site was established using a randomized blocked design with plots of shrub willow and corn blocked on lowland floodplains in Comfrey loam soil (fine-loamy, mixed, superactive, mesic Cumulic Endoaquolls) with 0% to 2% slope (northern or N-plots) and on upland plains in Symerton silt loam soil (fine-loamy, mixed, superactive, mesic Oxyaquic Argiudolls) with 2% to 10% slope (southern or S-plots). The N-plots served as a pseudo-control located in fertile, floodplain soil with limited nitrate leaching. Conversely, the S-plots with higher nitrate leaching were categorized as marginal areas. Nutrient cycling was assessed by sampling soil, soil water, groundwater, and vegetation for nutrient loss and uptake. Additionally, biomass, groundwater elevation, soil moisture, transpiration, and greenhouse gas flux were measured.

Results: Results show that, where the shrub willow was established, annual soil water nitrate concentrations were reduced by up to 87% as compared to where corn was established. Rates of water use and soil nutrient utilization in areas under the shrub willow were comparable to areas planted in corn. Furthermore, soils under shrub willow cover showed subsoil carbon sequestration potential and nitrous oxide (N_2O) emissions were reduced when compared to soils planted in corn.

Conclusions: Our study demonstrates that careful placement of perennial biomass crops such as shrub willow within intensively managed agricultural fields can provide multiple environmental benefits in addition to bioenergy feedstock.

2.23. Quantifying the Unknown: The Importance of Field Measurements and Genotype Selection in Mitigating the Atmospheric Impacts of Poplar Cultivation

Kiel, S.; Potosnak, M.J. and Rosenstiel, T.N.

Background: Many woody crops including *Populus* (poplar) and *Salix* (willow) emit significant quantities of volatile organic carbon (VOCs) especially the reactive hemiterpene isoprene to the atmosphere. The emission of VOCs from these woody crops play important roles in influencing regional and local atmospheric chemistry including influencing the formation of ground-level ozone and secondary organic aerosols. However, despite this knowledge, relatively few field-based studies have attempted to quantify the magnitude and diversity of VOC emissions from woody crops under actual field conditions or have evaluated how woody crop genotype selection and the cultivation method may affect overall VOC emissions. To address this general lack of understanding, the objectives of the present study were to deploy a novel field-based method to quantify VOC emissions in situ (relaxed eddy accumulation flux analysis) from a large-scale poplar biomass plantation established as part of the Advanced Hardwood Biofuels Northwest (AHB) Project. In addition, we evaluated leaf-scale variation in VOC emission from field-grown poplar genotypes over multiple growing seasons and we examined the impact of genotype selection on the potential for whole canopy VOC emission as a consequence of poplar cultivation.

Methods: We evaluated leaf-scale VOC emissions from eleven field-grown bioenergy poplar genotypes over multiple seasons as part of a large-scale field trial near Jefferson, Oregon, USA. In addition, we developed and deployed a field-based whole-canopy flux measurement system to quantify peak summer-time VOC emissions associated with the genotypes of this high-productivity coppiced poplar system.

Results: There was surprising and significant variation in leaf-level VOC emissions among the poplar genotypes examined. In addition, using our field-based relaxed eddy accumulation methods, we have observed some of the largest ecosystem fluxes of isoprene emission to date with whole canopy surface fluxes in excess of 100 nmol m^{-2} s^{-1}.

Conclusions: Results from this field-based study suggest that intensive woody crop cultivation is generally proposed for both poplar and willow and may represent a significant and poorly characterized, landscape-level source of VOC emissions with potentially significant implications on regional atmospheric chemistry. However, the observation of significant variation in VOC emissions among poplar genotypes suggests that both targeted breeding and genetic engineering may be an effective strategy for reducing the unintended atmospheric consequences associated with woody crop cultivation.

2.24. Greenhouse Gas and Energy Balance of Willow Biomass Crops Are Impacted by Prior Land Use and Distance from End Users

Volk, T.A.; Yang, S.; Fortier, M.-O. and Therasme, O.

Background: Few life cycle assessment (LCA) studies of willow (*Salix* spp.) biomass crops have investigated the effects and uncertainty of key parameters, e.g., soil organic carbon (SOC) changes associated with land use change and transportation distance with a spatially-explicit perspective.

Methods: This study uses a spatial LCA model that incorporates Geographic Information System (GIS) data in Central and Northern New York, USA to provide specific estimates of greenhouse gas (GHG) emissions and energy return on investment (EROI) of land use conversion and willow biomass production with SOC change in the model.

Results: Ninety-two percent of 9718 suitable parcels for willow biomass production had negative GHG emissions in willow biomass production indicating climate change mitigation potential. The average life cycle GHG emissions in the region were -126.8 kg CO_2eq Mg^{-1} biomass on cropland or pasture land and ranged from -53.2 kg to -176.9 kg CO_2eq Mg^{-1} biomass across the region. However, for grassland converted to willow biomass crops, there was a modeled decrease in SOC resulting in a slightly positive (27.7 kg CO_2eq Mg^{-1} biomass) GHG balance. Changes in SOC associated with land use change, willow crop yield and transportation distance had the greatest impact on GHG emissions. The uncertainty analysis showed large variations of probability distributions of GHG emissions in the five counties arising from differences in these key parameters. The average EROI of 19.2 was not affected by land use change but was influenced by transportation distance and yield.

Conclusions: The results showed substantial potential to reduce GHG emissions in the region by growing willow biomass crops as well as the potential to provide a woody biomass feedstock with a high EROI. Furthermore, these results suggest that willow biomass can serve as a low carbon and high EROI energy source in other regions of the world with similar infrastructure and soil conditions.

2.25. Growth Patterns and Productivity of Hybrid Aspen Clones in Northern Poland

Niemczyk, M.; Kaliszewski, A.; Wojda, T.; Karwański, M. and Liesebach, M.

Background: Rapid growth and favorable wood properties make aspen (*Populus tremula* L.) suitable for production of pulp and paper as well as reconstituted panels and pallets. Thus, plantations of fast-growing aspen and its hybrids seem to be a promising source of wood for satisfying the increasing need for wood-based products. In this study, we assessed the potential of tree improvement to enhance the productivity of aspen and its hybrids in northern Poland.

Methods: We studied a common garden trial that included 15 hybrid aspen clones of *Populus tremula* L. × *Populus tremuloides* Michaux, four hybrid clones of *P. tremula* × *Populus alba* L., and one *P. tremula* intraspecific cross. Hybrid clones had been crossed and selected in Poland and Germany and their growth was previously tested in the countries they were bred. For a comparison, clones of *P. tremula* plus trees selected in Białowieża Forest were studied. Tree height and diameter at breast height (DBH) were measured after four, five, six, and seven growing seasons. We calculated tree volume and the mean annual increment (MAI) for each clone at a given age. Data were analyzed using the Zero Inflated Generalized Linear Mixed Models (ZI-GLMM). Survival probability was calculated with the logistic model and the log-ANOVA model was used for trees that survived. Choice of the optimal model was based on the Akaike Information Criterion (AIC). All statistical analyses were performed with the Statistical Analysis Software (SAS version 9.4, Cary, North Carolina, USA).

Results: Survival probability for all clones was generally high (0.820 to 0.998). Biometric characteristics differed noticeably among clones but were significantly greater for all hybrids (inter- and intraspecific crosses) compared to pure species. Of the 21 analyzed clones, 'Białowieża' as a progeny of pure species had the poorest growth. In contrast, the clone 'Wä 13' (*P. tremula* × *P. tremuloides*) showed the highest DBH, height, volume production, and MAI (25.4 m^3 ha^{-1} year^{-1}) among all studied clones. The mean estimated stem volume of 'Wä 13' at age 7 was 4.6 times greater than pure *P. tremula*. Overall, significant clone by year interaction was observed for height and MAI. During the studied period, clones showed an increasing MAI that progressed over each analysis year.

Conclusions: All hybrid clones performed better than the offspring of pure *P. tremula*. Results of this study contribute to our understanding of aspen hybrid vigor (heterosis) and growth patterns. Experiments with aspen have been rarely completed in Poland. Therefore, our study has important practical implications since we demonstrated that aspen breeding can result in increased biomass productivity that could bring significant economic benefit.

2.26. The Economics of Rapid Multiplication of Elite Hybrid Poplar Biomass Varieties: Expediting the Delivery of Genetic Gains

Stanton, B.J.; Haiby, K.; Shuren, R.; Gantz, C. and Murphy, L.

Background: Poplar (*Populus* spp.) hybridization is key to advancing biomass yields and conversion efficiency. Once superior varieties are selected, there is a lag in commercial use while they are multiplied to scale. Considering the density of bioenergy plantations (3600 stems ha^{-1}) and the sizable area required for refinery operations, the length of the multiplication period impedes the expeditious delivery of genetic gains. The purpose of this study was to determine the level of gain in biomass yield and quality that justifies investment in advanced propagation techniques to speed the time to commercial deployment. Underlying objectives were: (1) quantify varietal variation in the efficiency of in vitro micro-propagation and greenhouse hedge and serial propagation and (2) determine the investment rate of return (IRR) into these propagation techniques at varying levels of genetic gains in biomass yield and quality throughout a 20-year rotation comprising a two-year establishment cycle and six three-year coppice cutting cycles.

Methods: Five varieties were studied by using the Finnish *Populus* × *wettsteinii* Hämet-Ahti model of in vitro shoot proliferation [3], which was followed by greenhouse propagation of succulent cuttings. Micro-cuttings were produced by in vitro culture and used in establishing greenhouse hedges that were sheared five times for succulent cuttings. Cutting yield and survival at each of the five hedgings were compared to five successive rounds of serial propagation. Data were analyzed as a 2 × 5 factorial analysis of variance.

Results: Variation in the time to initiate in in vitro culture (i.e., 26 to 47 days) and in the rate of laboratory shoot proliferation (i.e., 1.7× to 2.1×) was observed among the five varieties. Analysis of variance in the yield of succulent cuttings during greenhouse propagation showed that the main effects and the interaction of the propagation method and the timing of propagation were highly significant. The production of succulent cuttings during hedge propagation exceeded serial propagation by 72%. The seasonal effect was also quite strong with succulent cutting yield maximized during the late spring and the early summer. Economic modeling of returns from investment into the in vitro and greenhouse propagation system following a 20-year rotation of seven biomass harvests showed that attractive IRRs between 5% and 12% were achievable when revenues were increased due to: (1) premium pricing among refineries for improved feedstock that undergoes hydrolysis more cost effectively and (2) incremental biomass yield. The presentation demonstrated the relative change in IRR per-unit increases in yield and price.

Conclusions: Tacamahaca inter-specific hybrids adapted well to the Finnish propagation system. Variation during laboratory and greenhouse propagation can be exploited to improve propagation efficiency. Ultimately, increases in biomass yield and quality are required to justify investment in advanced propagation techniques to speed the deployment of elite varieties.

2.27. The Biomass Production Calculator: A Decision Tool for Hybrid Poplar Feedstock Producers and Investors

Shuren, R.A.; Busby, G. and Stanton, B.J.

Background: GreenWood Resources managed four large-scale hybrid poplar (*Populus* spp.) biomass demonstration farms under the United States Department of Agriculture-National Institute of Food and Agriculture, Agriculture and Food Research Initiative, Coordinated Agriculture Projects (AFRI CAP) program, Advanced Hardwood Biofuels Northwest (AHB) from 2012 to 2018. Each demonstration was managed for a five-year rotation comprised of a two-year establishment cycle and an ensuing three-year coppice cycle during which biomass yields and production activities and costs were compiled and documented. A principal AHB deliverable is an interactive decision tool—the biomass production calculator (BPC)—that growers and developers can use in evaluating hybrid poplar as a renewable energy feedstock crop or as a biomass investment opportunity. The BPC was also designed as a strategic planning tool to optimize feedstock production economics.

Methods: BPC employs a set of linked Excel worksheets that detail the full range of agronomic production activities, the costs of land, equipment, labor, chemicals, fuel, and the expense of biomass harvesting, and transportation. These are integrated with yields and anticipated biomass pricing in deriving internal rates of return specific to each of the main poplar production regions of the Pacific Northwest and northern California, USA. The calculator allows users to specify rotation lengths along with the regularity and duration of the constituent coppice cutting cycles. The frequency of each production activity, the extent of treated area, types of farming equipment, hours of operation, and manual labor needs can be tailored to each grower's situation. Harvesting assumptions are based on single-pass operations that cut, chip, and transfer biomass to off-loading equipment for farm gate delivery and reloading into trucks for bio-refinery transport. Yields are taken from actual inventories of mono-varietal production blocks within each demonstration farm. Users can adjust yields as percent increases or decreases from regional averages for specific varieties and growing conditions.

Results: Biomass yields of the top performing varieties varied among the four farms from 38 to 66 dry metric tons per ha (DMT ha^{-1}) and averaged 51 DMT ha^{-1} for the three-year coppice cycle. The combined costs of crop care and harvesting during the coppice cycle varied between \$1904 and \$2835 per ha. Delivered biomass cost for the coppice ranged from \$60 to \$119 per DMT with higher prices driven by crop care costs including irrigation at the California site.

Conclusions: The biomass production calculator is a valuable decision tool for organizations considering the advisability of adopting poplar as a woody energy crop once basic cost information has been determined. The calculator is used to determine the relative importance of cost components in optimizing production systems. It can be used nationwide to refine future assessments of biomass production potential.

2.28. Growth and Yield of Hybrid Poplar Mono-Varietal Production Blocks for Biofuel Production

Espinoza, J.; Shuren, R.; Zerpa, J. and Stanton, B.J.

Background: GreenWood Resources managed large-scale hybrid poplar (*Populus* spp.) farms ranging in size from 20 to 38 ha in Idaho, Oregon, California, and Washington, USA to demonstrate biomass yields, production costs, and harvesting technology in growing renewable feedstocks under the Advanced Hardwood Biofuel Northwest project (AHB). The purpose of this study was to quantify biomass yields and associate variation in productivity with leaf area indices for a range of clonal varieties suited to each of the four regions in which the demonstration farms were established.

Methods: Between seven and eleven mono-varietal production blocks of hybrid varieties were established at a density of 3588 trees ha^{-1} for each location. Mono-varietal production blocks varied from 2 to 4 ha at each farm. Inventory of permanent plots and destructive sampling of 1, 2, and 3-year-old coppice stands were conducted annually for each variety at all four locations. Yield equations were developed for each variety by least squares regression to predict individual tree dry weight from measurements of breast height diameter and tree height. Fertilizers were not applied during the coppice cycle. The leaf area index (LAI) was measured with a LAI-2200C Plant Canopy Analyzer (LI-COR, Lincoln, Nebraska, USA) for all varieties during the latter part of the third coppice season of the second production cycle.

Results: Yield of top-performing hybrid varieties during the three-year coppice production cycle at the Idaho, Oregon, California, and Washington farms was 15.7, 18.1, 12.9, and 22.2 dry metric tons per ha per year (DMT ha^{-1} year^{-1}), respectively. Varieties of the *Populus maximowiczii* Henry × *Populus deltoides* Marsh. taxon predominated at the Oregon farm while *Populus* × *generosa* Henry and *Populus trichocarpa* Torrey & Gray × *Populus nigra* L. taxa predominated at the Idaho farm. Top-rated varieties at the California farm belonged to the *Populus* × *canadensis* Moench taxon. Production at the Washington farm was highest for *P. deltoides* × *P. maximowiczii* and *P.* × *generosa* varieties. Linear correlations between yield and LAI consistently exhibited moderately good and positive associations at all four farms (e.g., *r* = 0.75–0.89). LAI generally ranged between 3.7 and 5.6. Taxa with the highest LAI were *P. deltoides* × *P. maximowiczii* followed by *P.* × *generosa*.

Conclusions: Important differences in biomass production were found among hybrid varieties at each location with the top-performing clones showing mean annual biomass increments of 13 to 22 DMT ha^{-1} year^{-1}. LAI exhibited a moderately high correlation with biomass production, which implies the potential of LAI to predict biomass production and to use as a selection criterion for highly productive varieties.

2.29. Poplar Productivity as Affected by Physiography and Growing Conditions in the Southeastern USA

Ghezehei, S.B.; Hazel, D.W.; Nichols, E.G. and Maier, C.

Background: Poplars (*Populus* spp.) have high productivity potential as Short Rotation Woody Crops (SRWC) provided that site-suitable clones are planted. The Coastal Plain, Piedmont, and Blue Ridge Mountain physiographic regions make up a significant part of the eastern and the southeastern USA and the assessment of productivity and adaptability of poplars in these regions provides valuable information to support successful large-scale implementation of poplar stands. Our objectives were to examine how the physiography and growing conditions in the Eastern and Southeastern USA can affect green wood productivity of poplar clones.

Methods: Woody biomasses of four-year-old poplar clones were estimated by using a volume equation derived by destructively sampling trees obtained from three physiographic regions (Coastal Plain, Piedmont and Blue Ridge Mountains). Clonal rankings of wood biomass were compared across growing conditions (including irrigation, marginality of lands) and the three physiographic provinces.

Results: Productivity of poplars varied among sites due to differences in site conditions, presumably fertility. There were variations in clonal rankings of wood biomass due to physiography and growing conditions. Changes in wood productivity by poplar due to physiography and growing conditions were more structured at the genotype level than the clonal level.

Conclusions: Even though clones that showed greater biomass variation to growing conditions generally belong to the same genotype, clone-level selection could produce greater wood biomass gains than selection at the genotype-level.

2.30. Hybrid Poplar Stock Type Impacts Height, DBH, and Estimated Total Dry Weight at Eight Years in a Hybrid Poplar Plantation Network

Hillard, S.C. and Froese, R.E.

Background: There are a number of plantation initiation decisions a potential hybrid poplar (*Populus* spp.) producer must decide upon. Major considerations include clone selection as well as planting stock, which impact initiation costs, growth, and yield. The long-term consequences for the initial choices of clone and planting stock on productivity are not fully understood. The objectives for the study were to utilize an eight-year annual growth data set from a hybrid poplar plantation network in northern Michigan, USA, to evaluate: (1) tip dieback in the out-planted stock, (2) growth models for height and diameter at breast height (DBH) using a mixed model approach, (3) performance of the clone × stock combinations using the growth models, and (4) assess potential financial outlook.

Methods: Eight years of growth data for three plantations of hybrid poplar clones (*Populus* × *canadensis* Moench 'DN34' and *Populus nigra* L. × *Populus maximowiczii* Henry 'NM6') and three stock types (poles, rooted stock and cuttings) were utilized to develop mixed effects models describing height and DBH growth for each stock type and clone combination. Models used nested random effects for individual tree, site, and treatment type. Planting stock (poles and rooted stock) tip dieback were evaluated by using mixed models with random effects for treatment type and site. Differences were tested by using the general linear hypothesis within species while between species differences were evaluated with *t*-tests and the Tukey's honestly significant difference test. The DBH model was used to inform an allometric model for hybrid poplar biomass and a financial model developed based on recent economic parameters.

Results: Differences in dieback were significant between clones for rooted stock but not pole stock and not significant between 'DN34' rooted stock and 'NM6' pole stock. Cutting stock differences were not significant in the height models while pole and rooted stock heights showed significant differences with 'DN34' yielding greater growth compared to 'NM6' on these sites. All growing stocks showed significant differences in height yields. DBH models were only significant between cuttings for the clones, while within the clone, there was no difference in DBH for cuttings and rooted stock. Significant differences in DBH were found between poles and rooted stock and poles and cuttings within clone. The estimated productivity of these treatments produced plantations with varying financial outlook at estimated peak mean annual increment. Overall 'NM6' pole stock was the closest to achieving a positive financial outlook.

Conclusion: This study demonstrates an evaluation of many initial choices available to a potential grower and how they impact growth as well as financial outlook of the operation. Poles had the most positive height growth outlook. Due to the high cost of poles, cuttings that showed DBH growth similar to rooted stock may be a viable option to reduce initiation costs. Efforts to reduce costs of inputs and harvest as well as increase productivity would improve financial outlook for biomass production systems.

2.31. Developing Phytoremediation Technology Using Pseudomonas Putida and Poplar for Restoring Petroleum-Contaminated Sites

Dewayani, A.A.; Zalesny, R.S., Jr.; Jose, S.; Nagel, S.C. and Lin, C.H.

Background: The majority of energy sources used for daily life in Indonesia and the United States are derived from petroleum and natural gas. There are more than 750 chemicals used for mining processes and many of these chemicals behave as endocrine disruptors for humans and animals. The organic pollutants, benzene, toluene, ethylbenzene, and xylene (BTEX) have been found in contaminated soils and water at and/or near mining sites. Due to the increasing number of petroleum mining operations and the potential increase of shale gas exploitation, it is necessary to develop a cost-effective and environmentally-friendly restoration strategy to reduce exposure of the remaining toxic chemicals into ecosystems. The objective of this study was to develop methods to degrade BTEX compounds especially toluene by introducing the bacterium degrader *Pseudomonas putida* into the rhizosphere of *Populus* spp. seedlings.

Methods: *Pseudomonas putida* has the todC1 gene that is responsible for the production of toluene dioxygenase enzyme, which enables *P. putida* to use toluene as the sole carbon source and take up toluene from the soil ecosystem. The *P. putida* could be introduced to the rhizosphere of poplar for restoring petroleum-contaminated sites. Solid-phase micro extraction (SPME) followed by gas chromatography mass spectrometry (GCMS) analyses were used to determine the degradation rates of toluene by the bacterium in a liquid phase experiment. The persistence of the introduced degraders was assessed by quantitative real-time polymerase chain reaction (*q*RT-PCR).

Results: During the 24-h incubation time, the concentration of toluene in the system was reduced 97% from 412 to 8.7 ppm. Current efforts are to evaluate the persistence of the introduced degraders in the poplar system and calculate the degradation kinetics of the pollutants.

Conclusions: The overarching outcome of this research is that such methods show the potential for refinement and application to help protect public health and restore ecological balance at mining sites in Indonesia, the USA, and other oil producing countries.

2.32. Comparison of Statistical Techniques for Evaluating the Fiber Composition of Early Rotation Pine and Hardwood Trees for the Production of Cellulose

Foust, A.M. and Headlee, W.L.

Background: With characteristics such as high strength and flexibility, cellulose has garnered much attention for its potential use in producing nano-material for the biomedical and engineering fields. Because cellulose naturally occurs in wood, small diameter trees from early rotational thinnings could be prime source material for nano-cellulose products. Ideal source material for nano-cellulose production would contain high cellulose and low lignin concentrations to expedite the cellulose extraction process. The selection process for source material would be assisted by a statistical technique that employs a single test for differences in recalcitrance rather than separate tests for each fiber component. The purpose of this study was to determine the fiber composition (i.e., percentage of nonpolar extracts (NPE), hemicellulose, cellulose, and combined lignin and ash) of small diameter pine and hardwood trees, compare the results of a single cumulative logit model using ordinal multinomial regression (OMR) analysis with those of a nested analysis of variance (ANOVA) with type 3 sums of squares analysis for each fiber component, and to determine if the former could provide a simpler test for species and spacing effects on overall recalcitrance.

Methods: Tree cores were collected from three different sites that had undergone an early rotational thinning or were of similar size and age class that would benefit from such a thinning. Sites include a hardwood spacing trial near Monticello, Arkansas, USA, containing sweetgum (*Liquidambar styraciflua* L.) planted at two spacings that underwent two thinning treatments. A loblolly pine (*Pinus taeda* L.) spacing study near the University of Arkansas Pine Tree Research Station with three replications of four spacing arrangements and a study at the University of Arkansas Pine Tree Research Station with four replications each of three oak species: cherrybark (*Quercus pagoda* Raf.), Nuttall (*Quercus texana* Buckl.), and Shumard (*Quercus shumardii* Buckl.). The fiber composition of tree cores from each study was analyzed using the ANKOM 2000 Automated Fiber Analyzer (ANKOM Technology, Macedon, New York, USA). Results were analyzed cumulatively using OMR analysis and by individual fiber components using ANOVA.

Results: In general, *p*-values from most of the OMR analyses for total fiber composition fell within the range of *p*-values from the ANOVA analyses for individual fiber components. The only instance that the OMR *p*-value did not fall within the span of ANOVA *p*-values was the spacing effect for the loblolly pine. There was a significant shift in the fiber composition for the plot within-spacing effect of the loblolly pine, which indicates differences in fiber composition between plots that may be due to tree survival differences. There was also a significant change in fiber composition among the three oak species with cherrybark oak showing a significant difference in fiber composition compared to Shumard oak.

Conclusions: Fiber composition can be analyzed more efficiently by conducting one OMR analysis compared to an ANOVA for each individual component. Since the OMR analysis is able to assess overall recalcitrance, it can more readily track significant shifts in fiber composition in response to treatment variables and has the added benefit of its confidence intervals being constrained within values of 0% to 100%. Thus, using the OMR analysis is a more convenient and biologically sound method for evaluating ideal source material for cellulose production.

2.33. Freshkills Anthropogenic Succession Study: Phase I: Deer Cafeteria Study

Hallett, R.A.; Piana, M.; Johnson, M.; Simmons, B. and Zalesny, R.S., Jr.

Background: Freshkills Park is an 890-ha public park being built on top of a landfill reclamation project on Staten Island in New York City, New York, USA. The Freshkills anthropogenic succession study is a designed experiment that was installed "off cap" but is on top of a "legacy dump." The experimental goal is to test three planting palettes for their ability to out compete exotic invasive plant species and become established on a site with highly disturbed soils.

Methods: Palette 1 included willow (*Salix* spp.) and poplar (*Populus* spp.) genotypes selected using phyto-recurrent selection techniques. Palette 2 included a mix of 18 species of shrubs and trees selected by the New York City Department of Parks and Recreation for the site. Palette 3 was a 50/50 mix of Palettes 1 and 2. The study area was not fenced and there are 52 white-tailed deer (*Odocoileus virginianus* Zimmermann) per km^2 on Staten Island. Tree growth and diameter were recorded along with mortality and deer browse severity.

Results: Early successional species had the highest mortality and were preferentially browsed. In the spring of the second growing season after planting, there was over 82% mortality on the site.

Conclusions: The process of creating a designed experiment in New York will be described along with future plans for what will hopefully be a long-term study on how to grow a forest in the city.

2.34. Adapting an Aspen Short Rotation Yield Model to Represent Hybrid Poplar Yield for Regional Hybrid Poplar Production Estimation

Hillard, S.C. and Froese, R.E.

Background: The upper Great Lakes Region (i.e., Michigan, Minnesota, and Wisconsin, USA) has abundant underutilized agricultural land that is suitable for hybrid poplar (*Populus* spp.) production. Interest in primary woody crop production systems is often hampered by low energy prices. Equally important, however, is the lack of targeted policies to encourage deployment. Current regional production estimates could benefit from improvements in scale. Improving the scale of regional productivity estimates could provide needed context to local legislators, policy makers, and stakeholders, which reduces barriers to increased production and improves targeted policies. Hybrid poplar generally lacks specific site index curves and growth and yield models that tie into remotely-sensed data to estimate productivity. The objectives for this study were to: (1) adapt a short rotation aspen (*Populus tremuloides* Michaux) growth and yield model to represent hybrid poplar and developing a link between aspen site index and hybrid poplar productivity, (2) apply the new growth model to a regional site index model for aspen, and (3) compare the new regional production estimates to previous regional productivity estimates.

Methods: Hybrid poplar yield was adapted to estimated aspen yield by tuning an existing aspen growth and yield model for the Great Lakes Region. Tuning used observed hybrid poplar heights from a plantation network and modeled aspen heights. Linear regression was used to develop a scaling function between aspen yield and hybrid poplar yield, which resulted in the development of a link between aspen site index and poplar biomass. A regional site index model developed for aspen was used to apply the model to extrapolate predicted hybrid poplar yield across the region. Spatial outputs were compared to previous regional production estimates via gridded raster outputs.

Results: Linear models linking aspen and hybrid poplar biomass had R^2 values ranging from 0.72 to 0.91 with standard errors of 0.03 to 0.10 Mg ha^{-1}. Using a Moderate Resolution Imaging Spectroradiometer (MODIS) derived site index model for aspen improved resolution of poplar yield estimates from 32 km, as reported in previous efforts, to 250 m. Model-estimated productivity for the hybrid poplar clone *Populus* × *canadensis* Moench 'DN34' ranged from <2 Mg ha^{-1} year^{-1} to 9 Mg ha^{-1} year^{-1} at age 10 while that of clone *Populus nigra* L. × *Populus maximowiczii* Henry 'NM6' ranged from <2 Mg ha^{-1} year^{-1} to 15 Mg ha^{-1} year^{-1} at age 10. For both models, this was approximately 5 Mg ha^{-1} year^{-1} less on average compared to previous models. The new regional estimates captured the same longitudinal gradient in productivity found previously and improved upon more localized yield estimates, which may have been impacted before by data limitations. Modeled yield estimates were compared to literature values using equivalence tests with no evidence to reject the null hypothesis of similarity.

Conclusions: The results indicated that an intragenus yield model may be adapted effectively to represent a species lacking a yield model. This technique can be applied to take advantage of remotely sensed variables and improve spatial resolution of production estimations. Improving resolution of estimates is vital for more cost-effective targeting of potentially productive areas, which improves prospects for short rotation woody crop deployment.

2.35. Potential Biomass Production of Four Cottonwood Clones Planted at Two Densities in the Arkansas River Valley, USA

Liechty, H.O. and Headlee, W.L.

Background: Eastern cottonwood (*Populus deltoides* Marsh.) is frequently planted in afforestation projects on retired agricultural fields in the Arkansas and Mississippi Alluvial Valleys, USA. Interest in cellulosic bioenergy feedstock production has increased the need to better understand the biomass production potential of various cottonwood clones grown at high densities with short rotations within these regions. The purpose of this study was to: (1) assess potential biomass production of four eastern cottonwood clones ('S7C20', 'S7C8', 'S7C4', 'F2') grown on a six-year rotation and (2) evaluate differences in growth, stem, and crown dimensional characteristics with different planting densities (8442 and 16,885 trees ha^{-1}).

Methods: Cuttings were planted in eight double rows (0.76 m between rows, 1.83 m between double rows) on either a 0.46 m or 0.91 m spacing in 2011. This planting design was replicated at three locations (row lengths of 345 m) in a retired sod farm field adjacent to the Arkansas River levee in Central Arkansas. In 2016, survival was visually assessed along the six double rows at each of the three clone, density replications. A total of eight plots (6.4 m long) encompassing the two middle double rows were located within the three clone, density replications where 75% or more of the planted trees survived. Each plot include 28 or 56 planted cutting locations depending on the initial establishment spacing. Diameter at breast height (DBH), total tree height, height to live crown, crown dieback, and crown class were determined. Tree biomass was determined from DBH and height measurement using equations developed by Dipesh et al. [4].

Results: Total dry woody biomass accumulation averaged between 26.3 and 65.5 Mg ha^{-1} (4.38 and 10.91 Mg ha^{-1} year^{-1}) for the eight clone, density combinations. Mean annual dry wood production for the highest and lowest establishment densities was 9.78 ± 1.59 and 7.81 ± 1.71 Mg ha^{-1} year^{-1}, respectively. Production significantly differed between high and low planting densities (10.79 and 4.39 Mg ha^{-1} year^{-1}) for only the 'S7C20' clone. Production was similar among clones except within the low density treatment where 'S7C20' had significantly lower total wood accumulation than the other three clones. The lower production of the 'S7C20' clone at this density treatment appeared to be related to the site conditions at which the plots were located rather than any inherent survival or characteristics associated with the clone.

Conclusions: Potential production in this study represented a best case scenario where survival of the planting stock was optimal. Under these conditions, wood production after six growing seasons was similar for the four selected clones and two planting densities. Thus, any productivity advantage associated with higher planting density had dissipated by the sixth year. Rotations of less than six years may result in higher production rates with the higher planting density relative to the lower planting density.

2.36. Growth Performance and Stability of Hybrid Poplar Clones in Simultaneous Tests on Six Sites in Minnesota, USA

Nelson, N.D.; Berguson, W.E.; McMahon, B.G.; Cai, M. and Buchman, D.

Background: Commercial adoption of poplar (*Populus* spp.) plantations requires genetic improvements for increased and consistent yield, disease resistance, and broadened adaptivity across a range of climate and soil types. Genotype × environment interaction (G × E) is a limitation that relates directly to adaptability, which complicates growth performance testing and reduces overall genetic gains. The objectives of this study were to: (1) identify clones that are superior in growth rate, disease resistance, and had acceptable genetic stability and G × E across different environments as candidates for moving into field tests approximating commercial conditions (i.e., yield blocks), (2) investigate clonal stability and G × E of the clones tested, and (3) discover general principles for understanding, controlling, and using clonal stability and G × E in hybrid poplar genetic improvement programs.

Methods: Growth, stability, and G × E were investigated for sixty-nine clones after five years at six agricultural sites in Minnesota, USA. Fifty-three of the clones were *Populus deltoides* Marsh. × *Populus nigra* L. (D × N) crosses, nine *P. deltoides* × *Populus maximowiczi* Henry, and seven other crosses of which most were previously screened in Minnesota. The experiment was a randomized complete block design with six blocks within each site and a clone trial, which is part of our sequential breeding and testing system.

Results: Five-year diameter (DBH) and basal area (BA) at 1.38-m height averaged 93.5 mm and 72.11 cm^2, respectively, over the six sites. DBH site means varied from 79.4 mm to 109.0 mm. The fastest-growing clone BA was 64% and 49% larger than the mean of the two commercial standards and the mean of the population, respectively. Clone, clone × site, and error were highly significant in the analysis of variance (ANOVA). The variance component for clone was over twice that of clone × site (G × E), which was promising for achieving genetic gains. Clonal rank did not change among sites. G × E interaction was dominated by relative performance differences of clones on the different sites with 26% of clones being stable (little change in growth among sites) and 74% unstable. Stability coefficients of the unstable clones varied over a 99% range. Only 15% of clones were both stable and fast growing. Seven D × N clones were selected for future testing in yield blocks based on growth, stability, and disease occurrence.

Conclusions: There was a strong and significant clone effect and it predominated over G × E. Clone stability and G × E are separate parameters. A collection of clones can be unstable, but exhibit low variance for stability, which results in low G × E. Most clones we studied were unstable and stability was widely variable. Selecting clones within the 85th percentile of growth gave an estimated genetic gain of 53% versus commercial control clones. Continuing large genetic improvements in yield are expected in further generations of a breeding system like ours that utilizes clonal selection within a family structure, which captures both additive and non-additive inheritance. Intentionally reducing G × E does not appear to be a feasible objective for inter-specific poplars in an applied breeding program i.e., without sacrificing productivity gains. This is because the fastest-growing clones have large variability in clonal stability and this variation is the driver of G × E.

2.37. Genetic Development, Evaluation, and Outreach for Establishing Hybrid Poplar Biomass Feedstock Plantations in the Midwestern United States

Nelson, N.D.; Host, G.E.; Lazarus, W. and Reichenbach, M.R.

Background: Hybrid poplar (*Populus* spp.) represents a promising long-term biomass feedstock for biofuels and the emerging bioproducts industry with significant potential for improving rural economies of the Midwestern United States. The economic viability of hybrid poplar is limited, however, by the current scale of breeding, selection, and testing and varies with shifts in commodity and petroleum prices and energy policies. Commercial adoption of poplar plantations requires genetic improvements for increased and consistent yield, disease resistance, and broadened adaptability across a range of climate and soil types. This project, commencing in 2018, leverages the extensive collection of *Populus* genetic material developed over several decades at the University of Minnesota Duluth-Natural Resources Research Institute to produce the next generation of genotypes for the Midwest, which is an epicenter for anticipated large-scale adoption of poplar culture. Specifically, we will test the hypothesis based on a preceding test in Indiana, that *Populus* inter-specific parentages incorporating *Populus deltoides* Marsh. originating from southern Minnesota and *Populus nigra* L. from a pan-European collection. Both species screened in northern Minnesota will provide fast growth and acceptable genotype × environment interaction (G × E) for sites as far south as southern Indiana. This will be accomplished by: (1) selecting promising genotypes based on our past breeding and field trials from a 500+ clone germplasm bank, (2) establishing replicated clonal field trials in Minnesota, Iowa, and Indiana, (3) annually monitoring survival, growth, and disease resistance, and (4) assessing G × E to identify broadly adapted clones. Ancillary novel experiments will test whether: (1) within-clone variation within sites can be used to predict G × E among sites and (2) nitrogen fertilization changes clonal rankings and G × E. The genetics component will be complemented by an assessment, which is closely linked to the genetics results of economic trade-offs between poplar versus other commodity production and integrated into regional Extension outreach programs. An Extension innovation will be the targeting of emerging bioenergy and bioproducts companies who may be seeking a captive biomass source from a dedicated energy crop on agricultural land.

Conclusions: This project will focus on the economically-viable deployment of improved poplar clones across the Midwest. Fundamental project outcomes are quantified genetic advances to improve the economics of poplar feedstock plantations and reduced production and investment risk for farmers and industry stakeholders.

2.38. Citizen Science in the Greenhouse: A Phytoremediation Case Study

Rogers, E.R.; Zalesny, R.S., Jr.; Hallett, R.A. and Westphal, L.M.

Background: Phyto-recurrent selection has been implemented for over a decade to determine superior *Salix* and *Populus* genotypes for field-scale phytotechnologies. Citizen science is the collection of data by scientifically-trained members of the general population. This study compares data collected by different citizen science cohorts and experts to determine the potential for citizen scientists to help with the phyto-recurrent selection process. The primary goals of this research were to: (1) assess the similarity of citizen-collected data compared with expert-collected data and (2) evaluate the effectiveness of a specifically-designed tree health rubric in collecting accurate phyto-recurrent selection data.

Methods: Three groups of eight citizens (high school students, science teachers, and young professionals) collected tree health data on 48, 49-day-old *Populus* and *Salix* trees. One group of eight experts also collected tree health data on the same trees. The trees were grown in six soils (five landfill soils and on control soil) in a greenhouse in Rhinelander, Wisconsin, USA. A health rubric developed for urban afforestation projects was adapted for the current study. Assessment variables were: height, diameter, leaf count, discoloration, and live crown ratio. Each group received the same training session on how to use the rubric and how to collect data. Comparisons were made assuming an acceptable range of accuracy of 10%.

Results: The citizen-collected averages for height, diameter, leaf count, and live crown ratio were all within 7% of the corresponding averages from the expert-collected data. The greatest variability occurred in the diameter average reported by the science teacher and student groups. The averages were 6.53% larger (*Salix*) and 5.02% larger (*Populus*) than that of the expert group, respectively. The least variability occurred in the average live crown ratio in which all citizen groups were within 2.53% of the expert group's average.

Conclusions: From these results, citizen-collected data were within an acceptable range of accuracy when compared to expert-collected data, which indicates that citizen scientists could play a valuable role in the evaluation of genotypes to predict future success when planted in the field.

3. Affiliations

Arsenov, D., Department of Biology and Ecology, University of Novi Sad, Novi Sad, Serbia.

Bauer, E.O., Institute for Applied Ecosystem Studies, Northern Research Station, USDA Forest Service, Rhinelander, Wisconsin, USA.

Benzel, T., Benzel Soil Services, LLC, Mercer, Wisconsin, USA.

Berguson, W.E., Consultant, retired Natural Resources Research Institute, University of Minnesota Duluth, Duluth, Minnesota, USA.

Borišev, M., Department of Biology and Ecology, University of Novi Sad, Novi Sad, Serbia.

Buchman, D., Natural Resources Research Institute, University of Minnesota Duluth, Duluth, Minnesota, USA.

Buechel, L., Waste Management of Wisconsin, Inc., Menomonee Falls, Wisconsin, USA.

Bura, R., Biofuels and Bioproducts Laboratory, School of Environmental and Forest Sciences, University of Washington, Seattle, Washington, USA.

Burken, J.G., Civil, Architectural, and Environmental Engineering, Missouri University of Science and Technology, Rolla, Missouri, USA.

Busby, G., GreenWood Resources, Inc., Portland, Oregon, USA.

Cacho, J.F., Environmental Science Division, Argonne National Laboratory, Lemont, Illinois, USA.

Cai, M., Natural Resources Research Institute, University of Minnesota Duluth, Duluth, Minnesota, USA.

Campbell, P., Environmental Science Division, Argonne National Laboratory, Lemont, Illinois, USA.

Cave, R.D., University of Florida, Institute of Food and Agricultural Sciences, Indian River Research and Education Center, Fort Pierce, Florida, USA.

Chowyuk, A., Biofuels and Bioproducts Laboratory, School of Environmental and Forest Sciences, University of Washington, Seattle, Washington, USA.

Cole, C., Department of Entomology, University of Wisconsin-Madison, Madison, Wisconsin, USA.

Da Ros, L.M., Department of Wood Science, University of British Columbia, Vancouver, British Columbia, Canada.

Dao, C., Biofuels and Bioproducts Laboratory, School of Environmental and Forest Sciences, University of Washington, Seattle, Washington, USA.

DeBauche, B.S., Civil, Architectural, and Environmental Engineering, Missouri University of Science and Technology, Rolla, Missouri, USA.

de Souza, D.P., College of Environmental Science and Forestry, State University of New York, Syracuse, New York, USA.

Desrochers, V., Plant Biology Research Institute (IRBV), Montreal, Québec, Canada.

Dewayani, A.A., Center for Agroforestry, and School of Natural Resources, University of Missouri, Columbia, Missouri, USA.

Drekić, M., Institute of Lowland Forestry and Environment, University of Novi Sad, Novi Sad, Serbia.

Dvoržak, D., Department for EU Programs and Projects & Economy, Section for Preparation and Implementation of Programs and Projects, City of Osijek, Osijek, Croatia.

Ellis, M.F., Green Carbon Solutions, Pepper Pike, Ohio, USA.

Eisenbies, M.H., College of Environmental Science and Forestry, State University of New York, Syracuse, New York, USA.

Espinoza, J., GreenWood Resources, Inc., Portland, Oregon, USA.

Fabio, E.S., School of Integrative Plant Science, Horticulture Section, Cornell University, Cornell AgriTech, Geneva, New York, USA.

Fištrek, Ž., Energy Institute Hrvoje Požar, Department for Renewable Energy Sources, Energy Efficiency & Environmental Protection, Zagreb, Croatia.

Fortier, M.-O., College of Environmental Science and Forestry, State University of New York, Syracuse, New York, USA.

Foust, A.M., Arkansas Forest Resource Center, University of Arkansas, Monticello, Arkansas, USA.

Froese, R.E., School of Forest Resources and Environmental Science, Michigan Technological University, Houghton, Michigan, USA.

Gantner, R., Faculty of Agriculture, J.J. Strossmayer University of Osijek, Osijek, Croatia.

Gantz, C., GreenWood Resources, Inc., Portland, Oregon, USA.

Ghezehei, S.B., Department of Forestry and Environmental Resources, North Carolina State University, Raleigh, North Carolina, USA.

Glavaš, H., Faculty of Electrical Engineering, Computer Science and Information Technology, J.J. Strossmayer University of Osijek, Osijek, Croatia.

Gustafson, R., Biofuels and Bioproducts Laboratory, School of Environmental and Forest Sciences, University of Washington, Seattle, Washington, USA.

Haiby, K., GreenWood Resources, Inc., Portland, Oregon, USA.

Hallett, R.A., Urban Forests, Human Health, and Environmental Quality, Northern Research Station, USDA Forest Service, Durham, New Hampshire, USA.

Hart, N.M., Advanced Hardwood Biofuels Northwest, Washington State University Extension, Lynnwood, Washington, USA.

Hazel, D.W., Department of Forestry and Environmental Resources, North Carolina State University, Raleigh, North Carolina, USA.

He, Z., University of Florida, Institute of Food and Agricultural Sciences, Indian River Research and Education Center, Fort Pierce, Florida, USA.

Headlee, W.L., Arkansas Forest Resource Center, University of Arkansas, Monticello, Arkansas, USA.

Henderson, D., AECOM, Milwaukee, Wisconsin, USA.

Hillard, S.C., Division of Forestry, Minnesota Department of Natural Resources, St. Paul, Minnesota, USA.

Host, G.E., Natural Resources Research Institute, University of Minnesota Duluth, Duluth, Minnesota, USA.

Hu, Y., Department of Renewable Resources, University of Alberta, Edmonton, Alberta, Canada.

Isebrands, J.G., Environmental Forestry Consultants, LLC, New London, Wisconsin, USA.

Ivezić, V., Faculty of Agriculture, J.J. Strossmayer University of Osijek, Osijek, Croatia.

Johnson, M., Urban Forests, Human Health, and Environmental Quality, Northern Research Station, USDA Forest Service, Bayside, New York, USA.

Jose, S., Center for Agroforestry, and School of Natural Resources, University of Missouri, Columbia, Missouri, USA.

Kaliszewski, A., Department of Forest Resources Management, Forest Research Institute, Raszyn, Poland.

Karwański, M., The Faculty of Applied Informatics and Mathematics, Warsaw University of Life Sciences, Warsaw, Poland.

Katanić, M., Institute of Lowland Forestry and Environment, University of Novi Sad, Novi Sad, Serbia.

Kebert, M., Institute of Lowland Forestry and Environment, University of Novi Sad, Novi Sad, Serbia.

Kiel, S., Department of Biology, Portland State University, Portland, Oregon, USA.

Kulišić, B., Energy Institute Hrvoje Požar, Department for Renewable Energy Sources, Energy Efficiency & Environmental Protection, Zagreb, Croatia.

Lazarus, W., Department of Applied Economics, University of Minnesota Twin Cities, St. Paul, Minnesota, USA.

Liechty, H.O., Arkansas Forest Resource Center, University of Arkansas, Monticello, Arkansas, USA.

Liesebach, M., Thünen Institute of Forest Genetics, Grosshansdorf, Germany.

Lin, C.H., Center for Agroforestry, University of Missouri; and, School of Natural Resources, University of Missouri, Columbia, Missouri, USA.

Lindroth, R., Department of Entomology, University of Wisconsin-Madison, Madison, Wisconsin, USA.

Liu, R., University of Florida, Institute of Food and Agricultural Sciences, Indian River Research and Education Center, Fort Pierce, Florida, USA.

Maier, C., Southern Institute of Forest Ecosystem Biology, Southern Research Station, USDA Forest Service, Research Triangle Park, North Carolina, USA.

Mansfield, S.D., Department of Wood Science, University of British Columbia, Vancouver, British Columbia, Canada.

McIvor, I., Plant & Food Research, Palmerston North, New Zealand.

McMahon, B.G., Natural Resources Research Institute, University of Minnesota Duluth, Duluth, Minnesota, USA.

Morrow, C., Department of Forest and Wildlife Ecology, University of Wisconsin-Madison, Madison, Wisconsin, USA.

Murphy, L., GreenWood Resources, Inc., Portland, Oregon, USA.

Nagel, S.C., School of Medicine, University of Missouri, Columbia, Missouri, USA.

Negri, M.C., Environmental Science Division, Argonne National Laboratory, Lemont, Illinois, USA.

Nelson, N.D., Natural Resources Research Institute, University of Minnesota Duluth, Duluth, Minnesota, USA.

Nichols, E.G., Department of Forestry and Environmental Resources, North Carolina State University, Raleigh, North Carolina, USA.

Niemczyk, M., Department of Silviculture and Forest Tree Genetics, Forest Research Institute, Raszyn, Poland.

Nikolić, N., Department of Biology and Ecology, University of Novi Sad, Novi Sad, Serbia.

Orlović, S., Institute of Lowland Forestry and Environment, University of Novi Sad, Novi Sad, Serbia.

Pekeč, S., Institute of Lowland Forestry and Environment, University of Novi Sad, Novi Sad, Serbia.

Peterson, M., Waste Management of Wisconsin, Inc., Menomonee Falls, Wisconsin, USA.

Piana, M., Department of Ecology, Evolution and Natural Resources, Rutgers University, New Brunswick, New Jersey, USA.

Pilipović, A., Institute of Lowland Forestry and Environment, University of Novi Sad, Novi Sad, Serbia.

Pohajda, I., Croatian Agriculture & Forestry Advisory Service, Zagreb, Croatia.

Poljaković-Pajnik, L., Institute of Lowland Forestry and Environment, University of Novi Sad, Novi Sad, Serbia.

Potosnak, M.J., Department of Environmental Science and Studies, DePaul University, Chicago, Illinois, USA.

Quinn, J.J., Environmental Science Division, Argonne National Laboratory, Lemont, Illinois, USA.

Reichenbach, M.R., Cloquet Forestry Center, University of Minnesota Twin Cities, Cloquet, Minnesota, USA.

Riehl, J.F., Department of Entomology, University of Wisconsin-Madison, Madison, Wisconsin, USA.

Rockwood, D.L., Florida FGT, LLC, Gainesville, Florida, USA.

Rogers, E.R., Institute for Applied Ecosystem Studies, Northern Research Station, USDA Forest Service, Rhinelander, Wisconsin, USA.

Rosenstiel, T.N., Department of Biology, Portland State University, Portland, Oregon, USA.

Seegers, R., Waste Management of Wisconsin, Inc., Whitelaw, Wisconsin, USA.

Shuren, R., GreenWood Resources, Inc., Portland, Oregon, USA.

Simmons, B., New York City Urban Field Station, New York City Department of Parks & Recreation, Bayside, New York, USA.

Smart, L.B., School of Integrative Plant Science, Horticulture Section, Cornell University, Cornell AgriTech, Geneva, New York, USA.

Soolanayakanahally, R.Y., Indian Head Research Farm, Agriculture and Agri-Food Canada, Indian Head, Saskatchewan, Canada.

Ssegane, H., Climate Corporation, St. Louis, Missouri, USA.

Stanton, B.J., GreenWood Resources, Inc., Portland, Oregon, USA.

Therasme, O., College of Environmental Science and Forestry, State University of New York, Syracuse, New York, USA.

Thomas, B.R., Department of Renewable Resources, University of Alberta, Edmonton, Alberta, Canada.

Townsend, P.A., Advanced Hardwood Biofuels Northwest, Washington State University Extension, Lynnwood, Washington, USA.

Van Acker, J., Laboratory of Wood Technology, Ghent University, Ghent, Belgium.

Volk, T.A., College of Environmental Science and Forestry, State University of New York, Syracuse, New York, USA.

Westphal, L.M., People and Their Environments, and The Strategic Foresight Group, Northern Research Station, USDA Forest Service, Evanston, Illinois, USA.

Wiese, A.H., Institute for Applied Ecosystem Studies, Northern Research Station, USDA Forest Service, Rhinelander, Wisconsin, USA.

Wojda, T., Department of Silviculture and Forest Tree Genetics, Forest Research Institute, Raszyn, Poland.

Yang, S., College of Environmental Science and Forestry, State University of New York, Syracuse, New York, USA.

Zalesny, R.S., Jr., Institute for Applied Ecosystem Studies, Northern Research Station, USDA Forest Service, Rhinelander, Wisconsin, USA.

Zerpa, J., GreenWood Resources, Inc., Barranquilla, Atlántico, Colombia.

Zumpf, C.R., Environmental Science Division, Argonne National Laboratory, Lemont, Illinois, USA.

Župunski, M., Department of Biology and Ecology, University of Novi Sad, Novi Sad, Serbia.

Author Contributions: W.L.H., J.R., R.Y.S. and R.S.Z.J. conceptualized this Conference Report. S.B.G., E.S.G., W.L.H., J.R., B.J.S., R.Y.S. and R.S.Z.J. reviewed, selected, and edited abstract submissions. E.S.G. compiled the first draft of this Conference Report for coauthor review. S.B.G., W.L.H., J.R., B.J.S., R.Y.S. and R.S.Z.J. provided review and editing of the initial draft. Authors of presentations contributed their individual abstract submissions.

Acknowledgments: We are grateful to the professional and efficient international team of conference organizers who helped to make this conference possible. In addition, we thank our sponsors for supporting these efforts. Erik Schilling, Tracy Stubbs, and Tammerah Garren from NCASI deserve special recognition for their unwavering commitment to conference planning along with Tammy Booth who developed our logo. Likewise, we thank Bernie McMahon for developing a world-class pre-conference tour. In addition, we are grateful to Judy Heikkinen, Chad Lashua, and Hakim Salaam for assistance with hosting the conference at Nicolet College. Lastly, we thank the presenters and participants for contributing to the networking and technology transfer during the field tours and technical sessions.

Conflicts of Interest: The authors declare no conflict of interest.

Appendix A

Table A1. List of presentations delivered at the 2018 Woody Crops International Conference, Rhinelander, Wisconsin, USA, 22–27 July 2018.

1.	Current Trends and Challenges in North American Poplar Breeding
2.	Investigation of Phytoremediation Potential of Poplar and Willow Clones in Serbia: A Review
3.	Potential for the Agricultural Sector to Produce Poplar Wood as Contribution to the Forestry Wood Industry Chain
4.	Reaching Economic Feasibility of Short Rotation Coppice (SRC) Plantations by Monetizing Ecosystem Services: Showcasing the Contribution of SRCs to Long Term Ragweed Control in the City of Osijek, Croatia
5.	Genetic Parameter Estimates for Coppiced Hybrid Poplar Bioenergy Trials
6.	Genetic and Environmental Effects on Variability in First-rotation Shrub Willow Bark and Wood Elemental Composition
7.	Is Hybrid Vigor Possible in Native Balsam Poplar Breeding?
8.	Uncovering the Genetic Architecture of Growth-defense Tradeoffs in a Foundation Forest Tree Species
9.	The Great Lakes Restoration Initiative: Reducing Runoff from Landfills in the Great Lakes Basin, USA
10.	Evaluating Poplar Genotypes for Future Success: Can Phyto-Recurrent Selection Assessment Techniques be Simplified?
11.	Growth of Poplars in Soils Amended with Fiber Cake Residuals from Paper and Containerboard Production
12.	Growth and Physiological Responses of Three Poplar Clones Grown on Soils Artificially Contaminated with Heavy Metals, Diesel Fuel and Herbicides
13.	Mitigating Downstream Effects of Excess Soil Phosphorus through Cultivar Selection and Increased Foliar Resorption
14.	Survival and Growth of Poplars and Willows Grown for Phytoremediation of Fertilizer Residues
15.	Stakeholder Assessment of the Feasibility of Poplar as a Biomass Feedstock and Ecosystem Services Provider in Rural Washington, USA
16.	Barriers and Opportunities for use of Short Rotation Poplar for the Production of Fuels and Chemicals
17.	Short Rotation Eucalypts: Opportunities for Bio-Products
18.	Variability of Harvester Performance Depending on Phenotypic Attributes of Short Rotation Willow Crop in New York, USA
19.	Cover Protection Affects Fuel Quality and Natural Drying of Mixed Leaf-On Willow and Poplar Woodchip Piles
20.	Historical Perspective and Evolution of the Short Rotation Woody Crops Program at Rhinelander, Wisconsin, USA
21.	Tree Willow Root Growth in Sediments Varying in Texture
22.	Environmental Benefits of Shrub Willow as Bioenergy Strips in an Intensively Managed Agricultural System
23.	Quantifying the Unknown: The Importance of Field Measurements and Genotype Selection in Mitigating the Atmospheric Impacts of Poplar Cultivation
24.	Greenhouse Gas and Energy Balance of Willow Biomass Crops are Impacted by Prior Land Use and Distance From End Users
25.	Growth Patterns and Productivity of Hybrid Aspen Clones in Northern Poland
26.	The Economics of Rapid Multiplication of Elite Hybrid Poplar Biomass Varieties: Expediting the Delivery of Genetic Gains
27.	The Biomass Production Calculator: A Decision Tool for Hybrid Poplar Feedstock Producers and Investors
28.	Growth and Yield of Hybrid Poplar Mono-Varietal Production Blocks for Biofuel Production
29.	Poplar Productivity as Affected by Physiography and Growing Conditions in the Southeastern USA
30.	Hybrid Poplar Stock Type Impacts Height, DBH, and Estimated Total Dry Weight at Eight Years in A Hybrid Poplar Plantation Network
31.	Developing Phytoremediation Technology Using *Pseudomonas putida* and Poplar for Restoring Petroleum-Contaminated Sites
32.	Comparison of Statistical Techniques for Evaluating Fiber Composition of Early Rotation Pine and Hardwood Trees for the Production of Cellulose
33.	Freshkills Anthropogenic Succession Study: Phase I: Deer Cafeteria Study
34.	Adapting An Aspen Short Rotation Yield Model to Represent Hybrid Poplar Yield for Regional Hybrid Poplar Production Estimation
35.	Potential Biomass Production of Four Cottonwood Clones Planted at Two Densities in the Arkansas River Valley, USA
36.	Growth Performance and Stability of Hybrid Poplar Clones in Simultaneous Tests on Six Sites in Minnesota, USA
37.	Genetic Development, Evaluation, and Outreach for Establishing Hybrid Poplar Biomass Feedstock Plantations in the Midwestern United States
38.	Citizen Science in the Greenhouse: A Phytoremediation Case Study

References

1. Zalesny, R.S., Jr.; Stanturf, J.A.; Gardiner, E.S.; Perdue, J.H.; Young, T.M.; Coyle, D.R.; Headlee, W.L.; Bañuelos, G.S.; Hass, A. Ecosystem Services of Woody Crop Production Systems. *Bioenergy Res.* **2016**, *9*, 465–491. [CrossRef]
2. Zalesny, R.S., Jr.; Stanturf, J.A.; Gardiner, E.S.; Bañuelos, G.S.; Hallett, R.A.; Hass, A.; Stange, C.M.; Perdue, J.H.; Young, T.M.; Coyle, D.R.; et al. Environmental technologies of woody crop production systems. *Bioenergy Res.* **2016**, *9*, 492–506. [CrossRef]
3. Haapala, T.; Pakkanen, A.; Pulkkinen, P. Variation in survival and growth of cuttings in two clonal propagation methods for hybrid aspen (*Populus tremula* × *P. tremuloides*). *For. Ecol. Manag.* **2004**, *193*, 345–354. [CrossRef]
4. Dipesh, K.C.; Will, R.E.; Hennessey, T.C.; Penn, C.J. Evaluating performance of short-rotation woody crops for bioremediation purposes. *New For.* **2015**, *46*, 267–281. [CrossRef]

forests

MDPI

Article

Hormones and Heterosis in Hybrid Balsam Poplar (*Populus balsamifera* L.)

Yue Hu * and Barb R. Thomas

Department of Renewable Resources, 442 Earth Science Building, University of Alberta, Edmonton, AB T6G 2E3, Canada; bthomas@ualberta.ca
* Correspondence: yhu6@ualberta.ca; Tel.: +1-403-992-2640

Received: 18 January 2019; Accepted: 8 February 2019; Published: 10 February 2019

Abstract: Balsam poplar (*Populus balsamifera* L.) is a transcontinental tree species in North America, making it an ideal species to study intra-specific hybrid vigour as a tool for increasing genetic gain in growth. We tested the hypothesis that intra-specific breeding of disparate populations of balsam poplar would lead to the expression of hybrid vigour and we determined the role of endogenous hormones linked to ecophysiological and growth performance. In September 2009, three field trials were established in Canada (two in Alberta (AB), i.e., Fields AB1 and AB2, and one in Quebec (QC), i.e., Field QC1) in conjunction with Alberta-Pacific Forest Industries Inc. and the Ministry of Forests, Wildlife and Parks, Quebec. Five male parents from each province as well as five female parents from QC and four female parents from AB were used for breeding intra-regional and inter-regional crosses. Based on a significant difference at year six for height and diameter, from the AB1 and AB2 field trials, the AB × QC cross-type was selected for further study. Cuttings from the AB × QC cross-type were grown in a randomized complete block design under near-optimal greenhouse conditions. Families were identified as slow- or fast-growing, and the relationship between hormone levels and growth performance of the genotypes within the families were examined. In late June, after 34 days of growth, internode tissue samples collected from each progeny were analyzed for gibberellic acids, indole-3-acetic acid, and abscisic acid content. Stem volume of two-month-old rooted cuttings, grown under optimal greenhouse conditions, was positively and significantly correlated with the photosynthetic rate, greenhouse growth, and stem volume of 8-year-old field-grown trees (Fields AB1 values: $r = 0.629$ and $p = 0.012$; AB2 values: $r = 0.619$ and $p = 0.014$, and QC1 values: $r = 0.588$ and $p = 0.021$, respectively). We determined that disparate and native populations of balsam poplar can be bred to produce superior progeny with enhanced stem growth traits.

Keywords: balsam poplar; disparate populations; hybrid vigour; plant physiology

---

1. Introduction

Heterosis, or hybrid vigour, refers to the phenomenon in which hybrids outperform their parents in yield, biomass, biotic and abiotic stress tolerance, or other traits [1]. Hybrid vigour, typically achieved through the controlled crossing of two species, or pure genetic lines of the same species, has long been exploited in agriculture [2–4] and in some tree species including *Populus* [5]. For poplars, including the aspens, two or more species are typically crossed to produce hybrid progeny, some of which can be expected to yield growth performance far superior than either parent (i.e., hybrid vigour or heterosis). Balsam poplar (*Populus balsamifera* L.) is a transcontinental species in North America ranging from Alaska to Newfoundland [6]. This species occupies a wide range of climatic and site conditions and often grows in mixed stands with conifers or other broadleaf trees, contributing to stand and landscape level diversity [6]. This wide range makes it an ideal species to study within-species hybrid vigour as a tool for increasing genetic gain in volume for a given population and rotation age [7,8]. Because of the clonal nature of this species and its ease of propagation from cuttings, any volume gain achieved through hybrid vigour can be rapidly exploited in a tree improvement program through cloning the superior individuals [9].

Knowledge regarding the physiological basis of heterosis is sporadic and has mainly focused on specific traits, such as freezing tolerance in *Arabidopsis* [10]. In addition, however, there have also been studies on the role of gibberellic acids (GAs) in the regulation of heterosis (e.g., [11]). The majority of GA metabolism genes in various plant species have been identified and characterized [12,13]. Gibberellins, a group of tetracyclic diterpenoid compounds, function as plant hormones that play an important role in the regulation of many aspects of plant growth and development, such as seed germination, stem elongation, leaf expansion, flower and fruit development, and wood formation [14]. Gibberellins residing in three different cellular compartments (plastid, endoplasmic reticulum, and cytoplasm), being synthesized via the terpenoid pathway, require terpene synthase (TPSs), cytochrome P450 mono-oxygenase (P450s), and 2-oxoglutarate dependent dehydrogenase (2 ODDs) for the biosynthesis of bioactive GA from geranylgeranyl diphosphate (GGDP) in plants [15]. The GA 20-oxidases (*GA20ox*) and GA 3-oxidases (*GA3ox*) convert the early GA structures (GA_{12} or GA_{53}) to the growth-active structures, GA_4 and GA_1, via the early non-13 hydroxylation pathway (leading to GA_4) and the early 13-hydroxylation pathway (leading to GA_1), respectively [11]. The inactivation of these growth-activating GAs and many of their early precursors is achieved by GA 2-oxidases (*GA2ox*) [16].

Recent work in rice has revealed evidence that endogenous gibberellins play a key role regulating heterosis [11]. Moreover, Rood et al. [17] analyzed the differences in responsiveness to the exogenous application of GA_3 and endogenous levels of GAs between F1 hybrids and their inbred parents of diallele combinations in maize (*Zea mays*). They concluded that the increased endogenous concentration of GA in the hybrids could provide a phytohormone basis for heterosis for shoot growth. Additionally, Park et al. [14] compared the growth of young conifer seedlings under optimal conditions with the field performance of the same seedlings via a retrospective approach and found that endogenous GA levels may explain much of the natural variation seen in tree stem size in even-aged pine forests.

Indole-3-acetic acid (IAA), the most common plant hormone of the auxin class, has been linked to tree stem radial growth both for conifers [18] and deciduous trees [19]. IAA regulates various aspects of plant growth and development and acts as a positional signal that controls cambial growth rate by regulating the radial number of dividing cells in the cambial meristem, which is an important component in determining cambial growth rate [20]. Overall, IAA promotes cell division, elongation, and differentiation, whereas abscisic acid (ABA) regulates IAA biosynthesis and activity [21]. Abscisic acid, a sesquiterpene, plays an important role in seed development and maturation and induces dormancy in buds, underground stem, and seeds [22]. ABA is also called a stress hormone because the production of hormones is stimulated by drought, water logging, and other adverse environmental conditions by regulating stomatal closure, inducing the expression of stress responsive genes and the accumulation of osmo-compatible solutes [23].

In order to test the role of endogenous hormones including links to physiological and growth performance in balsam poplar produced from breeding disparate populations, a greenhouse study was designed to grow progeny from selected families under near-optimal growing conditions. The stem tissue was harvested, prior to any senescence or corresponding hormone degradation, for GA, IAA, and ABA analysis in order to examine whether there is a causal relationship between hormone concentration in the elongating stem (internodes) tissue and growth rate of the vegetatively propagated progeny from selected families of balsam poplar.

2. Materials and Methods

2.1. Selection of Families and Progeny

In order to test the hypothesis that within-species breeding will lead to the expression of hybrid vigour, a series of controlled crosses were completed including local × local and local × distant parental types from both Alberta (AB) and Quebec (QC) sources of balsam poplar. Five male parents from each province (Abitibi, QC and Athabasca, AB regions) as well as five female parents from QC and four female parents from AB were used for breeding, both for within-region and between-region crosses. Parent trees were identified in AB and QC and bred in the winter of 2005, the seedlings were grown in stool-beds, and cuttings were taken to establish three field trials in September 2009. In each field trial trees were planted in 4-tree family plots, with 10 blocks at a 2.5 × 2.5 m spacing on two sites (Fields AB1 and AB2) in AB (Alberta-Pacific Forest Industries Inc. (Al-Pac) millsite, 54° N, 112° W, 575 m, mean annual precipitation of 458 mm [24]), and a single site (Field QC1) in QC located at Trécesson, (48° N, 78° W, 348 m, mean annual precipitation of 890 mm [25]). The crosses produced a total of 33 families of AB × AB, AB × QC, QC × AB, and QC × QC cross-types, respectively. Each progeny from each family is represented once in each trial across the 10 blocks. In summer 2016, a greenhouse trial using families identified as slow- and fast-growing in cross-type AB × QC were selected in order to use extremes of performance for examining the relationship between hormone levels and growth performance of the selected genotypes within specific families.

Stem volumes were calculated using the 6-year-old tree data from AB1 and AB2 field sites. The relationship between individual progeny and stem volume for the AB × QC cross-types was then plotted (Figure 1). Across the eight families in this cross-type, there were distinctly different patterns of stem volume in the progeny. Three groups with three progeny per family were selected as follows: (1) a fast-growing (FG) group (families AP5396, AP5402, and AP 5416); (2) a slow-growing (SG) group (families AP5401, AP5411, and AP5414); and (3) three slow-growing progeny selected from the fast-growing families (SFG) (families AP5396, AP5402, and AP 5416). The selection of these progeny and families allowed for the following comparisons: slow-growing vs. fast-growing within the same family group, and slow-growing vs. slow-growing between different family groups. In total, 27 individual progeny were selected for study within each family and group type within the AB × QC cross-type (Table 1).

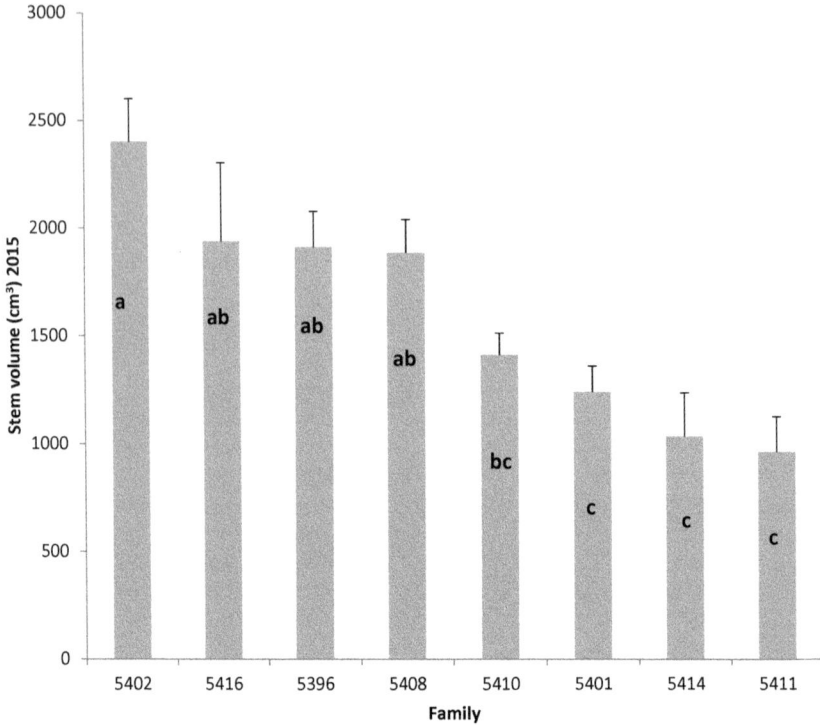

Figure 1. Mean stem volumes (cm^3) (+SE) for 6-year-old Alberta × Quebec cross-type families grown at two Alberta sites (AB1 and AB2). Significant differences between family means are indicated by different letters at $p \leq 0.05$.

Table 1. Family codes, and selected progeny for each group type, fast-growing (FG), slow-growing (SG), and slow fast-growing (SFG), based on 6-year-old growth from two field sites (AB1 and AB2).

Family	Fast-Growing Progeny	Group (Fast-Growing (FG))	Slow-Growing Progeny	Group (Slow-Growing in Fast-Growing Group (SFG) or Slow-Growing (SG) Group)
AP5396	147071	FG	147043	SFG
	147083	FG	147051	SFG
	147072	FG	147041	SFG
AP5401			178092	SG
			178073	SG
			178104	SG
AP5402	180071	FG	180093	SFG
	180103	FG	180063	SFG
	180081	FG	180094	SFG
AP5411			255081	SG
			255011	SG
			255102	SG
AP5414			270051	SG
			270101	SG
			270061	SG
AP5416	272084	FG	272024	SFG
	272102	FG	272071	SFG
	272091	FG	272023	SFG

2.2. Greenhouse Propagation

Dormant branch cuttings that were 40–50 cm long were collected on March 31, 2016 from the selected 27 progeny located in the AB1 field at the Al-Pac mill site, placed in black plastic bags and stored in the freezer for eight days at −4 °C, until the greenhouse experiment started. Cuttings were soaked in cold water for two days at room temperature in the lab without the use of any additional rooting hormone [26], with water replaced daily with fresh cold water. On April 11, eight stem cuttings 6–9 cm long with a minimum of two buds for each of the 27 progeny were rooted in Format 360 Hillsons Rootrainers trays (Beaver Plastics Ltd., Acheson, AB, Canada) filled with Sunshine Mix #3 (Sun Gro Horticulture, Vancouver, BC, Canada). Budburst was scored and recorded three times over the course of one week. Upon bud flush, and after height growth of approximately 10 cm was reached (~Day 40 after striking), cuttings were transplanted directly into 2 L pots filled with Sunshine Mix #4 (Sun Gro Horticulture, Vancouver, BC, Canada). The rooted cuttings (stecklings) were grown in the greenhouse at the University of Alberta and Day 1 of the experiment started on 24 May 2016 under natural light supplemented by cool-white fluorescent lamps to provide a 21 h long photoperiod and a minimum photosynthetic photon flux density (PPFD) of 400 μmol m^{-2} s^{-1} at plant level. Maximum day and night temperatures were maintained at approximately 25 °C and 18 °C, respectively, throughout the experimental period. The stecklings were kept well-watered and fertilized using a 20-20-20 commercial water-soluble fertilizer (20:20:20 plus micronutrients (Fe 0.1%; Mn 0.05%; Zn 0.05%; Cu 0.05%; B 0.02%; Mo 0.0005%)) (Plant Products Co. Ltd. Brampton, ON, Canada) at a pH of 5.8–6.3, adjusted by adding phosphoric acid (H_3PO_4). The greenhouse was well ventilated and PPFD, humidity, and air temperature were continuously monitored and recorded using a HOBO U12-012 data logger (Onset Computer Cooperation, Pocasset, MA, USA). The potted stecklings were rotated weekly to minimize position effect in the greenhouse [27].

2.3. Measurements, Harvest and Selection for Hormone Analysis

Caliper, measured at the base of the new stem, and height growth were measured every 10 days after transplanting, starting on Day 1 of the experiment (24 May 2016). Gas exchange measurements including photosynthetic rate (A), stomatal conductance (Gs), and intrinsic water use efficiency (iWUE) were made using a CIRAS-3 infrared gas analyzer (IRGA) (PP Systems, Amesbury, MA, USA) and a broad leaf cuvette (PLC4 (B) Broad Leaf Cuvette, PP Systems, Amesbury, MA, USA), the cuvette window being 18 mm diameter in size with a total area of 2.5 cm^2, just prior to harvesting on Day 34, 27 June 2016 (near the longest day of the year). In all, 15 progeny were selected (six from FG, six from SFG, and three from SG) from three families for further hormone analysis.

At harvest, 8 ramets of each of the 15 progeny (genotypes) were grouped based on a visual assessment of vigour, which in turn was based on height and diameter, to select: (a) the largest, (b) the second largest, and (c) the third largest ramet for use in the hormone level analysis in the elongating internode stem tissue [14].

For each of the selected ramets, leaves were snipped off at the base of their petiole, along the chosen length of stem while the tree was still intact. The upper 20–30% (all internodes which were still elongating plus two lower internodes which had ceased to elongate) of the stem were harvested, wrapped in a double-layer of aluminum foil forming a package, and placed onto dry ice for storage prior to preparation for the hormone analysis. The roots were then carefully washed and put into paper bags. The biomass components (leaves, remaining stem tissue, and roots) were stored in paper bags before drying for two days at 65 °C, then measured using a model AV53 scale (readability 0.001 g, OHAUS Adventurer Pro, Melrose, MA, USA). After drying, the leaves were ground in a ball grinder (Model MM200, Retsch Inc., Haan, North Rhine-Westphalia, Germany) and stored in 20 mL plastic scintillation vials (Fisher Scientific, Hampton, NH, USA) in preparation for $\delta^{13}C$ analysis at the University of Alberta's Natural Resources Analytical Laboratory (NRAL). After five days in the −80 °C freezer, the stem samples, which were harvested earlier for hormone analysis, were freeze-dried in a FreeZone®2.5 L Benchtop freeze dry system (Labconco Corporation, Kansas City, MO, USA) for three days.

2.4. Analysis of GAs, IAA, and ABA

One gram dry weight (DW) of each tissue sample was ground with liquid N_2 and washed sea sand (Fisher Scientific, Fair Lawn, NJ, USA), then extracted in 80% MeOH (H_2O:MeOH = 20:80, v/v). Following this, 250 ng [$^{13}C_6$] IAA (gift from Dr. J. Cohen, available from Cambridge Isotope Laboratories, Inc., Tewksbury, MA, USA), 200 ng [$^2H_6$] ABA (a gift from Drs. L. Rivier and M. Saugy, University of Lausanne, Lausanne, Switzerland), and 20–40 ng each of [$^2H_2$] GA_{15}, GA_{24}, GA_9, GA_{20}, GA_4, GA_1, GA_8, and GA_{34} (deuterated GAs were obtained from Professor L.N. Mander, Research School of Chemistry, Australian National University, Canberra, Australia) were added to the aqueous MeOH extraction solvent as internal standards. Subsequent purification, separation, and stable isotope dilution analysis by gas chromatography–mass spectrometry (GC-MS) selected ion monitoring (SIM) were accomplished as described by Kurepin et al. [28].

2.5. Data Analysis

All the growth data (height, caliper, and stem volume) were analyzed by ANOVA using SAS 9.4 [29]. Following significant main effects, multiple comparisons among means were completed using the Student–Newman–Keuls test. A result of $p \leq 0.05$ was considered significant. The correlation coefficient (r) and probability (p) were determined by Pearson's correlation analysis using the Statistical Package in SigmaPlot 13.0 (Systat Software Inc., San Jose, CA, USA).

3. Results

3.1. Hybrid Vigour in Intra-Specific Hybrids

Preliminary analysis of the 6-year-old tree height, diameter, and stem volume data from the Al-Pac field sites, AB1 and AB2 only, indicated differences in family performance among the different cross-types. In order to detect the intra-specific hybrid vigour, intra-regional crosses (AB♀× AB♂and QC♀× QC♂) were used as a control to compare with inter-regional (AB♀× QC♂and QC♀× AB♂) crosses. In general, AB♀× AB♂(the female is described first) families were the slowest growing, whereas AB × QC crosses ranked first and significantly better than intra-regional crosses in terms of growth performance (height, diameter at breast height (DBH), and stem volume) (Table 2). These results were used to make further selections for families and progeny based on cross-type for the greenhouse experiment.

Table 2. Mean height (± SE), diameter at breast height (DBH), and stem volume at age six from two Alberta sites (AB1 and AB2) for the four cross-types as indicated by Alberta (AB) and Quebec (QC) with the female parent listed first. Significant differences between cross-type means are indicated by different letters.

Cross-Types	Height (m)	DBH (mm)	Stem Volume (cm^3)
AB♀× AB♂	3.67 ± 0.05 [c]	33.70 ± 0.76 [b]	1375.94 ± 65.88 [b]
AB♀× QC♂	4.06 ± 0.04 [a]	37.77 ± 0.54 [a]	1786.78 ± 59.48 [a]
QC♀× AB♂	3.78 ± 0.03 [bc]	34.73 ± 0.40 [b]	1389.78 ± 35.27 [b]
QC♀× QC♂	3.80 ± 0.05 [b]	34.52 ± 0.63 [b]	1433.09 ± 55.63 [b]

3.2. Greenhouse Growth at Two Months (34 Days after Transplanting)

Initial growth of cuttings after striking, for all families in the FG, SG, and SFG groups, showed similar heights during the first seven days of growth, early in the growing season. However, after the emergence of primary leaves, cuttings of the fast-growing group (FG) grew much faster than the cuttings of the intermediate (SFG) or the SG group selected from the slow-growing families. By Day 34 after transplanting (about two months old), the FG group performed better than both the SG and SFG groups in stem volume (Figure 2) and biomass (Figure 3). Additionally, within the same family (i.e., family AP5416), fast-growing progeny (272084) performed better than the slow-growing progeny (272023 and 272024) under near-optimal greenhouse conditions, thus indicating the wide range of performance variability within a single family (Figure 4).

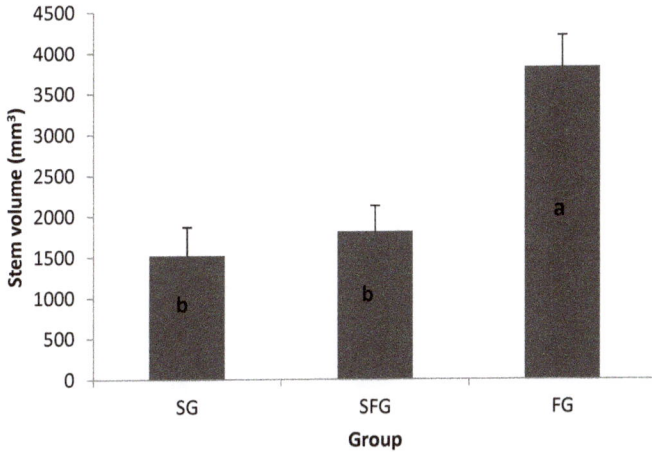

Figure 2. Mean stem volume (mm^3) (+SE) of three groups at Day 34 under near-optimal greenhouse conditions. Note: SG = slow-growing progeny; SFG = slow-growing progeny from a fast-growing family; and FG = fast-growing progeny. Significant differences between group means are indicated by different letters at $p \leq 0.05$.

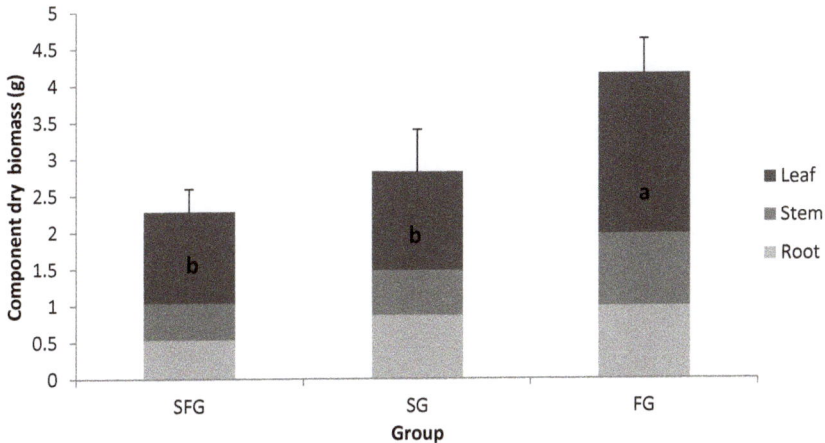

Figure 3. Mean component dry biomass (g) (+SE) of three groups at Day 34 under near-optimal greenhouse conditions. Note: SG = slow-growing progeny; SFG = slow-growing progeny from a fast-growing family; and FG = fast-growing progeny. Significant differences between group means of total dry mass are indicated by different letters at $p \leq 0.05$.

Stem volume (mm^3) at Day 34

6000
5000
4000
3000
2000
1000
0

| a |
| b |
| bc |
| cd |
| cd |
| cd | cd |
| cde | cd |
| cd |
| de | de | de |
| e | e | e |

272024 272023 178092 272091 178073 180063 272071 272102 180081 180093 178104 180094 180103 180071 272084

Group SFG SFG SG FG SG SFG SFG FG FG SFG SG SFG FG FG FG

Progeny

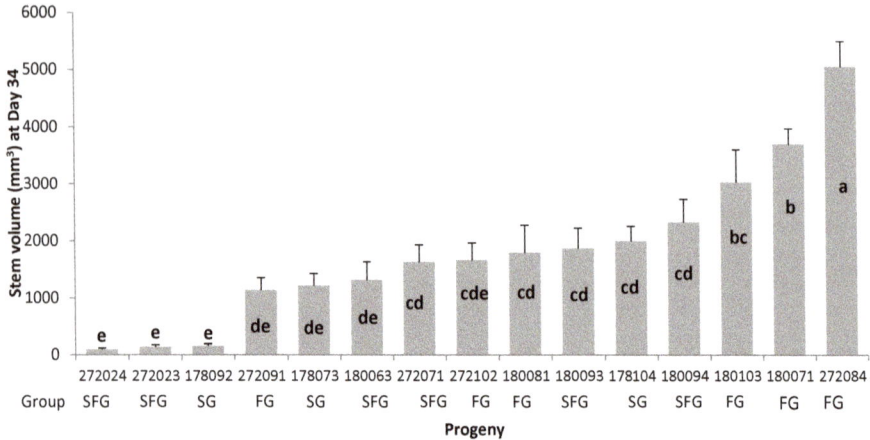

Figure 4. Mean stem volume (mm^3) (+ SE) of 15 selected balsam poplar progeny at Day 34 under near-optimal greenhouse conditions. Note: SG = slow-growing progeny; SFG = slow-growing progeny from a fast-growing family; and FG = fast-growing progeny. Significant differences between progeny means are indicated by different letters at $p \leq 0.05$.

3.3. Greenhouse Gas Exchange

Table 3 shows that higher rates of photosynthesis were correlated with growth (height, caliper, and stem volume) at Day 34 (after transplanting, about 2-months old), just prior to harvest. Additionally, increased photosynthetic demand was supported by an increased supply of CO_2 through an increase in stomatal conductance (g_s). It was also found that $\delta^{13}C$ in the leaf tissue showed a significant negative correlation with height which supports the g_s results, indicating that the stomates were open, promoting an increase in gas exchange and carbohydrate production, resulting in an increase in height growth.

Table 3. Pearson's correlations (r) analysis among physiological variables under optimal growth condition in greenhouse for selected progeny growth at two months.

	Height (cm)	Caliper (mm)	Stem Volume (mm^3)	A	g_s	iWUE	$\delta^{13}C_{leaf}$ (‰)
Height (cm)	1	0.933 **	0.861 **	0.751 *	0.532 *	0.606 *	−0.545 **
Caliper (mm)		1	0.890 **	0.743 *	0.614 *	0.517 *	−0.502
Stem volume (mm^3)			1	0.556 *	0.327	0.344	−0.495
A				1	0.596 *	0.796 **	−0.461
g_s					1	0.205	−0.563 *
iWUE						1	−0.505 *

* Significant at $p \leq 0.05$: (1) ** Significant at $p \leq 0.01$. A: Photosynthetic rate (μmol CO_2 m^{-2} s^{-1}); g_s: stomatal conductance (mol H_2O m^{-2} s^{-1}); iWUE: intrinsic water use efficiency (μmol CO_2 mmol^{-1} H_2O); $\delta^{13}C$ leaf: carbon isotope composition.

3.4. Comparisons of Greenhouse Growth at Day 34 and Field Growth Performance at Age Eight Years

The Pearson's correlation analysis showed that stem volumes calculated from height and caliper of Day 34 stecklings grown under near-optimal greenhouse conditions are positively and significantly correlated with stem volumes of 6-year- and 8-year-old field-grown trees of the same genotypes (Figure 5). Additionally, positive and significant correlations were also obtained when stem dry biomass and stem volumes of Day 34 stecklings grown under near-optimal greenhouse conditions (Table 4) were regressed against each of 6- and 8-year-old stem diameters and stem volume of field-grown trees of the same genotypes.

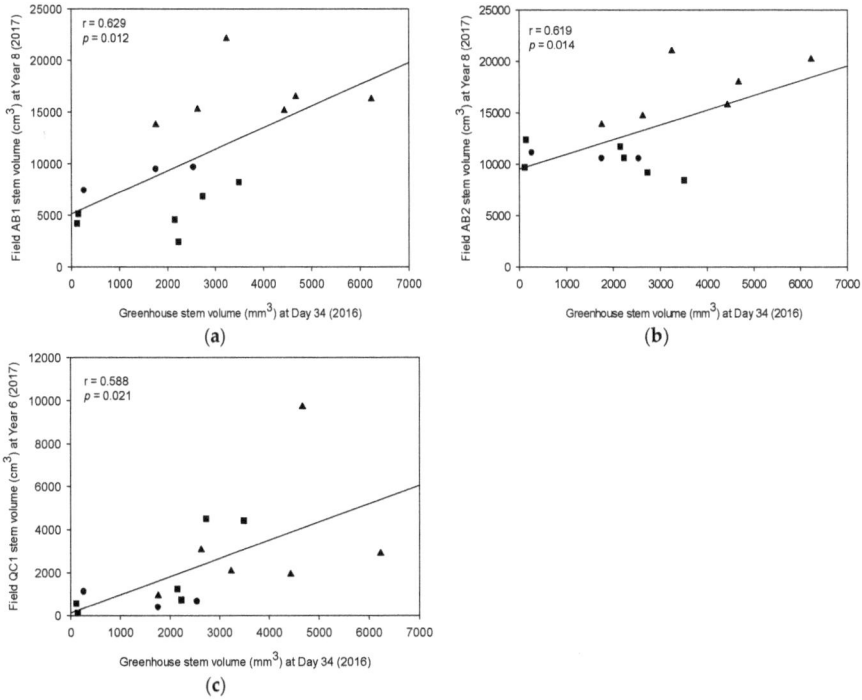

Figure 5. Pearson's correlation (r) between field stem volume and greenhouse growth at Day 34 (after transplanting) for 15 selected progeny in 2017 for: (**a**) Year 8, Field AB1, AB; (**b**) Year 8, Field AB2, AB; and (**c**) Year 6, Field QC1, QC. Symbols represent individual progeny that had been previously grouped with FG (▲), SFG (■), and SG (●). Note: SG = slow-growing progeny; SFG = slow-growing progeny from a fast-growing family; and FG = fast-growing progeny. AB = Alberta and QC = Quebec.

Table 4. Pearson's correlation (r) analysis between phenotypic characteristics of 8-year-old field-grown hybrid *Populus balsamifera* trees and Day 34 greenhouse-grown stecklings for the same 15 genotypes. AB = Alberta and QC = Quebec.

Field AB1		
Stem parameters of 8-year-old trees at Field AB1	**Day 34 stem volume (mm³)**	**Day 34 stem dry weight (g)**
Height (cm)	0.550 *[1]	0.516 *
DBH (mm)	0.608 *	0.584 *
Stem volume (cm³)	0.629 *	0.601 *
Field AB2		
Stem parameters of 8-year-old trees at	**Day 34 stem volume (mm³)**	**Day 34 stem dry weight (g)**
Height (cm)	0.654 **	0.686 **
DBH (mm)	0.537 *	0.604 *
Stem volume (cm³)	0.619 *	0.671 **
Field QC1		
Stem parameters of 6-year-old trees at	**Day 34 stem volume (mm³)**	**Day 34 stem dry weight (g)**
Height (cm)	0.758 **	0.739 **
DBH (mm)	0.666 **	0.705 **
Stem volume (cm³)	0.588 *	0.684 **

[1] All values represent the correlation coefficient (r) from Pearson's correlation analysis. DBH = diameter at breast height (1.3 m). ** Significant at $p \leq 0.01$. * Significant at $p \leq 0.05$.

3.5. Hormone Analysis

Since the gibberellin (GA) profiles are likely to be very different between the active growth phase and growth cessation (bud set) phase, we grew the poplar trees under near-optimal conditions and harvested the tissue near the longest day of the year to avoid the deficiency and degradation of bioactive GAs [30].

The concentrations of three endogenous plant hormone classes, namely, ABA, IAA, and GAs, were quantified in stem tissues of 15 selected FG, SFG, and SG progeny. Stem IAA concentration in Day 34 stecklings was positively and significantly correlated with stem biomass and stem volume (Figure 6a,b). In contrast, stem ABA in Day 34 stecklings was negatively and significantly correlated with stem volume and stem dry biomass (Figure 6c,d).

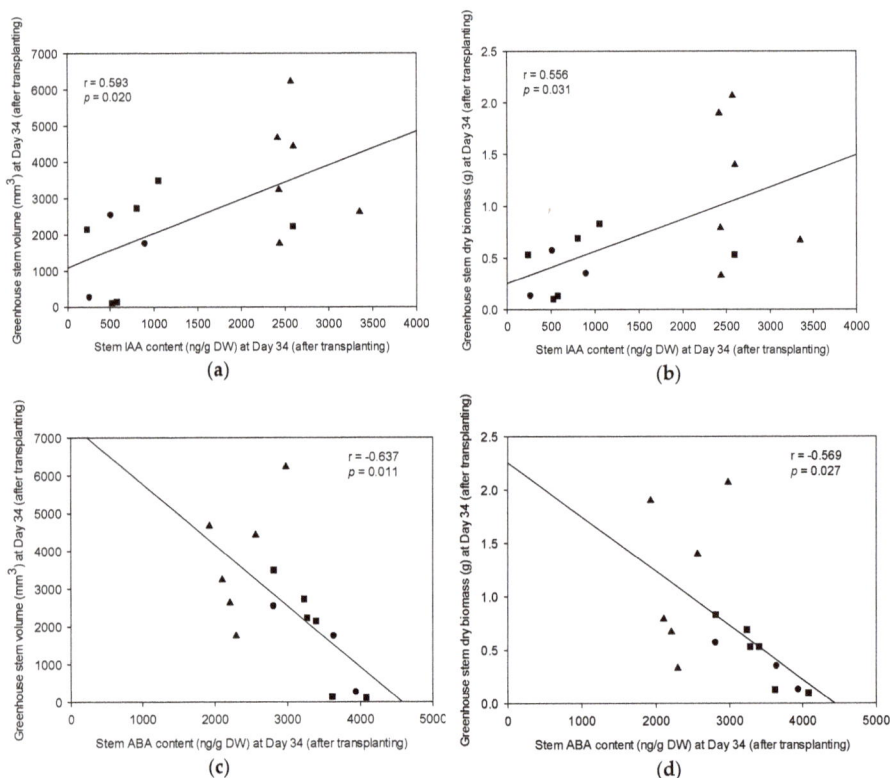

Figure 6. Pearson's correlations (r) of tissue concentrations of indole-3-acetic acid (IAA) and abscisic acid (ABA), versus stem dry biomass or stem volume of nursery-grown cuttings (new growth) at Day 34 (after transplanting). (**a**) New growth stem volume at Day 34 *versus* IAA in stem tissue at Day 34; (**b**) New growth stem dry biomass at Day 34 versus IAA in stem tissue at Day 34; (**c**) New growth stem volume at Day 34 *versus* ABA in stem tissue at Day 34; and (**d**) New growth stem dry biomass at Day 34 *versus* ABA in stem tissue at Day 34. Symbols represent individual progeny that had been previously grouped with FG (▲), SFG (■), and SG (●). Note: SG = slow-growing progeny; SFG = slow-growing progeny from a fast-growing family; and FG = fast-growing progeny.

3.6. Endogenous Plant Growth Hormone in Greenhouse-Grown Stecklings versus Field Growth Performance

The relationships between stem GA levels and stem volume were analyzed through Pearson's correlation (Figure 7). Our results showed that stem GA_{19} and GA_{20} content were all significant and negatively correlated with greenhouse stem volume, which indicates that GA_{19} and GA_{20} serve as persecutors for GA_8 (Figure 7a,b). In addition, we confirmed that both GA_{19} and GA_{20} play an important role in the early *GA20ox* portion of the GA biosynthesis pathway as found by Ma et al. [11]. In the elongated stems of the Day 34 stecklings, a significant and positive correlation between stem volume and the content of GA_8 was detected, with an r of 0.840 ($p < 0.001$) (Figure 7c). Moreover, a positive and significant correlation was also detected between field growth (Fields AB1 and AB2) trees and the content of GA_8 in the greenhouse-grown trees for the 15 selected progeny (r = 0.749 ($p = 0.001$) and r = 0.734 ($p = 0.002$), respectively) (see Figure 8).

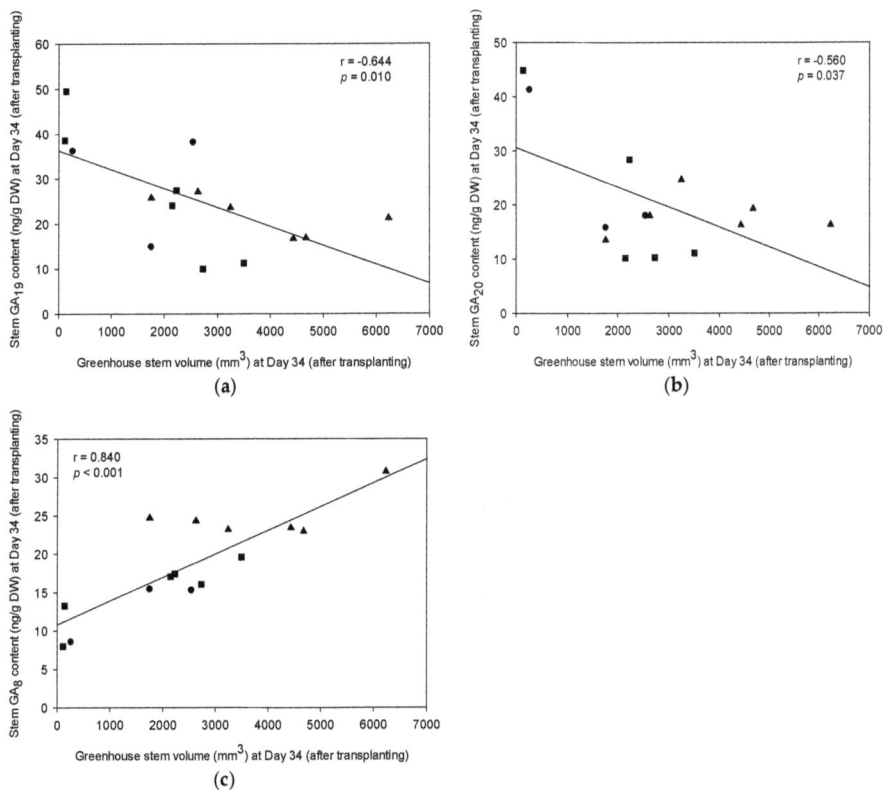

Figure 7. Pearson's correlation (r) between greenhouse stem volume at Day 34 (after transplanting) and plant hormones for 15 selected progeny; (**a**) GA_{19} (ng/g DW); (**b**) GA_{20} (ng/g DW); and (**c**) GA_8 (ng/g DW). Symbols represent individual progeny that had been previously grouped with FG (▲), SFG (■), and SG (●). Note: SG = slow-growing progeny; SFG = slow-growing progeny from a fast-growing family; and FG = fast-growing progeny.

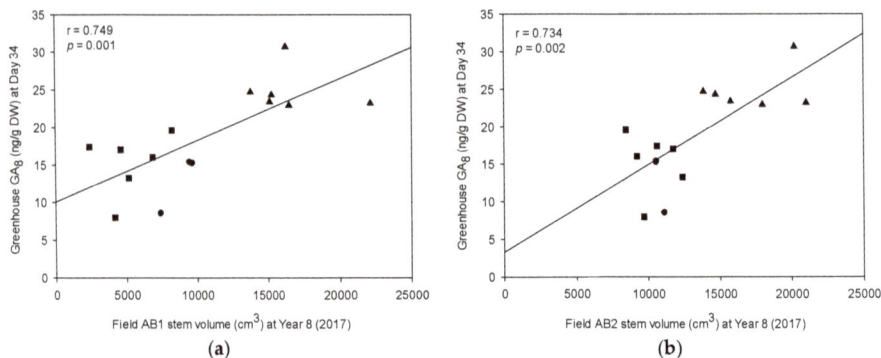

Figure 8. Pearson's correlation (r) between greenhouse GA_8 (ng/g DW) at Day 34 (after transplanting) and field stem volume for 15 selected progeny; (**a**) Field AB1, AB; and (**b**) Field AB2, AB. Symbols represent individual progeny that had been previously grouped with FG (▲), SFG (■), and SG (●). Note: SG = slow-growing progeny; SFG = slow-growing progeny from a fast-growing family; and FG = fast-growing progeny.

4. Discussion

Phenology is the study of the timing of yearly events such as spring bud flush, growth cessation, autumn bud set, and leaf senescence. The two major environmental cues which are primarily responsible for the induction of phenological traits are temperature and photoperiod (day length) [31]. There is an apparent trade-off between photosynthetic carbon assimilation rates (A), growth, and phenology in *P. balsamifera* [27]. If a transfer south is associated with warmer than optimal temperatures, growth cessation will be delayed and bud formation will proceed more slowly. If a transfer north is associated with colder than optimal temperatures, growth cessation will also be delayed and additionally bud formation will take longer [32]. Trees from the AB genotypes have higher A but accomplish far less growth than the trees from the QC genotypes because of the latitude differences that drive the photoperiod differences. However, if no intrinsic physiological constraints exist which might prevent the combination of high A from the AB parent and the longer growing period of the QC parent, the progeny of intra-specific crosses between AB and QC populations may accomplish more growth than local crosses (i.e., heterosis). From our results, the crossing of AB and QC *P. balsamifera* (Table 2) showed superior growth and heterosis (hybrid vigour), the phenomenon where hybrids outperform their parents in yield, biomass, and other traits [1]. Additionally, our findings showed that intra-specific hybridizations between geographically distant populations may lead to heterosis [33,34] through bringing together a new combination of alleles. However, there were no significant differences between the QC × AB cross-type compared with the local crosses, and other intra-regional crosses may indicate that the maternal effect played a role in the growth of the F_1 hybrids, with offspring from AB mothers exhibiting higher fitness than those from QC mothers [35,36]. This may explain why the AB × QC cross-type showed better performance than the corresponding QC × AB families.

Several past studies have indicated that morphological measures of young conifer trees grown under near-optimal conditions can be predictive of inherently rapid stem growth, at the family level, at older ages (i.e., 9 years or 32 years) [14,37]. However, our study is the first study to show a significant and positive correlation between greenhouse-grown (Day 34) and field-grown (age eight) balsam poplars, while similar findings have been found in the literature for conifers, namely, 12 open-pollinated families of *Pinus densiflora* Sieb. et Zucc., where seedling stem volume at age six months was significantly correlated with field performance at age 32 years [14]. In addition, Pharis et al. [37] found that full-sibling families *of P. radiata*, that is, seedling stem volume at age 138 days, gave reliable estimates of field performance at any age measured over nine years. Additionally, six-month heights of full-sibling families of black spruce (*P. mariana* Mill.) which were grown in a greenhouse environment were significantly correlated with age 13 field heights [38]. From the above discussion, two common characteristics were identified. First, for all the retrospective approaches studied, the cuttings were grown under near-optimal environmental conditions in greenhouses. Second, morphological measurements were made prior to the setting of the terminal bud. Therefore, these findings imply that the early stem and shoot growth of young rooted cuttings (two months) that are raised under near-optimal conditions can be a useful trait for identifying inherently rapid stem growth in mature balsam poplar.

Leaf $\delta^{13}C$ is largely related to the ratio of CO_2 partial pressure inside the leaf and ambient air (ci/ca) [39], which is driven by stomatal conductance and photosynthetic processes. Several studies have shown a strong positive correlation between $\delta^{13}C$ and plant water use efficiency (WUE) via ci/ca [40–42], which suggests that leaf $\delta^{13}C$ can be measured as a proxy for plant WUE. Water use efficiency reflects the balance between carbon fixation and the amount of water released by plants. Water use efficiency trends were assessed through $\delta^{13}C$ values of leaf tissue, with higher $\delta^{13}C$ values generally associated with greater WUE, whereas more negative $\delta^{13}C$ values were linked to reduced WUE and greater water loss [43,44]. Trees displaying both high WUE and high productivity, therefore, likely indicate that WUE is primarily associated with variation in photosynthetic capacity, as opposed to variation in stomatal conductance, which would theoretically lead to a decrease in productivity. In conifer species, positive correlations between growth traits and $\delta^{13}C$ values have consistently been found [43,45], and our results, shown in Table 3, also indicate the same trend in a deciduous species, suggesting variation in WUE is driven primarily by photosynthetic capacity. A high sink demand resulting from increased growth leads to an increase in photosynthetic rate, and, subsequently, more positive $\delta^{13}C$ values and higher WUE.

High IAA levels have been associated with enhanced levels of expression of two GA biosynthesis genes, *GA20ox* and *GA3ox* [46–48]. Thus, biosynthesis of the GAs required for rapid stem growth may be dependent on an adequate supply of the auxin IAA. In our study, stem tissue ABA concentrations in Day 34 stecklings showed a significant negative correlation with stem volume and dry biomass (Figure 6c,d); thus, an ABA:IAA "balance" may control very early height growth in rooted cuttings. Recent evidence suggests that IAA acts downstream of the ABA response [49], and there is also the potential for high ABA concentrations to negatively influence the biosynthesis of GAs [50], including inhibition at the gene expression level.

Elevated levels of bioactive GA usually suppress the expression of *GA20ox* and *GA3ox* while stimulating the expression of *GA2ox*; conversely, a drop in the bioactive GA level usually up-regulates the expression of *GA20ox* and *GA3ox* and down-regulates the expression of *GA2ox* [11,51]. As the concentration of growth-activating GAs in growing tissues is controlled by the transcriptional control of biosynthetic (*GA20ox* and *GA3ox*) or catabolic (*GA2ox*) genes [13], the modification of the expression of these genes alters plant growth, which has been successfully demonstrated in various plant species including poplars [52–54]. The trend shown in Figure 8 indicates the possibility of using endogenous GA levels to accelerate the early selection of balsam poplars that possess traits for inherently rapid stem growth. Moreover, the correlations found between stem volume and GA content in our study confirm the findings of others such as Zhang et al. [55] who found that GA content was correlated with heterosis in plant height in wheat hybrids where it was reported that increased elongation of the uppermost internode contributed most to heterosis for plant height. In the case of *Populus*, fiber and vessel lengths increase in the transition zone between juvenile and mature wood [56]. Therefore, the time it takes for a cell to mature within the different differentiation zones will ultimately determine its size and cell wall thickness. It is possible that the action of GAs like GA_1 and/or GA_4 extends this transition time, and thereby increases the fiber length in hybrid aspen (*Populus tremula × P. tremuloides*) compared to pure *P. tremula* plants [18]. Additionally, these results also provide new insights into the mechanisms whereby GAs control growth and development in trees. Overexpression of *Arabidopsis* GA 20-oxidase (*AtGA20ox1*) in transgenic hybrid aspen (*Populus tremula × P. tremuloides*) resulted in a substantial increase in plant height and fiber length, nearly doubling stem dry weight relative to the wild-type control. Additionally, *GA20ox* appears to play an important role in wood development through tree growth. Ko et al. [57] showed that poplar GA20-oxidase (*PtGA20ox*) has the highest expression in developing xylem, where several GA receptors were highly expressed in the tissue by using tissue-specific transcriptome analysis.

5. Conclusions

In conclusion, the data presented in this research confirms the widespread occurrence of heterosis typically found in interspecific crosses, with disparate population breeding of intra-specific hybrids in balsam poplar. The combined analyses of endogenous GA content and growth data revealed that GAs play a regulatory role in heterosis for hybrid balsam poplar at the physiological level. Larger scale investigations of multiple plant hormones at more developmental stages of hybrid balsam poplar are anticipated to confirm and extend these findings. Moreover, the results obtained from selected progeny are validated by retrospective comparisons with the stem growth performance of the corresponding field-grown trees at eight years of age. These results with *P. balsamifera* also point toward the successful early selection of progeny from genetic crosses of other deciduous tree species, that is, the identification of progeny which are inherently fast growing and have high stem biomass production. Additionally, the relationship between GAs and growth performance (both field and greenhouse) may open up new ways to manipulate trees to grow faster and produce more biomass by increasing endogenous GA levels. Moreover, screening for high levels of expression of GA20ox genes may be a useful technique for future tree-breeding programs. Future RT-PCR work should be used to test *GA20oxs*, *GA3oxs*, and *GA2oxs* which are the main targets of regulation by GA signaling to establish homeostasis.

Author Contributions: Y.H. and B.R.T. conceived and designed the experiments; performed the experiments; analyzed the data; and wrote the paper.

Funding: This research was funded by a Natural Sciences and Engineering Research Council of Canada (NSERC) Industrial Research Chair in Tree Improvement to B.R. Thomas, grant number 474636-13.

Acknowledgments: We are grateful to Pierre Périnet (QC) who conducted the breeding, David Kamelchuk (Al-Pac), Michael Thomson, Briana Ledic, Gregg Hamilton, and Ho-Chun (Aaron) Chen for their assistance in collecting and processing plant material, establishing field trials, and offering assistance with data collection. We also thank Steven Williams (retired), University of Alberta, and Loeke Janzen for the hormone analysis, University of Calgary. Funding for this manuscript was provided through the Industrial Research Chair in Tree Improvement held by B.R. Thomas supported by the Natural Sciences and Engineering Research Council (NSERC), Alberta-Pacific Forest Industries Inc., Alberta Newsprint Company Timber Ltd., Canadian Forest Products Ltd., Millar Western Forest Products Ltd., Huallen Seed Orchard Company Ltd., West Fraser Mills Ltd. (including Alberta Plywood, Blue Ridge Lumber Inc., Hinton Wood Products, and Sundre Forest Products Inc.), and Weyerhaeuser Company Ltd., (Pembina Timberlands and Grande Prairie Timberlands). Additional funding was also provided by the Government of Alberta. Finally, we want to offer our special thanks and dedicate this paper to (Dick) R.P. Pharis who, although no longer with us, continues to inspire us by his knowledge and love of research.

Conflicts of Interest: The authors declare no conflict of interest.

References

1. Sinha, S.K.; Khanna, R. Physiological, biochemical, and genetic basis of heterosis. *Adv. Agron.* **1975**, *27*, 123–174.
2. Wehrhahn, C.; Allard, R.W. The detection and measurement of the effects of individual genes involved in the inheritance of a quantitative character in wheat. *Genetics* **1965**, *51*, 109–119. [PubMed]
3. Stuber, C.W. Heterosis in plant breeding. *Plant Breed Rev.* **1994**, *12*, 227–251.
4. Baranwal, V.K.; Mikkilineni, V.; Zehr, U.B.; Tyagi, A.K.; Kapoor, S. Heterosis: Emerging ideas about hybrid vigour. *EXBOTJ* **2012**, *63*, 6309–6314. [CrossRef] [PubMed]
5. Stettler, R.F.; Zsuffa, L.; Wu, R. The role of hybridization in the genetic manipulation of Populus. In *Biology of Populus and its Implications for Management and Conservation*; Part I, Chapter 4; Stettler, R., Bradshaw, T., Heilman, P., Hinckley, T., Eds.; NRC (National Research Council of Canada) Research Press: Ottawa, ON, Canada, 1996; pp. 87–112.

6.   Zasada, J.C.; Phipps, H.M. *Populus balsamifera* L. Balsam Poplar. In *Silvics of North America Volume 2. Hardwoods; Agriculture Handbook 654*; Burns, R.M., Honkala, B.H., Eds.; U.S. Department of Agriculture, Forest Service: Washington, DC, USA, 1990; Volume 2, p. 877.

7.   Hewitt, G. The genetic legacy of the Quaternary ice ages. *Nature* **2000**, *405*, 907–913. [CrossRef] [PubMed]

8.   Excoffier, L.; Ray, N. Surfing during population expansions promotes genetic revolutions and structuration. *Trends Ecol. Evol.* **2008**, *23*, 347–351. [CrossRef] [PubMed]

9.   Arnold, M.L.; Cornman, R.S.; Martin, N.H. Hybridization, hybrid fitness and the evolution of adaptations. *Plant Biosyst.* **2008**, *142*, 166–171. [CrossRef]

10. Korn, M.; Gärtner, T.; Erban, A.; Kopka, J.; Selbig, J.; Hincha, D.K. Predicting Arabidopsis Freezing Tolerance and Heterosis in Freezing Tolerance from Metabolite Composition. *Molecular Plant* **2010**, *3*, 224–235. [CrossRef] [PubMed]

11. Ma, Q.; Hedden, P.; Zhang, Q. Heterosis in Rice Seedlings: Its Relationship to Gibberellin Content and Expression of Gibberellin Metabolism and Signaling Genes. *Plant Physiol.* **2011**, *156*, 1905–1920. [CrossRef] [PubMed]

12. Olszewski, N.; Sun, T.-P.; Gubler, F. Gibberellin signaling: biosynthesis, catabolism, and response pathways. *Plant Cell* **2002**, *14*, S61–S80. [CrossRef] [PubMed]

13. Thomas, S.G.; Hedden, P. Gibberellin biosynthesis and its regulation. *Biochem. J.* **2012**, *444*, 11–25.

14. Kurepin, L.V.; Janzen, L.; Park, E.-J.; Lee, W.-Y.; Zhang, R.; Pharis, R.P. Plant hormone-assisted early family selection in *Pinus densiflora* via a retrospective approach. *Tree Physiol.* **2014**, *35*, 86–94.

15. Ueguchi-Tanaka, M.; Fujisawa, Y.; Kobayashi, M.; Ashikari, M.; Iwasaki, Y.; Kitano, H.; Matsuoka, M. Rice dwarf mutant d1, which is defective in the alpha subunit of the heterotrimeric G protein, affects gibberellin signal transduction. *Proc. Natl. Acad. Sci. USA* **2000**, *97*, 11638–11643. [CrossRef] [PubMed]

16. Rieu, I.; Eriksson, S.; Powers, S.J.; Gong, F.; Griffiths, J.; Woolley, L.; Benlloch, R.; Nilsson, O.; Thomas, S.G.; Hedden, P.; et al. Genetic analysis reveals that C19-GA 2-oxidation is a major gibberellin inactivation pathway in Arabidopsis. *Plant Cell.* **2008**, *20*, 2420–2436. [CrossRef] [PubMed]

17. Rood, S.B.; Buzzell, R.I.; Mander, L.N.; Pearce, D.; Pharis, R.P. Gibberellins: a phytohormonal basis for heterosis in maize. *Science* **1988**, *241*, 1216–1218. [CrossRef] [PubMed]

18. Uggla, C.; Moritz, T.; Sandberg, G.; Sundberg, B. Auxin as a positional signal in pattern formation in plants. *Proc. Natl. Acad. Sci. USA* **1996**, *93*, 9282–9286. [CrossRef] [PubMed]

19. Tuominen, H.; Puech, L.; Fink, S.; Sundberg, B. A radial concentration gradient of indole-3-acetic acid is related to secondary xylem development in hybrid aspen. *Plant Physiol.* **1997**, *115*, 577–585. [CrossRef]

20. Uggla, C. Indole-3-Acetic Acid Controls Cambial Growth in Scots Pine by Positional Signaling. *Plant Physiol.* **1998**, *117*, 113–121. [CrossRef]

21. Fu, S.-F.; Wei, J.-Y.; Chen, H.-W.; Liu, Y.-Y.; Lu, H.-Y.; Chou, J.-Y. Indole-3-acetic acid: A widespread physiological code in interactions of fungi with other organisms. *Plant Signal. Behav.* **2015**, *10*, e1048052. [CrossRef]

22. Zeevaart, J.A.D.; A Creelman, R. Metabolism and physiology of abscisic acid. *Annu. Rev. Plant Biol.* **1988**, *39*, 439–473. [CrossRef]

23. Lim, C.W.; Baek, W.; Jung, J.; Kim, J.-H.; Lee, S.C. Function of ABA in stomatal defense against biotic and drought stresses. *IJMS* **2015**, *16*, 15251–15270. [CrossRef] [PubMed]

24. Climate Edmonton. Available online: https://en.climate-data.org/north-america/canada/alberta/edmonton-610/ (accessed on 16 January 2019).

25. Climate Trécesson. Available online: https://en.climate-data.org/north-america/canada/quebec/trecesson-719862/ (accessed on 16 January 2019).

26. DesRochers, A.; Thomas, B.R. A comparison of pre-planting treatments on hardwood cuttings of four hybrid poplar clones. *New Forests* **2003**, *26*, 17–32. [CrossRef]

27. Drewes, E.C.; Soolanayakanahally, R.Y.; Guy, R.D.; Silim, S.N.; Schroeder, W.R. Enhanced assimilation rate and water use efficiency with latitude through increased photosynthetic capacity and internal conductance in balsam poplar (*Populus balsamifera* L.). *Plant Cell Environ.* **2009**, *32*, 1821–1832.

28. Kurepin, L.V.; Emery, R.J.N.; Pharis, R.P.; Reid, D.M. Uncoupling light quality from light irradiance effects in *Helianthus annuus* shoots: putative roles for plant hormones in leaf and internode growth. *EXBOTJ* **2007**, *58*, 2145–2157. [CrossRef]

29. SAS Institute Inc. *Base SAS® 9.4 Procedures Guide: Statistical Procedures*, 2nd ed.; SAS Institute Inc.: Cary, NC, USA, 2013.

30. Zawaski, C.; Busov, V.B. Roles of gibberellin catabolism and signaling in growth and physiological response to drought and short-day photoperiods in *Populus* trees. *PLoS ONE* **2014**, *9*, e86217. [CrossRef]

31. Ford, K.R.; Harrington, C.A.; Clair, J.B.S. Photoperiod cues and patterns of genetic variation limit phenological responses to climate change in warm parts of species' range: modeling diameter-growth cessation in coast Douglas-fir. *Glob. Change Biol.* **2017**, *23*, 3348–3362. [CrossRef]

32. Rohde, A.; Bastien, C.; Boerjan, W.; Thomas, S. Temperature signals contribute to the timing of photoperiodic growth cessation and bud set in poplar. *Tree Physiol.* **2011**, *31*, 472–482. [CrossRef]

33. Schmidtling, R.C.; Nelson, C.D. Interprovenance crosses in loblolly pine using selected parents. *Forest Genet.* **1996**, *3*, 53–66.

34. Harfouche, A.; Bahrman, N.; Baradat, P.; Guyon, J.P.; Petit, R.J.; Kremer, A. Provenance hybridization in a diallel mating scheme of maritime pine (*Pinus pinaster*). II. Heterosis. *Can. J. For. Res.* **2000**, *30*, 10–16. [CrossRef]

35. Kirk, H.; Vrieling, K.; Klinkhamer, P.G.L. Maternal effects and heterosis influence the fitness of plant hybrids. *New Phytol.* **2005**, *166*, 685–694. [CrossRef]

36. Bräutigam, K.; Soolanayakanahally, R.; Champigny, M.; Mansfield, S.; Douglas, C.; Campbell, M.M.; Cronk, Q. Sexual epigenetics: gender-specific methylation of a gene in the sex determining region of *Populus balsamifera*. *Sci. Rep.* **2017**, *7*, 45388. [CrossRef] [PubMed]

37. Pharis, R.P.; Yeh, F.C.; Dancik, B.P. Superior growth potential in trees: What is its basis, and can it be tested for at an early age? *Can. J. For. Res.* **1991**, *21*, 368–374. [CrossRef]

38. Williams, D.J.; Dancik, B.P.; Pharis, R.P. Early progeny testing and evaluation of controlled crosses of black spruce. *Can. J. For. Res.* **1987**, *17*, 1442–1450. [CrossRef]

39. Farquhar, G.D.; Ehleringer, J.R.; Hubick, K.T. Carbon isotope discrimination and photosynthesis. *Annu. Rev. Plant Biol.* **1989**, *40*, 503–537. [CrossRef]

40. Körner, C.; Farquhar, G.D.; Wong, S.C. Carbon isotope discrimination by plants follows latitudinal and altitudinal trends. *Oecologia* **1991**, *88*, 30–40. [CrossRef]

41. Monclus, R.; Dreyer, E.; Villar, M.; Delmotte, F.M.; DeLay, D.; Petit, J.-M.; Barbaroux, C.; Le Thiec, D.; Bréchet, C.; Brignolas, F. Impact of drought on productivity and water use efficiency in 29 genotypes of *Populus deltoides* × *Populus nigra*. *New Phytol.* **2006**, *169*, 765–777. [CrossRef]

42. Chamaillard, S.; Fichot, R.; Vincent-Barbaroux, C.; Bastien, C.; Depierreux, C.; Dreyer, E.; Villar, M.; Brignolas, F. Variations in bulk leaf carbon isotope discrimination, growth and related leaf traits among three *Populus nigra* L. populations. *Tree Physiol.* **2011**, *31*, 1076–1087. [CrossRef]

43. Guy, R.D.; Holowachuk, D.L. Population differences in stable carbon isotope ratio of *Pinus contorta* Dougl. ex Loud.: relationship to environment, climate of origin, and growth potential. *Can. J. Bot.* **2001**, *79*, 274–283. [CrossRef]

44. Diefendorf, A.F.; Mueller, K.E.; Wing, S.L.; Koch, P.L.; Freeman, K.H. Global patterns in leaf ^{13}C discrimination and implications for studies of past and future climate. *Proc. Natl. Acad. Sci. USA* **2010**, *107*, 5738–5743. [CrossRef]

45. Eilmann, B.; de Vries, S.M.G.; den Ouden, J.; Mohren, G.M.J.; Sauren, P.; Sass-Klaassen, U. Origin matters! Difference in drought tolerance and productivity of coastal Douglas-fir (*Pseudotsuga menziesii* (Mirb.)) provenances. *Forest Ecol. Manag.* **2013**, *302*, 133–143. [CrossRef]

46. Wolbang, C.M.; Ross, J.J. Auxin promotes gibberellin biosynthesis in decapitated tobacco plants. *Planta* **2001**, *214*, 153–157. [CrossRef] [PubMed]

47. Ozga, J.A.; Yu, J.; Reinecke, D.M. Pollination-, development-, and auxin-specific regulation of gibberellins 3β-hydroxylase gene expression in pea fruit and seeds. *Plant Physiol.* **2003**, *131*, 1137–1146. [CrossRef] [PubMed]

48. Nemhauser, J.L.; Hong, F.; Chory, J. Different plant hormones regulate similar processes through largely non-overlapping transcriptional responses. *Cell* **2006**, *126*, 467–475. [CrossRef] [PubMed]

49. Rinaldi, M.A.; Liu, J.; Enders, T.A.; Bartel, B.; Strader, L.C. A gain-of-function mutation in IAA16 confers reduced responses to auxin and abscisic acid and impedes plant growth and fertility. *Plant Mol. Biol.* **2012**, *79*, 359–373. [CrossRef] [PubMed]

50. Kurepin, L.V.; Walton, L.J.; Pharis, R.P.; Emery, R.J.N.; Reid, D.M. Interactions of temperature and light quality on phytohormone-mediated elongation of *Helianthus annuus* hypocotyls. *Plant Growth Regul.* **2010**, *64*, 147–154. [CrossRef]
51. Hedden, P.; Phillips, A.L. Gibberellin metabolism: new insights revealed by the genes. *Trends Plant Sci.* **2000**, *5*, 523–530. [CrossRef]
52. Eriksson, M.E.; Israelsson, M.; Olsson, O.; Moritz, T. Increased gibberellin biosynthesis in transgenic trees promotes growth, biomass production and xylem fiber length. *Nat. Biotechnol.* **2000**, *18*, 784–788. [CrossRef] [PubMed]
53. Busov, V.B.; Meilan, R.; Pearce, D.W.; Ma, C.; Rood, S.B.; Strauss, S.H. Activation tagging of a dominant gibberellin catabolism gene (GA 2-oxidase) from Poplar that regulates tree stature. *Plant physiol.* **2003**, *132*, 1283–1291. [CrossRef]
54. Reinecke, D.M.; Wickramarathna, A.D.; Ozga, J.; Kurepin, L.V.; Jin, A.L.; Good, A.G.; Pharis, R.P. Gibberellin 3-oxidase gene expression patterns influence gibberellin biosynthesis, growth, and development in Pea. *Plant physiol.* **2013**, *163*, 929–945. [CrossRef]
55. Yao, Y.; Nie, X.; Zhang, Y.; Ni, Z.; Sun, Q. Gibberellins and heterosis of plant height in wheat (*Triticum aestivum* L.). *BMC Genet.* **2007**, *8*, 40.
56. Park, E.J.; Kim, H.T.; Choi, Y.I.; Lee, C.H.; Nguyen, V.P.; Jeon, H.W.; Cho, J.S.; Pharis, R.P.; Kurepin, L.V.; et al. Overexpression of gibberellin 20-oxidase1 from *Pinus densiflora* results in enhanced wood formation with gelatinous fiber development in a trangenic hybrid poplar. *Tree Physiol.* **2015**, *35*, 1264–1277. [PubMed]
57. Ko, J.-H.; Kim, H.-T.; Han, K.-H.; Ko, J.; Kim, H.; Hwang, I.; Han, K. Tissue-type-specific transcriptome analysis identifies developing xylem-specific promoters in poplar. *Plant Biotechnol. J.* **2012**, *10*, 587–596. [CrossRef] [PubMed]

forests

Article

Survival, Height Growth, and Phytoextraction Potential of Hybrid Poplar and Russian Olive (*Elaeagnus Angustifolia* L.) Established on Soils Varying in Salinity in North Dakota, USA

Ronald S. Zalesny Jr. [1,*], Craig M. Stange [2] and Bruce A. Birr [1]

[1] USDA Forest Service, Northern Research Station, Institute for Applied Ecosystem Studies, 5985 Highway K, Rhinelander, WI 54501, USA

[2] USDA Natural Resources Conservation Service, Bismarck Plant Materials Center, 3308 University Drive, Bismarck, ND 58504, USA

*   Correspondence: ron.zalesny@usda.gov; Tel.: +1-715-362-1132; Fax: +1-715-362-1166

Received: 24 May 2019; Accepted: 28 July 2019; Published: 9 August 2019

Abstract: Salt-affected soils in the Northern Great Plains, USA, can impact the long-term survival and growth of trees recommended for agroforestry systems, with Russian olive (*Elaeagnus angustifolia* L.) being one of few options that survives on these sites. Similarly, hybrid poplars have been used for phytotechnologies on high-salinity soils throughout the world. The objective of this study was to test the survival, height growth, and phytoextraction potential of eight hybrid poplar clones (*Populus deltoides* Bartr. ex Marsh. × *P. nigra* L. 'Robusta', 'DN17', 'DN182', 'DN5'; *P. deltoides* × *P. maximowiczii* A. Henry 'NC14104', 'NC14106'; *P. nigra* × *P. maximowiczii* 'NM2', 'NM6') versus Russian olive grown on soils categorized according to initial salinity levels: low (0.1 to 3.9 dS m^{-1}), medium (4.0 to 5.9 dS m^{-1}), and high (6.0 to 10.0 dS m^{-1}). Seven trees per genotype were grown in each salinity treatment at a spacing of 3 × 3 m for four years in Burleigh County, North Dakota. Survival and height were determined following the first four growing seasons, and leaf phytoextraction potential of Al, Ca, Cd, Fe, K, Mg, Mn, Na, and Zn was measured for one-year-old trees. Soil salinity decreased over time, reflecting the phytoextraction potential of the trees. Russian olive did not survive as well as expected, having lower overall survival than three of the hybrid poplar clones ('DN17', 'DN5', 'NM6'). At the end of three years when trees were removed per a landowner maintenance agreement, 86%, 71%, and 43% of the Russian olive trees were alive in the low-, medium-, and high-salinity soils, respectively. At this time, 'NM2' was the only hybrid poplar clone with similar survival to Russian olive in the high-salinity soils. Russian olive had greater Na, Cd, and Fe leaf concentrations than the hybrid poplar clones, but it also had the worst uptake of Ca and Mg of all genotypes. For hybrid poplar, the *P. deltoides* × *P. nigra* genomic group had the broadest clonal variability among all traits, with 'Robusta' and 'DN182' exhibiting great potential for establishment on high-salinity soils. 'Robusta' and 'DN17' are the same genotype but they came from different nursery sources (i.e., hence their different nomenclature), and they did not differ for height nor leaf phytoextraction. *Populus deltoides* × *P. maximowiczii* clones were not suitable for the soil conditions and silvicultural applications (e.g., tree shelters) of the current study, while *P. nigra* × *P. maximowiczii* clones exhibited the most stable performance across all years and salinity treatments. Both 'NM2' and 'NM6' had superior fourth-year survival and height, as well as average or above average phytoextraction of all elements tested.

Keywords: agroforestry; metal uptake; phytoremediation; phytotechnologies; *Populus*; salt uptake

1. Introduction

Agroforestry practices such as planting trees and shrubs in agricultural areas are useful for moderating winds, managing snow distribution, and reducing energy demands for snow removal, livestock feed, and heating buildings [1]. However, salt-affected soils such as those in the Northern Great Plains of North Dakota, USA, can prevent or reduce the long-term survival and growth of trees and shrubs recommended for agroforestry systems [2]. Russian olive (*Elaeagnus angustifolia* L.) is one of the few tree species that is salt tolerant and able to withstand these high-salinity soil conditions such as those in the region [3–5], yet the species is highly invasive and has spread into riparian habitats throughout much of western North America [6]. As a result, landowners are often faced with the choice between dealing with Russian olive and the potential for its invasive biology to impact the environment, or not being able to have viable options for afforestation of their lands.

To address the need for other tree options that can be established on these high-salinity soils, researchers at the US Department of Agriculture (USDA) Northern Research Station and their collaborators have been developing genotypes of hybrid poplars (*Populus* species and their hybrids) for decades to provide biomass feedstocks for biofuels, bioenergy, and bioproducts [7–9], as well as "green tools" for phytoremediation and associated phytotechnologies [10,11]. Two advancements in these regional hybrid poplar development programs have provided the potential for identification of pure species or hybrids that can survive and thrive on these salty soils. First, phyto-recurrent selection was developed as a data-driven methodology for choosing hybrid poplar clones to appropriately match the particular climatic and soil conditions of field sites [12,13]. Second, based on the use of phyto-recurrent selection in phytotechnologies related to highly-salinity wastewater irrigation, certain hybrid poplar genotypes been identified that can tolerate salinities as high as 9 dS m^{-1} [14–16], which may provide tree options for planting in the Northern Great Plains.

Similar activities by the USDA Agricultural Research Service (USDA ARS) in the southwestern United States have corroborated this potential application for hybrid poplars [11,17]. For example, Bañuelos et al. [18] and Shannon et al. [19] reported broad genetic variation in salt tolerance among eight hybrid poplar genotypes belonging to three genomic groups. Genotype-specific biomass impacts occurred at salinities ranging from 3.3 to 7.6 dS m^{-1}, with each salinity unit decreasing biomass by 10 to 15%. More recently, Bañuelos et al. [20] tested hybrid poplar growth and uptake of boron (B) following irrigation with high salinity and high B waters, and they reported seasonal fluctuations in B and chloride (Cl^{-}) uptake along with tolerance and lack of biomass impacts of some clones.

Although less salinity research has been conducted for trees than agricultural crops, select studies have classified salt tolerance of poplars. For example, Mirck and Zalesny [21] classified poplars as sensitive to moderately salt tolerant depending on genomic groups and specific genotypes within groups. Regardless of plant type, classifying salinity tolerance depends on salt concentrations, measurement methods, and source of salinity treatments [21]. In particular, for treatment sources, salinity may be: (1) introduced via wastewaters such as landfill leachate, (2) present in groundwaters, or (3) naturally occurring in soils. In addition to those cited above, studies testing salt tolerance of poplars have included soil salinities ranging from 0.1 to 15.5 dS m^{-1} [22–25], with tolerance thresholds ranging from 2.3 to 5.0 dS m^{-1} and electrical conductivity values of root zone salinity resulting in 50% biomass reductions (i.e., C_{50} values) ranging from 3.3 to 5.8 dS m^{-1} [26]. Salinity studies of poplars have occurred in short durations under controlled environments [25], and through entire rotations with practical field applications [27], thus resulting in opportunities for combining poplar phytoremediation and agroforestry [28].

Recently, Limmer et al. [29] reported results of using hybrid poplar clone 'HP-308' (*Populus charkowiensis* Schröder × *P. incrassata* Dode) and Russian olive for phytoremediation of benzene, toluene, and chlorobenzene, which is the first field-scale report that we know of combining these genera for phytotechnologies. The objective of the current study was to compare the performance of known saline-tolerant hybrid poplar genotypes compared with Russian olive at a field site in Burleigh County, North Dakota that had soils naturally varying in salinity. Specifically, based on initial salinity values, we categorized the soils as low (0.1 to 3.9 dS m^{-1}), medium (4.0 to 5.9 dS m^{-1}), and high (6.0 to 10.0 dS m^{-1}) salinity treatments and grew eight hybrid poplar genotypes plus a Russian olive accession for four growing seasons. We tested for differences among salinity treatments, genotypes, and their interactions for tree survival, height growth, and phytoextraction potential of salts and metals (i.e., the uptake of salts and metals into leaf tissue in the current study, where greater concentrations equal higher potential). This information is important for researchers, land managers, and landowners when making decisions about species selection for agroforestry systems in the region.

2. Materials and Methods

2.1. Site and Soils Description

The study site was located in Burleigh County, near Bismarck, North Dakota, USA (46.82707° N, −100.59790° W) (Figure 1). The soils were classified as Harriet loam, and soil properties of the study area are shown in Tables 1 and 2. Given heterogeneity in soil salinity across the site, initial salinity levels were mapped 21 days before planting on 20 May 2014 (Figure 2), and then at seven additional dates throughout the study (Table 1). Electrical conductivity (EC) readings were measured with a FieldScout Direct Soil EC Meter at 23 cm depth (Spectrum Technologies, Inc., Plainfield, Illinois, USA). Salinity values are reported as 1:1 measurements. The initial salinity ranges were categorized into three soil treatments: low (0.1 to 3.9 dS m^{-1}), medium (4.0 to 5.9 dS m^{-1}), and high (6.0 to 10.0 dS m^{-1}). Figure 2 illustrates the locations of these treatments across the site.

On 27 August 2014, six soil samples from each salinity treatment were collected to a depth of 30 cm (Figure 2) and brought back to the USDA Forest Service, Institute for Applied Ecosystem Studies in Rhinelander, Wisconsin, USA for laboratory analyses of pH, EC, chloride (Cl$^-$), and the following elements: aluminum (Al), cadmium (Cd), calcium (Ca), cobalt (Co), chromium (Cr), copper (Cu), iron (Fe), potassium (K), magnesium (Mg), manganese (Mn), sodium (Na), nickel (Ni), lead (Pb), and zinc (Zn) (Table 2).

A Fisher Scientific Accumet XL-50 m (Thermo Fisher Scientific, Carlsbad, California, USA) was used to measure pH, EC, and Cl$^-$. For pH, 4 g of air-dried, crushed, and sieved soil samples (≤2 mm particle size) were weighed into 5-dram clear polystyrene vials (Thornton Plastics, Salt Lake City, Utah, USA), and 5 mL of 0.01 M calcium chloride (CaCl$_2$) was added. The vials were capped, placed horizontally on an orbital shaker, and shaken at 80 rpm for 40 min. The vials were then placed in a tray, caps removed, and pH measured using a Fisherbrand^{TM} AccuCap^{TM} capillary junction pH electrode (Thermo Fisher Scientific, Carlsbad, California, USA). The meter was calibrated with buffer standards prepared from Fisher Buffer Salt of pH 4.01 and 7.41 (Thermo Fisher Scientific, Carlsbad, California, USA).

For EC and Cl$^-$, 4 g of dried, crushed, and sieved soil samples (≤2 mm particle size) were weighed into 5-dram clear polystyrene vials (Thornton Plastics, Salt Lake City, Utah, USA), and 5 mL of water purified with a RODI-T2 reverse osmosis plus Type II DI system (Aqua Solutions, Jasper, Georgia, USA) was added [15,21]. The vials were handled and shaken in the same manner as for pH. Electrical conductivity was measured using a Fisherbrand^{TM} Accumet^{TM} temperature-compensated two-cell conductivity probe using standards prepared from 5000 dS m^{-1} conductivity standard (Ricca Chemical, Arlington, Texas, USA). After the EC measurements, 0.1 mL of 5-M sodium nitrate (NaNO$_3$) solution was added as an ionic strength adjustor to each vial. The solution was mixed thoroughly, and Cl$^-$ was measured using a Fisherbrand^{TM} Accumet^{TM} solid-state half-cell Cl$^-$-specific electrode using standards prepared from 1000 mg L^{-1} Cl$^-$ standard (Ricca Chemical, Arlington, Texas, USA).

Figure 1. (**A**) Landscape of the Northern Great Plains in North Dakota, USA, with the study plot and trees planted into tree shelters in the center of the photo. (**B**) Surface salt crust of the soils at the study plot; note trees planted into tree shelters in the background. (**C**) Black weed barrier fabric protecting each tree from competition. (**D**) FieldScout Direct Soil EC (electrical conductivity) Meter with 61-cm T-handle probe used to measure soil salinity levels (Spectrum Technologies, Inc., Plainfield, Illinois, USA). (**E**) Trees growing at the study plot. (**F**) Poplar clone 'DN5' (*Populus deltoides* × *P. nigra*) growing out of its 1.5-m tree shelter. All photos were taken by Ron Zalesny (USDA Forest Service) at 77 days after planting.

An Agilent AA240FS fast sequential atomic absorption spectrometer (Agilent Technologies, Santa Clara, California, USA) was used to analyze for soil concentrations of Al, Cd, Ca, Co, Cr, Cu, Fe, K, Mg, Mn, Na, Ni, Pb, and Zn (Table 2). In particular, 4 g of dried, crushed, and sieved soil samples (≤2 mm particle size) were weighed into 50-mL polypropylene Digestion Cups (Environmental Express, Charleston, South Carolina, USA), and 45 mL of Mehlich-3 extracting solution was added. The cups were handled and shaken in the same manner as for pH. Filtermate® filters (Environmental Express, Charleston, South Carolina, USA) were inserted and carefully pressed through the liquid to leave the clear supernatant above the filter. These samples were presented to the Agilent AA240FS and analyzed for the panel of elements presented above.

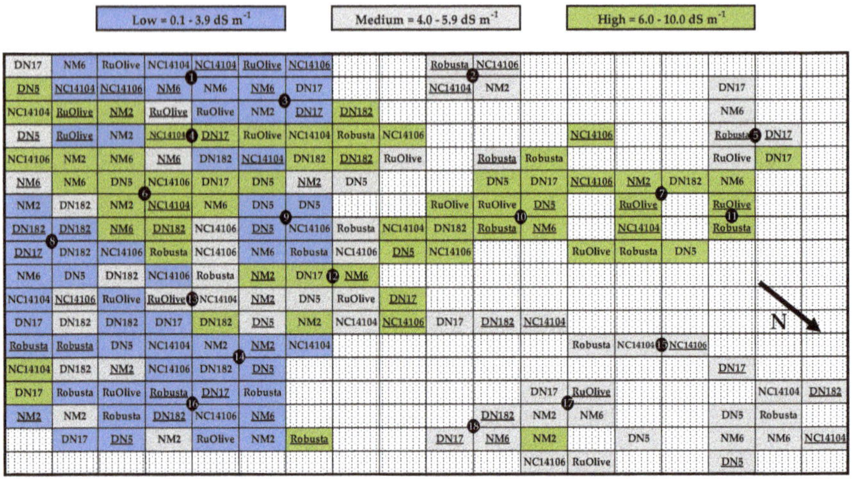

Low = 0.1 - 3.9 dS m⁻¹   Medium = 4.0 - 5.9 dS m⁻¹   High = 6.0 - 10.0 dS m⁻¹

Figure 2. Field map showing locations of eight poplar clones and Russian Olive (*Elaeagnus angustifolia* L.) established on low- (blue), medium- (gray), and high-salinity (green) soils in North Dakota, USA (n = 7 trees per soil treatment × genotype combination). Black circles with numbers indicate soil sampling points (n = 6 samples per soil treatment). Leaves were sampled for phytoextraction potential from genotypes that are underlined (n = 3 trees per soil treatment × genotype combination). See Table 1 for genotype information. Dotted cells indicate areas where planting was not possible given the presence of aboveground debris and/or where salinity of the dotted cell was not needed as a planting site (see Figure 1).

Table 1. Mean value (± standard error; n = 63) for soil salinity (dS m⁻¹) during eight sampling dates in a study testing the survival, growth, and phytoextraction potential of hybrid poplar in North Dakota, USA. Initial soil salinity values were categorized into three treatment levels for data analyses: low (0.1 to 3.9 dS m⁻¹), medium (4.0 to 5.9 dS m⁻¹), and high (6.0 to 10.0 dS m⁻¹). The column labeled '*Overall*' represents the overall mean averaged across all treatment levels (n = 189).

	Salinity Treatment			
Date	Low	Medium	High	*Overall*
20 May 2014	1.89 ± 0.12	4.87 ± 0.07	7.73 ± 0.15	*4.83 ± 0.19*
3 July 2014	1.79 ± 0.13	4.26 ± 0.12	6.31 ± 0.19	*4.12 ± 0.16*
15 July 2014	1.67 ± 0.12	3.61 ± 0.13	5.87 ± 0.16	*3.71 ± 0.15*
27 August 2014	1.35 ± 0.10	3.08 ± 0.11	4.72 ± 0.14	*3.05 ± 0.12*
22 October 2014	1.19 ± 0.09	2.77 ± 0.09	3.89 ± 0.17	*2.62 ± 0.11*
8 July 2015	1.22 ± 0.09	2.74 ± 0.11	4.02 ± 0.17	*2.66 ± 0.11*
21 June 2016	1.21 ± 0.09	2.14 ± 0.11	3.46 ± 0.16	*2.27 ± 0.10*
12 October 2016	1.11 ± 0.09	1.43 ± 0.11	2.53 ± 0.12	*1.47 ± 0.07*

Table 2. Mean value (± standard error; $n = 6$) for soil pH, electrical conductivity (EC; dS m^{-1}), and metals in a study testing the survival, growth, chloride, and phytoextraction potential of hybrid poplar in North Dakota, USA. Initial soil salinity values were categorized into three treatment levels for data analyses: low (0.1 to 3.9 dS m^{-1}), medium (4.0 to 5.9 dS m^{-1}), and high (6.0 to 10.0 dS m^{-1}). The column labeled 'Overall' represents the overall mean averaged across all treatment levels ($n = 18$). Soils were classified as Harriet loam. See Materials and Methods for details about analytical methods.

	Salinity Treatment			
Parameter	Low	Medium	High	*Overall*
pH	7.01 ± 0.23	7.68 ± 0.17	8.02 ± 0.15	*7.57 ± 0.14*
EC	0.96 ± 0.28	2.98 ± 0.54	3.56 ± 0.97	*2.50 ± 0.45*
		g kg^{-1}		
Calcium (Ca)	3.56 ± 0.18	4.09 ± 0.70	5.30 ± 0.53	*4.32 ± 0.33*
Chloride (Cl$^-$)	0.27 ± 0.04	0.76 ± 0.18	1.33 ± 0.28	*0.79 ± 0.15*
Magnesium (Mg)	1.20 ± 0.08	1.99 ± 0.18	1.89 ± 0.17	*1.69 ± 0.12*
Sodium (Na)	1.28 ± 0.40	3.91 ± 0.39	5.86 ± 0.72	*3.68 ± 0.54*
		mg kg^{-1}		
Aluminum (Al)	533.07 ± 44.86	407.26 ± 70.97	256.94 ± 58.99	*399.09 ± 42.21*
Cadmium (Cd)	0.08 ± 0.03	0.06 ± 0.03	0.10 ± 0.02	*0.08 ± 0.01*
Cobalt (Co)	1.69 ± 0.11	2.09 ± 0.23	1.75 ± 0.20	*1.84 ± 0.11*
Chromium (Cr)	0.54 ± 0.12	0.50 ± 0.08	0.52 ± 0.10	*0.52 ± 0.06*
Copper (Cu)	2.48 ± 0.79	2.49 ± 0.58	2.77 ± 0.56	*2.58 ± 0.36*
Iron (Fe)	374.21 ± 19.44	368.29 ± 8.83	344.41 ± 23.33	*362.30 ± 10.38*
Potassium (K)	711.86 ± 74.87	572.25 ± 22.67	474.99 ± 24.69	*586.37 ± 34.87*
Manganese (Mn)	187.01 ± 21.23	227.24 ± 24.82	181.12 ± 20.03	*198.46 ± 12.99*
Nickel (Ni)	22.62 ± 2.08	28.06 ± 2.76	24.79 ± 2.97	*25.16 ± 1.53*
Lead (Pb)	2.01 ± 0.60	2.33 ± 0.70	3.12 ± 0.67	*2.49 ± 0.37*
Zinc (Zn)	4.83 ± 0.58	4.16 ± 0.30	3.64 ± 0.23	*4.21 ± 0.25*

2.2. Plant Material and Experimental Design

Eight hybrid poplar clones [*Populus deltoides* Bartr. ex Marsh. × *P. nigra* L. 'Robusta', 'DN17', 'DN182', 'DN5'; *P. deltoides* × *P. maximowiczii* A. Henry 'NC14104', 'NC14106'; *P. nigra* × *P. maximowiczii* 'NM2', 'NM6'] and one Russian olive (*Elaeagnus angustifolia* L.) accession were tested (Table 3). For consistency with hybrid poplar clones, Russian olive is referred to as 'RuOlive' in the Results. 'Robusta' and 'DN17' are the same genotype but they came from different nursery sources (i.e., hence their different nomenclature), and they were analyzed as two different clones to evaluate performance of planting stock from these two sources. 'Robusta' was acquired from Big Sioux Nursery, Watertown, South Dakota, USA, while all other hybrid poplar clonal material (including 'DN17') came from the USDA Forest Service, Institute for Applied Ecosystem Studies in Rhinelander, Wisconsin, USA from one-year-old coppice shoots collected from stoolbeds at Hugo Sauer Nursery. All hybrid poplar consisted of 25.4 cm dormant, unrooted cuttings that were grown in 10.16- × 10.16- × 35.56-cm TP414 Treepots (Stuewe and Sons, Inc., Tangent, Oregon, USA) for one growing season at the USDA Natural Resources Conservation Service, Bismarck Plant Materials Center in Bismarck, North Dakota, USA. Russian olive was bare root conservation grade stock purchased the year of establishment from Lincoln Oakes Nursery in Bismarck, North Dakota, USA.

On 10 June 2014, seven trees of each genotype were planted from the TP414 Treepots (hybrid poplars) and as bare root stock (Russian Olive) in a completely random experimental design at 3.05- × 3.05-m spacing within each soil salinity treatment (i.e., 189 trees were tested). Given the soil heterogeneity described above, some trees did not have full tree competition on all sides, resulting in spacing that was greater than 9.3 m^2 per tree (Figure 2). The depth to the bottom of each root ball was 30.48 cm. Due to the sticky nature of the soils, 0.007 m^3 of Mandan silt loam soil was used to backfill each planting hole. Once planted, 1.52 m Tubex® tree shelters (Tubex, Old Hickory, Tennessee, USA) were installed on each tree for protection from white-tailed deer (*Odocoileus virginianus* Zimmermann) (Figure 1). A 1.83 × 1.83 m square of 90.7 g DeWitt Sunbelt woven ground cover (DeWitt Company, Sikeston, Missouri, USA) was installed around each tree to eliminate competing vegetation (Figure 1).

Table 3. *Populus* genomic groups and clones tested on low-, medium-, and high-salinity soils in North Dakota, USA. Russian olive (*Elaeagnus angustifolia* L.) (accession number 9019582) was grown as an experimental control.

Accession [a]	Clone	Genomic Group [b]
9094432	Robusta	'DN' *P. deltoides* × *P. nigra*
9094423	DN17	"
9094422	DN182	"
9094421	DN5	
9094424	NC14104	'DM' *P. deltoides* × *P. maximowiczii*
9094425	NC14106	"
9094426	NM2	'NM' *P. nigra* × *P. maximowiczii*
9094427	NM6	"

[a] Accession numbers from the USDA-Natural Resources Conservation Service Bismarck Plant Materials Center, Bismarck, North Dakota, USA. [b] Sections and authorities are: *Aigeiros* Duby—*P. deltoides* Bartr. ex Marsh, *P. nigra* L. *Tacamahaca* Spach—*P. maximowiczii* A. Henry.

2.3. Data Collection and Analysis

Survival and tree height (m) were determined following first, second, third, and fourth year budset. Height was measured from the base of the tree to the tip of the terminal bud. Per a landowner maintenance agreement, Russian olive was removed after the third growing season to prevent its spread onto the landscape. Thus, survival and height data were not available for Russian olive during year four.

On 27 August 2014 leaves from leaf plastochron index (LPI) 5, 10, and 15 [30] were collected from three trees per soil treatment × genotype combination (i.e., 81 trees were tested) (Figure 2). The leaves were brought back to the USDA Forest Service, Institute for Applied Ecosystem Studies in Rhinelander, Wisconsin, USA, for evaluation of phytoextraction potential via laboratory analyses of Al, Ca, Cd, Fe, K, Mg, Mn, Na, and Zn using an Agilent AA240FS fast sequential atomic absorption spectrometer (Agilent Technologies, Santa Clara, California, USA). In particular, leaf tissue was ground with a Cyclotec 1093 sample mill (Foss North America, Inc., Eden Prairie, Minnesota, USA) using a 20-mesh screen. For each tree, 100 to 400 mg of ground leaf tissue was placed in 50-mL polypropylene Digestion Cups (Environmental Express, Charleston, South Carolina, USA), and 7 mL of 10.6-N nitric acid (HNO_3) was added. The cups were placed at room temperature under an operating fume hood for at least 30 min to initiate pre-digestion. One or more empty tubes were included in each digestion set to serve as "blanks". After a minimum of 30 min, under an operating fume hood, the digestion cups were placed into a Smartblock 125® digester (Columbia Analytical Instruments, Inc., Irmo, South Carolina, USA) that was pre-heated to 125 °C. The samples were digested for at least 45 min, watching from that point to be sure they were not reduced to dryness. When all samples were reduced to 2 to 5 mL and were clear, the cups were removed from the block and cooled to room temperature. After a minimum of 15 min, 45 mL of water purified with a RODI-T2 reverse osmosis plus Type II DI system (Aqua Solutions, Jasper, GA, USA) was added to the digestion cups that were then capped and inverted several times to thoroughly mix the contents before being transferred to new, clean digestion cups. The new cups were brought to 50 mL volume with purified water, capped, and inverted to mix. Filtermate® filters (Environmental Express, Charleston, SC, USA) were inserted and carefully pressed through the liquid to leave the clear supernatant above the filter. These samples were presented to the Agilent AA240FS and analyzed for the panel of elements presented above.

All height and phytoextraction data were subjected to analyses of variance (ANOVA) and analyses of means (ANOM) according to SAS® (PROC GLM; PROC ANOM; SAS INSTITUTE, INC., Cary, North Carolina, USA) assuming the aforementioned completely random design including the main effects of soil salinity treatment (fixed), genotype (fixed), and their interactions. Fisher's protected least significant difference (LSD) was used to separate means of main effects at a probability level of $p < 0.05$.

3. Results

3.1. Survival

Annual survival rates across all soil salinity treatments and genotypes decreased approximately 20% each year, with overall means ranging from 97% after the first growing season to 40% at four years after planting (Figure 3). First-year survival rates were relatively stable across treatment × genotype interactions; six of the nine genotypes exhibited 100% survival in all three soil treatments (Figure 3). Soil salinity impacted hybrid poplar clone 'NC14106' (*P. deltoides* × *P. maximowiczii*) the most, with survival rates decreasing from 100% in low-salinity soils to 86% and 71% in medium- and high-salinity treatments, respectively. Fourteen percent of the two remaining clones, 'DN182' (*P. deltoides* × *P. nigra*) and 'DN5' (*P. deltoides* × *P. nigra*), died during the first year when planted in the high-salinity soils, while 'DN182' also had this mortality rate for the medium salinity treatment.

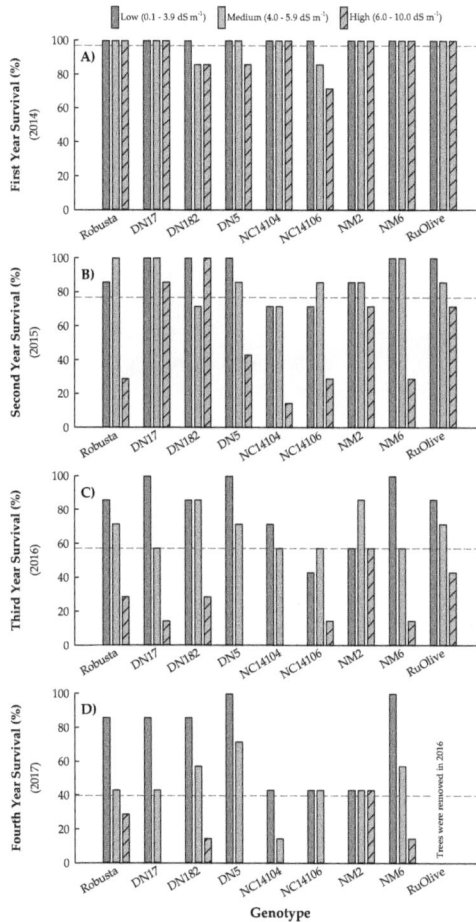

Figure 3. Mean survival (*n* = 7) after the first (**A**), second (**B**), third (**C**), and fourth (**D**) growing seasons of eight hybrid poplar clones and Russian Olive (*Elaeagnus angustifolia* L.) grown on soils varying in salinity in North Dakota, USA. Soil treatments were based on initial values and included low (0.1 to 3.9 dS m^{-1}), medium (4.0 to 5.9 dS m^{-1}), and high (6.0 to 10.0 dS m^{-1})] salinity ranges. The dashed line represents the overall mean. Per a landowner maintenance agreement, all Russian olive trees were removed in 2016.

After the second growing season, survival ranged from 52% (high salinity) to 90% (low salinity) for salinity treatments and 52% ('NC14104'; *P. deltoides* × *P. maximowiczii*) to 95% ('DN17'; *P. deltoides* × *P. nigra*) for genotypes (Figure 3). Survival for Russian olive (*Elaeagnus angustifolia* L. 'RuOlive') in year two was similar to 'NC14106' in year one, with 100% of its trees surviving in low-salinity soil followed by 86% and 71% in medium- and high-salinity treatments, respectively. There was an inverse relationship between survival and salinity levels, with most hybrid poplar clones having substantially lower survival in the high-salinity soils, though some genotypes were less impacted. For example, 'DN182' had 100% survival in the high-salinity treatment despite 29% of its trees dying in the medium-salinity soils. All trees planted in low- and medium-salinity soils survived for clone 'DN17', and its trees of the high-salinity treatment had 86% survival, which was the best performance across all genotypes and soil treatments in year two.

Furthermore, noticeable mortality occurred during the third growing season, with 43% of the trees dying across treatments and genotypes during that year (Figure 3). Survival ranged from 22% (High treatment) to 81% (Low) for salinity treatments, and 38% ('NC14106') to 67% ('DN182', 'RuOlive', 'NM2' (*P. nigra* × *P. maximowiczii*)) for genotypes. All of the 'DN5' and 'NC14104' trees planted in the high-salinity soils died, and survival decreased to 14% for 'DN17' and 'NC14106' for this treatment. In contrast, trees of 'NM2' were not impacted by the high-salinity soils relative to those with low salinity, having survival rates of 57% in both treatments.

After the fourth growing season, survival ranged from 11% (High) to 65% (Low) for salinity treatments and 19% ('NC14104') to 57% ('DN5', 'NM6' (*P. nigra* × *P. maximowiczii*)) for genotypes (Figure 3). Distinct survival patterns emerged during year four. First, in addition to 'DN5' and 'NC14104' that exhibited complete mortality in high-salinity soils during year 3, all trees of clones 'DN17' and 'NC14106' died in this treatment during the fourth growing season. Second, two clones, 'Robusta' (*P. deltoides* × *P. nigra*) and 'NM6' had stable survival rates from years two through four, with those in the last two years of study being nearly identical (within each genotype). Third, of all the genotypes tested, 'NM2' exhibited the greatest survival in high-salinity soils (i.e., 43%), which was equal to its survival rate in both the low- and medium-salinity soils at the end of the study.

3.2. Height Growth

The treatment × genotype interaction was negligible for first- and second-year height ($p = 0.2405$, $p = 0.5738$, respectively), yet the treatment ($p < 0.0001$) and genotype ($p < 0.0001$) main effects were both significant for these heights (Table 4). After the first growing season, low-salinity soils had the significantly largest trees that were 9% and 16% taller than trees in the medium and high salinity treatments (which differed from each other), respectively (Table 5). Height of the low-treatment trees was 8% greater than the overall mean, while high-treatment trees were 9% shorter than the overall mean. Similarly, second-year height was significantly greatest for the low salinity treatment, where its trees were 9% and 36% taller than trees in the medium and high salinity treatments (which differed from each other), respectively (Table 5). For the genotype main effect, one-year-old trees ranged in height from 0.81 ± 0.03 ('RuOlive') to 1.72 ± 0.05 m ('DN5'), with an overall mean of 1.50 ± 0.02 m (Figure 4). There were no differences in height within hybrid poplar genomic groups, and clones exhibited similar values, which were all 49% significantly taller than for 'RuOlive'. Height was much more variable in year two, ranging from 1.44 ± 0.13 ('RuOlive') to 2.13 ± 0.15 m ('DN5'), with an overall mean of 1.79 ± 0.04 m (Figure 4). Clones 'Robusta', 'DN5', and 'NM6' had the greatest height among clones, and there were differences among genotypes within genomic groups. For example, 'DN182' trees were 20% shorter than for the other *P. deltoides* × *P. nigra* clones. Likewise, 'NM6' had 28% taller trees than 'NM2' within the *P. nigra* × *P. maximowiczii* genomic group. In contrast, 'NC14014' and 'NC14016' exhibited identical second-year height (Figure 4).

Table 4. Probability values from analyses of variance comparing height and leaf phytoextraction potential of eight hybrid poplar clones and Russian olive (*Elaeagnus angustifolia* L.) (*G* = nine genotypes) established on low-, medium-, and high-salinity soils (*T* = three soil treatments) in North Dakota, USA. Significant values listed in bold were compared in the Results and illustrated in Table 5 and Figures 4–6.

Parameter	Source of Variation		
	T	G	T × G
Height			
(2014) First Year	**<0.0001**	**<0.0001**	0.2405
(2015) Second Year	**<0.0001**	**<0.0001**	0.5738
(2016) Third Year	<0.0001	<0.0001	**0.0486**
(2017) Fourth Year	<0.0001	0.0002	**<0.0001**
Leaf Phytoextraction			
Aluminum (Al)	**0.0011**	**<0.0001**	0.6113
Calcium (Ca)	**0.0004**	**<0.0001**	0.3609
Cadmium (Cd)	0.0827	**<0.0001**	0.8508
Iron (Fe)	**0.0239**	**<0.0001**	0.2088
Potassium (K)	**0.0247**	0.8734	0.1237
Magnesium (Mg)	**<0.0001**	**<0.0001**	0.3731
Manganese (Mn)	**0.0010**	**0.0019**	0.7696
Sodium (Na)	<0.0001	<0.0001	**0.0044**
Zinc (Zn)	0.4731	**0.0008**	0.2629

Table 5. Mean height (*n* = 33 to 63) and leaf phytoextraction (*n* = 27) (± standard error) across eight hybrid poplar clones and Russian Olive (*Elaeagnus angustifolia* L.) for initial soil salinity treatment main effects [low (0.1 to 3.9 dS m^{-1}), medium (4.0 to 5.9 dS m^{-1}), and high (6.0 to 10.0 dS m^{-1})] in a study testing the survival, growth, and phytoextraction potential of hybrid poplar in North Dakota, USA. Means with different letters within rows were different at $p < 0.05$, while those labeled with an asterisk (*) were different than the overall mean at $p < 0.05$. Leaves were sampled on 27 August 2014. See Figure 2 for locations of trees from which leaves were sampled.

Parameter	Salinity Treatment						Overall
	Low		Medium		High		
Height (m)							
(2014) First Year	1.63 ± 0.04	a *	1.48 ± 0.04	b	1.37 ± 0.04	c *	*1.50 ± 0.02*
(2015) Second Year	2.03 ± 0.07	z *	1.84 ± 0.06	y	1.31 ± 0.05	x *	*1.79 ± 0.04*
Leaf phytoextraction (g kg^{-1})							
Calcium (Ca)	10.41 ± 0.65	b *	12.77 ± 0.87	a	14.16 ± 0.92	a *	*12.44 ± 0.50*
Magnesium (Mg)	3.80 ± 0.20	y *	5.56 ± 0.38	z	5.63 ± 0.31	z	*5.00 ± 0.20*
Leaf phytoextraction (mg kg^{-1})							
Aluminum (Al)	198.94 ± 20.11	b *	278.87 ± 21.09	a	271.37 ± 21.13	a	*249.72 ± 12.51*
Iron (Fe)	67.50 ± 5.30	yx	92.36 ± 11.22	z	75.21 ± 5.73	zy	*78.35 ± 4.65*
Potassium (K)	28.12 ± 1.01	b	28.53 ± 1.20	b	32.79 ± 1.70	a *	*29.81 ± 0.80*
Manganese (Mn)	53.38 ± 5.13	y *	76.55 ± 4.69	z	77.25 ± 6.06	z	*69.06 ± 3.28*

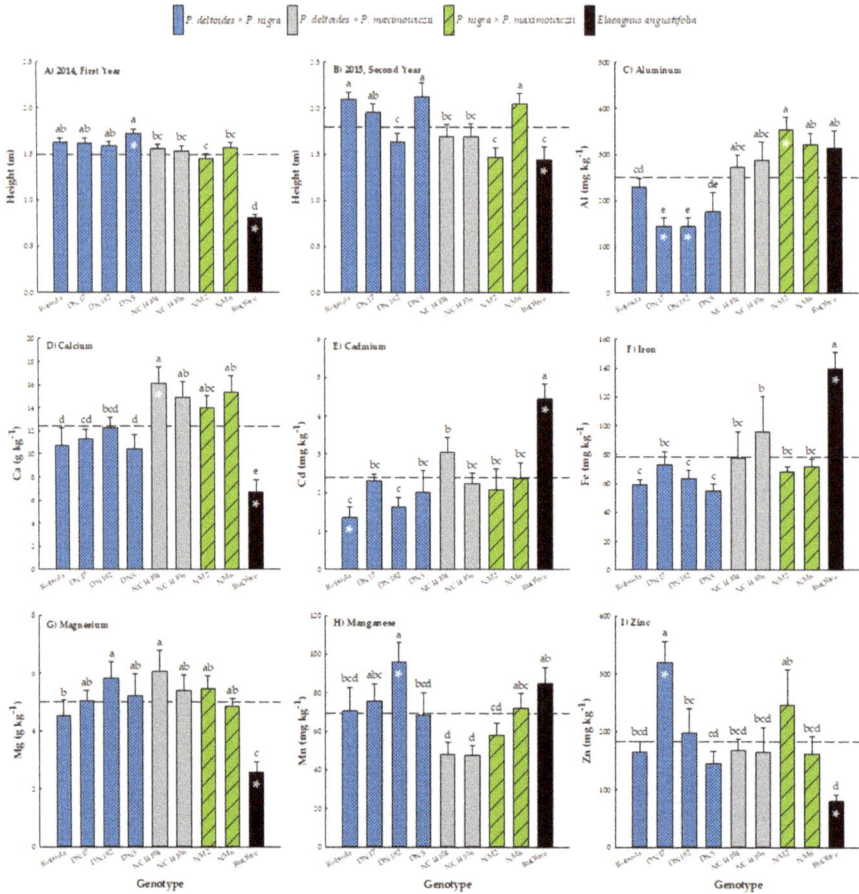

Figure 4. Mean height (± standard error) after the first (**A**) and second (**B**) growing seasons, as well as leaf phytoextraction (± standard error) of aluminum (Al) (**C**), calcium (Ca) (**D**), cadmium (Cd) (**E**), iron (Fe) (**F**), magnesium (Mg) (**G**), manganese (Mn) (**H**), and zinc (Zn) (**I**) of eight hybrid poplar clones and Russian Olive (*Elaeagnus angustifolia* L.) grown on soils varying in salinity in North Dakota, USA. The dashed line represents the overall mean, while bars with asterisks indicate means that differ from the overall mean at *p* < 0.05. Bars with different letters were different according to Fisher's protected least significant difference (LSD) at *p* < 0.05.

Treatment × genotype interactions were significant for height at three (*p* = 0.0486) and four (*p* < 0.0001) years after planting (Table 4). After the third growing season, height ranged from 0.91 ± 0.00 ('NC14106' High) to 3.04 ± 0.16 m ('DN5' Low), with an overall mean of 2.23 ± 0.04 m (Figure 5). Although all of the high-salinity trees died for 'DN5' and 'NC14104', height did not differ for the other treatments within either of these clones. Similarly, there were no differences among salinity treatments for 'DN182', 'NC14106', 'NM2', 'NM6' nor 'RuOlive'. In contrast, trees of 'DN17' grew 53% taller in low- versus high-salinity soils. Likewise, trees in both low and medium salinity treatments were significantly taller than those in high-salinity soils, having 53% and 50% greater height, respectively (Figure 5). In the fourth growing season, height ranged from 0.61 ± 0.16 ('Robusta' High) to 3.28 ± 0.20 m ('DN5' Low), with an overall mean of 2.44 ± 0.09 m (Figure 5). As noted above, only four of the eight clones tested (i.e., 'RuOlive' was removed after the third growing season per a landowner maintenance agreement) were still alive on the high salinity treatment in year four: 'Robusta', 'DN182',

'NM2', and 'NM6'. Of those, height was only significantly different for 'Robusta', wherein trees with the same height for low and medium treatments grew 78% taller than those in the high salinity soils.

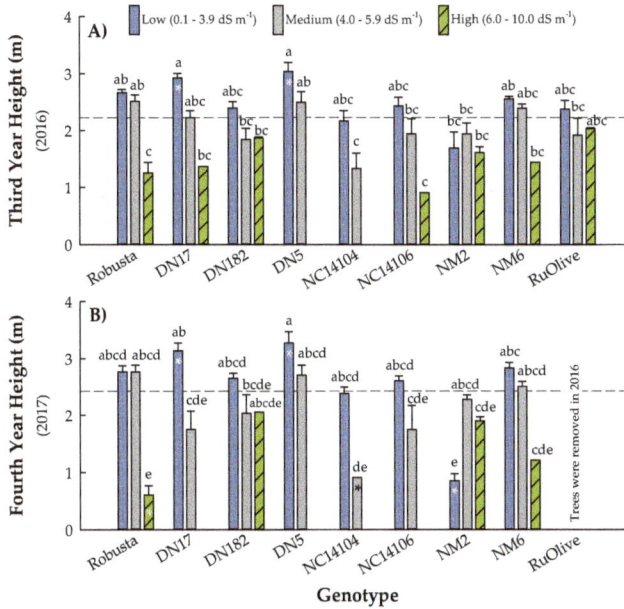

Figure 5. Mean height (± standard error) after the third (**A**) and fourth (**B**) growing seasons of eight hybrid poplar clones and Russian Olive (*Elaeagnus angustifolia* L.) grown on soils varying in salinity in North Dakota, USA. Soil treatments included low (0.1 to 3.9 dS m^{-1}), medium (4.0 to 5.9 dS m^{-1}), and high (6.0 to 10.0 dS m^{-1}) initial salinity ranges. The dashed line represents the overall mean, while bars with asterisks indicate means that differ from the overall mean at $p < 0.05$. Bars with different letters were different according to Fisher's protected LSD at $p < 0.05$. Per a landowner maintenance agreement, all Russian olive trees were removed in 2016.

3.3. Leaf Phytoextraction

The treatment × genotype interaction was negligible for phytoextraction of Al, Ca, Cd, Fe, K, Mg, Mn, and Zn, yet the treatment main effect was significant for all of these metals ($p < 0.05$) except Cd and Zn (Table 4). Uptake of Al was greatest on medium- and high-salinity soils that were not different from one another but were 28% greater than the low salinity treatment, which was 20% less than the overall mean (Table 5). Similar trends were shown for Ca, Mg, and Mn, where the medium and high salinity treatments had 26%, 33%, and 31% more leaf uptake than the low salinity treatment, respectively. Iron uptake was greatest for medium-salinity soils, which were not different from Fe phytoextraction from high-salinity treatments but were 27% greater than in the low-salinity soils. Lastly, K uptake was 15% greater in high-salinity soils than either of the other treatments (that did not differ from one another) and 9% more than the overall mean (Table 5).

Clones differed in their leaf phytoextraction of Al, Ca, Cd, Fe, Mg, Mn, and Zn ($p < 0.05$) (Table 4). The leaf concentration of Al ranged from 144.41 ± 19.30 ('DN182') to 353.12 ± 28.94 mg Al kg^{-1} ('NM2'), with a mean of 249.73 ± 12.51 mg Al kg^{-1} (Figure 4). The *P. deltoides* × *P. nigra* clones exhibited 78% less Al uptake than the *P. deltoides* × *P. maximowiczii* genotypes, *P. nigra* × *P. maximowiczii* genotypes, and 'RuOlive', which were not different from one another. Clones 'DN17' and 'DN182' had the lowest Al uptake, which was 73% less than the overall mean. Calcium phytoextraction ranged from 6.71 ± 1.11 ('RuOlive') to 16.14 ± 1.42 g Ca kg^{-1} ('NC14104'), with a mean of 12.44 ± 0.50 g Ca kg^{-1} (Figure 4). Genotypes segregated into three response groups based on genomic group: 1) clones belonging to the *P. deltoides* × *P. maximowiczii* and *P. nigra* × *P. maximowiczii* genomic groups exhibited the greatest Ca phytoextraction, 2) *P. deltoides* × *P. nigra* clones had the second highest Ca levels, and 3) 'RuOlive', which had 85% less Ca uptake than the overall mean, also had significantly less Ca than either of the other response groups. For Cd, leaf concentrations ranged from 1.36 ± 0.27 ('Robusta') to 4.45 ± 0.40 mg Cd kg^{-1} ('RuOlive'), with a mean of 2.39 ± 0.15 mg Cd kg^{-1} (Figure 4). In contrast to Ca where Russian olive had the lowest phytoextraction potential, Cd in the leaves of 'RuOlive' was 31% ('NC14104') to 69% ('Robusta') greater than the other clones and 46% greater than the overall mean. For Fe, leaf concentrations ranged from 55.31 ± 4.56 ('DN5') to 140.14 ± 10.91 mg Fe kg^{-1} ('RuOlive'), with a mean of 78.35 ± 4.65 mg Fe kg^{-1} (Figure 4). Similar to the results for Cd, Fe in the leaves of 'RuOlive' was 31% ('NC14106') to 61% ('DN5') greater than the other clones and 44% greater than the overall mean. The leaf concentration of Mg ranged from 2.57 ± 0.36 ('RuOlive') to 6.06 ± 0.72 g Mg kg^{-1} ('NC14104'), with a mean of 5.00 ± 0.20 g Mg kg^{-1} (Figure 4), which was the opposite trend than that for Cd and Fe. In particular, despite 'DN182' and 'NC14104' having 23% and 26% more Mg in their leaves than 'Robusta', respectively, all of the hybrid poplar genotypes exhibited significantly higher Mg concentrations than 'RuOlive', which had 95% less Mg than the overall mean. Manganese phytoextraction ranged from 47.79 ± 4.66 ('NC14106') to 95.82 ± 10.66 mg Mn kg^{-1} ('DN182'), with a mean of 69.06 ± 3.28 mg Mn kg^{-1} (Figure 4). Clones within genomic groups were consistent in their uptake of Mn, with *P. deltoides* × *P. nigra* genotypes having the greatest overall concentrations (that were similar to 'RuOlive') followed by *P. nigra* × *P. maximowiczii* clones and then *P. deltoides* × *P. maximowiczii* genotypes. For Zn, leaf concentrations ranged from 80.02 ± 11.43 ('RuOlive') to 320.29 ± 36.09 mg Zn kg^{-1} ('DN17'), with a mean of 183.29 ± 12.98 mg Zn kg^{-1} (Figure 4). With the exception of 'DN17' having the greatest Zn uptake that was 43% higher than the overall mean and 'RuOlive' with the least Zn uptake that was 129% lower than the overall mean, genotypes exhibited relatively uniform phytoextraction of Zn.

The treatment × genotype interaction was significant for phytoextraction of Na ($p = 0.0044$) (Table 4). Uptake of Na into the leaves ranged from 0.03 ± 0.01 ('NC14106' Low) to 5.88 ± 0.77 g Na kg^{-1} ('RuOlive' High), with an overall mean of 0.93 ± 0.17 g Na kg^{-1} (Figure 6). Sodium concentration in the leaves did not differ among treatments within individual hybrid poplar genotypes (Figure 6), yet general trends were similar as with height during years three and four. In particular, the *P. deltoides* × *P. nigra* clones exhibited substantial within-group variability, with 'DN5' having 70% lower overall Na uptake than the other clones. In addition, the *P. deltoides* × *P. maximowiczii* clones had the lowest overall Na uptake, that being only 70% of the overall mean. For *P. nigra* × *P. maximowiczii* clones, 'NM2' had some of the highest Na concentrations for medium- and high-salinity soils of all treatment × genotype combinations, while 'NM6' had consistent Na for low and medium treatments that were 125% less than that for the high-salinity soils, albeit not statistically different. 'RuOlive' exhibited the greatest Na phytoextraction, with trees grown in the high-salinity soils taking up to 78% more Na than those of the low salinity treatment.

4. Discussion

In this study, survival, height, and phytoextraction potential of known saline-tolerant hybrid poplar genotypes were compared with that of Russian olive at a field site in Burleigh County, North Dakota, USA that had soils naturally varying in salinity. Overall, hybrid poplars exhibited greater survival and height than Russian olive, and phytoextraction potential depended on specific element × genotype combinations. It is worth noting, however, that Russian olive, by nature, is substantially shorter than any of the hybrid poplar genotypes tested, regardless of soil conditions. Hybrid poplar survival rates of the current study were consistent with previous reports during the first growing season yet lower than expected during years two through four, wherein survival decreased 20% per year to an overall survival rate of 40% at the end of the study. Expected survival rates throughout 10- to 12-year rotations of poplars are 90% to 95% for biomass applications [31,32] and 80% to 90% for phytotechnologies [33,34]. For example, in applications that are most similar to the current study with respect to salinity, Zalesny et al. [14] reported 78% survival for eight poplar genotypes that were irrigated with high-salinity landfill leachate (i.e., 6.2 to 9.4 dS m^{-1}) for two growing seasons. In a long-term component of that study, 8-year-old companion trees of *P. nigra* × *P. maximowiczii* 'NM2' had 90% survival [35], which was similar to 14 *P. deltoides* clones receiving tertiary treated municipal wastewater at 27 months after planting in Florida, USA [36].

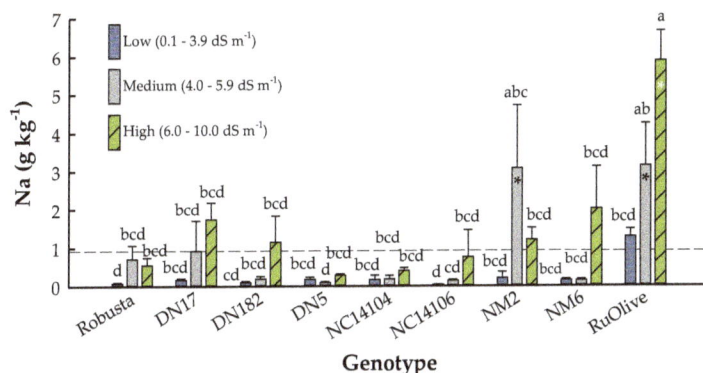

Figure 6. Leaf phytoextraction of sodium (Na) (± standard error) of eight hybrid poplar clones and Russian olive (*Elaeagnus angustifolia* L.) grown on soils varying in salinity in North Dakota, USA. Soil treatments included low (0.1 to 3.9 dS m^{-1}), medium (4.0 to 5.9 dS m^{-1}), and high (6.0 to 10.0 dS m^{-1}) initial salinity ranges. The dashed line represents the overall mean, while bars with asterisks indicate means that differ from the overall mean at $p < 0.05$. Bars with different letters were different according to Fisher's protected LSD at $p < 0.05$.

The three most plausible silvicultural causes of the lower-than-expected survival rates in the current study included high soil salinities, extremes of soil moisture availability, and tree shelter effects. First, and the most obvious, is that hybrid poplars are not typically planted in soils with salinity levels in excess of 5 dS m^{-1} [37]. Although high-biomass clones have been identified that have tolerated such salinity thresholds [17,20], some genotypes in the current study were not adapted to these local site conditions (i.e., *P. deltoides* × *P. maximowiczii* 'NC14104', 'NC14106'). Nevertheless, using methods such as phyto-recurrent selection is warranted for future efforts, wherein genotypes adapted to such soil conditions could be selected before field deployment, thereby increasing the probability of long-term survival and growth [13,18,19]. Second, trees in the current study were not irrigated, and they underwent periods of excessive moisture and prolonged drought, both of which likely stressed the trees and limited their response potential to the high-salinity soils [38,39]. While most industrial plantations of hybrid poplars in the Midwest are not irrigated [40], very few have experienced these combinations of concurrent water and salinity stress. Third, hybrid poplars are not typically planted in tree shelters, which in the current study resulted in altered aboveground morphology (i.e., tall and thin trees with reduced branch/leaf development) [41] that likely contributed to confounding of the trees' stress responses from the moisture and salinity mentioned above. While tree shelters protect the trees from deer, they abrade the tree boles and branches where they emerge from the shelter, resulting in deformity. In addition, shelters may affect the trees' ability to harden off, thus leading to winter dieback. Both of which occurred in the current study.

These silvicultural stress responses associated with adaptability, water use mechanisms, and external impacts for survival likely contributed to the observed variability in height, as well. Overall, the inverse relationship between growth and salinity (i.e., growth declined with increasing salinity levels) corroborated those of previous studies [18,22,24]. The magnitude of growth responses of the trees in the current study was similar yet slightly higher than that reported for other hybrid poplars used for salinity-related phytotechnologies. For example, Zalesny and Bauer [42] irrigated three *P.*

deltoides × *P. nigra* clones ('DN34', 'DN17', 'DN182') and two *P. nigra* × *P. maximowiczii* genotypes ('NM2', 'NM6') with low-salinity landfill leachate for one growing season and reported mean height across clones of 99.74 cm, which was 63% of that reported here for the low initial soil salinity treatment (i.e., 0.1 to 3.9 ds m^{-1}). Similarly, six hybrid poplar genotypes ('DN34', 'DN5', 'NC14104', 'NM2', 'NM6', *P. deltoides* × *P. nigra* 'I4551') were irrigated every other day with high salinity (i.e., 8.7 ds m^{-1}) landfill leachate from 42 to 70 days after planting, and trees receiving leachate had 15% greater height than those irrigated with non-saline well water [43]. Four clones ('DN5', 'NC14104', 'NM2', 'NM6') were used in both studies, and their associated height was consistently higher in the current study: 'DN5' (+66%), 'NC14104' (+65), 'NM2' (+63%), and 'NM6' (+60%). After two years, Zalesny et al. [14] reported that eight hybrid poplar genotypes irrigated with high-salinity landfill leachate (i.e., 6.2 to 9.4 dS m^{-1}) were 0.98 to 1.81 m tall compared with 1.44 to 2.13 m in the current study, resulting in their mean height being 20% less than that of the current study. Five clones ('DN5', 'NC14104', 'NC14016', 'NM2', 'NM6') were used in both studies, and height comparisons varied among these genotypes. Trees were taller in the current study for 'DN5' (+31%), 'NC14106' (+26), and 'NM6' (+22%) yet shorter for 'NC14104' (-7%) and 'NM2' (-14%). While not grown in the field, Steppuhn et al. [25] tested first-year stem length of four *P. deltoides* × *P. nigra* clones ('Assiniboine', 'CanAm', 'Manitou', 'Walker') grown in high-salinity solutions ranging in electrical conductivity from 2 to 31 dS m^{-1}, which translated to equivalent electrical conductivity (EC_e) of saturated soil paste extracts of 1.0 to 15.5 dS m^{-1} (i.e., this lower range should be used in comparisons with the current study) [44]. At EC_e of 6.9 dS m^{-1}, they reported that 'Assiniboine' exhibited 14% greater stem length than the other three clones, resulting in an 8.4% stem length reduction per EC_e unit in their study [25]. Similar clonal trends were observed for their salinity treatments of 2.3 dS m^{-1} and 4.2 dS m^{-1} [25], which were equivalent to the low and medium levels of the current study, respectively.

Leaf phytoextraction in the current study was element- and genotype-specific. Russian olive is known to be highly salt tolerant [3], and it is one of few options for establishing trees on the highly saline soils of North Dakota [5,6]. As such, Russian olive exhibited greater Na, Cd, and Fe leaf concentrations than any of the hybrid poplar genotypes. However, despite expectations, Russian olive also had the worst uptake of Ca and Mg of all genotypes tested. Soil concentrations of these elements were analyzed at the time of leaf sampling (Table 2), and conducting a mass balance assessment of the ultimate fate of these metals was beyond the scope of this study. Nevertheless, such experimentation would be a meaningful next step for future projects. While minimal information is available about the phytoextraction potential of Russian olive, leaf uptake of Cd in the current study was 99% higher than that for a study testing the concentration of heavy metals in leaves of Russian olive trees growing on industrial brownfields [45]. In contrast, leaf phytoextraction of Fe was 43% lower in the current study, thus corroborating the need to match genotypes with elements of interest and specific soil conditions. Arık and Yaldız [45] also reported Russian olive leaf concentrations of Al, Mn, and Zn that were 52%, 48%, and 78% lower than in the current trees, respectively. Similar results were shown for Cd in a study testing Russian olive as a bioindicator of heavy metal pollution [46], wherein Russian olive trees had 72% less Cd concentrations than in the current study. In contrast to comparisons with Arik and Yaldiz [45], Zn concentrations of the current study were 28% lower (rather than 78% higher) than those of Aksoy and Şahin [46].

Trees of the current study exhibited moderate levels of salt toxicity symptoms consisting of premature leaf abscission, burned leaf margins, and winter dieback [18]. Leaf phytoextraction of hybrid poplar showed similar salinity stress response trends as for survival and growth. In particular, with the exception of 'DN182' and 'DN17' having much higher concentrations of Mn and Zn than most other clones, respectively, the hybrid poplar genomic groups had two general responses. First, *P. nigra* × *P. maximowiczii* genotypes exhibited superior phytoextraction of all elements. Second, *P. deltoides* × *P. nigra* and *P. deltoides* × *P. maximowiczii* clones each had leaf concentrations that were similar to both of the other genomic groups, with two exceptions. The *P. deltoides* × *P. nigra* clones had 78% less Al and 37% less Ca uptake than the other genomic groups, while the *P. deltoides* × *P. maximowiczii* genotypes extracted 53% less Mn and 155% less Na relative to the other groups. Furthermore, Bañuelos et al. [20] reported similar variability among seven hybrid poplar genomic groups for leaf concentrations of Cl⁻ and boron (B) following high-salinity irrigation, while Bañuelos et al. [18] and Shannon et al. [19] showed a range in selenium (Se) and B uptake among eight clones belonging to three genomic groups that were irrigated similarly. Although uptake of Cl⁻, B, and Se were not tested in the current study, phytoextraction potential of other parameters corroborated previous results for hybrid poplar [47,48]. For example, Chen et al. [24] tested the effect of NaCl on growth and physiological parameters of *P. euphratica* Oliv. and *P. tomentosa* Carrière, and they reported that 40% and 36% of total Na from their treatments was taken up and distributed to the leaves, respectively. Their substantially less Na uptake into hybrid poplar leaves relative to Russian olive was likely due to known physiological responses of hybrid poplar to Na stress [24]. Specifically, hybrid poplars typically restrict Na transport from roots to leaves [19,23], which has been shown in field trials in the North Central, USA [15]. Testing five of the same genotypes as the current study, Zalesny et al. [15] irrigated the trees with high-salinity landfill leachate for two growing seasons and reported Na phytoextraction values that were 5% lower for 'NM6' yet 42% ('NM2') to 79% ('DN5') greater than the current study for the other clones. Comparisons between current results and those from the same genotypes irrigated with high-salinity landfill leachate have shown similar outcomes to those reported above. In particular, phytoextraction values were typically higher for leachate-irrigated trees versus those grown on the soils of the current study, yet responses of individual genotypes to specific soil conditions governed phytoextraction potential [49,50]. This was especially important in the current study given that soil salinity levels decreased over time. Although it was not directly tested, higher mortality rates in subsequent growing seasons suggested that the trees are extracting and accumulating salts up to and beyond their biological thresholds, thus resulting in impacts to their internal physiological processes. Overall, this is a worthwhile indication that phytoextraction occurred.

5. Conclusions

The primary objective in the current study was to test whether any of the hybrid poplar genotypes performed as well or better than Russian olive, to offer landowners a tree-planting alternative to Russian olive on their high-salinity soils. While this objective was to find one or more hybrid poplar genotypes to replace Russian olive, it is worth noting that shrubs and grasses may also have potential for establishing vegetation on these high-salinity soils in North Dakota, USA. As a result, it would be beneficial in future studies for researchers and land managers to test a suite of species belonging to these plant functional groups. Nevertheless, in the current study Russian olive did not survive nor grow as well as expected, having lower survival rates and smaller trees than many of the hybrid poplar clones tested. At the end of three years, 86%, 71%, and 43% of the Russian olive trees were alive in the low-, medium-, and high-salinity soils, respectively. Despite the underperformance of Russian olive, hybrid poplar genomic groups exhibited broad genetic variability, which has led to their use in studies of this nature [13,18,19]. The *P. deltoides* × *P. nigra* genomic group had the broadest clonal variability among all traits, with 'Robusta' and 'DN182' exhibiting great potential for the establishment on high-salinity soils, based on survival, growth, and leaf phytoextraction. An interesting observation (as noted in the Materials and Methods) is that 'Robusta' and 'DN17' are the same genotype that came from different nursery sources in the current study (i.e., hence their different nomenclature). Based on statistical contrasts (data not reported), 'Robusta' and 'DN17' did not differ for height within nor across salinity treatments. The same was true for leaf phytoextraction, with two exceptions. First, across treatments, 'Robusta' had 37% more Al uptake than 'DN17'. Second, across treatments, 'DN17' had 49% more Zn uptake than 'Robusta'. These results are difficult to interpret, and we ascertain that the differences were due to individual responses to microsites or quality of the planting stock. Nevertheless, as expected, these clones exhibited similar responses overall. In contrast to the *P. deltoides* × *P. nigra* clones, the *P. deltoides* × *P. maximowiczii* genomic group was not suitable for the soil conditions nor silvicultural applications (e.g., the tree shelters) of the current study. Based on previous results [13,14], these clones have been highly successful for phytotechnologies, and their response in the current study supports the need for phyto-recurrent selection to monitor and choose genotypes under controlled conditions before advancing them to the field. Lastly, the *P. nigra × P. maximowiczii* clones exhibited the most stable performance across all years and salinity treatments, when considering all the traits simultaneously. Both 'NM2' and 'NM6' had the greatest fourth-year survival and height, as well as average or above average phytoextraction of all elements tested. Overall, our results suggest that *P. deltoides × P. nigra* and *P. nigra × P. maximowiczii* clones that are advanced from phyto-recurrent selection (or a similar selection method), grown in the open air without the use of tree shelters, and monitored for moisture demands (and irrigated during establishment, if necessary) to eliminate the potential for drought-induced stresses are best suited for similar systems of this nature, especially in North Dakota, USA.

Author Contributions: C.M.S. and R.S.Z. conceived and designed the study; C.M.S. established and maintained the field trial; C.M.S., R.S.Z., and B.A.B. conducted measurements; C.M.S. and B.A.B. performed sample laboratory analyses; R.S.Z. and B.A.B. performed the data analyses and interpreted the results; R.S.Z., C.M.S., and B.A.B. wrote the paper.

Funding: This research received no external funding.

Acknowledgments: We are grateful to Elizabeth Rogers and Joe Zeleznik for reviewing earlier versions of this manuscript.

Conflicts of Interest: We used the products mentioned in this manuscript because they met our research needs. Use of specific equipment is left to the discretion of the researcher. Endorsement is not intended by the USDA Forest Service nor USDA Natural Resources Conservation Service.

References

1. Schultz, R.C.; Colletti, J.P.; Faltonson, R.R. Agroforestry opportunities for the United States of America. *Agrofor. Syst.* **1995**, *31*, 117–132. [CrossRef]
2. Franzen, D. Managing saline soils in North Dakota. *North Dak. State Univ. Ext. Bull.* **2013**, SF1087, 11.
3. Monk, R.W.; Wiebe, H.H. Salt tolerance and protoplasmic salt hardiness of various woody and herbaceous ornamental plants. *Plant Physiol.* **1961**, *36*, 478–482. [CrossRef] [PubMed]
4. Khamzina, A.; Lamers, J.P.; Vlek, P.L. Nitrogen fixation by *Elaeagnus angustifolia* in the reclamation of degraded croplands of Central Asia. *Tree Physiol.* **2009**, *29*, 799–808. [CrossRef] [PubMed]
5. Enescu, C.M. Russian olive (*Elaeagnus angustifolia* L.): A multipurpose species with an important role in land reclamation. *Curr. Trends Nat. Sci.* **2018**, *7*, 54–60. Available online: http://www.natsci.upit.ro/media/1642/paper-8.pdf (accessed on 5 April 2019).
6. Katz, G.L.; Shafroth, P.B. Biology, ecology and management of *Elaeagnus angustifolia* L. (Russian olive) in western North America. *Wetlands* **2003**, *23*, 763–777. [CrossRef]
7. Riemenschneider, D.E.; Berguson, W.E.; Dickmann, D.I.; Hall, R.B.; Isebrands, J.G.; Mohn, C.A.; Stanosz, G.R.; Tuskan, G.A. Poplar breeding and testing strategies in the north-central U.S.: Demonstration of potential yield and consideration of future research needs. *For. Chron.* **2001**, *77*, 245–253. [CrossRef]
8. Zalesny, R.S., Jr.; Hall, R.B.; Zalesny, J.A.; Berguson, W.E.; McMahon, B.G.; Stanosz, G.R. Biomass and genotype × environment interactions of *Populus* energy crops in the Midwestern United States. *BioEnergy Res.* **2009**, *2*, 106–122. [CrossRef]
9. Nelson, N.D.; Berguson, W.E.; McMahon, B.G.; Cai, M.; Buchman, D.J. Growth performance and stability of hybrid poplar clones in simultaneous tests on six sites. *Biomass Bioenergy* **2018**, *118*, 115–125. [CrossRef]
10. Zalesny, R.S., Jr.; Stanturf, J.A.; Gardiner, E.S.; Perdue, J.H.; Young, T.M.; Coyle, D.R.; Headlee, W.L.; Bañuelos, G.S.; Hass, A. Ecosystem services of woody crop production systems. *BioEnergy Res.* **2016**, *9*, 465–491. [CrossRef]
11. Zalesny, R.S., Jr.; Stanturf, J.A.; Gardiner, E.S.; Bañuelos, G.S.; Hallett, R.A.; Hass, A.; Stange, C.M.; Perdue, J.H.; Young, T.M.; Coyle, D.R.; et al. Environmental technologies of woody crop production systems. *BioEnergy Res.* **2016**, *9*, 492–506. [CrossRef]
12. Zalesny, R.S., Jr.; Bauer, E.O. Selecting and utilizing *Populus* and *Salix* for landfill covers: Implications for leachate irrigation. *Int. J. Phytoremed.* **2007**, *9*, 497–511. [CrossRef] [PubMed]

13. Zalesny, J.A.; Zalesny, R.S., Jr.; Wiese, A.H.; Hall, R.B. Choosing tree genotypes for Phytoremedion of landfill leachate using phyto-recurrent selection. *Int. J. Phytoremed.* **2007**, *9*, 513–530. [CrossRef] [PubMed]

14. Zalesny, J.A.; Zalesny, R.S., Jr.; Coyle, D.R.; Hall, R.B. Growth and biomass of *Populus* irrigated with landfill leachate. *For. Ecol. Manag.* **2007**, *248*, 143–152. [CrossRef]

15. Zalesny, J.A.; Zalesny, R.S., Jr.; Wiese, A.H.; Sexton, B.; Hall, R.B. Sodium and chloride accumulation in leaf, woody, and root tissue of *Populus* after irrigation with landfill leachate. *Environ. Pollut.* **2008**, *155*, 72–80. [CrossRef] [PubMed]

16. Zalesny, J.A.; Zalesny, R.S., Jr. Chloride and sodium uptake potential over an entire rotation of *Populus* irrigated with landfill leachate. *Int. J. Phytoremed.* **2009**, *11*, 496–508. [CrossRef] [PubMed]

17. Bañuelos, G.S.; Dhillon, K.S. Developing a sustainable phytomanagement strategy for excessive selenium in western United States and India. *Int. J. Phytoremed.* **2011**, *13*, 208–228. [CrossRef] [PubMed]

18. Bañuelos, G.S.; Shannon, M.C.; Ajwa, H.; Draper, J.H.; Jordahl, J.; Licht, L. Phytoextraction and accumulation of boron and selenium by poplar (*Populus*) hybrid clones. *Int. J. Phytoremed.* **1999**, *1*, 81–96. [CrossRef]

19. Shannon, M.C.; Banuelos, G.S.; Draper, J.H.; Ajwa, H.; Jordahl, J.; Licht, L. Tolerance of hybrid poplar (*Populus*) trees irrigated with varied levels of salt, selenium and boron. *Int. J. Phytoremed.* **1999**, *1*, 273–288. [CrossRef]

20. Bañuelos, G.S.; LeDuc, D.; Johnson, J. Evaluating the tolerance of young hybrid poplar trees to recycled waters high in salinity and boron. *Int. J. Phytoremed.* **2010**, *12*, 419–439. [CrossRef]

21. Mirck, J.; Zalesny, R.S., Jr. Mini-review of knowledge gaps in salt tolerance of plants applied to willows and poplars. *Int. J. Phytoremed.* **2015**, *17*, 640–650. [CrossRef] [PubMed]

22. Fung, L.E.; Wang, S.S.; Altman, A.; Hütterman, A. Effect of NaCl on growth, photosynthesis, ion and water relations of four poplar genotypes. *For. Ecol. Manag.* **1998**, *107*, 135–146. [CrossRef]

23. Chen, S.; Li, J.; Fritz, E.; Wang, S.; Hüttermann, A. Sodium and chloride distribution in roots and transport in three poplar genotypes under increasing NaCl stress. *For. Ecol. Manag.* **2002**, *168*, 217–230. [CrossRef]

24. Chen, S.; Li, J.; Wang, S.; Fritz, E.; Hüttermann, A.; Altman, A. Effects of NaCl on shoot growth, transpiration, ion compartmentation, and transport in regenerated plants of *Populus euphratica* and *Populus tomentosa*. *Can. J. For. Res.* **2003**, *33*, 967–975. [CrossRef]

25. Steppuhn, H.; Kort, J.; Wall, K.G. First year growth response of selected hybrid poplar cuttings to root zone salinity. *Can. J. Plant Sci.* **2008**, *88*, 473–483. [CrossRef]

26. Steppuhn, H.; Genuchten, M.T.; Grieve, C.M. Root zone salinity. *Crop Sci.* **2005**, *45*, 221–232. [CrossRef]

27. Smesrud, J.K.; Duvendack, G.D.; Obereiner, J.M.; Jordahl, J.L.; Madison, M.F. Practical salinity management for leachate irrigation to poplar trees. *Int. J. Phytoremed.* **2012**, *14*, 26–46. [CrossRef] [PubMed]

28. Rockwood, D.L.; Naidu, C.V.; Carter, D.R.; Rahmani, M.; Spriggs, T.A.; Lin, C.; Alker, G.R.; Isebrands, J.G.; Segrest, S.A. Short-rotation woody crops and Phytoremedion: Opportunities for agroforestry? *Agrofor. Syst.* **2004**, *61*, 51–63. [CrossRef]

29. Limmer, M.A.; Wilson, J.; Westenberg, D.; Lee, A.; Siegman, M.; Burken, J.G. Phytoremedion removal rates of benzene, toluene, and chlorobenzene. *Int. J. Phytoremed.* **2018**, *20*, 666–674. [CrossRef]

30. Larson, P.R.; Isebrands, J.G. The plastochron index as applied to developmental studies of cottonwood. *Can. J. For. Res.* **1971**, *1*, 1–11. [CrossRef]

31. Headlee, W.L.; Zalesny, R.S., Jr.; Donner, D.M.; Hall, R.B. Using a process-based model (3-PG) to predict and map hybrid poplar biomass productivity in Minnesota and Wisconsin, USA. *BioEnergy Res.* **2013**, *6*, 196–210. [CrossRef]

32. Lazarus, W.; Headlee, W.L.; Zalesny, R.S., Jr. Impacts of supplyshed-level differences in productivity and land costs on the economics of hybrid poplar production in Minnesota, USA. *BioEnergy Res.* **2015**, *8*, 231–248. [CrossRef]

33. Zalesny, R.S., Jr.; Bauer, E.O.; Hall, R.B.; Zalesny, J.A.; Kunzman, J.; Rog, C.J.; Riemenschneider, D.E. Clonal variation in survival and growth of hybrid poplar and willow in an in situ trial on soils heavily contaminated with petroleum hydrocarbons. *Int. J. Phytoremed.* **2005**, *7*, 177–197. [CrossRef] [PubMed]

34. Zalesny, R.S., Jr.; Bauer, E.O. Genotypic variability and stability of poplars and willows grown on nitrate-contaminated soils. *Int. J. Phytoremed.* **2019**. [CrossRef] [PubMed]

35. Zalesny, R.S., Jr.; Headlee, W.L.; Gopalakrishnan, G.; Bauer, E.O.; Hall, R.B.; Hazel, D.W.; Isebrands, J.G.; Licht, L.A.; Negri, M.C.; Guthrie-Nichols, E.; et al. Ecosystem services of poplar at long-term Phytoremedion sites in the Midwest and Southeast, United States. *WIREs Energy Environ.* **2019**, in press. [CrossRef]

36. Minogue, P.J.; Miwa, M.; Rockwood, D.L.; Mackowiak, C.L. Removal of nitrogen and phosphorus by *Eucalyptus* and *Populus* at a tertiary treated municipal wastewater sprayfield. *Int. J. Phytoremed.* **2012**, *14*, 1010–1023. [CrossRef] [PubMed]

37. Neuman, D.S.; Wagner, M.; Braatne, J.H.; Howe, J. Part II: Stress physiology—Abiotic. In *Biology of Populus and its Implications for Management and Conservation*; Stettler, R.F., Bradshaw, H.D., Jr., Heilman, P.E., Hinckley, T.M., Eds.; NRC Research Press, National Research Council of Canada: Ottawa, ON, Canada, 1996; Chapter 17; pp. 423–458. Available online: www.nrcresearchpress.com/doi/book/10.1139/9780660165066#.XKUA8JAUmRs (accessed on 5 April 2019).

38. Dickmann, D.I.; Liu, Z.; Nguyen, P.V.; Pregitzer, K.S. Photosynthesis, water relations, and growth of two hybrid *Populus* genotypes during a severe drought. *Can. J. For. Res.* **1992**, *22*, 1094–1106. [CrossRef]

39. Liu, Z.; Dickmann, D.I. Responses of two hybrid *Populus* clones to flooding, drought, and nitrogen availability. I. Morphology and growth. *Can. J. Bot.* **1992**, *70*, 2265–2270. [CrossRef]

40. Zalesny, R.S., Jr.; Donner, D.M.; Coyle, D.R.; Headlee, W.L. An approach for siting poplar energy production systems to increase productivity and associated ecosystem services. *For. Ecol. Manag.* **2012**, *284*, 45–58. [CrossRef]

41. Zalesny, J.A.; Zalesny, R.S., Jr.; Coyle, D.R.; Hall, R.B.; Bauer, E.O. Clonal variation in morphology of *Populus* root systems following irrigation with landfill leachate or water during 2 years of establishment. *BioEnergy Res.* **2009**, *2*, 134–143. [CrossRef]

42. Zalesny, R.S., Jr.; Bauer, E.O. Evaluation of *Populus* and *Salix* continuously irrigated with landfill leachate II. Soils and early tree development. *Int. J. Phytoremed.* **2007**, *9*, 307–323. [CrossRef] [PubMed]

43. Zalesny, R.S., Jr.; Wiese, A.H.; Bauer, E.O.; Riemenschneider, D.E. *Ex situ* growth and biomass of *Populus* bioenergy crops irrigated and fertilized with landfill leachate. *Biomass Bioenergy* **2009**, *33*, 62–69. [CrossRef]

44. Ayers, R.S.; Westcot, D.W. *Water Quality for Agriculture*; FAO Irrigation and Drainage Paper 29 (Revision 1); Food and Agriculture Organization of the United Nations: Rome, Italy, 1985; p. 174. Available online: http://www.fao.org/3/t0234e/t0234e00.htm (accessed on 5 April 2019).

45. Arık, F.; Yaldız, T. Heavy metal determination and pollution of the soil and plants of southeast Tavşanlı (Kütahya, Turkey). *Clean Soil Air Water* **2010**, *38*, 1017–1030. [CrossRef]

46. Aksoy, A.; Şahin, U. *Elaeagnus angustifolia* L. as a biomonitor of heavy metal pollution. *Turk. J. Bot.* **1999**, *23*, 83–87.

47. Burges, A.; Alkorta, I.; Epelde, L.; Garbisu, C. From Phytoremedion of soil contaminants to phytomanagement of ecosystem services in metal contaminated sites. *Int. J. Phytoremed.* **2018**, *20*, 384–397. [CrossRef]

48. Pilipović, A.; Zalesny, R.S., Jr.; Rončević, S.; Nikolić, N.; Orlović, S.; Beljin, J.; Katanić, M. Growth, physiology, and phytoextraction potential of poplar and willow established in soils amended with heavy-metal contaminated, dredged river sediments. *J. Environ. Manag.* **2019**, *239*, 352–365. [CrossRef] [PubMed]

49. Zalesny, R.S., Jr.; Bauer, E.O. Evaluation of *Populus* and *Salix* continuously irrigated with landfill leachate I. Genotype-specific elemental Phytoremedion. *Int. J. Phytoremed.* **2007**, *9*, 281–306. [CrossRef]

50. Zalesny, J.A.; Zalesny, R.S., Jr.; Wiese, A.H.; Sexton, B.T.; Hall, R.B. Uptake of macro- and micro-nutrients into leaf, woody, and root tissue of *Populus* after irrigation with landfill leachate. *J. Sustain. For.* **2008**, *27*, 303–327. [CrossRef]

![forests logo] *forests*

MDPI

Article

Relationships among Root–Shoot Ratio, Early Growth, and Health of Hybrid Poplar and Willow Clones Grown in Different Landfill Soils

Elizabeth R. Rogers [1], Ronald S. Zalesny, Jr. [1,*], Richard A. Hallett [2], William L. Headlee [3] and Adam H. Wiese [1]

[1] USDA Forest Service, Northern Research Station, Rhinelander, WI 54501, USA; errogers@fs.fed.us (E.R.R.); awiese@fs.fed.us (A.H.W.)
[2] USDA Forest Service, Northern Research Station, Durham, NH 03824, USA; rhallett@fs.fed.us
[3] UA Division of Agriculture, Arkansas Forest Resources Center, Monticello, AR 71656, USA; headlee@uamont.edu
* Correspondence: rzalesny@fs.fed.us; Tel.: +1-715-362-1132

Received: 23 November 2018; Accepted: 22 December 2018; Published: 10 January 2019

Abstract: Root–shoot allocation of biomass is an underrepresented criterion that could be used for tree selection in phytoremediation. We evaluated how root–shoot allocations relate to biomass production and overall health of poplar and willow clones grown in landfill soil treatments. Fifteen poplar clones and nine willows were grown in a greenhouse for 65 days in soils from five Wisconsin landfills and one greenhouse control. We tested for treatment, clone, and interaction differences in root–shoot ratio (RSR), health, and growth index, along with relationships between RSR with diameter, health, height, total biomass, and growth index. Treatments, clones, and their interactions were not significantly different for poplar RSR, but willow clones differed ($p = 0.0049$). Health significantly varied among willow clones ($p < 0.0001$) and among the clone $\times$ treatment interaction for poplars ($p = 0.0196$). Analysis of means showed that willow clones 'Allegany' and 'S365' exhibited 28% and 21% significantly greater health scores than the overall mean, respectively. Root–shoot ratio was not significantly correlated with health in either genus but was positively correlated with growth index for poplars, which was corroborated via regression analyses. Selecting clones based on a combination of biomass allocation, health, and growth indices may be useful for using phyto-recurrent selection to satisfy site-specific ecosystem services objectives.

Keywords: phytoremediation; hybrids; health; biomass production; phytotechnologies

1. Introduction

There are numerous phytotechnologies available for remediating contaminated sites. For example, through the use of woody plants like trees, phytoremediation aims to remediate many forms of soil and water pollution [1,2]. Common targets of phytoremediation include metals, metalloids, petroleum hydrocarbons, pesticides, explosives, chlorinated solvents, and industrial byproducts [3]. Remediation occurs through decontamination, the removal of pollutants through plant uptake, or stabilization, which alters the soil chemistry to stabilize the pollutant(s) [4]. Phytoextraction is a mode of decontamination in which plants accumulate contaminants and are later harvested. This removal of pollutants rather than stabilization prevents transfer to nearby groundwater, streams, and lakes. The degree of phytoextraction possible is specific to site conditions, pollutant bioavailability and plant characteristics [5,6]. Soil texture, pH, and pollutant concentration must all be within the limits of the plants used to enhance productivity and growth, which in turn will promote maximum phytoextraction. Because of the range in performance and tolerance among tree species, methods of species selection are vital to maximize potential remediation.

A recent development in phytotechnologies, phyto-recurrent selection is the screening and selection of readily available clones based on performance in experimental trials [7,8]. Superior clones are selected for their survivability, rooting ability, yield, and pest/disease resistance. Multiple testing cycles are employed to eliminate less desirable clones. As the number of clones left in a trial decreases, the complexity of the data increases. The data collected from the last cycle of a trial determines which clones to plant at a specific site. In one study, phyto-recurrent selection was used to select *Populus* clones for phytoremediation of landfill leachate [7]. The clones responded differently to leachate treatments for all traits tested in the first growing cycle. In the third growing cycle rooting differences were evident between leachate treatments, which was corroborated in the field [9]. This highlights the need for testing and selecting clones for site-compatible characteristics, which has been a focus of phytotechnologies research throughout the past decade [8].

Phytoremediation research has focused on ways to simultaneously maximize yield and root growth. *Populus* and *Salix* genera are ideal short rotation woody crops (SRWC) due to their high yield and fast growth [10]. *Populus* is also one of the most popular genera used for phytoremediation because of its rapid juvenile growth, ease of hybridization, and vegetative propagation [11]. Rapid growth and proper establishment at the start of a plant's life, especially when grown in contaminated soils, is essential to successful phytoremediation. Poplars can also regrow after multiple harvests and effectively remove inorganics from contaminated soil [12,13].

Like *Populus*, the genus *Salix* is a main source of clones for phytoremediation [4]. Willows and poplars are similar in that they both have the potential for high biomass productivity, effective nutrient uptake, and clone-specific capacity for taking up heavy metals [14]. Willows also exhibit high evapotranspiration, a key factor determining rapid chemical uptake in plants [15]. Such uptake of contaminants is useful in a variety of situations on different contaminated sites. For example, willows planted on land contaminated with dredged sediment containing mineral oil and polyaromatic hydrocarbons decreased the mineral oil composition almost four times more than decreases on a paired site where all vegetation was removed [16]. Willows can also experience enhanced growth when immersed in contaminated conditions. Higher yield, increased number of shoots, and increased plant dry mass have all occurred in willows treated with wastewater compared to control plants [14,17].

Currently, phytoremediation research on poplars and willows is focusing on understanding the role of biomass distribution, across genera and even among specific hybrid clones in response to contaminated soils [18]. The relationship between below- and above-ground biomass of a plant is known as the root–shoot ratio (RSR). Understanding the RSR provides insight to a plant's ability to perform across different sites or contaminant levels. Biomass ratios can vary among species, between plants of different sizes, and in differing moisture conditions [19,20]. In addition, nutrient availability, oxygen, light, and temperature can impact the biomass ratio in plants [21–23]. These factors emphasize the need to better understand RSRs and the genotype × environment interactions that influence them.

Biomass allocation is important to phytoremediation because it constitutes criteria for selecting superior site-specific clones. Research to date has shown immense variability in above- and below-ground biomass production among clones [24–26]. Not only does variation exist among clones but between clones grown in different soils as well. Willows specifically have been shown to have a near twofold increase in number of leaves and coarse roots when grown in clay versus sand substrates [27]. In particular, cuttings grown in the clay substrate exhibited higher net primary production, dry mass distribution, shoot height, or fine root number than those grown in sand. Marginal sites such as landfills exhibit similar soil heterogeneity that will likely result in variable biomass production and phytoremediation capability among clones. Across this diversity of site types, there is a need to determine which clones can survive, maximize growth and are most effective at remediating the site. For example, species with greater belowground biomass allocations may provide the greatest benefit in sites where the primary objective involves the roots, such as phytostabilization, rhizofiltration, or rhizodegredation. On the other hand, if the objective involves the aboveground processes of the plant, such as phytovolatilization or phytodegredation, species with greater aboveground biomass allocations should be implemented.

In addition to the variation in poplar and willow biomass allocation and health across different soil treatments, differences among clones have been shown to dictate success of these production systems [13]. For example, Wullschleger et al. [26] found that biomass allocation in poplars varies with genotype and age. They also conclude that root–shoot relationships should be investigated because of the importance of root distribution to carbon sequestration and phytoremediation. In a similar study, Barigah et al. [28] tested the differences in aboveground biomass production among five hybrid poplar clones after one year of growth and reported that all clones differed from one another, with the best clone producing over 3.5 times more aboveground biomass and aboveground biomass per area than the poorest-performing clone. Variation was attributed to phenotypic traits of the individual clones such as large individual leaf size, total leaf area, and photosynthetic rates. Results were similar for another small-scale study comparing four poplar clones, though the variation was not as distinct [29]. The range in RSRs was 0.26 to 0.36, with significant shifts in clonal biomass rankings across the two-year study. Thus, not only was there variation among same-aged clones in biomass production and allocation, but clonal variation over time as well. Knowledge of such variation in biomass allocations among clones, especially in regard to temporal variation, is important to the success of a phytoremedation site. Clones that exhibit greater belowground biomass allocations early on have the potential for better initial establishment in harsh field conditions (i.e., contamination, available nutrients, moisture levels) when compared to clones that do not. Successful establishment can lead to better tree health, and increased overall effectiveness of the phytoremediation at a site.

More research must be done to determine which clones are best for each allocation strategy, be it higher mass dedicated to shoots or roots, a balance between the two, or adaptability among all three. Scientists can then use biomass allocation data to choose clones that meet site-specific remediation goals like high root allocations to stabilize riparian buffer zones, high shoot allocations to allow for maximum evapotranspiration, or the ability to change between both at highly dynamic sites. In the current study, we tested the hypothesis that poplar and willow clones will vary in their root–shoot allocations when grown in different landfill soil treatments, both among clones and treatments. Furthermore, we assessed the relationships among RSR and the early growth and health of the clones. Methods and results will provide researchers with another criterion for the phyto-recurrent selection process regardless of specific genotypes or geographic locations.

2. Materials and Methods

2.1. Soil Collection and Site Description

Soils were collected from five closed solid waste municipal landfills in eastern Wisconsin, USA; Bellevue, Caledonia, Menomonee Falls, Slinger, and Whitelaw (Figure 1). Annual mean temperature across the landfills ranges from 6.8 to 8.8 °C while precipitation ranges from 749 to 876 mm [30]. Soil was collected down to a depth of 0.8 m from between 5 and 16 randomly selected points from each site. The total amount of soil collected from each site was 3.8 m^3. Soil samples were sieved (0.6 cm screen) and homogenized by site. The sieved soils were loaded into Agrimaster poly stock tanks (Behlen Manufacturing Company, Columbus, NE, USA) with a resulting individual tree soil volume of 0.02 m^3.

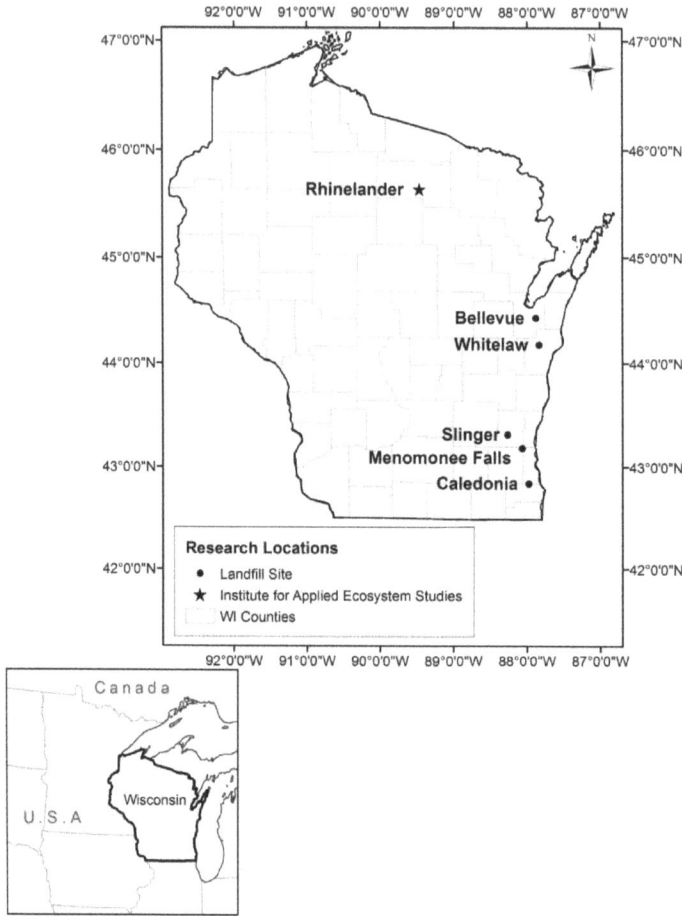

Figure 1. Locations of the five landfill sites where soil was collected and the phytotechnologies greenhouse where phyto-recurrent selection was conducted (Rhinelander, WI, USA) in a study assessing the relationships among root–shoot allocations, early growth, and health of hybrid poplar and willow clones when grown for 65 days in the landfill soils.

2.2. Soil Properties

Six 25-mg subsamples of soil were collected from the sieved and homogenized soil of each site, processed, and subsequently sent to Northern Lake Service, Inc. (Crandon, WI, USA). The subsamples were analyzed for volatile organic compounds, including vinyl chloride, which were later determined to be undetectable at soil harvesting depths. Soil pH, texture, and chemical properties are listed in Table 1. The aim of the current study was to determine if RSR varied among clones, soils, or their interaction, and therefore we did not evaluate the uptake of particular soil components, nor their effect on tree growth.

Table 1. pH, texture, and chemical properties of soils collected from five landfills in eastern Wisconsin, USA in a study assessing the relationships among root–shoot allocations, early growth, and health of hybrid poplar and willow clones when grown in the landfill soils for 65 days.

	Soil Treatment					
	Bellevue	**Caledonia**	**Menomonee Falls**	**Slinger**	**Whitelaw**	**Control**
pH	7.64 ± 0.02	7.20 ± 0.01	7.52 ± 0.02	7.37 ± 0.01	7.74 ± 0.02	4.49 ± 0.01
Texture	Clay Loam	Clay	Clay Loam	Clay Loam	Sandy Loam	-
	Percent					
Clay	38	48	28	28	7	-
Sand	37	21	33	25	55	-
Silt	35	31	39	47	38	-
C	0.024 ± 0.001	0.029 ± 0.004	0.064 ± 0.000	0.024 ± 0.001	0.041 ± 0.003	0.310 ± 0.005
N	0.000 ± 0.000	0.001 ± 0.000	0.001 ± 0.000	0.001 ± 0.000	0.001 ± 0.000	0.007 ± 0.000
	$g \, kg^{-1}$					
Al	41.82 ± 0.56	52.46 ± 4.21	31.68 ± 0.30	45.71 ± 0.32	35.43 ± 4.62	22.75 ± 0.67
Ca	43.19 ± 1.78	30.51 ± 8.27	97.14 ± 1.12	23.22 ± 0.24	62.74 ± 8.14	11.24 ± 0.20
Fe	22.19 ± 0.30	29.56 ± 2.21	20.34 ± 0.18	26.29 ± 0.27	21.14 ± 2.11	16.61 ± 0.38
K	20.81 ± 0.24	19.57 ± 1.27	11.58 ± 0.36	13.88 ± 0.20	14.78 ± 1.69	4.43 ± 0.40
Na	5.40 ± 0.05	4.45 ± 0.12	4.14 ± 0.03	6.21 ± 0.05	4.95 ± 0.16	2.71 ± 0.15
Ti	2.56 ± 0.02	3.39 ± 0.24	2.03 ± 0.01	3.23 ± 0.01	2.35 ± 0.24	2.48 ± 0.06
Si	255.47 ± 3.59	234.58 ± 6.19	163.75 ± 1.42	263.10 ± 1.00	209.66 ± 3.23	76.98 ± 2.51
	$mg \, kg^{-1}$					
Mg	17.68 ± 0.63	21.14 ± 4.06	47.25 ± 0.31	15.08 ± 0.17	37.21 ± 3.85	25.53 ± 0.65
Mn	396.80 ± 8.50	648.96 ± 66.74	512.31 ± 4.55	790.67 ± 5.27	392.88 ± 66.65	257.87 ± 5.76
P	435.47 ± 8.75	426.05 ± 4.79	357.47 ± 5.89	445.51 ± 3.14	381.91 ± 9.96	380.88 ± 12.91
Sr	114.21 ± 0.58	101.06 ± 1.55	100.34 ± 0.27	103.01 ± 0.43	101.50 ± 1.36	56.84 ± 1.60
Zr	16.58 ± 0.15	16.55 ± 0.51	11.92 ± 0.38	23.77 ± 0.26	13.30 ± 0.94	2.00 ± 0.13

2.3. Genotype Selection and Experimental Design

Twenty-four clones were used in this study, fifteen poplar clones and nine willow clones. *Populus* and *Salix* cuttings were obtained from the USFS NRS Hugo Sauer Nursery (Rhinelander, WI, USA), Iowa State University Clonal Orchard at the Iowa Department of Natural Resources State Nursery (*Populus* only) (Ames, IA, USA), Michigan State University Clonal Orchard at the Tree Research Center (*Populus* only) (Lansing, MI, USA), University of Minnesota Natural Resources Research Institute Clonal Orchard at the North Central Research and Outreach Center Nursery (*Populus* only) (Grand Rapids, Minnesota, USA), and Double A Willow (*Salix* only) (Fredonia, NY, USA). These sources grew whips of the clones for one growing season in stool beds. Dormant, unrooted cuttings were processed to a length of 20.32 cm. Cuts were made to position at least one primary bud within 2.54 cm from the top of each cutting. Cuttings were then stored in polyethylene bags at 5 °C and subsequently soaked in water to a height of 10.16 cm for 48 h before planting. The trees were grown in the greenhouse with a 16-h photoperiod. Temperatures were held at 24 °C in the daytime and 20 °C in the nighttime. Trees received irrigation with unfiltered well water based on water demand. The trees of each selection cycle were arranged in a split-plot design with random block effects, fixed soil treatment whole plots, and fixed clone sub-plots. Three blocks and six soil treatments (soil from each of the five landfills plus a control of Jolly Gardner Pro-Line C/G Custom Growing Mix (a mix of Canadian sphagnum peat, medium perlite, and vermiculite) (Atlanta, GA, USA) were tested. For the purposes of this study, "soil treatment" is defined as these six different types of soil; soils were not treated or amended in any way. Clones themselves were arranged in randomized complete blocks in order to account for potential greenhouse environmental gradients. A total of 432 cuttings were planted, with 18 cuttings per clone, and three cuttings per clone per soil treatment. Trees were established in Agrimaster poly stock tanks with individual tank volumes of 0.26 m³.

This project is part of a larger experiment that consisted of three phyto-recurrent selection growing cycles. The 24 clones used in this study were those that were selected for cycle 3 trials using weighted summation indices (described below). There were 15 hybrid poplar clones from four genomic groups and nine willow clones from six genomic groups (Table 2).

Table 2. *Populus* and *Salix* genomic groups and clones tested in phyto-recurrent selection cycle 3 in a study assessing how root–shoot allocations contribute to biomass production and overall health when grown for 65 days in soils from five landfills in eastern Wisconsin, USA.

Genomic Group	Clone
Populus [a]	
P. deltoides × *P. maximowiczii* 'DM'	313.55, DM111, NC14106
P. deltoides × *P. nigra* 'DN'	9732-32, 9732-36, 21700, 99038022, 99038026, BR 3960, DN5, DN34, DN177
P. nigra × *P. maximowiczii* 'NM'	NM2, NM6
(*P. trichocarpa* × *P. deltoides*) × *P. deltoides* 'TDD'	NC13820
Salix [b]	
S. caprea hybrid 'C'	S365
S. miyabeana 'M'	SX67
S. purpurea 'P'	Allegany
S. purpurea × *S. miyabeana* 'PM'	Millbrook
S. sachalinensis × *S. miyabeana* 'SM'	Canastota
S. viminalis × *S. miyabeana* 'VM'	Fabius, Owasco, Tully Champion
S. viminalis × (*S. sachalinensis* × *S. miyabeana*) 'VSM'	Preble

[a] Sections and authorities for *Populus* are: *Aigeiros* Duby—*P. deltoides* Bartr. ex Marsh, *P. nigra* L.; *Tacamahaca* Spach—*P. maximowiczii* A. Henry, *P. trichocarpa* Torr. & Gray. [b] Sections and authorities for *Salix* are: *Cinerella* Seringe—*S. caprea* L.; *Helix* Dumortier—*S. purpurea* L., *S. miyabeana* Seemen; *Vimen* Dumortier—*S. sachalinensis* F. Schmidt; *Viminella* Seringe—*S. viminalis* L.

2.4. Weighted Summation Indices

In all cycles, weighted summation indices were used for phyto-recurrent selection [7]. Allometric traits (root dry mass; combined leaf and stem dry mass; number of roots; height; diameter; total leaf number; stem dry mass; leaf dry mass) were given weights (sum of weight = 1) based on their relative contribution to initial survival and early establishment. The weights were then multiplied by the adjusted or unadjusted means for each trait, and the values were added across all traits. Clones that exhibited greater relative index scores were selected to move on to the next cycle. The present study involves clones that advanced to cycle 3 through superior index scores in the previous two cycles.

2.5. Data Collection

Tree health was measured at 58 days after planting using a four-category qualitative scale ranging from 0 to 3, where 0 = dead, 1 = poor health, 2 = moderate health, and 3 = optimal health (Figure 2). Two researchers measured the health of all the clones to promote consistency in ratings. At 65 days after planting, height, measured from the point of attachment of the primary stem to the original cutting, and diameter, taken 1.5 cm from the point of attachment to avoid stem swell, were measured. Trees were harvested, washed, and dissected into roots, stems, leaves, and cuttings. All components were oven dried at 70 °C until constant mass was obtained. Root–shoot ratio (RSR) was calculated as the ratio of root dry mass to dry mass of stems + leaves. Cutting dry mass was not included in the RSR calculations.

Figure 2. Photographs of tree health at 58 days after planting using a four-category qualitative scale ranging from 0 to 3, where 0 = dead, 1 = poor health, 2 = moderate health, and 3 = optimal health in a study assessing the relationships among root–shoot allocations, early growth, and health of hybrid poplar (P) and willow (W) clones grown in soils from five landfills in eastern Wisconsin, USA.

2.6. Growth Index

A simplified growth index was created by calculating z-scores [31] for diameter, height, and number of leaves and then averaging the scores for each tree. This index was created separately for each genus. To make the index more intuitive we made all values positive by adding a constant equal to the absolute value of the lowest score in the data set. Higher values represented more robust trees.

2.7. Data Analysis

Height, diameter, total dry mass (above- plus below-ground), RSR, and growth index data were analyzed using analyses of variance, analyses of means, and correlation analyses according to SAS® (PROC GLM; PROC ANOM; PROC CORR; SAS Institute, Inc., Cary, NC, USA). These analyses assumed the split plot design with a random block effect and fixed main effects for soil treatment and clone. Analyses of covariance (ANCOVA) were conducted to test for the effect of cutting dry mass on all traits because of its broad variation at 65 days after planting (0.79 to 22.97 g). Cutting dry mass was a significant covariate for leaf dry mass, stem dry mass, and aboveground dry mass ($p < 0.05$) but did not have a significant effect on root dry mass ($p > 0.05$). Main effects and their interactions were considered significantly different at $p < 0.05$ according to comparison analyses using Tukey's adjustment. The Kruskal-Wallis test was used to test for differences in tree health scores. Again, significant effects were differentiated using Tukey's adjustment. Clones were classified as generalists or specialists based on their mean health scores. Generalists were clones whose mean health score (Section 2.5) was 2.7 or better in four or more of the six soil treatments. Specialist clones exhibited mean health scores of 2.7 or better in three or less of the treatments. All analyses were performed to assess the relationships among root–shoot allocations, early growth, and health of the trees for soil treatment × clone interactions, within genera.

To evaluate whether the relationships with RSR from the correlation analyses differed significantly by treatment or clone, ANCOVAs were carried out using PROC GLM in SAS®. Linear equations were fit for each response variable (diameter, height, total dry mass, and growth index) with experimental effects of treatment or clone, as well as the covariate RSR and interactions between RSR and the experimental effects. This approach allows testing for differences in the intercepts (via treatment or clone), whether the overall relationship with RSR differs from zero (via RSR), and whether the relationship with RSR differs by treatment or clone (via interactions with RSR). In all cases, statistical significance ($\alpha = 0.05$) was tested using F-ratios based on Type III sums of squares, and the coefficient of determination (r^2) for each model was recorded. When RSR interactions were significant, slope estimates for each treatment or clone were generated and t-tests were conducted to determine which parameter estimate(s) differed significantly from zero. When justified by the presence of significant experimental effects, the slope estimates were generated using treatment- or clone-specific intercepts.

3. Results

3.1. Analyses of Variance

Soil treatment, clone, and the treatment × clone interaction were not significant for RSR in poplars (Table 3). However, there were differences among clones for willows ($p = 0.0049$), with RSR ranging from 0.006 ± 0.011 (clone 'SX67') to 0.061 ± 0.009 ('Owasco') and an overall mean of 0.035 ± 0.009 (Figure 3). Root–shoot ratios among willow genomic groups differed without any noticeable patterns. In general, RSR was not significantly different among the top six clones. Although mean RSR for willow clones varied, they were not significantly different from the overall mean RSR (Figure 3). Despite a significant clone main effect ($p < 0.0001$), the treatment × clone interaction governed tree health for poplar ($p = 0.0196$) (Table 3). Of the poplars, 48% had optimal health (score = 3), 34% had moderate health (score = 2), 2% had poor health (score = 1), and 16% were dead (score = 0). Across clones, mean health scores were stable in individual soil treatments, with values varying by 0.2 health points (Table 4). Similarly, nine of the 15 poplar clones tested in cycle 3 performed as generalists, with mean health scores not significantly differing across soil treatments. The remaining six clones were specialists according to health scores. Clone 'DN34' had significantly greater health in soils from Menomonee Falls, Bellevue, and Slinger than soils from the other three sites, which were not different from one another. The remaining five specialist clones had significantly better health in one or two of the soil treatments.

Table 3. Probability values from analyses of variance in a study assessing the relationships among root–shoot allocations, early growth, and health of hybrid poplar and willow clones when grown for 65 days in soils from five landfills in eastern Wisconsin, USA.

Source of Variation	Root–Shoot Ratio	Tree Health	Growth Index
	Populus		
Soil treatment	0.1427	0.2877	**<0.0001**
Clone	0.0726	**<0.0001**	**0.0003**
Soil treatment × clone	0.9778	**0.0196**	0.1750
	Salix		
Soil treatment	0.0697	0.1842	**<0.0001**
Clone	**0.0049**	**<0.0001**	**0.0029**
Soil treatment × clone	0.3145	0.9550	0.0755

Means for root–shoot ratio of both genera were adjusted for the variation in cutting dry mass, which was a significant covariate (*Populus*, $p = 0.0027$; *Salix*, $p = 0.0048$). Significant effects are shown in bold.

Table 4. Mean tree health scores ($n = 3$) for each combination of soil treatment and hybrid poplar clone in a study assessing the relationships among root–shoot allocations, early growth, and health of hybrid poplar and willow clones when grown for 65 days in soils from five landfills in eastern Wisconsin, USA.

Performance Group	Genomic Group	Clone	Soil Treatment					
			Bellevue	Caledonia	Menomonee Falls	Slinger	Whitelaw	Control
Generalist	DM	313.55	2.7 ± 0.3 a	3.0 ± 0.0 a	3.0 ± 0.0 a	3.0 ± 0.0 a	2.3 ± 0.3 ab	2.7 ± 0.3 a
Generalist	DM	DM111	3.0 ± 0.0 a	3.0 ± 0.0 a	ne	2.3 ± 0.3 ab	2.0 ± 0.0 bc	2.3 ± 0.7 ab
Generalist	DM	NC14106	2.7 ± 0.3 a	3.0 ± 0.0 a	ne	3.0 ± 0.0 a	3.0 ± 0.0 a	3.0 ± 0.0 a
Specialist	DN	9732-32	2.5 ± 0.4 ab	2.0 ± 0.0 bc	2.0 ± 0.0 bc	**2.7 ± 0.3** a	2.3 ± 0.3 ab	2.0 ± 0.0 bc
Specialist	DN	9732-36	**3.0 ± 0.0** a	2.0 ± 0.0 bc	2.0 ± 0.0 bc	2.5 ± 0.4 ab	**3.0 ± 0.0** a	1.7 ± 0.3 c
Specialist	DN	21700	2.0 ± 0.0 bc	2.0 ± 0.0 bc	**3.0 ± 0.0** a	2.3 ± 0.3 a	**3.0 ± 0.0** a	2.0 ± 0.6 bc
Specialist	DN	99038022	2.0 ± 0.0 bc	**2.7 ± 0.3** a	2.0 ± 0.0 bc	**2.7 ± 0.3** a	2.0 ± 0.0 bc	2.0 ± 0.0 bc
Generalist	DN	99038026	2.0 ± 0.0 bc	2.5 ± 0.4 ab	2.0 ± 0.6 bc	1.7 ± 0.3 c	2.0 ± 0.0 bc	2.0 ± 0.0 bc
Specialist	DN	BR 3960	**2.7 ± 0.3** a	2.3 ± 0.3 ab	2.0 ± 0.0 bc	2.3 ± 0.3 ab	2.0 ± 0.0 bc	2.0 ± 0.6 bc
Generalist	DN	DN5	3.0 ± 0.0 a	2.7 ± 0.3 a	3.0 ± 0.0 a	3.0 ± 0.0 a	2.5 ± 0.4 ab	3.0 ± 0.0 a
Specialist	DN	DN34	**3.0 ± 0.0** a	2.0 ± 0.0 bc	**3.0 ± 0.0** a	**2.7 ± 0.3** a	2.0 ± 0.0 bc	2.0 ± 0.0 bc
Generalist	DN	DN177	2.7 ± 0.3 a	2.5 ± 0.4 ab	3.0 ± 0.0 a	3.0 ± 0.0 a	2.3 ± 0.3 ab	2.7 ± 0.3 a
Generalist	NM	NM2	3.0 ± 0.0 a	3.0 ± 0.0 a	3.0 ± 0.0 a	3.0 ± 0.0 a	2.5 ± 0.4 ab	3.0 ± 0.0 a
Generalist	NM	NM6	3.0 ± 0.0 a	3.0 ± 0.0 a	3.0 ± 0.0 a	3.0 ± 0.0 a	2.7 ± 0.3 a	3.0 ± 0.0 a
Generalist	TDD	NC13820	2.5 ± 0.4 ab	3.0 ± 0.0 a	3.0 ± 0.0 a	2.7 ± 0.3 a	3.0 ± 0.0 a	2.7 ± 0.3 a
		Across	2.7 ± 0.1	2.6 ± 0.1	2.6 ± 0.1	2.7 ± 0.1	2.4 ± 0.1	2.4 ± 0.1

The control treatment was a standard greenhouse potting mix. Clones were categorized into performance groups based on their stability for tree health across sites. Means with the same letter were not significantly different at $p = 0.05$. ne = not estimable. Bold values indicate soil treatments wherein specialist clones performed better than with other landfill soils. Means labelled as 'across' represent the overall soil treatment means averaged across clones within a particular soil treatment. See Table 2 for genomic group descriptions.

Willows, on the other hand, only exhibited a significant clone main effect ($p < 0.0001$), with mean health scores ranging from 1.9 ± 0.2 ('Canastota') to 2.8 ± 0.1 ('Allegany') and an overall mean of 2.20 ± 0.05 (Figure 3). Of the willows, 30% had optimal health, 56% had moderate health, 10% had poor health, and 4% were dead. All genomic groups containing *Salix viminalis* showed below average health, in addition to those with *S. sachalinensis* × *S. miyabeana* parentage (i.e., 'Canastota' and 'Preble'). In contrast, analysis of means showed that clones 'Allegany' and 'S365' exhibited 28% and 21% significantly greater tree health than the overall mean, respectively (Figure 3).

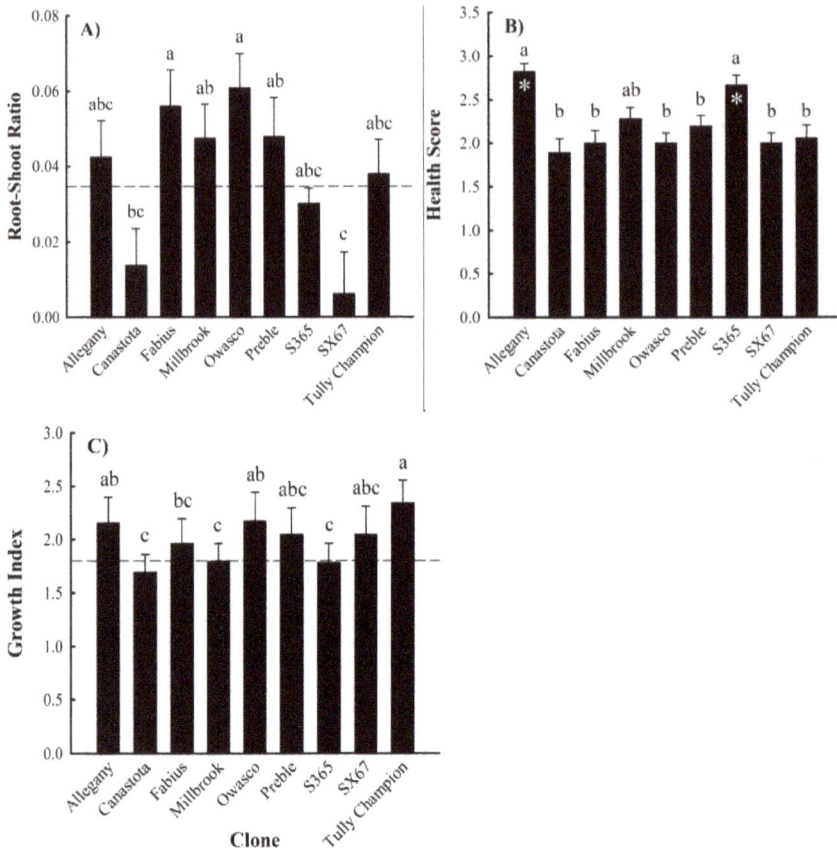

Figure 3. Root–shoot ratio (**A**), health score (**B**), and growth index (**C**) for the *Salix* clone main effect in a study assessing the relationships among root–shoot allocations, early growth, and health of hybrid poplar and willow clones when grown for 65 days in soils from five landfills in eastern Wisconsin, USA. The dashed line represents the overall mean, while bars with asterisks indicate means that differ from the overall mean at $p < 0.05$. Bars with the same letters were not different according to Tukey's adjustment at $p > 0.05$.

Soil treatment and clone main effects were significant for growth indices in poplars and willows ($p < 0.05$), while their interaction was not significant for both genera ($p > 0.05$) (Table 3). Trends in growth indices for soil treatments were similar for poplars and willows, with the control exhibiting nearly twice the magnitude of any other treatment and 72% to 75% percent higher indices than the overall mean, respectively (Figure 4). For both genera, the growth index for Bellevue, Caledonia, and Slinger was the same, and Whitelaw and Menomonee Falls had the lowest indices of all soil treatments. Furthermore, the variability in growth index for willow clone main effects was intermediate of that for RSR and health score, and no clone exhibited a distinct advantage nor were any clones significantly different than the overall mean (Figure 3). In contrast, for poplars, clone 'BR 3960' exhibited 30% to 117% higher growth indices than the second best ('NM6') and worst ('313.55') clones, respectively (Figure 5). Also, the growth index of clone 'BR 3960' was 53% greater than the overall mean. Overall, trends for individual poplar genomic groups were non-existent for growth indices.

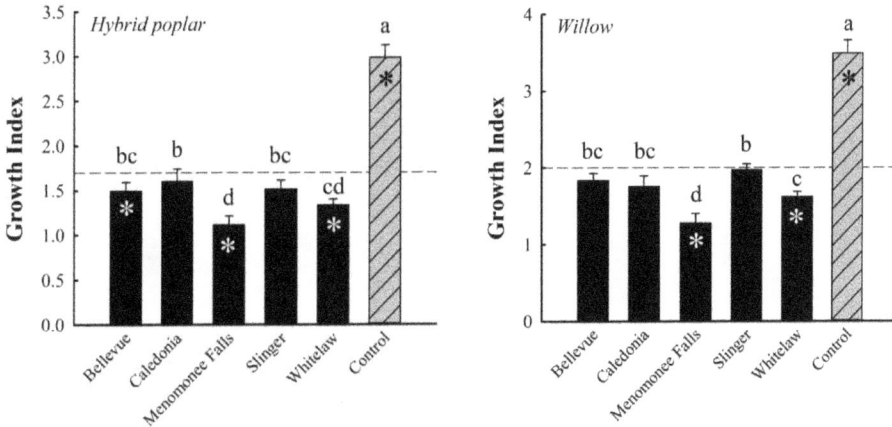

Figure 4. Growth index for the soil treatment main effect in a study assessing the relationships among root–shoot allocations, early growth, and health of hybrid poplar and willow clones when grown for 65 days in soils from five landfills in eastern Wisconsin, USA. The dashed line represents the overall mean, while bars with asterisks indicate means that differ from the overall mean at $p < 0.05$. Bars with the same letters were not different according to Tukey's adjustment at $p > 0.05$.

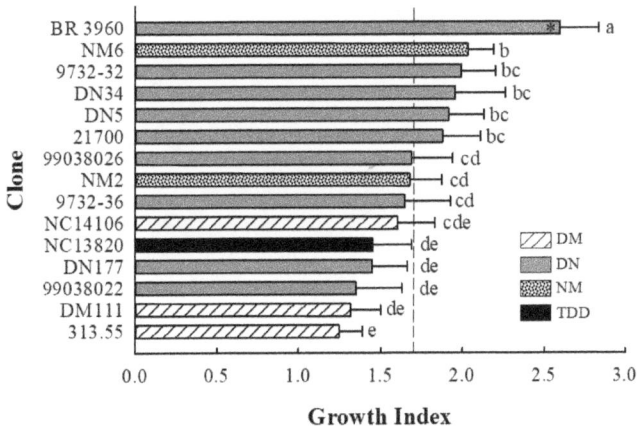

Figure 5. Growth index for the hybrid poplar clone main effect in a study assessing the relationships among root–shoot allocations, early growth, and health of hybrid poplar and willow clones when grown for 65 days in soils from five landfills in eastern Wisconsin, USA. The dashed line represents the overall mean, while bars with asterisks indicate means that differ from the overall mean at $p < 0.05$. Bars with the same letters were not different according to Tukey's adjustment at $p > 0.05$. Shading of bars indicates different genomic groups; see Table 2 for their descriptions.

3.2. Correlations

Poplar phenotypic correlations ranged from -0.17 (health with diameter) to 0.95 (height with growth index) (Table 5). Root–shoot ratio was not significantly correlated with health ($r = -0.08$, $p = 0.2215$) but was correlated with height ($r = 0.36$, $p < 0.0001$), diameter ($r = 0.45$, $p < 0.0001$), total biomass ($r = 0.56$, $p < 0.0001$), and growth index ($r = 0.50$, $p < 0.0001$). All correlations not including RSR were significant except health with height ($r = -0.12$, $p = 0.0687$). For willows, phenotypic correlations ranged from -0.18 (health with diameter) to 0.96 (diameter with growth index) (Table 5). Root–shoot ratio was not correlated with health ($r = -0.09$, $p = 0.2826$), height ($r = 0.12$, $p = 0.1381$), nor growth index ($r = 0.14$, $p = 0.0865$), but was significantly correlated with diameter ($r = 0.20$, $p = 0.0143$) and total biomass ($r = 0.22$, $p = 0.0068$). All correlations not including RSR were significant ($p < 0.05$), except the correlation between health and growth index ($r = -0.13$, $p = 0.1155$).

Table 5. Phenotypic correlations among six traits in a study assessing the relationships among root–shoot ratio (RSR), early growth, and health of hybrid poplar (above diagonal) and willow (below diagonal) clones when grown for 65 days in soils from five landfills in eastern Wisconsin, USA.

	Diameter	Health	Height	RSR	Total Biomass	Growth Index
Diameter		−0.17 **0.0120**	0.91 **<0.0001**	0.45 **<0.0001**	0.89 **<0.0001**	0.94 **<0.0001**
Health	−0.18 **0.0245**		−0.12 0.0687	−0.08 0.2215	−0.15 **0.0213**	−0.18 **0.0066**
Height	0.91 **<0.0001**	−0.16 **0.0439**		0.36 **<0.0001**	0.85 **<0.0001**	0.95 **<0.0001**
RSR	0.20 **0.0143**	−0.09 0.2826	0.12 0.1381		0.56 **<0.0001**	0.50 **<0.0001**
Total biomass	0.80 **<0.0001**	−0.17 **0.0390**	0.75 **<0.0001**	0.22 **0.0068**		0.88 **<0.0001**
Growth index	0.96 **<0.0001**	−0.13 0.1155	0.94 **<0.0001**	0.14 0.0865	0.91 **<0.0001**	

Probability values are shown below the correlations. Significant correlations are shown in bold.

3.3. Analyses of Covariance

The ANCOVA results indicated that the relationship with RSR differed significantly by treatment for diameter, height, total dry mass, and growth index, but not for health (Table 6). This was true both for poplars and for willows. In general, the fit of treatment models was weak for health ($r^2 = 0.07$ to 0.10) but relatively strong for the other variables ($r^2 = 0.57$ to 0.81). Clones differed significantly in their relationship with RSR only for diameter and height, and even then only for willows and not for poplars. The fit of clone models was relatively weak in all cases ($r^2 = 0.16$ to 0.48), even for the willow models in which the relationship with RSR was found to differ significantly among clones ($r^2 = 0.19$ to 0.21) (Table 6).

For the models in which the relationship with RSR differed significantly by treatment or clone, the estimates of the slopes for each treatment or clone are shown in Tables 7 and 8, respectively. For poplars, the Whitelaw soils only showed a significant positive relationship of RSR with total dry mass, whereas all other soil treatments showed significant positive relationships with diameter, height, and growth index. All but one treatment (i.e., Menomonee Falls) also showed a significant positive relationship between RSR and total dry mass (Table 7). For willows, three treatments (Caledonia, Control, and Menomonee Falls) showed significant positive relationships between RSR and diameter, total dry mass, and growth index; two of those treatments (Caledonia and Menomonee Falls) also showed significant positive relationships between RSR and height (Table 7). In addition, four willow clones ('Allegany', 'Owasco', 'SX67', and 'Tully Champion') showed significant positive relationships between RSR and height; three of those clones ('Owasco', 'SX67', and 'Tully Champion') also showed significant positive relationships for diameter (Table 8).

Table 6. Results of analyses of covariance (ANCOVA) (i.e., *p*-values from *F*-tests of model effects and model r^2 values) for diameter, height, total dry biomass, and growth index in a study assessing the relationships among root–shoot allocations, early growth, and health of hybrid poplar and willow clones when grown for 65 days in soils from five landfills in eastern Wisconsin, USA.

Genus	Model	Diameter	Height	Total Biomass	Growth Index
Populus	T	<0.0001	<0.0001	0.0451	<0.0001
	RSR	<0.0001	<0.0001	<0.0001	<0.0001
	T × RSR	<0.0001	0.0009	<0.0001	0.0003
	r^2	0.76	0.64	0.81	0.66
	C	0.6784	0.0233	0.5592	0.0473
	RSR	<0.0001	<0.0001	<0.0001	<0.0001
	C × RSR	0.0597	0.0770	0.0553	0.0630
	r^2	0.43	0.40	0.48	0.42
Salix	T	<0.0001	<0.0001	0.0008	<0.0001
	RSR	<0.0001	<0.0001	<0.0001	<0.0001
	T × RSR	<0.0001	<0.0001	<0.0001	<0.0001
	r^2	0.69	0.57	0.76	0.72
	C	0.1344	0.0613	0.5170	0.2447
	RSR	0.0016	0.0103	0.0004	0.0159
	C × RSR	0.0341	0.0202	0.0572	0.0505
	r^2	0.19	0.21	0.20	0.16

Significant interactions with root–shoot ratio (RSR) (bold) indicate that soil treatments (T) or clones (C) differed in their relationship with RSR.

Table 7. Slope parameter estimates and associated *p*-values for the relationship between root–shoot ratio (RSR) and diameter, height, total dry biomass, and growth index in a study assessing the relationships among root–shoot allocations, early growth, and health of hybrid poplar and willow clones when grown for 65 days in soils from five landfills in eastern Wisconsin, USA.

Genus	Slope (Treatment)	Diameter		Height		Total Biomass		Growth Index	
		Est	*p*-Value	Est	*p*-Value	Est	*p*-Value	Est	*p*-Value
Populus	Menomonee Falls	45.4	0.0006	709.9	0.0035	106.9	0.0696	19.9	0.0057
	Control	57.1	<0.0001	628.7	<0.0001	419.8	<0.0001	19.2	<0.0001
	Bellevue	44.7	<0.0001	631.3	0.0001	161.2	<0.0001	23.0	<0.0001
	Slinger	32.2	0.0056	497.2	0.0202	128.1	0.0143	16.7	0.0087
	Caledonia	73.6	<0.0001	1011.7	0.0003	312.3	<0.0001	33.9	<0.0001
	Whitelaw	7.8	0.1527	105.9	0.2957	61.0	0.0141	4.0	0.1809
Salix	Menomonee Falls	39.9	<0.0001	992.3	<0.0001	112.9	0.0029	19.1	<0.0001
	Control	30.3	<0.0001	360.9	0.0573	353.6	<0.0001	19.2	<0.0001
	Bellevue	−1.1	0.6915	−115.4	0.1426	14.3	0.2771	−1.7	0.2933
	Slinger	4.6	0.4018	86.7	0.5639	43.8	0.0828	1.5	0.6220
	Caledonia	25.3	0.0014	698.1	0.0013	98.7	0.0062	12.9	0.0034
	Whitelaw	11.0	0.0584	112.3	0.4797	48.9	0.0675	3.1	0.3404

Significant slopes (bold) indicate that the relationship with RSR for a given soil treatment differs from zero.

Table 8. Slope parameter estimates and associated *p*-values for the relationship between root–shoot ratio (RSR) with diameter and height in a study assessing the relationships among root–shoot allocations, early growth, and health of hybrid poplar and willow clones when grown for 65 days in soils from five landfills in eastern Wisconsin, USA.

Genus	Slope (Clone)	Diameter		Height	
		Est	*p*-Value	Est	*p*-Value
Salix	Allegany	17.4	0.0597	497.6	**0.0199**
	Canastota	12.8	0.3139	267.5	0.3615
	Fabius	0.5	0.9155	−41.4	0.6776
	Millbrook	12.4	0.0833	212.3	0.1955
	Owasco	17.3	**0.0045**	307.8	**0.0273**
	Preble	12.9	0.1058	126.3	0.4906
	S365	10.4	0.4382	−382.9	0.2166
	SX67	29.4	**0.0212**	663.7	**0.0237**
	Tully Champion	23.2	**0.0079**	376.9	0.0590

Significant slopes (bold) indicate that the relationship with RSR for a given clone differs from zero.

4. Discussion

Understanding biomass allocation among hybrid clones is important for phytoremediation because it can be used as a selection criterion in phyto-recurrent selection to determine which clones best match site-specific goals. The high variation in previous studies in biomass allocation among treatments, moisture conditions and even individual clones of poplars and willows highlights the need for incorporating biomass allocation into the phyto-recurrent selection process. Results of the present study were consistent with those of Rytter [27] regarding biomass allocations and health across different soil treatments (Table 3). As with Rytter [27], biomass production was significantly different across soil treatments, but not biomass allocations. The variation in RSR across treatments was marginally significant for willow clones (*p* = 0.0697), but the variation was less apparent among poplars (*p* = 0.1427). Similarly, soil treatments did not influence health of either genus, which may have resulted from the smaller sample sizes of cycle 3, short growing cycle (i.e., 65 days) relative to an entire field growing season (i.e., 110 days), and uniform climatic conditions in the greenhouse. Field-simulating moisture conditions in greenhouse trials have been linked with considerable variation of RSR in black willow cuttings [32]. Our consistent, adequate watering of trees prevented this heterogeneity from becoming evident.

Results of this study corroborated previous findings of variation among clones. In poplars, tree health varied significantly among clones, while in willows, both tree health and RSR varied significantly among clones (Table 3). Nevertheless, despite these univariate results, health was not significantly related to RSR in general, and the relationship did not differ significantly among treatments or clones. Thus, tree health in the current study does not appear to be strongly related to any sort of imbalance in above- versus below-ground partitioning of resources. In contrast, most metrics of poplar size (i.e., diameter, height, total dry mass, and growth index) showed significant positive relationships with RSR under most treatments. The two exceptions were Menomonee Falls and Whitelaw; the former showed significant positive relationships for all metrics except total dry mass, and the latter showed a significant positive relationship only for total dry mass (Table 6). For willows, the metrics of plant size showed significant positive relationships with RSR for only half the treatments. Thus, it appears the different soil conditions were mediating the extent to which RSR influenced enhanced growth, particularly for willow (Table 6). Furthermore, when accounting for treatment differences, total dry mass had the strongest relationship with RSR of all the variables considered in this study, as demonstrated by the associated models having the best fit for both poplar (r^2 = 0.81) and willow (r^2 = 0.76). Notably, the highest slope for total dry mass in both poplars and willows was observed for the control soil, which suggests that the levels and/or types of contaminants in the other soils may serve to "dull" the relationship between RSR and total biomass production to varying degrees.

The weaker fit of the clone models (Table 8) compared to the treatment models (Table 7) suggested that genotype differences were less important than soil differences in explaining the relationship between RSR and plant size metrics. However, significant differences among clones were observed for some size metrics in willows, with about a third of the willow clones having slopes that differ significantly from zero. This suggested that genotype can also play a role at times in determining the extent to which higher RSR is associated with improved performance. Overall, treatment and clone main effects were examined independently for the ANCOVA in this study, without regard for their potential interactions. While the evaluation of such interactions may provide further insights (e.g., whether treatment-specific slopes differ among clones), the replication within each treatment × clone combination ($n = 3$) in the current study was insufficient to justify fitting unique slopes for each of these combinations. Future studies with larger sample sizes are therefore recommended, so that these potential interactions can be explored.

Moreover, the soil treatment × clone interaction for health scores in poplars exhibited enough variation to classify specialist and generalist clones (Table 4). Average health scores for individual clones were compared across treatments, with nine clones being classified as generalists and six as specialists. Variation among clones was to be expected, due to genotypic and phenotypic differences. However, one rather surprising finding was that 'DN34' was classified as a specialist not a generalist, as anticipated. Clone 'DN34' is a classic generalist; it was bred over 100 years ago and has been used since then by countless researchers across many continents [33,34]. As such, 'DN34' is known across the poplar community to not only thrive but also outperform other clones in a multitude of conditions. In the present study, 'DN34' defied expectations and did not perform equally well in all of the six soil treatments. Instead, 'DN34' only had significantly superior health in three out of the six landfill soils. Similarly, clear underperforming and superior performing willows were also made apparent through the analyses of variance and analyses of means. Clones 'Canastota' and 'Preble' along with all genomic groups containing *Salix viminalis* had below average health, while clones 'Allegany' and 'S365' had greater than average health. These results may lead researchers to select 'Allegany' and 'S365' if clone health is a priority. Further research needs to be done to quantify the differences among clones so that best-performing clones may be selected based on RSR, health score, and growth index.

Despite promising results from the analyses of variance and analyses of means, correlation analyses did not reveal anticipated results. Although root biomass was significantly related to growth indices such that the indices explained 52% of its variability in poplars and 39% in willows (data not reported), RSR itself was not significantly correlated with health in either poplars or willows (Table 5). Two correlations were of particular interest: (1) RSR with total biomass and (2) health with total biomass, as they were both significant in poplars and willows. This study may not directly prove that balanced root–shoot ratios lead to increased health in poplar and willow hybrids, but the "health" component should be interpreted with caution. The health scores used in this study were not all-inclusive by any means; they did not take into account biomass production among other variables. Often in weighted summation indices of phyto-recurrent selection, total biomass production is heavily weighted. Researchers use total biomass production as an indicator of health. Clones with high production are viewed as strong, healthy, and superior, whereas those that produce relatively less biomass are considered weak or unfit. Keeping that in mind, results of this study related the biomass component of health to RSR. Since it is shown that a higher RSR is correlated to higher total biomass production, it can be concluded that RSR indirectly affects health.

5. Conclusions

Root–shoot ratio (RSR) is a valuable selection criterion that can be incorporated into phyto-recurrent selection of poplars and willows used for phytoremediation and associated phytotechnologies, particularly when used in weighted summation indices. In willows, RSR has been shown to vary among individual clones. This suggests that researchers should implement RSR as a selection criterion in weighted summation indices for the phyto-recurrent selection of willows.

The weight given to RSR will depend on site objectives, i.e., if soil stabilization is the main objective, clones with higher RSR may result in the greatest benefit. On the other hand, poplars did not show any variation in RSR among treatment, clone, or their interaction. RSR in poplars should not be discounted for use as a selection criterion, however. Only 15 poplar clones, out of the hundreds that exist, were used in this study, so there should be continued research on a larger array of poplar clones. Furthermore, these 15 clones were already selected for performance in these soils in the first two selection cycles; a relationship may exist among the clones that had been rejected. Both poplars and willows showed correlations between RSR and total biomass. These results indicate that to some degree, aboveground biomass production can be used to predict belowground biomass production but should be tested further in order to develop models that relate the two. Results of this study can serve as a baseline for similar analyses relating health, growth index, and RSR of these short rotation woody crops. It is worth noting that RSR is not intended to be a stand-alone metric for evaluating clone performance. Root–shoot ratio is most useful when weighed with and against other factors, as in the weighted summation indices mentioned above. Further research is needed to address the shortcomings depicted here, namely: cycle length and simplicity of clone health assessments. Furthermore, analyses should be advanced to the field where there will likely be a marked difference in biomass allocation, especially given varying climatic conditions and soil moisture regimes. Whether or not a correlation exists between the RSR of final-cycle greenhouse trees and biomass production of trees established in the field also warrants additional study. Scientists and site managers can use the methodologies and results of this study as a means for selecting the best clones to fulfill their individual objectives.

Author Contributions: E.R.R., R.S.Z.J., and A.H.W. conceived and designed the experiments; E.R.R., R.S.Z.J., and A.H.W. performed the experiments; R.S.Z.J., E.R.R., R.A.H., and W.L.H. analyzed the data and wrote the paper.

Funding: This work was funded by the Great Lakes Restoration Initiative (GLRI; Template #738 Landfill Runoff Reduction) and the USDA Forest Service, Northern Research Station. Support was also provided by the USDA National Institute of Food and Agriculture (McIntire Stennis Project 1009221).

Acknowledgments: We are grateful to Larry Buechel, Jim Forney, Mike Peterson, and Ray Seegers of Waste Management of Wisconsin, Inc. for information about and access to the five landfills used in this study. Also, we thank Andrej Pilipović and Lorenzo Rossi for reviewing earlier versions of this manuscript.

Conflicts of Interest: The authors declare no conflict of interest.

References

1. Schnoor, J.L.; Licht, L.A.; McCutcheon, S.C.; Wolfe, N.L.; Carreira, L.H. Phytoremediation of organic and nutrient contaminants. *Environ. Sci. Technol.* **1995**, *29*, 318–323. [CrossRef] [PubMed]
2. Suresh, B.; Ravishanker, G.A. Phytoremediation: A novel and promising approach for environmental clean-up. *Crit. Rev. Biotechnol.* **2004**, *24*, 97–124. [CrossRef]
3. Robinson, B.H.; Green, S.R.; Mills, T.M.; Clothier, B.E.; van der Velde, M.; Laplane, R.; Fung, L.; Deurer, M.; Hurst, S.; Thayalakumaran, T.; et al. Phytoremediation: Using plants as biopumps to improve degraded environments. *Aust. J. Soil Res.* **2003**, *41*, 599–611. [CrossRef]
4. Mirck, J.; Isebrands, J.G.; Verwijst, T.; Ledin, S. Development of short-rotation willow coppice systems for environmental purposes in Sweden. *Biomass Bioenergy* **2005**, *28*, 219–228. [CrossRef]
5. Cunningham, S.D.; Shann, J.R.; Crowley, D.E.; Anderson, T.A. Phytoremediation of contaminated water and soil. In *Phytoremediation of Soil and Water Contaminants*; Kruger, E.L., Anderson, T.A., Coats, J.R., Eds.; American Chemical Society: Washington, DC, USA, 1997; pp. 2–17, ISBN 9780841235038.
6. Cunningham, S.D.; Ow, D.W. Promises and prospects of phytoremediation. *Plant Physiol.* **1996**, *110*, 715–719. [CrossRef] [PubMed]
7. Zalesny, J.A.; Zalesny, R.S., Jr.; Wiese, A.H.; Hall, R.B. Choosing tree genotypes for phytoremediation of landfill leachate using phyto-recurrent selection. *Int. J. Phytorem.* **2007**, *9*, 513–530. [CrossRef] [PubMed]
8. Zalesny, R.S., Jr.; Stanturf, J.A.; Gardiner, E.S.; Bañuelos, G.S.; Hallett, R.A.; Hass, A.; Stange, C.M.; Perdue, J.H.; Young, T.M.; Coyle, D.R.; et al. Environmental technologies of woody crop production systems. *BioEnergy Res.* **2016**, *9*, 492–506. [CrossRef]

9.   Zalesny, J.A.; Zalesny, R.S., Jr.; Coyle, D.R.; Hall, R.B.; Bauer, E.O. Clonal variation in morphology of *Populus* root systems following irrigation with landfill leachate or water during two years of establishment. *BioEnergy Res.* **2009**, *2*, 134–143. [CrossRef]

10.   Netzer, D.; Tolsted, D.; Ostry, M.E.; Isebrands, J.G.; Riemenschneider, D.; Ward, K. *Growth, Yield, and Disease Resistance of 7- to 12-Year-Old Poplar Clones in the North Central United States*; General Technical Report NC 229; U.S. Department of Agriculture, Forest Service, North Central Research Station: St. Paul, MN, USA, 2002; pp. 1–40.

11.   Zalesny, R.S., Jr.; Stanturf, J.A.; Gardiner, E.S.; Perdue, J.H.; Young, T.M.; Coyle, D.R.; Headlee, W.L.; Bañuelos, G.S.; Hass, A. Ecosystem services of woody crop production systems. *BioEnergy Res.* **2016**, *9*, 465–491. [CrossRef]

12.   Baldantoni, D.; Cicatelli, A.; Bellion, A.; Castiglione, S. Different behaviours in phytoremediation capacity of two heavy metal tolerant poplar clones in relation to iron and other trace elements. *J. Environ. Manag.* **2014**, *146*, 94–99. [CrossRef]

13.   Laureysens, I.; Blust, R.; De Temmerman, L.; Lemmens, C.; Ceulemans, R. Clonal variation in heavy metal accumulation and biomass production in a poplar coppice culture: I. Seasonal variation in leaf, wood and bark concentrations. *Environ. Pollut.* **2004**, *131*, 485–494. [CrossRef] [PubMed]

14.   Perttu, K.; Kowalik, P. *Salix* vegetation filters for purification of waters and soils. *Biomass Bioenergy* **1997**, *12*, 9–19. [CrossRef]

15.   Dietz, A.C.; Schnoor, J.L. Advances in phytoremediation. *Environ. Health Perspect.* **2001**, *109*, 163–168. [CrossRef]

16.   Vervaeke, P.; Luyssaert, S.; Mertens, J.; Meers, E.; Tack, F.; Lust, N. Phytoremediation prospects of willow stands on contaminated sediment: A field trial. *Environ. Pollut.* **2003**, *126*, 275–282. [CrossRef]

17.   Holm, B.; Heinsoo, K. Municipal wastewater application to short rotation coppice of willows—Treatment efficiency and clone response in an Estonian case study. *Biomass Bioenergy* **2013**, *57*, 126–135. [CrossRef]

18.   Isebrands, J.G.; Aronsson, P.; Carlson, M.; Ceulemans, R.; Coleman, M.; Dickinson, N.; Dimitriou, J.; Doty, S.; Gardiner, E.; Heinsoo, K.; et al. Environmental applications of poplars and willows. In *Poplars and Willows: Trees for Society and the Environment*; Isebrands, J.G., Richardson, J., Eds.; Food and Agriculture Organization (FAO) of the United Nations: Rome, Italy; CAB International, Inc.: Boston, MA, USA, 2014; Chapter 6; pp. 258–336, ISBN 978-1-78064-108-9.

19.   Monk, C. Ecological importance of root/shoot Ratios. *Bull. Torrey Bot. Club* **1966**, *93*, 402–406. [CrossRef]

20.   Lloret, F.; Casanovas, C.; Penuelas, J. Seedling survival of Mediterranean shrubland species in relation to root: Shoot ratio, seed size and water and nitrogen use. *Funct. Ecol.* **1999**, *13*, 210–216. [CrossRef]

21.   Van der Werf, A.; Visser, A.J.; Schieving, F.; Lambers, H. Evidence for optimal partitioning of biomass and nitrogen at a range of nitrogen availabilities for a fast- and slow-growing species. *Funct. Ecol.* **1993**, *7*, 63–74. [CrossRef]

22.   Gedroc, J.J.; Mcconnaughay, K.D.; Coleman, J.S. Plasticity in root/shoot partitioning: Optimal, ontogenetic, or both? *Funct. Ecol.* **1996**, *10*, 44–50. [CrossRef]

23.   Friend, A.L.; Coleman, M.D.; Isebrands, J.G. Carbon allocation to root and shoot systems of woody plants. In *Biology of Adventitious Root Formation*; Davis, T.D., Haissig, B.E., Eds.; Plenum Press: New York, NY, USA, 1994; pp. 245–273.

24.   Adler, A.; Karacic, A.; Weih, M. Biomass allocation and nutrient use in fast growing woody and herbaceous perennials used for phytoremediation. *Plant Soil* **2008**, *305*, 189–206. [CrossRef]

25.   Weih, M.; Nordh, N. Characterising willows for biomass and phytoremediation: Growth, nitrogen and water use of 14 willow clones under different irrigation and fertilisation regimes. *Biomass Bioenergy* **2002**, *23*, 397–413. [CrossRef]

26.   Wullschleger, S.D.; Yin, T.M.; Difazio, S.P.; Tschaplinski, T.J.; Gunter, L.E.; Davis, M.F.; Tuskan, G.A. Phenotypic variation in growth and biomass distribution for two advanced-generation pedigrees of hybrid poplar. *Can. J. For. Res.* **2005**, *35*, 1779–1789. [CrossRef]

27.   Rytter, R. Biomass production and allocation, including fine-root turnover, and annual N uptake in lysimeter-grown basket willows. *For. Ecol. Manag.* **2001**, *140*, 177–192. [CrossRef]

28.   Barigah, T.; Saugier, B.; Mousseau, M.; Guittet, J.; Ceulemans, R. Photosynthesis, leaf area and productivity of 5 poplar clones during their establishment year. *Ann. Sci. For.* **1994**, *51*, 613–625. [CrossRef]

29. Pallardy, S.; Gibbins, D.; Rhoads, J. Biomass production by two-year-old poplar clones on floodplain sites in the Lower Midwest, USA. *Agroforest. Syst.* **2003**, *59*, 21–26. [CrossRef]

30. U.S. Climate Data. Available online: http://www.usclimatedata.com/ (accessed on 23 April 2017).

31. Green, R. *Sampling Design and Statistical Methods for Environmental Biologists*; Wiley: New York, NY, USA, 1979; ISBN 9780471039013.

32. Pezeshki, S.R.; Anderson, P.H.; Shields, F.D. Effects of soil moisture regimes on growth and survival of black willow (*Salix nigra*) posts (cuttings). *Wetlands* **1998**, *18*, 460–470. [CrossRef]

33. Zalesny, R.S., Jr.; Hall, R.B.; Zalesny, J.A.; McMahon, B.G.; Berguson, W.E.; Stanosz, G.R. Biomass and genotype × environment interactions of *Populus* energy crops in the Midwestern United States. *BioEnergy Res.* **2009**, *2*, 106–122. [CrossRef]

34. Marmiroli, M.; Pietrini, F.; Maestri, E.; Zacchini, M.; Marmiroli, N.; Massacci, A. Growth, physiological and molecular traits in *Salicacae* trees investigated for phytoremediation of heavy metals and organics. *Tree Physiol.* **2011**, *31*, 1319–1334. [CrossRef]

forests

MDPI

Article

Phytoremediation Efficacy of *Salix discolor* and *S. eriocephela* on Adjacent Acidic Clay and Shale Overburden on a Former Mine Site: Growth, Soil, and Foliage Traits

Alex Mosseler * and John E. Major

Natural Resources Canada, Canadian Forest Service—Atlantic Forestry Centre, 1350 Regent St., P.O. Box 4000, Fredericton, NB E3B 5P7, Canada; john.major@canada.ca
* Correspondence: alex.mosseler@canada.ca; Tel.: +1-506-452-2440; Fax: +1-506-452-3525

Received: 27 October 2017; Accepted: 28 November 2017; Published: 2 December 2017

Abstract: Plants regularly experience suboptimal environments, but this can be particularly acute on highly-disturbed mine sites. Two North American willows—*Salix discolor* Muhl. (DIS) and *S. eriocephala* Michx. (ERI)—were established in common-garden field tests on two adjacent coal mine spoil sites: one with high clay content, the other with shale overburden. The high clay content site had 44% less productivity, a pH of 3.6, 42% clay content, high water holding capacity at saturation (64%), and high soil electrical conductivity (EC) of 3.9 mS cm^{-1}. The adjacent shale overburden site had a pH of 6.8, and after removing 56.5% stone content, a high sand content (67.2%), low water holding capacity at saturation (23%), and an EC of 0.9 mS cm^{-1}. The acidic clay soil had significantly greater Na (20×), Ca (2×), Mg (4.4×), S (10×), C (12×) and N (2×) than the shale overburden. Foliar concentrations from the acidic clay site had significantly greater Mg (1.5×), Mn (3.3×), Fe (5.6×), Al (4.6×), and S (2×) than the shale overburden, indicating that these elements are more soluble under acidic conditions. There was no overall species difference in growth; however, survival was greater for ERI than DIS on both sites, thus overall biomass yield was greater for ERI than DIS. Foliar concentrations of ERI were significantly greater than those of DIS for N (1.3×), Ca (1.5×), Mg (1.2×), Fe (2×), Al (1.5×), and S (1.5×). There were no significant negative relationships between metal concentrations and growth or biomass yield. Both willows showed large variation among genotypes within each species in foliar concentrations, and some clones of DIS and ERI had up to 16× the Fe and Al uptake on the acidic site versus the adjacent overburden. Genetic selection among species and genotypes may be useful for reclamation activities aimed at reducing specific metal concentrations on abandoned mine sites. Results show that, despite having a greater water holding capacity, the greater acidity of the clay site resulted in greater metal mobility—in particular Na—and thus a greater EC. It appears that the decline in productivity was not due to toxicity effects from the increased mobility of metals, but rather to low pH and moisture stress from very high soil Na/EC.

Keywords: acidic soil; foliar nutrient and metal concentration; *Salix*; site reclamation; species variation

---

1. Introduction

There has been longstanding interest in Europe, and more recently in North America, in using willows (*Salix* spp.) as a source of biomass for energy purposes [1–5]. More recently, there has also been a growing interest in using willows for various environmental applications and land reclamation, including phytoremediation and phytoextraction of contaminated soils [6–17]. Highly disturbed mine sites with very low pH and high levels of metals and electrical conductivity (EC) can be challenging for reclamation and revegetation [18]. Although reclamation and phytoremediation of heavy-metal-contaminated sites using willows have been investigated in Europe, these studies have been restricted to a limited number of species and a limited number of genotypes (or clones) within species [11,12,19–32]. Of special concern for bioenergy production in the context of land reclamation is the identification of well-adapted, native plant species that can tolerate highly acidic soil conditions that may contain potentially toxic levels of metals and high EC [8,13,14,21,33–36]. Soil pH changes the solubility and mobility of metals, thereby increasing potential toxicity as soil acidity increases [37–41]. For instance, soil acidification can induce zinc (Zn) phytotoxicity [37] and aluminum (Al) toxicity in plants [42,43], and although trees are generally tolerant of high Al concentrations, irreversible damage to tree roots can occur at pH < 4.2, resulting in decreased growth and biomass yields [30,41,44,45].

With more than 350 species worldwide, willows are widespread across the northern hemisphere, and Canada has 76 native willows [46]. Willows are among the first woody plants to colonize former mine sites [9,46–48]. Although they are most often associated with wetlands and riparian zones, some willows are also adapted to drier upland sites. *Salix discolor* Muhl. (DIS) and *S. eriocephala* Michx. (ERI) are widespread across eastern and central Canada and are being field tested in common-garden studies because both species show promise as fast-growing sources of biomass for bioenergy production [1,49]. Both DIS and ERI are shrub willows that occur in natural populations on disturbed wetland sites across eastern and central North America, but DIS can also be found colonizing drier upland sites throughout its botanical range and is a common, naturally occurring colonizer of the Salmon Harbour (SH) coal mine site near Minto, New Brunswick (NB), Canada (Lat. 46.07° N; Long. 66.05° W). Several other willows, including *S. bebbiana* Sarg. and *S. lucida* Muhl., as well as aspen (*Populus tremuloides* Michx.), balsam poplar (*P. balsamifera* L.), birches (*Betula papyrifera* Marsh. and *B. populifolia* Marsh.), pin cherry (*Prunus pensylvanica* L.), and black locust (*Robinia pseudoacacia* L.), the latter introduced from eastern USA seed sources, can also be found colonizing recently disturbed shale rock overburden at the former SH coal mine. *Salix eriocephala*, however, is more commonly associated with deeper, more fertile soils along stream banks [49] and is only rarely found as a natural colonizer of the SH coal mine spoils.

In a previous paper, we detailed the coppice biomass yield components, which ultimately showed 44% less productivity on acidic clay deposits versus adjacent shale rock overburden at the SH mine site [45]. In this paper, our objective was to assess soil properties in relation to foliar concentration of nutrients and metals in 1-year-old coppice regrowth from DIS and ERI clones and assess these traits in light of productivity differences between sites. In addition, we wanted to quantify these growth differences for selection of superior clones for phytoremediation, biomass production, and land reclamation purposes. In order to assess adaptation and utility of willows for growth on reclaimed mine sites and biomass production for bioenergy purposes, we hypothesized that there would be both species and clonal differences in response to the different soil conditions and that foliar concentrations of some of the absorbed metals would negatively affect aboveground growth response.

2. Materials and Methods

Periodic flash flooding and ponding of mine drainage water on the former SH coal mine has formed a patchwork of small clay deposits of varying depths scattered over the exposed shale rock overburden that dominates these coal mine spoils. A common-garden field test and a clonal gene bank of selected genotypes of DIS and ERI containing 15 genotypes in common (Table 1) were established in 2008 and 2009, respectively, on two adjacent sites separated by several hundred meters at the SH coal mine spoils near Minto, NB, on a property operated by NB Power, the local electrical power utility. These adjacent mine spoil sites consisted of recently landscaped shale rock overburden resulting from coal strip-mining operations. However, prior to our study or any reclamation activities, one of these sites had existed for approximately 3 years as a settling pond that was part of a watercourse draining both surface runoff and water pumped from flooded coal seams. This pond deposited a thick layer of clay, covering a uniformly flat rectangular area measuring approximately 70 m × 40 m over the shale rock overburden that dominates the mine site (see contour map in Figure 1). The depth of the clay deposit was determined by digging 44 holes with a soil auger in a grid pattern at 6 m × 6 m intervals across the area of the clone bank and measuring depth to the rock overburden. The clay layer ranged in depth from 11 cm to 48 cm, with an average depth of 35 cm. Six soil samples were taken by shovel to a depth of 15 cm from each of these two sites in August 2013. Soil analyses were conducted according to McKeague [50] as follows: available P-#TP-CSS-MSSA 4.41 (sodium bicarbonate extraction), exchangeable cations-#TP-CSS-MSSA 4.5, FCMM 15 (ammonium acetate extraction), pH-#TP-CSS-MSSA3.13 (pH in 1:1 water), texture-#TP-CSSMSSA 2.12 (Hydrometer method), organic matter, N, C and S-#TP-LFIM (total C by LECO induction furnace) by the Laboratory for Forest Soils and Environmental Quality at the University of New Brunswick (UNB) in Fredericton, NB. The water content at saturation parameter was measured using the disturbed soil sample, which provides an estimate for this quantitative value.

Figure 1. A map of the former Salmon Harbour coal mine site showing the locations of the acidic clay site (red square) and the three sampled sites (red dots) on the adjacent shale overburden. The insert shows the location of the coal mine site in New Brunswick, Canada.

The common garden experiment on the shale overburden site was established in May 2008 using unrooted stem cuttings of 15 DIS and ERI clones and was located approximately 300 m from the clone bank (acidic clay deposit), extending for a distance of approximately 180 m along the watercourse that drained the mine site (see Figure 1, Table 1)). This experiment was established along a gentle slope that had been landscaped to minimize soil erosion and surface runoff into the adjacent watercourse. Each genotype in the common-garden test was established as a linear, five-ramet (clonal) row plot running perpendicular to the watercourse, with the five ramets spaced at 0.5 m within the plot and 2 m between plots. Each clonal row plot was replicated three times in a randomized complete block test design.

The clay deposit site was established in May 2009 with unrooted stem cuttings to maintain 31 of the best-performing genotypes of DIS and ERI that had been selected from a common-garden field test established at the Montreal Botanical Gardens (MBG) in Montreal, PQ, Canada, in 2005. Also included on both the acidic clay deposit and the shale overburden sites was a check clone of *S. viminalis* clone #5025 (VIM), introduced from the Swedish willow breeding program [51], and which has been widely used in short-rotation willow biomass trials across Canada [3,52]. On the acidic clay site, each genotype was randomly established as a single linear row plot consisting of 21 plants (ramets) per genotype (clone) spaced at 1 m between ramets within the row plot and 2 m between adjacent clonal row plots, giving each plant approximately 2 m^2 of growing space.

Table 1. Origins of 15 *Salix discolor* and *S. eriocephala* genotypes used for measurements of foliar uptake of nutrients and metals from two common-garden studies at the Salmon Harbour mine site near Minto, New Brunswick (NB).

Species	Population Location *	Latitude North	Longitude West	Selected Clones
	Levis, QC	46°78′	71°18′	LEV-D3, LEV-D6
	Lower Anfield, NB	46°92′	67°49′	ANF-D1
S. discolor	Hawkesbury, ON	45°39′	74°75′	HAW-D5
	Montmagny, QC	46°94′	70°60′	MON-D1
	Mud Lake, ON	45°88′	76°78′	MUD-D4
	Richmond Fen, ON	45°13′	75°82′	RIC-D2
	Ste. Anne de la Perade, QC	46°56′	72°20′	ANN-E6
	Bristol, NB	46°47′	67°58′	BRI-E2
	Fosterville, NB	45°78′	67°76′	FOS-E1
S. eriocephala	Fredericton, NB	45°94′	66°62′	FRE-E1
	Green River, NB	47°34′	68°19′	GRE-E1
	Norton, NB	45°67′	65°81′	NOR-E10
	Shepody Creek, NB	45°71′	64°77′	SHE-E3
	Rivière au Saumon, QC	47°21′	70°35′	SAU-E3

* province of QC (Quebec), ON (Ontario), NB (New Brunswick).

The linear clonal plots of each species were randomly assigned on each site. However, on the acidic clay site, each of the 15 clonal plots of interest was established as a clone bank consisting of a single linear row plot containing 21 ramets per clone. Therefore, the clones on the acidic clay did not represent a true block effect, but the site was flat, uniform, and confined to a small area. The 21-ramet linear row plot on the clay deposit was divided into three *post hoc* blocks, with each block assigned a single, seven-ramet sample plot per clone per block. Thus, clonal variation could not be precisely determined due to this constraint on randomization, and estimates of within-clone variation should be interpreted with caution. Nevertheless, the clay site consisted of a 62 m by 22 m area of clay deposit, which differed greatly from the adjacent shale overburden (Table 2).

On both study sites, willow clones were established using 20 cm long, rootless stem cuttings collected during the dormant season from vigorous 1- and 2-year-old stem sections (as per [53]) from coppiced plants in a common garden established at the MBG. Each plant on the shale overburden had 1 m^2 of growing space versus 2 m^2 of growing space on the acidic clay deposit. However, the 1-year-old coppice growth of the willow clones on both sites was quite poor relative to more fertile sites such as the MBG from where these clones had been collected [53], and none of the plants on the mine site fully occupied the growing space available to them nor were they in competition with each other for growing space. The aboveground biomass was harvested from both sites in the fall of 2011 as 2-year-old coppice growth [45]. The green mass of the biomass of each harvested plant was measured to the nearest 10 g using an electronic weigh scale (Electronic Infant Scale, model ACS-20A-YE, Peoples' Republic of China). Green mass in t ha^{-1} was calculated by converting the harvested biomass per plant to biomass production per hectare to a standard 1 m^2 by multiplying by 10 (e.g., multiplying by 10,000 plants per ha divided by 1000 kg per ton). Overall yield was the product of green mass and survival.

Table 2. (**a,b**) Soil properties (Mean ± SE) from two sites, an acidic clay deposit and a shale rock overburden, at the Salmon Harbour mine site. Sites with different letters are significantly different at $p = 0.05$.

(**a**)

Site	Carbon (g kg⁻¹)	Nitrogen (g kg⁻¹)	Potassium (ppm)	Calcium (ppm)	Magnesium (ppm)	Phosphorus (ppm)	Sodium (%)
Acidic clay	53.1 ± 2.3a	2.07 ± 0.17a	116.1 ± 14.5a	2968 ± 116a	352.4 ± 14.6a	10.75 ± 0.6a	0.443 ± 0.031a
Shale overburden	4.6 ± 2.3b	1.02 ± 0.11b	91.1 ± 14.5a	1466 ± 116b	80.2 ± 14.6b	3.98 ± 0.65b	0.022 ± 0.031b

Exchangeable cations extracted using ammonium acetate, phosphorus was extracted by sodium bicarbonate.

(**b**)

Site	Sand (%)	Silt (%)	Clay (%)	pH	C:N ratio	Sulfur (%)	EC [1] (mS cm⁻¹)	WC at Sat. [2] (%)
Acidic clay	12.9 ± 2.4b	44.9 ± 2.5a	42.3 ± 1.9a	3.6 ± 0.2b	25.9 ± 1.4b	0.079 ± 0.012a	3.89 ± 0.12a	64.0 ± 1.4a
Shale overburden	67.2 ± 2.4a	23.4 ± 2.5b	9.4 ± 1.9b	6.8 ± 0.2a	4.6 ± 1.4a	0.008 ± 0.012b	0.88 ± 0.12b	22.9 ± 1.4b

EC [1] = electrical conductivity; WC at Sat. [2] = water content at saturation.

Healthy foliage samples were harvested on 8 September 2014 from 1-year-old coppice stems of up to five ramets within each clonal row plot from three blocks (replications) on each of the two sites. Foliage from the five plants (ramets) per plot was bulked for analysis, resulting in a total of 90 foliage samples as experimental units. The foliage was stored in paper sampling bags, and placed in drying ovens at 65 °C for 48 h. Foliage was ground to a fine powder, and the grinder was washed with ethanol between samples. The Laboratory for Forest Soils and Environmental Quality at the University of New Brunswick used standard protocols (e.g., Method numbers TP-SSMA 15.3.1, 15.3.3, and 15.4 from [50]), for foliage analysis of C, N, P, K, Ca, Mg, Mn, Fe, Na, Zn, Al , and S. Foliage dry ashing is done by muffle furnace 500 °C for 4 h (slow ramp up), and extracted with 8N HCl at 90 °C for 30 min. and filtered. K, Ca, Mg, Mn, Al, Fe and Zn (cations) were run on the Varian SpectrAA 400 (Varian Techtron Pty. Limited, Mulgrave, Victoria, Australia) P is run on the Technicon Traaccs Autoanalyser (Technicon Instruments Corp., Tarrytown, New York, USA). Method #TP-LFIM (Total carbon by LECO induction furnace (LECO Corp. St. Joseph, Michigan, USA)) was used for N, C, and S [46].

Statistical Analysis

Foliar analysis data from both study sites were subjected to analyses of variance (ANOVA) in which site, species, and clones were considered fixed effects. The ANOVA model used was as follows:

$$Y_{ijklm} = \mu + B_{i(j)} + T_j + S_k + C_{l(k)} + S_kT_j + C_{l(k)}T_j + e_{ijklm},$$

where Y_{ijklm} is the dependent ramet trait of the ith block, of the jth site, of the kth species, of the lth clone of the mth ramet and where μ is the overall mean. $B_{i(j)}$ is the effect of the ith block ($i = 1, \ldots 3$) nested within the jth site , T_j is the effect of the jth site ($j = 1, 2$), S_k is the effect of the kth species ($k = 1, 2$), $C_{l(k)}$ is the effect of the lth clone ($l = 1, \ldots 8$) nested within the kth species , and S_kT_j is the effect of the species by site interaction, $C_{l(k)}T_j$ is the effect of the clone nested within species by site interaction, and e_{ijklm} is the random error component.

The impact of nutrient and metal concentrations (mean values) on growth, using total aboveground green mass (mean values) from [45] were analyzed using analysis of covariance (ANCOVA). In these analyses, three sources of variation were studied: (1) covariate (i.e., nutrient or metal), (2) independent effect (species), and (3) independent effect × covariate. The analyses were based on the following model:

$$Y_{ij} = B_0 + B_{0i} + B_1X_{ij} + B_{1i}X_{ij} + e_{ij}$$

where Y_{ij} is the dependent trait of the jth tree of the ith site or species, B_0 and B_1 are average regression coefficients, B_{0i} and B_{1i} are the site or species-specific coefficients, X_{ij} is the independent variable, and e_{ij} is the error term. Results were considered statistically significant at $p = 0.05$. The data had satisfied normality and equality of variance assumptions. The general linear model from Systat (Version 12, Chicago, IL, USA) was used for analysis.

3. Results

3.1. Growth

The clay site had 44% less green mass productivity than the overburden site, with 1.27 and 1.83 t ha^{-1}, respectively ($p = 0.012$, Figure 2a), and there were no differences in productivity by species on either site. However, there was a significant species, site, and species × site interaction for survival. Overall, *ERI* and *DIS* had 93% and 65% survival, respectively (Figure 2b). The significant species × site interaction was a result of rank change with site. *Salix eriocephala* had greater survival on overburden vs. the clay site, at 99% and 86%, respectively; whereas, *DIS* had lower survival on the overburden vs. clay at 62% and 68%, respectively. Overall yield, which includes a survival factor, showed that the overburden had 51% greater overall yield than the clay site, with 1.54 and 1.01 t ha^{-1}, respectively

(Figure 2c). In contrast to plant green mass yields, overall yield (yield × survival) had a species effect with 48% greater yield for *ERI* compared with *DIS*, at 1.52 and 1.03 t ha⁻¹, respectively.

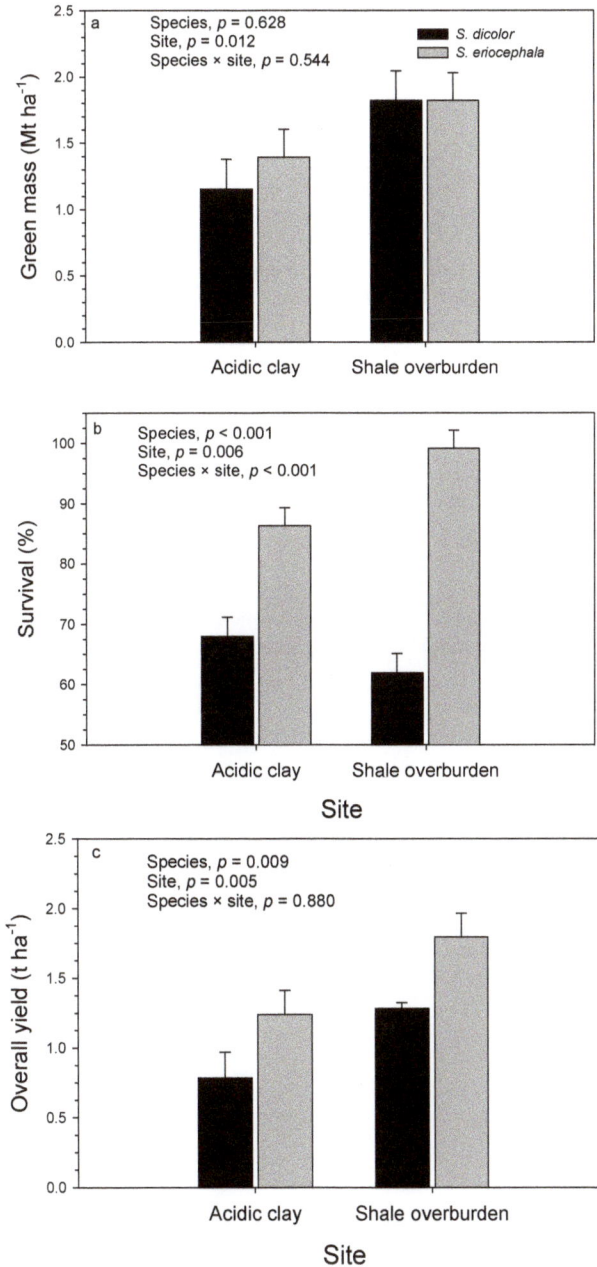

Figure 2. (a) Green mass, (b) survival, and (c) overall yield (mean ± SE) of *Salix discolor* and *S. eriocephala* on acidic clay and shale overburden sites at the Salmon Harbour coal mine near Minto, NB. Reproduced from Mosseler and Major [45].

3.2. Soil

The two sites showed significant differences in soil characteristics (Table 2). Although the soils of both study sites were composed from similar shale rock overburden, the clay deposit (42% clay content) was highly acidic (pH = 3.6), had high water holding capacity at saturation (64%) and a high EC of 3.9 mS cm^{-1}. The adjacent shale overburden site had a pH of 6.8, and after removing 56.5% stone content following sieving using a 2 mm sieve, the soil had a high sand content (67.2%), low water holding capacity at saturation (23%), and EC of 0.9 mS cm^{-1}, with comparatively little clay (9.4%) (Table 2). The acidic clay soil had significantly greater Na (20×), Ca (2×), Mg (4.4×), S (10×), C (12×) and N (2×) than the shale overburden site.

3.3. Foliage

The acidic clay site had significantly greater foliar concentrations of Mg (1.5×), Mn (3.3×), Fe (5.6×), Al (4.6×), and S (2×) than the shale overburden site (Table 3). Foliar concentrations of ERI were significantly greater than those of DIS for N (1.3×), Ca (1.5×), Mg (1.2×), Fe (2×), Al (1.5×), and S (1.5×). Differences in both foliar C and N concentrations were small but significantly lower on the acidic clay soil despite having significantly greater values in the soil samples (Table 2). Foliar N concentration was significantly higher in ERI clones on both site types, particularly on the shale rock overburden (Table 4, species × site $p < 0.001$; Figure 3a). Foliar P did not differ significantly by site or species, but there was a near significant species × site interaction ($p = 0.088$) showing that ERI had greater foliar P on the acidic clay, whereas DIS had greater foliar P on the shale rock overburden (Table 4; Figure 3b). Foliar K$^+$ had a significant species effect (DIS > ERI) but no site or species × site interaction (Figure 3c). Foliar Ca concentration had significant species (ERI > DIS) and site (overburden > acidic clay) effects (Figure 4a). Foliar Mg concentration was again significantly greater for ERI than DIS on both sites (Figure 4b). Foliar Mn concentration did not differ significantly between the two species, but site differences were highly significant (acidic clay > overburden, 3.3×; Figure 4c). Similarly, foliar Fe concentration was 5.6× greater on the acidic clay than on the shale overburden, and ERI had significantly greater foliar Fe than DIS, particularly on the acidic clay site (species × site $p = 0.020$; Figure 5a). Differences for foliar Zn concentrations were not significant for species or site effects (not shown, Table 4). Foliar Al concentration was significantly greater (4.6×) on the acidic clay site than on overburden, and there were also significant species differences, particularly on the acidic clay site where ERI foliage had twice the Al concentration than DIS (Figure 5b). The same basic pattern and significant differences were found for foliar S concentration (Figure 5c). The foliar C:N ratio did not differ significantly between site types but was significantly greater in DIS than ERI on both the acidic clay and the shale overburden (not shown). At the species level, patterns of foliar metal concentrations varied as Mn > Zn > Al > Fe for DIS, and for ERI, Mn > Fe > Al > Zn.

Table 3. (a,b) Foliage concentrations of nutrients and metals (Mean ± SE) from two sites, an acidic clay deposit and a shale rock overburden, at the Salmon Harbour mine site. Sites with different letters are significantly different at p = 0.05.

(a)

Site	Carbon (%)	Nitrogen (%)	Carbon: Nitrogen Ratio	Phosphorus (%)	Potassium (%)	Calcium (%)	Magnesium (%)
Acidic clay	49.95 ±0.06b	1.66 ±0.03b	30.8 ± 0.45a	0.210 ± 0.004a	1.01 ± 0.02a	1.73 ± 0.04b	0.292 ± 0.007a
Shale overburden	50.26 ±0.06a	1.75 ± 0.03a	29.9 ± 0.45a	0.208 ± 0.004a	0.99 ± 0.02a	1.89 ± 0.04a	0.195 ± 0.007b

(b)

Site	Manganese (ppm)	Iron (ppm)	Sodium (%)	Zinc (ppm)	Aluminum (ppm)	Sulfur (%)
Acidic clay	609.5 ± 30.3a	250.8 ± 26.4a	0.89 ± 0.04a	243.3 ± 8.9a	258.4 ± 19.8a	0.433 ± 0.014a
Shale overburden	185.4 ± 30.3b	44.5 ± 26.4b	0.90 ± 0.04a	257.3 ± 8.9a	56.1 ± 19.8b	0.243 ± 0.014b

Table 4. (a–c) Analyses of variance (ANOVA) for foliar concentrations of nutrients and metals, including source of variation, degrees of freedom (df), mean square values (MS), p values, and coefficient of determination (R^2). p values < 0.05 are in bold print.

(a)

Source of Variation	df	Carbon (%)		Nitrogen (%)		C:N ratio		Phosphorus (%)	
		MS	p Value	MS	p Value	MS	p Value	MS	p Value
Block (site)	4	0.98	**0.001**	0.18	**<0.001**	56.3	**<0.001**	1.8×10^{-3}	0.096
Species	1	2.20	**0.001**	2.87	**<0.001**	899.3	**<0.001**	0.4×10^{-3}	0.508
Site	1	2.20	**0.001**	0.18	**0.021**	19.5	0.149	0.1×10^{-3}	0.849
Species × site	1	1.16	**0.017**	0.54	**<0.001**	142.0	**<0.001**	2.6×10^{-3}	0.088
Genotype (species)	13	0.06	0.991	0.12	**<0.001**	39.5	**<0.001**	2.5×10^{-3}	**0.002**
Genotype (species) × site	13	0.17	0.567	0.02	0.816	7.5	0.636	1.3×10^{-3}	0.155
Error	56	0.19		0.03		9.1		0.8×10^{-3}	
R^2			0.541		0.773		0.789		0.554

Table 4. *Cont.*

(b)

Source of Variation	df	Iron (ppm)		Zinc (ppm)		Aluminum (ppm)		Sulfur (%)	
		MS	p Value	MS	p Value	MS	p Value	MS	p Value
Block (site)	4	10.3×10^4	**0.017**	17.0×10^3	**0.002**	10.3×10^4	**0.039**	0.014	0.164
Species	1	24.8×10^4	**0.007**	2.1×10^3	0.443	24.8×10^4	**0.005**	0.202	**<0.001**
Site	1	95.4×10^4	**<0.001**	4.4×10^3	0.270	95.4×10^4	**<0.001**	0.812	**<0.001**
Species × site	1	17.9×10^4	**0.020**	1.4×10^3	0.533	17.9×10^4	**0.006**	0.095	**0.001**
Genotype (species)	13	8.5×10^4	**0.005**	24.5×10^3	**<0.001**	8.5×10^4	**<0.001**	0.011	0.268
Genotype (species) × site	13	8.8×10^4	**0.004**	5.5×10^3	0.120	8.8×10^4	**<0.001**	0.015	0.079
Error	56	3.1×10^4		3.5×10^3		3.1×10^4		0.008	
R^2		0.700		0.703		0.752		0.764	

(c)

Source of Variation	df	Potassium (%)		Sodium (%)		Calcium (%)		Magnesium (%)		Manganese (ppm)	
		MS	p Value	MS	p Value	MS	p Value	MS	p Value	MS	p Value
Block (site)	4	0.060	**0.010**	0.021	0.880	0.682	**<0.001**	0.014	**<0.001**	25.5×10^4	**<0.001**
Species	1	0.396	**<0.001**	0.082	0.288	3.679	**<0.001**	0.029	**0.001**	0.9×10^4	0.647
Site	1	0.015	0.341	0.002	0.878	0.575	**0.007**	0.210	**<0.001**	402.8×10^4	**<0.001**
Species × site	1	<0.001	0.955	0.134	0.176	0.004	0.822	0.002	0.358	0.4×10^4	0.757
Genotype (species)	13	0.109	**<0.001**	0.143	**0.038**	0.242	**0.001**	0.012	**<0.001**	7.2×10^4	0.074
Genotype (species) × site	13	0.029	0.067	0.131	0.061	0.109	0.159	0.002	0.463	1.9×10^4	0.938
Error	56	0.016		0.072		0.074		0.002		4.1×10^4	
R^2		0.728		0.491		0.767		0.783		0.731	

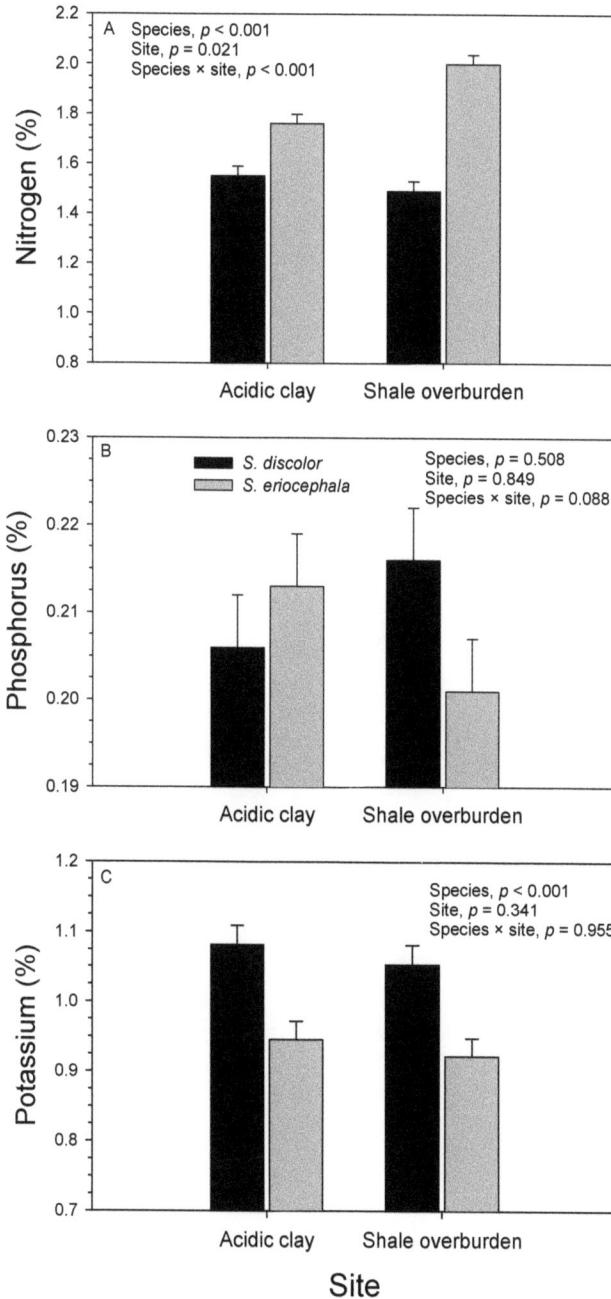

Figure 3. Foliar (**a**) nitrogen, (**b**) phosphorus, and (**c**) potassium concentrations (mean ± SE) of *Salix discolor* and *S. eriocephala* on acidic clay and shale overburden sites.

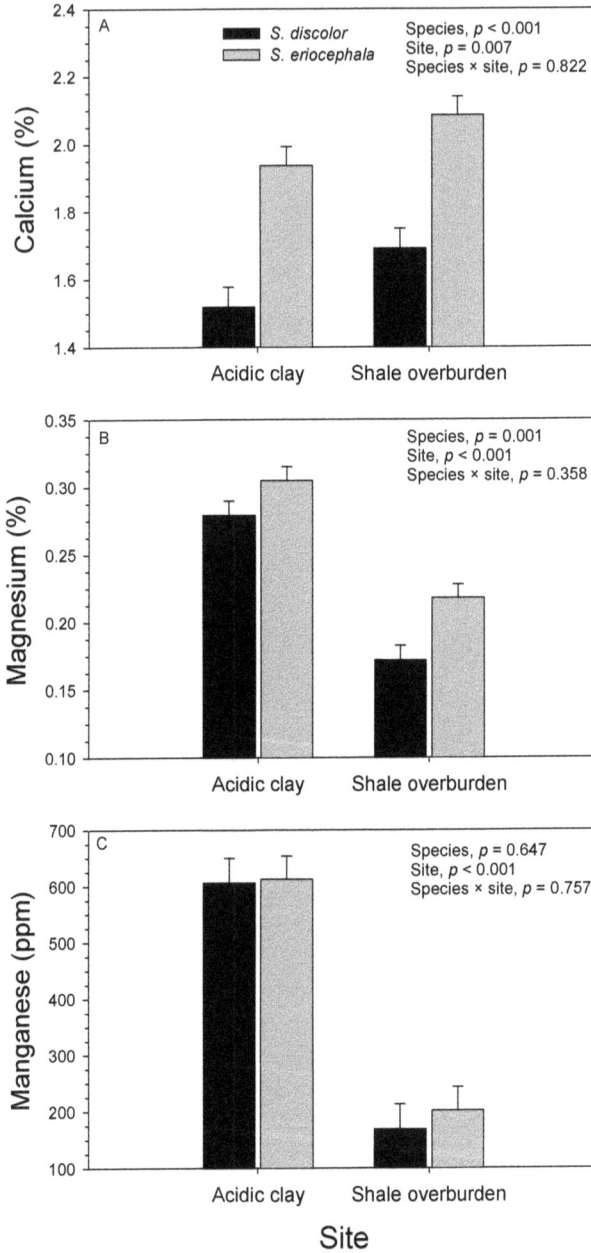

Figure 4. Foliar (**a**) calcium, (**b**) magnesium, and (**c**) manganese concentrations (mean ± SE) of *Salix discolor* and *S. eriocephala* on acidic clay and shale overburden sites.

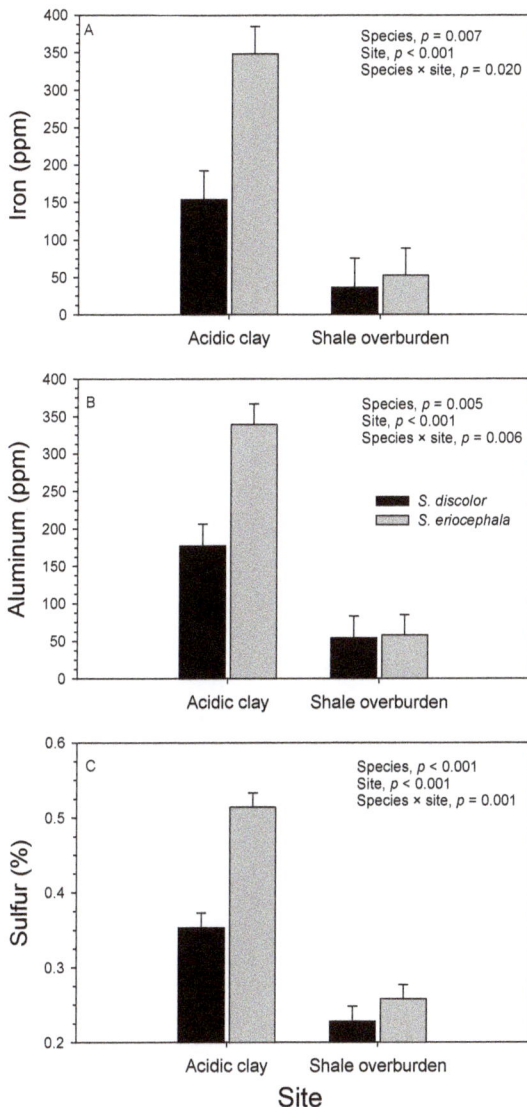

Figure 5. Foliar (**a**) iron, (**b**) aluminum and (**c**) sulfur concentrations (mean ± SE) of *Salix discolor* and *S. eriocephala* on acidic clay and shale overburden sites.

At the genotype level (clones within species), ANOVA showed significant differences among clones in foliar concentrations for most macronutrients and metals, including N, P, K, Ca, Mg, Na, Fe, Zn, and Al (Table 4), but only Fe and Al showed significant clone × site interactions. Foliar Fe and Al concentrations were consistently greater on the acidic clay site and showed some very strong genotypic variation up to eight times greater for Fe and Al in some DIS clones (e.g., clones ANF-D1, RIC-D2) (Figures 6a and 7a). For ERI, some clones also had up to eight times the Fe and Al concentrations than other ERI clones (e.g., clones FOS-E1, FRE-E1) (Figures 6b and 7b). Metal concentrations in certain ERI clones were up to 16× those of some DIS clones.

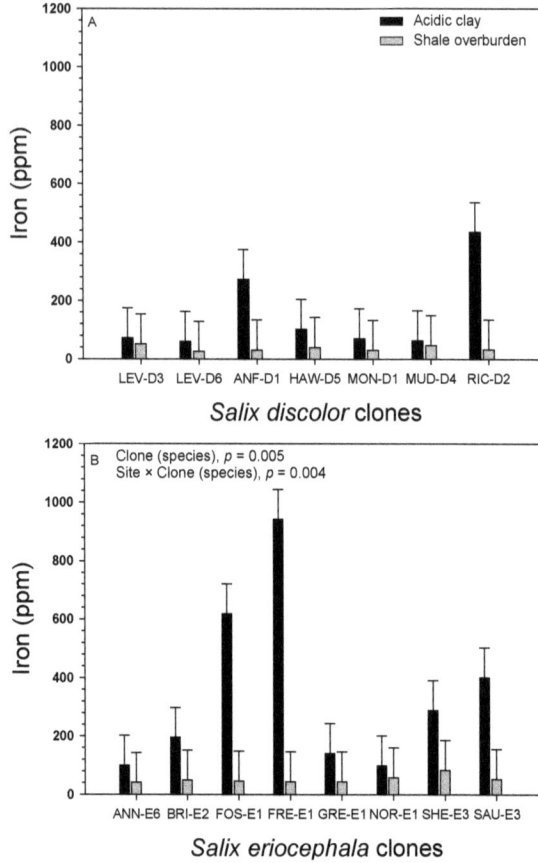

Figure 6. Foliar iron concentrations (mean ± SE) from 15 genotypes of (**a**) *Salix discolor* (seven) and (**b**) *S. eriocephala* (eight) on acidic clay and shale overburden sites.

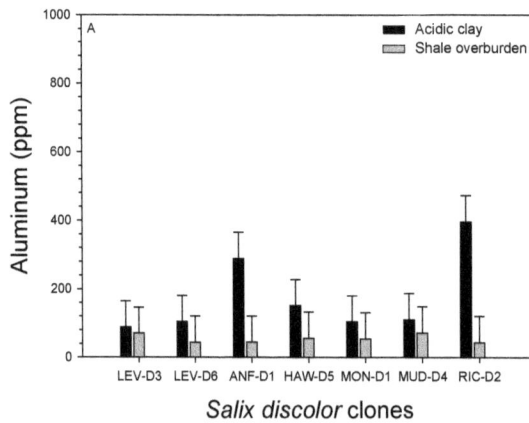

Figure 7. *Cont.*

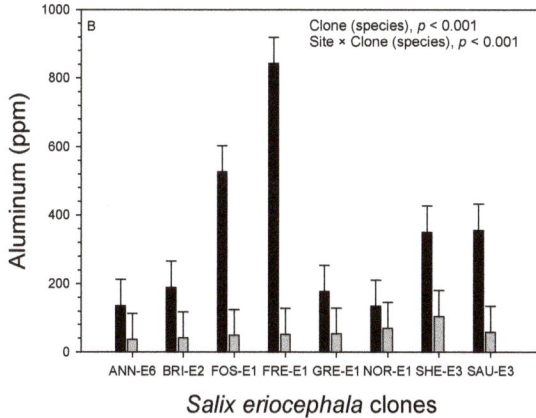

Figure 7. Foliar aluminum concentrations (mean ± SE) from 15 genotypes of (**a**) *Salix discolor* (seven) and (**b**) *S. eriocephala* (eight) on acidic clay and shale overburden sites.

Covariate analysis of total aboveground green mass, using macronutrients or metals as the covariate and testing species effect, showed no significant nutrient/metal effects, species effects, or species × nutrient/metal interaction for any of the 12 foliar nutrients, metals or C:N ratio (Table 3), except for Ca. Green mass showed a significant Ca × species interaction ($p = 0.019$), species ($p = 0.027$), and Ca ($p = 0.008$) effects. Green mass had a different positive slope in relation to increasing Ca concentration, with a steeper slope for DIS than ERI (Figure 8).

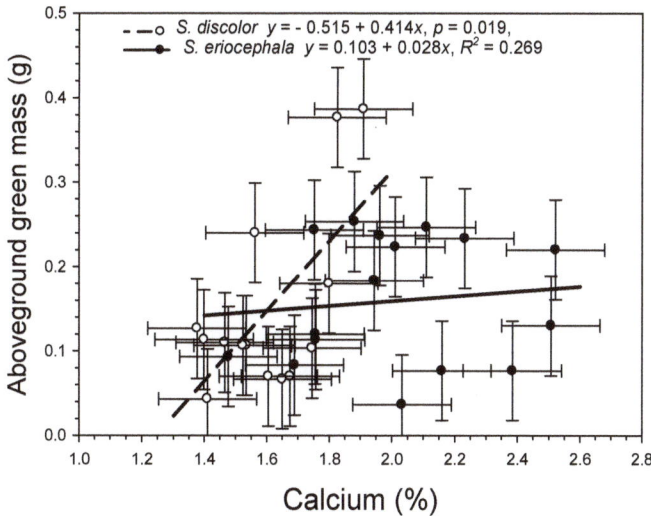

Figure 8. Covariate analysis of green mass vs. calcium concentration for *Salix discolor* and *S. eriocephala*.

4. Discussion

Willows are recognized for their phytoremediation potential and for phytoextraction of soils contaminated by heavy metals [11,22,23,25,26,31,32,54–56]. Both species and clonal differences have been observed for heavy metal concentrations, but these higher concentrations have sometimes had little apparent effect on growth and biomass accumulation in willows [30,54–57]. A limited number of European willows have been assessed for genetic variation both among and within species in their capacity for absorption of various soil contaminants under field conditions [19,21,23,25,27,58–60]. However, much less is known about such responses for North American willows [6,8,9,13,14], or their potential use in phytoremediation.

Both low and high pH soils can have high EC. Generally, acidic soils have lower EC compared with high pH soils. However, high soil EC under acidic conditions can be related to mobility of elements under acidic conditions, including some nutrients, Na and other metals [37,38,61]. In addition, clay itself has greater EC than sandy sites, generally due to greater cation exchange capacity [62]. The higher clay content of the acidic clay site could also result in increased binding of metals to clay mineral particles. Moreover, the acidic clay site had 20× the Na than the shale overburden, which presents another unfavorable condition, saline soils. The clay site that was formed as a settling pond and is often inundated after winter snow melt has a greater water content at saturation. Although high EC/saline soils may be physically wet, they can be a physiologically dry habitat for plants depending on the saline concentrations or EC. Increases in soil EC create lower soil water potential, which makes it more difficult for plants to extract soil water, creating an osmotic "drought" effect [63]. Salinity is customarily measured using EC, and the EC on the clay site was 3.9 mS cm^{-1}, which is considered high salinity for typical glycophytic plants. In a *Salix* species × salinity experiment examining response to control (0.45 mS cm^{-1}), "moderate" (1.5 mS cm^{-1}), and "high" salinity (3.0 mS cm^{-1}), DIS and ERI did not survive the high EC/saline conditions to the end of the experiment (day 160); however, *S. interior* Rowlee (INT) did show a 35% survival [64]. At moderate salinity, survival was 67%, 68% and 100%, for DIS, ERI and INT, respectively. Thus, despite the highly acidic, high EC/Na soil conditions, DIS and ERI managed to survive and even showed modest growth under these conditions. Note, the potential productivity of DIS and ERI on a fertile site such as the MBG was on average 24 and 45 Mt ha^{-1}, respectively. The only other species growing on this site naturally is coltsfoot (*Tussilago farfara* L.), despite a considerable diversity of invasive species surrounding the site and occupying the shale overburden site [53].

Despite the 20× higher Na concentration on the acidic clay compared with the shale overburden, there was no difference in foliar Na between sites. In almost all plant salinity studies, Na appears to reach toxic concentrations before Cl does, and so most studies have concentrated on Na exclusion and control of Na within plants [65]. In an analysis of foliar Na and other nutrients in a controlled salinity × willow species study, DIS and ERI had similar foliar Na; however, INT had a far greater capacity to incorporate the excess Na [66]. In a review of the physiological mechanisms of plant salinity/Na tolerance, there are generally three distinct adaptation mechanisms to salinity: cellular osmotic stress adjustment; Na or Cl exclusion; and tolerance of tissue to accumulated Na or Cl [65]. Salinity can cause the accumulation of high concentrations of ions such as Na, Cl, and SO$_4$ [67]. Some species do not show much increase in salt concentration over time and thus can avoid leaf mortality, whereas other species exclude or store salt in vacuoles, thereby controlling uptake and compartmentalization of ions such as Na, Cl, Fe, and S [68,69]. Among these ions, Na is considered the main ion negatively affecting plant growth. Thus, Na must be effectively partitioned and confined within cell vacuoles. Exactly how plants direct Na to the vacuole is unclear (69), but by doing so, plants osmotically adjust so they can extract water from a saline site.

Although we did not assess total metal concentrations in the soil, the soils on these sites were derived largely from the same shale rock parent material and total metal concentrations were assumed to be reasonably similar across our study sites. Foliar concentrations were greater on the acidic clay site for Mn (3.3×), Fe (5.6×), and Al (4.6×) compared with the adjacent shale overburden. The significantly greater foliar concentrations for metals such as Al, Fe, and Mn in plants on the acidic (pH = 3.6) clay site compared with an adjacent shale overburden (pH = 6.8), indicate that these metals were most likely mobilized by low pH [37,38,70] and taken up in higher concentrations by willow foliage. Whereas Zn concentrations did not vary significantly either by species or site (250 ppm), apparently unaffected by acidic conditions. Baseline concentrations of Zn in eight European willows assessed by Nissen and Lepp [21] (Table 3) ranged from 82 to 296 ppm. On shale soils, Alloway [70] suggested that Zn concentrations greater than 300 ppm approached phytotoxic levels. Rosselli et al. [36] showed that both *Salix* and *Betula* growing on contaminated sites had higher foliar Zn concentrations in their foliage than other tree species tested (*Alnus*, *Fraxinus*, and *Sorbus*).

Plants growing in highly acidic soil conditions can experience aluminum (Al) toxicity [43,44,71], of which the most easily recognized symptom is inhibition of root growth [72]. Foliage analysis indicated significantly greater Al (4.6×) levels on the acidic clay deposit than the shale overburden (Table 2), and green mass was significantly lower (44%) on the acidic clay than on the shale overburden, but increased levels of foliar Al did not influence growth on either site. From a land reclamation perspective, perhaps the more important finding was that willow species can grow roots from rootless stem cuttings on these sites and still maintain comparatively good growth under low pH and high EC and potentially phytotoxic soil conditions.

On the acidic clay site, Mn was 610 ppm; whereas, mean values of Mn at 850 ppm have been reported [68]. Bourret et al. [9] found foliar Mn concentrations ranging from approximately 350 ppm in *Salix geyeriana* Andersson, to 150 ppm in *S. monticola* Bebb growing in unsaturated mine tailings. At foliar Mn concentrations of 610 ppm on the acidic clay site, neither DIS nor ERI showed any foliar symptoms of phytotoxicity on the SH coal mine spoils. Manganese is an essential plant nutrient, but toxic levels of Mn have been reported in arid areas under irrigation and from metal-rich mine surface runoff and drainage waters [35]. Shanahan et al. [13] established toxicity thresholds of 2791 and 3117 mg L^{-1} for Mn, and 556 and 623 mg L^{-1} for Zn, in two willow species native to western North America. Bioavailability of both Mn and Zn in soils increases dramatically as soil pH declines, and this was also evident in foliar Mn concentrations in our study, which showed 3.3 times the concentration on the acidic clay site compared with the shale overburden. High levels of both Mn and Zn can result in nutrient deficiencies by interfering in the uptake of other essential elements, including C, Fe, and Mg [13], but no such effects were evident in the foliage of either DIS or ERI on the two sites assessed in our study. Other plants have shown a wide tolerance in foliar concentrations of Mn [73].

In Swedish field trials, VIM generally showed poorer growth on acidic clay soils [30], where low soil pH may have mobilized metals to an extent that led to increased foliar concentrations. Mine tailings from base metal mining operations can be highly acidic [33], resulting in potentially phytotoxic levels for metals such as Al, Mn, and Zn [9,14,37,38,70]. Although, the pH of the clay deposit at the SH coal mine was very low, none of the visible symptoms of foliar toxicity, such as leaf chlorosis or necrosis, bleaching, decreases in leaf sizes, leaf curling, etc. (see Marschner [74]), were evident in plants on either site. There was a small but significant site effect on survival, with 77.1% and 80.5% survival for the acidic clay and overburden sites, respectively, and biomass yield was 44% greater on the overburden compared with the acidic clay site [45]. On both sites, ERI had significantly greater survival than DIS but both species grew comparatively well, despite low pH on the clay site, demonstrating that both species have a reasonably high tolerance for such harsh site conditions. On the acidic clay site, mean biomass growth for the best DIS and ERI genotypes were comparable to the highly selected VIM check clone #5027 [45]. On the shale overburden site, the best DIS genotype exceeded the biomass of VIM by 12%, whereas the best ERI genotype underperformed VIM by 27%. As Pulford et al. [27] noted in their phytoremediation field trials, European willows have the ability to grow and produce high biomass yields capable of absorbing high metal concentrations, and this can be important where the aim of phytoextraction is the removal of metals from moderately contaminated sites.

In field trials of VIM in Switzerland that compared phytoextraction of Cd and Zn on an acidic site (pH = 5.2) versus a calcareous site (pH = 7.3), Hammer et al. [29] reported higher levels of Zn in the VIM foliage on the acidic site. In our study, significantly greater foliage concentrations of Al, Fe, Mn, S, and Mg were also found on the acidic clay site, but no differences between sites were found for foliage concentration of K, P, and Zn. Through application of elemental S to a calcareous site, Hammer et al. [29] were able to solubilize heavy metals, thereby increasing metals uptake by VIM. The highly acidic clay site in our study already had a very high S concentration, probably due to the presence of metal sulfides in floodwaters associated with the high S concentration of coal seams at the SH mine site. This high S concentration and related acidity may also have promoted high foliar concentrations of metals on the clay site. Both studies show that low pH and presence of high S concentration are associated with heavy metal uptake and lower biomass yields in willow [45].

Soil C was more than 10× greater on the acidic clay than shale overburden site and at first glance might indicate greater organic matter in the soil. However, there was only a scattering of coltsfoot on the site, and as this was a coal mine site, it would appear more likely that the greater C was from the settling of fine coal particles on the acidic clay site. Soil N was comparatively low (see Hansen et al. [75]) for the shale rock overburden and acidic clay sites (0.1% and 0.2%, respectively). However, foliar N concentrations were relatively high in both DIS and ERI considering the infertile site conditions and were greatest on the shale overburden despite low soil N. The relatively high foliar N concentration is puzzling in view of the low soil N concentration and supports recent research demonstrating the presence of endophytes in the stems of willows capable of fixing N on nutrient-poor sites [76,77].

There were some strong and significant differences in species responses with respect to uptake of both macronutrients (e.g., N, K, Ca, Mg, and S) and metals (e.g., Al, Fe, and Zn) (see also Pulford et al. [27]). The higher foliar concentrations of N, Ca, and Mg in ERI on both sites may reflect a higher nutrient demand in ERI. In natural habitats, ERI is generally associated with more fertile sites, whereas DIS is a natural colonizer of highly disturbed, drier, and rockier upland sites and is commonly found as a natural invader and colonizer of the shale overburden of the SH mine site [49]. ERI also had significantly greater foliar concentrations of metals such as Fe and Al, especially under the acidic conditions of the clay site, and the results suggest that selection of specific ERI genotypes may be useful in areas contaminated by these metals.

Landberg and Greger [19,59] described clonal variation in two European willows, VIM and *S. dasyclados*, for cadmium (Cd) uptake and noted variation in clonal tolerance for specific heavy metals, as well as a general tolerance in certain willow clones. Greger and Landberg [40] also showed that accumulation, transport, and tolerance for specific heavy metals can be highly clone specific and that this specificity can vary with site conditions. Our results showed strong clonal differences with respect to uptake of metals in both DIS and ERI, with a remarkable consistency in the capacity of certain willow clones to take up several different metals simultaneously. For instance, clones of either DIS or ERI able to take up the largest quantities of Al were the same clones that took up the largest quantities of Fe. These results suggest that certain genotypes may have a greater capacity for uptake or tolerance for metals and that clonal selection for increased metal uptake on contaminated sites looks promising.

Despite large variation in metal concentrations among species and genotypes, there were no significant negative (or positive) relationships with productivity, as might have been expected. In addition, there was no relationship between nutrient concentrations and productivity except for a positive relationship of increasing Ca concentration on productivity, particularly for DIS. Calcium may help neutralize very acidic soil conditions. The genotype with the greatest productivity and Ca concentration was the DIS clone MON-D1.

5. Conclusions

Phytoremediation efforts often require plant tolerance to a number of suboptimal conditions. On a former coal mine spoil, neither *Salix discolor* nor *S. eriocephala* showed any negative relationships between metal concentrations and plant growth, nor any visible effects on foliage health. Genetic differences among species and genotypes within these species in their biological capacity to take up and tolerate metals suggests that selection and breeding for willow clones with special capacities for uptake of heavy metals from moderately contaminated soils looks promising. *Salix eriocephala* showed particular promise for use in phytoextraction of heavy metals such as Fe and Al. Results showed that despite having a greater water holding capacity, the clay site was very acidic, resulting in greater metal mobility and EC, and it would appear that the decline in productivity on a highly acidic site was not due to toxicity of mobile metals but to low pH and moisture stress from very high soil Na/EC.

Acknowledgments: We are grateful to Moira Campbell, Ted Cormier, John Malcolm, and Peter Tucker for their assistance in collecting and processing plant material, establishment of field tests, and assistance with data collection. We also thank Michele Coleman at Mine Restoration Inc., a subsidiary of NB Power, for providing space for common-garden field tests, Michel Labrecque and the Montreal Botanical Gardens for hosting the clonal willow gardens used to obtain plant material, Ian DeMerchant for preparing the contour map of the mine site study locations, Jim Estey of the Laboratory for Forest Soils and Environmental Quality at the University of New Brunswick for analysis of soil and foliage samples, and to three anonymous reviewers, one of which provided an insightful and constructive critique of an earlier version of our manuscript. Funding for this study was provided by the Canadian Forest Service and the New Brunswick Wildlife Trust.

Author Contributions: A. M. and J. E. M. conceived and designed the experiments; performed the experiments; analyzed the data; and wrote the paper.

Conflicts of Interest: The authors declare no conflict of interest.

Abbreviations

DIS	*Salix discolor*
EC	electrical conductivity
ERI	*Salix eriocephala*
MBG	Montreal Botanical Garden
Al	Aluminum
C	Carbon
Ca	Calcium
Fe	Iron
INT	*Salix interior*
K	Potassium
Mg	Magnesium
Mn	Manganese
Na	Sodium
N	Nitrogen
NB	New Brunswick
ON	Ontario
P	Phosphorous
QC	Quebec
S	Sulfur
VIM	*Salix viminalis*
Zn	Zinc

References

1. Mosseler, A.; Zsuffa, L.; Stoehr, M.U.; Kenney, W.A. Variation in biomass production, moisture content, and specific gravity in some North American willows (*Salix* L.). *Can. J. For. Res.* **1988**, *18*, 1535–1540. [CrossRef]
2. Zsuffa, L. Genetic improvement of willows for energy plantations. *Biomass* **1990**, *22*, 35–47. [CrossRef]
3. Labrecque, M.; Teodorescu, T.I. Field performance and biomass production of 12 willow and poplars in short-rotation coppice in southern Quebec (Canada). *Biomass Bioenergy* **2005**, *29*, 1–9.
4. Volk, T.A.; Abrahamson, L.P.; Nowak, C.A.; Smart, L.B.; Tharakan, P.J.; White, E.H. The development of short-rotation willow in the northeastern United States for bioenergy and bio-products, agroforestry and phytoremediation. *Biomass Bioenergy* **2006**, *30*, 715–727. [CrossRef]
5. Zalesny, R.S.; Stanturf, J.A.; Gardiner, E.S.; Perdue, J.H.; Young, T.M.; Coyle, D.R.; Headlee, W.L.; Banuellos, G.S.; Hass, A. Ecosystem services of woody crop production systems. *Bioenergy Res.* **2016**, *9*, 465–491. [CrossRef]
6. Labrecque, M.; Teodorescu, T.I.; Daigle, S. Effect of wastewater sludge on growth and heavy metal bioaccumulation of two *Salix* species. *Plant Soil* **1994**, *171*, 303–316. [CrossRef]

7.   Labrecque, M.; Teodorescu, T.I. Influence of plantation site and wastewater sludge fertilization on the performance and foliar nutrient status of two willow species grown under SRIC in southern Quebec (Canada). *For. Ecol. Manag.* **2001**, *150*, 223–239. [CrossRef]
8.   Kuzovkina, Y.A.; Knee, M.; Quigley, M.F. Cadmium and copper uptake and translocation in five willow species. *Int. J. Phytorem.* **2004**, *6*, 269–287. [CrossRef] [PubMed]
9.   Bourret, M.M.; Brummer, J.E.; Leininger, W.C.; Heil, D.M. Effect of water table on willows grown in amended mine tailing. *J. Environ. Qual.* **2005**, *34*, 782–792. [CrossRef] [PubMed]
10. Kuzovkina, Y.A.; Quigley, M.F. Willows beyond wetlands: Uses of *Salix* L. species for environmental projects. *WASP* **2005**, *162*, 183–204. [CrossRef]
11. Meers, E.; Lamsal, S.; Vervaeke, P.; Hopgood, M.; Lust, N.; Tack, F.M.G. Availability of heavy metals for uptake by *Salix viminalis* on a moderately contaminated dredged sediment disposal site. *Environ. Pollut.* **2005**, *137*, 354–364. [CrossRef] [PubMed]
12. Meers, E.; Vandecasteele, B.; Ruttens, A.; Vangronsveld, J.; Tack, F.M.G. Potential of five willow species (*Salix* spp.) for phytoextraction of heavy metals. *Environ. Exp. Bot.* **2007**, *60*, 57–68. [CrossRef]
13. Shanahan, J.O.; Brummer, J.E.; Leininger, W.C.; Paschke, M.W. Manganese and zinc toxicity thresholds for mountain and Geyer willow. *Int. J. Phytorem.* **2007**, *9*, 437–452. [CrossRef] [PubMed]
14. Boyter, M.J.; Brummer, J.E.; Leininger, W.C. Growth and metal accumulation of Geyer and mountain willow in topsoil versus amended mine tailings. *WASP* **2009**, *198*, 17–29. [CrossRef]
15. Kuzovkina, Y.A.; Volk, T.A. The characteristics of willow (*Salix* L.) varieties for use in ecological engineering applications: Co-ordination of structure, function and autecology. *Ecol. Eng.* **2009**, *35*, 1178–1189. [CrossRef]
16. Witters, N.; van Slycken, S.; Ruttens, A.; Adriaensen, K.; Meers, E.; Meiresonne, L.; Tack, F.M.G.; Thewys, T.; Laes, E.; Vangronsveld, J. Short-rotation coppice of willow for phytoremediation of a metal-contaminated agricultural area: A sustainability assessment. *Bioenergy Res.* **2009**, *2*, 144–152. [CrossRef]
17. Dimitriou, I.; Mola-Yudego, B.; Aronsson, P.; Eriksson, J. Changes in organic carbon and trace elements in the soil of willow short-rotation coppice plantations. *Bioenergy Res.* **2012**, *5*, 563–572. [CrossRef]
18. Tordoff, G.M.; Baker, A.J.M.; Willis, A.J. Current approaches to the revegetation and reclamation of metalliferous mine wastes. *Chemosphere* **2000**, *41*, 219–228. [CrossRef]
19. Landberg, T.; Greger, M. Differences in uptake and tolerance to heavy metals in *Salix* from unpolluted and polluted areas. *Appl. Geochem.* **1996**, *11*, 175–180. [CrossRef]
20. Landberg, T.; Greger, M. Differences in oxidative stress in heavy metal resistant and sensitive clones of *Salix viminalis*. *J. Plant Physiol.* **2002**, *159*, 69–75. [CrossRef]
21. Nissen, L.R.; Lepp, N.W. Baseline concentrations of copper and zinc in shoot tissues of a range of *Salix* species. *Biomass Bioenergy* **1997**, *12*, 115–120. [CrossRef]
22. Perttu, K.L.; Kowalik, P.J. *Salix* vegetation filters for purification of water and soils. *Biomass Bioenergy* **1997**, *12*, 9–19. [CrossRef]
23. Punshon, T.; Dickinson, N.M. Acclimation of *Salix* to metal stress. *New Phytol.* **1997**, *137*, 303–314. [CrossRef]
24. Punshon, T.; Dickinson, N.M. Mobilization of heavy metals using short-rotation coppice. *Asp. Appl. Biol.* **1997**, *49*, 285–292.
25. Punshon, T.; Dickinson, N.M. Heavy metal resistance and accumulation characteristics in willows. *Int. J. Phytorem.* **1999**, *4*, 361–385. [CrossRef]
26. Aronsson, P.; Perttu, K. Willow vegetation filters for wastewater treatment and soil remediation combined with biomass production. *For. Chron.* **2001**, *77*, 293–298. [CrossRef]
27. Pulford, I.D.; Riddell-Black, D.M.; Stewart, C. Heavy metal uptake by willow clones from sewage sludge-treated soil: The potential for phytoremediation. *Int. J. Phytorem.* **2002**, *4*, 59–72. [CrossRef]
28. Pulford, I.D.; Watson, C. Phytoremediation of heavy metal-contaminated land by trees—A review. *Environ. Int.* **2003**, *29*, 529–540. [CrossRef]
29. Hammer, D.; Kayser, A.; Keller, C. Phytoextraction of Cd and Zn with *Salix viminalis* in field trials. *Soil Use Manag.* **2002**, *19*, 187–192. [CrossRef]
30. Klang-Westin, E.; Eriksson, J. Potential of *Salix* as phytoextractor for Cd on moderately contaminated soils. *Plant Soil* **2003**, *249*, 127–137. [CrossRef]
31. Keller, C.; Hammer, D.; Kayser, A.; Richner, W.; Brodbeck, M.; Sennhauser, M. Root development and heavy metal phytoextraction efficiency: Comparison of different species in the field. *Plant Soil* **2003**, *249*, 67–81. [CrossRef]

32. Vervaeke, P.; Luyssaert, S.; Mertens, J.; Meers, E.; Tack, F.M.G.; Lust, N. Phytoremediation prospects of willow stands on contaminated sediments: A field trial. *Environ. Pollut.* **2003**, *126*, 275–282. [CrossRef]

33. Bagatto, G.; Shorthouse, J.D. Biotic and abiotic characteristics of ecosystems on acid metalliferous mine tailings near Sudbury, Ontario. *Can. J. Bot.* **1999**, *77*, 410–425.

34. Berti, W.R.; Cunningham, S.D. Phytostabilization of metals. In *Phytoremediation of Toxic Metals—Using Plants to Clean Up the Environment*; Raskin, I., Ensley, B.D., Eds.; John Wiley and Sons: New York, NY, USA, 2000; pp. 71–88.

35. Green, C.H.; Heil, D.M.; Cardon, G.E.; Butters, G.L.; Kelly, E.F. Solubilization of manganese and trace metals in soils affected by acid mine runoff. *J. Environ. Qual.* **2003**, *32*, 1323–1334. [CrossRef] [PubMed]

36. Rosselli, W.; Keller, C.; Boschi, K. Phytoextraction capacity of trees growing on a metal contaminated soil. *Plant Soil* **2003**, *256*, 265–272. [CrossRef]

37. Chaney, R.L. Zinc phytotoxicity. In Proceedings of the International Symposium on Zinc in Soils and Plants, Perth, Australia, 27–28 September 1993; Robson, A.D., Ed.; Kluwer Academic: London, UK, 1993; pp. 135–149.

38. Kahle, H. Response of roots of trees to heavy metals. *Environ. Exp. Bot.* **1993**, *33*, 99–119. [CrossRef]

39. Tack, F.M.G.; Callewaert, O.W.; Verloo, M.G. Metal solubility as a function of pH in contaminated, dredged sediment affected by oxidation. *Environ. Pollut.* **1996**, *91*, 199–208. [CrossRef]

40. Greger, M.; Landberg, T. Use of willow in phytoextraction. *Int. J. Phytorem.* **1999**, *1*, 115–123. [CrossRef]

41. Rout, G.R.; Samantaray, S.; Das, P. Aluminium toxicity in plants: A review. *Agronomie* **2001**, *21*, 3–21. [CrossRef]

42. Mariano, E.D.; Pinheiro, A.S.; Garcia, E.F.; Keltjens, W.G.; Jorge, R.A.; Menossi, M. Differential aluminium-impaired nutrient uptake along the root axis of two maize genotypes contrasting in resistance to aluminium. *Plant Soil* **2015**, *388*, 323–335. [CrossRef]

43. Rehmus, A.; Bigalke, M.; Valarezo, C.; Castillo, J.M.; Wilcke, W. Aluminum toxicity to tropical montane forest tree seedlings in southern Ecuador: Response of nutrient status to elevated Al concentrations. *Plant Soil* **2015**, *388*, 87–97. [CrossRef]

44. Andersson, M. Toxicity and tolerance of aluminium in vascular plants. *WASP* **1988**, *39*, 439–462.

45. Mosseler, A.; Major, J.E. Coppice growth responses of two North American willows (*Salix* spp.) in acidic clay deposits on coal mine overburden. *Can. J. Plant Sci.* **2014**, *94*, 1269–1279. [CrossRef]

46. Argus, G.W. *Salix* L. In *Flora of North America North of Mexico, Volume 7: Magnoliophyta: Salicaceae to Brassicaceae*; Editorial Committee, Ed.; Oxford University Press: Oxford, UK; New York, NY, USA, 2010; pp. 23–162.

47. Russell, W.B.; La Roi, G.H. Natural vegetation and ecology of abandoned coal-mined land, Rocky Mountain Foothills, Alberta, Canada. *Can. J. Bot.* **1986**, *64*, 1286–1298. [CrossRef]

48. Strong, W.L. Vegetation development on reclaimed lands in the Coal Valley Mine of western Alberta, Canada. *Can. J. Bot.* **2000**, *78*, 110–118.

49. Mosseler, A.; Major, J.E.; Labrecque, M. Growth and survival of seven native willow species on highly disturbed coal mine sites in eastern Canada. *Can. J. For. Res.* **2014**, *44*, 1–10. [CrossRef]

50. McKeague, J.A. (Ed.) *Manual on Soil Sampling and Methods of Analysis*, 2nd ed.; Canadian Society of Soil Science: Ottawa, ON, Canada, 1978; 212p.

51. Gullberg, U. Towards making willows pilot species for coppicing production. *For. Chron.* **1993**, *69*, 721–726. [CrossRef]

52. Guidi-Nissim, W.G.; Pitre, F.E.; Teodorescu, T.I.; Labrecque, M. Long-term biomass productivity of willow bioenergy plantations maintained in southern Quebec, Canada. *Biomass Bioenergy* **2013**, *56*, 361–369. [CrossRef]

53. Mosseler, A.; Major, J.E.; Labrecque, M. Genetic by environment interactions of two North American *Salix* species assessed for coppice yield and components of growth on three sites of varying quality. *Trees* **2014**, *28*, 1401–1411. [CrossRef]

54. Klang-Westin, E.; Perttu, K. Effects of nutrient supply and soil cadmium concentration on cadmium removal by willow. *Biomass Bioenergy* **2002**, *23*, 415–426. [CrossRef]

55. Mleczek, M.; Rutkowski, P.; Rissmann, I.; Kaczmarek, Z.; Golinski, P.; Szentner, K.; Strazynska, K.; Stachowiak, A. Biomass productivity and phytoremediation potential of *Salix alba* and *Salix viminalis*. *Biomass Bioenergy* **2010**, *34*, 1410–1418. [CrossRef]

56. Syc, M.; Pohorely, M.; Kamenikova, P.; Habart, J.; Svoboda, K.; Puncochar, M. Willow trees from heavy metals phytoextraction as energy crops. *Biomass Bioenergy* **2012**, *37*, 106–113. [CrossRef]

57. Utmazian, M.N.; Wieshammer, G.; Vega, R.; Wenzel, W.W. Hydroponic screening for metal resistance and accumulation of cadmium and zinc in twenty clones of willows and poplars. *Environ. Pollut.* **2007**, *148*, 155–165. [CrossRef] [PubMed]

58. Riddell-Black, D.M. Heavy metal uptake by fast growing willow species. In *Willow Vegetation Filters for Municipal Wastewaters and Sludges: A Biological Purification System*; Aronsson, P., Perttu, K., Eds.; Sveriges Lantbruksuniversitet: Uppsala, Sweden, 1994; pp. 133–144.

59. Landberg, T.; Greger, M. Interclonal variation of heavy metal interactions in *Salix viminalis*. *Environ. Toxicol. Chem.* **2002**, *21*, 2669–2674. [CrossRef] [PubMed]

60. Vyslouzilova, M.; Tlustos, P.; Szakova, J. Cadmium and zinc phytoextraction of seven clones of *Salix* spp. planted on heavy metal contaminated soils. *Plant Soil Environ.* **2003**, *49*, 542–547.

61. Zottl, H.W. Heavy metal levels and cycling in forest ecosystems. *Experientia* **1985**, *41*, 1104–1113. [CrossRef]

62. Domsch, H.; Giebel, A. Estimation of soil textural features from soil electrical conductivity recorded using the EM38. *Precis. Agric.* **2004**, *5*, 389–409. [CrossRef]

63. Sheldon, A.R.; Dalal, R.C.; Kirchhof, G.; Kopittke, P.M.; Menzies, N.W. The effect of salinity on plant-available water. *Plant Soil* **2017**, *418*, 477–491. [CrossRef]

64. Major, J.E.; Mosseler, A.; Malcolm, J.W.; Heartz, S. Salinity tolerance of three *Salix* species: Survival, biomass yield and allocation, and biochemical efficiencies. *Biomass Bioenergy* **2017**, *105*, 10–22. [CrossRef]

65. Munns, R.; Tester, M. Mechanisms of salinity tolerance. *Ann. Rev. Plant Biol.* **2008**, *59*, 651–681. [CrossRef] [PubMed]

66. Major, J.E.; Mosseler, A.; Malcolm, J.W. Salix species variation in leaf gas exchange, sodium, and nutrient parameters at three levels of salinity. *Can. J. For. Res.* **2017**, *47*, 1045–1055. [CrossRef]

67. Flowers, T.J.; Colmer, T.D. Salinity tolerance in halophytes. *New Phytol.* **2008**, *179*, 945–963. [CrossRef] [PubMed]

68. Cassaniti, C.; Romano, D.; Hop, M.E.C.M.; Flowers, T.J. Growing floricultural crops with brackish water. *Environ. Exp. Bot.* **2013**, *92*, 165–175. [CrossRef]

69. Dong, Y.; Ma, Y.; Wang, H.; Zhang, J.; Zhang, G.; Yang, M.-S. Assessment of tolerance of willows to saline soils through electrical impedance measurements. *For. Sci. Pract.* **2013**, *15*, 32–40.

70. Alloway, B.J. (Ed.) *Heavy Metals in Soils*, 2nd ed.; Blaikie Academic and Professional: London, UK, 1995; p. 368.

71. Delhaize, E.; Ryan, P.R. Aluminum toxicity in plants. *Plant Physiol.* **1995**, *107*, 315–321. [CrossRef] [PubMed]

72. Larcheveque, M.; Desrochers, A.; Bussiere, B.; Cartier, H.; David, J.-S. Re-vegetation of non-acid-generating, thickened tailings with boreal trees: A greenhouse study. *J. Environ. Qual.* **2013**, *42*, 351–360. [CrossRef] [PubMed]

73. Chen, C.; Zhang, H.; Wang, A.; Lu, M.; Shen, Z.; Lian, C. Phenotypic plasticity accounts for most of the variation in leaf manganese concentrations in *Phytolacca americana* growing in manganese-contaminated environments. *Plant Soil* **2015**, *396*, 215–227. [CrossRef]

74. Marschner, H. *Mineral Nutrition of Higher Plants*; Academic Press Inc.: London, UK, 1986; 674p.

75. Hansen, E.A.; McLaughlin, R.A.; Pope, P.E. Biomass and nitrogen dynamics of hybrid poplar on two different soils: Implications for fertilization strategy. *Can. J. For. Res.* **1988**, *18*, 223–230. [CrossRef]

76. Doty, S.L.; Oakley, B.; Xin, G.; Kang, J.W.; Singleton, G.; Khan, Z.; Vajzovic, A.; Staley, J.T. Diazotrophic endophytes of native black cottonwood and willow. *Symbiosis* **2009**, *47*, 23–33. [CrossRef]

77. Knoth, J.L.; Kim, S.-H.; Ettl, G.J.; Doty, S.L. Biological nitrogen fixation and biomass accumulation within poplar clones as a result of inoculation with diazotrophic consortia. *New Phytol.* **2013**, *201*, 599–609. [CrossRef] [PubMed]

Article

Stakeholder Assessment of the Feasibility of Poplar as a Biomass Feedstock and Ecosystem Services Provider in Southwestern Washington, USA

Noelle M. Hart [1], Patricia A. Townsend [1,*], Amira Chowyuk [2] and Rick Gustafson [2]

[1] Advanced Hardwood Biofuels Northwest, Washington State University Extension, 600 128th St SE, Everett, WA 98208, USA; noelle.hart@wsu.edu
[2] Biofuels and Bioproducts Laboratory, School of Environmental and Forest Sciences, University of Washington, Seattle, WA 98115, USA; anchowyu@uw.edu (A.C.); pulp@uw.edu (R.G.)
* Correspondence: patricia.townsend@wsu.edu

Received: 19 September 2018; Accepted: 18 October 2018; Published: 20 October 2018

Abstract: Advanced Hardwood Biofuels Northwest (AHB), a USDA NIFA-funded consortium of university and industry partners, identified southwestern Washington as a potential location for a regional bioproducts industry using poplar trees (*Populus* spp.) as the feedstock. In this qualitative case study, we present the results of an exploratory feasibility investigation based on conversations with agricultural and natural resources stakeholders. This research complements a techno-economic modelling of a hypothetical biorefinery near Centralia, WA, USA. Interviews and group discussions explored the feasibility of a poplar-based bioproducts industry in southwestern WA, especially as it relates to converting land to poplar farms and the potential for poplar to provide ecosystem services. Stakeholders revealed challenges to local agriculture, past failures to profit from poplar (for pulp/sawlogs), land-use planning efforts for flood mitigation and salmon conservation, questions about biorefinery operations, and a need for a new economic opportunity that "pencils out". Overall, if the business model is convincing, participants see chances for win-win situations where landowners could profit growing poplar on otherwise low-value acreage and achieve ecosystem services for wastewater or floodplain management.

Keywords: woody bioenergy crop; social acceptance; short rotation coppice; bioeconomy; biorefinery; land-use; agriculture; crop adoption; wastewater; floodplain

1. Introduction

Hybrid poplar trees (*Populus* spp.) have a long history of cultivation in the Pacific Northwest (PNW) for a variety of uses [1,2]. In the 1980s and 1990s, poplar was established for pulp and paper. In the 2000s, the poplar tree industry transitioned to saw log and peeler log production due to poor market prices for poplar chips. Looking toward future markets, poplar could be a feedstock for bioproducts, including fuels and chemicals [3]. "Poplar for biomass" trees would be grown as a perennial agricultural crop, harvested on two- to three-year rotations (i.e., coppice) [4].

Another use of poplar trees is to provide ecosystem services. Poplar is capable of phytoremediation (using plants and their associated microbes to remove contaminants), making it effective for environmental clean-up (e.g., land reclamation) [5–7]. Many municipalities in the PNW use poplar plantations to aid with wastewater and biosolids management [8–10]. Poplar can also be planted as buffers along degraded river systems to treat nonpoint-source pollution, provide shade, prevent erosion, sequester carbon, lower peak flood flows, and provide recreation and aesthetic benefits [9,11–13].

Since 2011, Advanced Hardwood Biofuels Northwest (AHB, a consortium of university and industry partners led by the University of Washington) has been laying the foundation for a poplar-based bioeconomy in the PNW (hardwoodbiofuels.org). Originally focused on poplar for biofuels, the project expanded into exploring poplar's joint potential to provide biomass (for various bioproducts) and ecosystem services. This could lead to win-win opportunities by improving the local environment, decreasing the cost of biomass production, and benefiting rural economies.

Based on models of land-use, crop prices, poplar growth, and other factors, AHB identified southwestern Washington as a potential location for a regional poplar bioproducts industry [14]. Agricultural and pasture land in southwestern WA could produce reasonably-priced hybrid poplar biomass. Techno-economic modelling indicates many possible sites in this region for a moderately-sized (250,000 dry tons of biomass/year), cost-effective biorefinery. If the biorefinery could be co-located with the TransAlta power plant in Centralia, WA, USA there is an opportunity for major capital savings. The biorefinery would save costs by utilizing the power plant's excess boiler capacity or low-pressure steam that has modest value for power production.

If poplar farms can simultaneously provide ecosystem services for southwestern WA, this would increase the area's potential for a poplar-based bioeconomy (Figure 1). In particular, wastewater and floodplain management applications show potential. The Chehalis Regional Water Reclamation Facility already utilizes poplar, irrigating a 176-acre poplar tree plantation with reclaimed water during times when the Chehalis River has low flow, typically April through November [15]. Flooding is a major concern in the region, as the Chehalis River Basin has experienced multiple devastating floods in the past 30 years. Researchers are investigating whether land suitable for poplar plantations exists near other wastewater treatment facilities and are exploring how poplar can be incorporated into floodplain modelling.

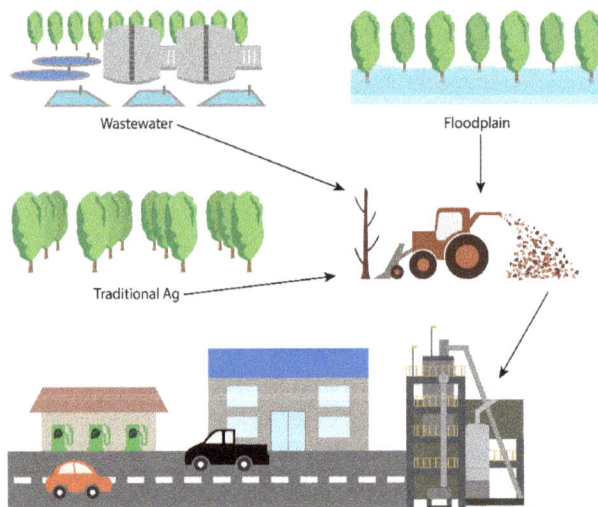

Figure 1. Potential sources of poplar biomass for a biorefinery, including traditional agriculture and ecosystem services plantings.

While techno-economic feasibility is critical, so is practicality and social acceptability. Our study takes a first step toward evaluating whether a poplar bioeconomy is practical and socially acceptable from the perspectives of people from the local agricultural/natural resource community in southwestern WA. We conducted a qualitative case study for a hypothetical poplar-based bioindustry centered around a biorefinery in Centralia, WA, USA. We used interviews and group discussions with agricultural and natural resources stakeholders to:

1. capture information about the specific context of the study region,
2. identify opportunities, challenges, and questions related to growing poplar for a bioproducts industry, and
3. explore the feasibility of incorporating ecosystem services.

Within the PNW, there are few publications on stakeholder views about poplar biomass [16,17]. A survey of WA landowners found that 36% of landowners are at least somewhat willing to grow poplar [17]. Willingness to grow poplar was connected to interest in new crops and general interest in bioenergy crops, which in turn was correlated to profit, tradition, and soil preservation.

Based on studies of other bioenergy crops in other regions, potential growers may be concerned about the status of the biomass market, failures of past introduction of new crops, unfamiliarity with the cropping method, and uncertainty about financial security [18,19]. For short rotation woody crops, farmers may not feel that the crops fit their identity, lifestyle, farming culture, and prioritization of food production [20]. Other major barriers to a new bioeconomy are likely to include concerns related to the biorefinery (e.g., increased traffic, water usage) [21] and economic challenges for commercial-scale biofuel production [22].

A new poplar bioeconomy needs a foundation of dedicated growers, receptive communities, interested biorefinery investors, and enthusiastic policy makers and community leaders. To build this foundation, we need to understand the values, needs, and concerns of those stakeholders. This stakeholder assessment is the first published qualitative study of perspectives on poplar as a biomass feedstock and ecosystem services provider, to the best of our knowledge. Specifically, we explore whether a poplar bioeconomy would fit the needs, values, and context of southwestern WA based on the knowledge of local agricultural and natural resources stakeholders.

2. Materials and Methods

2.1. Study Region

The study region (Figure 2) roughly matches a techno-economic assessment model of a hypothetical biorefinery adjacent to an existing power plant (TransAlta, Centralia Generation) in Centralia, WA, USA [23]. The biorefinery would be supplied by poplar trees produced within 100 km (62 mi) of the site. Although poplar tree farms existed in greater abundance in the past, poplar is still currently grown in the study region. Two example sites are the City of Chehalis plantation for the application of reuse water and a 4800-acre commercial plantation in Clatskanie, OR, USA, owned by GreenWood Resources Inc. (Portland, OR, USA).

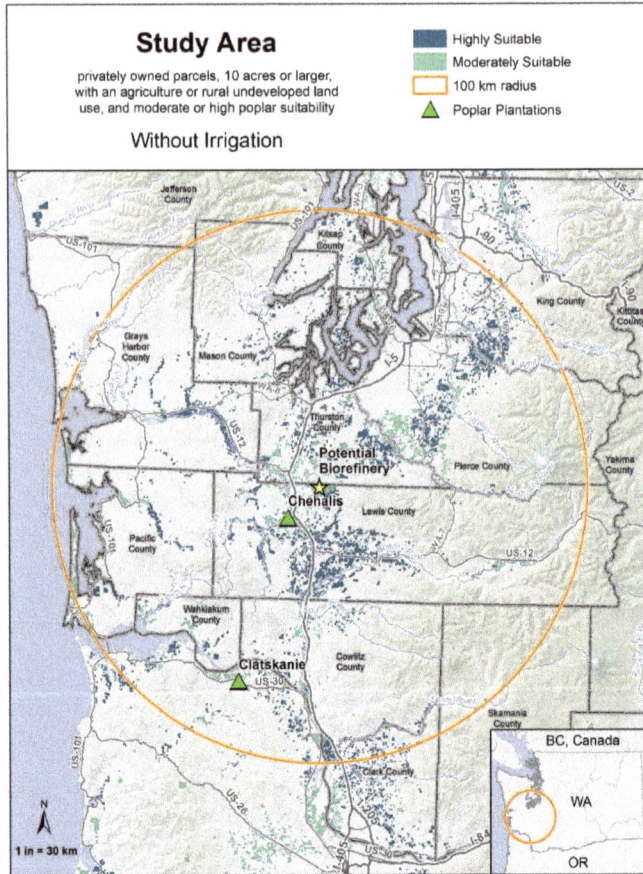

Figure 2. Parcels suitable for poplar growth without the use of irrigation within 100 km of a potential biorefinery in Centralia, WA, USA [24]. (Note: The techno-economic assessment used 100 km road distance between parcels and the potential refinery. We used a 100 km radius straight-line distance, making our study region larger.) Suitability is based on soil type, topography, and other geophysical site characteristics. Most parcels labelled as moderately suitable become highly suitable with irrigation. Map produced by Andrew Cooke, Natural Resources Spatial Informatics Group, Precision Forestry Cooperative, University of Washington.

2.2. Interviews and Group Discussions

We conducted 16 semi-structured interviews and two group discussions (one focus group and one short group discussion) with agricultural and natural resources stakeholders. By using qualitative research methods (e.g., interviews), we were able to explore a breadth of values and gain a deeper understanding of participants' rationales [25]. This is in contrast to quantitative techniques (e.g., surveys), for which participant responses are limited to preselected options or short answers.

Interviews took place in-person or over-the-phone between January–May 2018, and typically lasted 30–60 min. The focus group was conducted in January 2018 during a Chehalis Basin Partnership meeting and lasted 90 min (Appendix A). The short group discussion took place during 20 min of the February 2018 Lewis County Farm Bureau meeting. In total, we heard from approximately 33 participants.

Interviewees were recruited through purposeful sampling and snowball sampling [25]. Stakeholders came from: environmental consultanting agencies, a county conservation district, WSU Extension, farming operations, non-profit conservation/landowner organizations, WA Department of Ecology, economic/land development interests, and wastewater treatment plants (Table 1). Nearly all stakeholders resided or worked on projects in Lewis or Thurston County. To learn about some of the other counties in the study region, we spoke with one stakeholder about Pierce County, another about Cowlitz County, and some stakeholders in the Chehalis Basin Partnership spoke about Grays Harbor County. Although members of tribal nations are important stakeholders in the study region, we did not seek out direct participation from tribal nations due to time and budget constraints.

Table 1. Number of individuals interviewed in each stakeholder group *. These numbers do not include participants in the focus group or short group discussion unless they were also interviewed.

Stakeholder Group	Number of Interviewed Participants
Environmental Consultanting Agencies	2
County Conservation District	2
WSU Extension	3
Farming Operations	3
Non-Profit Conservation/Landowner Organizations	2
WA Department of Ecology	1
Economic/Land Development Interests	2
Wastewater Treatment Plants	3

* There are more individuals [18] than the number of interviews [16] because two interviews had two participants.

Interviewees were provided background information factsheets (Appendix B), and group discussion participants were given a presentation. An interview guide (Appendix C) structured the conversation with flexibility to explore ideas and expertise of the participants. We asked our participants to consider three different applications of poplar farms—traditional agriculture, wastewater management, and floodplain management. To explore the potential for an industry, we asked participants to assume that a market was available for the poplar. We focused on the acceptability and likelihood of poplar farming, and we also noted comments related to the potential biorefinery. We included questions about both challenges and opportunities to address potential bias toward optimistic or pessimistic views that may have otherwise arisen given our personal experience as AHB researchers.

The conversations were recorded and then fully transcribed (see Appendix A for exception). The transcripts were coded using descriptive and initial (i.e., open) coding methods to sort the data based on recurrent topics [26]. The codes were then further categorized based on our research objectives. Coding was conducted by a single researcher, using a word processor. The transcript data was reviewed and interpreted by a second researcher to ensure validity. Discussions with co-authors and a research assistant (Catherine Gowan, WSU Extension) refined the interpretation.

Participants voluntarily agreed to take part in the study and consented to an audio recording. The recordings were deleted upon completion of this manuscript. The WSU Office of Research Assurances found this project to be exempt from the need for IRB review.

2.3. Limitations

The information garnered is not generalizable to the entire population of the study region, or to all agricultural or natural resources stakeholders in the region. Given time and budget constraints, we were generally unable to interview more than one or two people from the same line of work. (This is especially relevant for the wastewater treatment case, as much of the pertinent information came from a single, highly-knowledgeable interviewee.) Instead, we looked for a wide assortment of natural resource and agricultural professionals to get a breadth of contextual information. This approach is justifiable for an exploratory study of variability in perspectives and as a precursor to further inquiry [25].

Our interviewees may have been more open-minded or optimistic compared to other stakeholders, as participants may have been more curious, interested individuals. We did not independently verify the veracity of information provided by participants; the results reflect their best understanding.

3. Results and Discussion

The following section presents what was learned from our interviews and group discussions. Participants' overall perspectives about supplying poplar biomass for a biorefinery (Table 2) are presented first. These thoughts on feasibility apply across the different applications, whether poplar is grown for agricultural, wastewater management, or floodplain management purposes. Quotes and discussion are inserted throughout this section. Quotes were selected for illustrative purposes and are not attributed to specific stakeholder categories because we often heard similar sentiments expressed across groups. We edited quotes for brevity when it was possible to do so without losing meaning.

Table 2. Summary of participants' thoughts on general benefits and concerns about the development of a poplar-based bioproducts industry in southwestern Washington, USA.

Benefits	Concerns
Could help the struggling agriculture community	Will this poplar-based industry actually "pencil out"
Faster turn-around than other tree crops	Can damage from wildlife be prevented
Could be easier to manage than alternative agriculture options	Will harvesting equipment be available
Could work on wet, marginal land	Who will make the upfront investment
Comparatively low soil disturbance	What is coming in and going out of the biorefinery

Following the overview, we detail specific results for poplar grown for traditional agricultural, wastewater management, and floodplain management. Results for each application are presented, with example quotes in three tables. Following each table, a discussion is presented on what was learned from participants about local context, opportunities, challenges, and questions.

3.1. Overall Perspectives on Supplying Poplar Biomass

The first reaction of many stakeholders was to tell us that the region has tried in the past to make money growing poplar and it did not work out for them. About 15 years ago, farmers planted poplar in anticipation of highly-profitable markets (e.g., pulp, sawlogs) that never emerged. Stakeholders questioned how this proposed poplar industry would be different: *"You know ... poplar is one of these things that won't go away. We've been doing poplar off-and-on for 30 years in the area, and unfortunately it's always ended up being a bust, a loser."*

However, stakeholders also said they would be interested in growing poplar if they knew they could make money doing so. The primary concern was whether the poplar-based industry "penciled out": *"I think it would be a very easy area to get people to convert whatever they have going if you could show them that they could make money."*

People acknowledged that poplar grew well in the area and that a new crop market could be very helpful for the agricultural community. Short rotation poplar was appealing as a crop because it:

- has a fast turn-around for a tree crop (3 years)

  "I think [short rotation poplar] would appeal to some people ... because the poplars before ... they were eight years, nine years, it was a really long timeline before you 'got your payout'."

- requires less labor and input than row crops and livestock production

  "Once that crop is in the ground, you have very little requirement for cultivation or no requirement for cultivation...So something that could be more planted and left just like an annual forage crop, hay or alfalfa. Those kinds of things could have a lot of possibilities."

- might provide income on marginal, wet farmland

  "There's lots of marginal areas in Lewis County, and it really can't crop anything. I don't know if you can even pasture it. It's just so wet ... and those are the areas something like that'd work."

- does not tear up the soil (after initial planting)

  "I really like this idea where you're not even taking it out of the ground. Especially from a land standpoint cause you're not destroying the soil, the ecosystem as far as the soil goes and everything."

Wildlife damage from elk, deer, and beaver is a major consideration. Some former poplar growers experienced severe devastation: *"[The elk and deer] ate 30,000 trees of 60,000 in the first year."* While elk fences may be an option, stakeholders commented that fences are an additional expense and will not hold up in areas with regular flooding: *"Winter's typically when we got our damage [in the poplar fields]. Any fence is going to get washed out with flooding."* Elk damage is a concern for agriculture more broadly, as farmers are also having trouble elsewhere in the PNW [27].

Participants wanted to know about the availability of harvest equipment. There is not equipment in the region currently and purchasing would be costly: *"Somebody's got to be financing it. ... you get a bunch of people to plant it and then nobody's got the money to buy the harvester, then you got a problem."* However, one person we spoke to saw this as an additional business opportunity. They felt that there would be interest in starting a harvesting contracting company: *"I know industrial people here that would just clamor to get an opportunity to buy the equipment that could do the harvesting and that's a whole industry, just harvesting."*

Many stakeholders seemed to feel that the biggest hurdle would be securing the initial investments necessary to start the industry: *" ... seems like it would take a major initial investment and whether that has any validity to it at this point ... I certainly don't know. I don't know if anyone knows."* Although there are investors interested in economic development, they would need to be sure of a return: *"This county's very, very conservative. You've got to show them the wares before they'll invest a nickel of their money."* Failures from past ventures make growers wary of poplar, and most would first need to see money invested in poplar-related infrastructure. That said, an agricultural consultant noted that farming is always a risky venture: *"Farmers are gamblers. They're the biggest gamblers in the world. Because you have not only weather, you've got markets and the government and all of those things can affect what happens to your crops."*

The stakeholders we spoke to did not think that poplar farming would influence the general public, or if it did, the reception would be positive. They informed us that the region is familiar with tree farms (e.g., Christmas tree farms) and has a long history of forestry and farming. They felt the biorefinery would be of greater concern to the local community. Poplar biomass can be made into a variety of bioproducts, and we did not specify a particular output for the biorefinery during our interviews and discussions. Participants felt that the local community would want to know what the end product is, as well as what is transported in-and-off the site, what by-products or wastes are generated, and any impacts on air or water quality: *"They're going to want to know what you're bringing into the site, what you're producing on the site, and what you're taking off the site."* A participant shared with us that public health is a priority in their particular community: *"I think it comes down to what are you doing to the community. As far as putting it in the air and the water, that's what they are concerned about."* These are common concerns for industrial-scale biomass facilities expressed by local communities [28,29]. In addition, perceptions about biorefinery impacts could affect growers' decisions: *" ... certain people won't grow things if it's going to create something that could be harmful."*

3.2. Poplar and Agriculture

Poplar could be an alternative crop available to farmers to support their livelihood. We asked stakeholders about: (a) the state of local agriculture, (b) where participants saw opportunities and challenges, and (c) the questions they would need answered. A summary of the main points and associated quotes are presented in Table 3, followed by a discussion of the results.

Table 3. Summary and quotes related to the potential for poplar farming to contribute to agriculture in southwestern Washington, from participant stakeholders in agriculture and natural resources.

Local Context	Quotes about Poplar and Agriculture
(a) Agriculture is in trouble (hard to make a living; expanding development pressures)	(a) *"Many of the folks that I knew and their families that farmed for a living twenty years ago ... farming is [now] secondary to some other line of work, the day to day"* (a) *"One thing that Pierce County has been inundated with are the increase [sic] in the number of warehouses and distribution centers."*
(b) Integral to region's rural history and character	(b) *"Like the small community feel ... the amount of land one person can have to themselves, agriculture opportunities, and closeness to family."*
(c) Past poplar ventures (for pulp/sawlogs) failed because of a lack of markets	(c) *"[Poplar] isn't news. This was tried on the harbor, there's a half a dozen different places in Grays Harbor that I know of, and it didn't prove profitable for the people."*
Opportunities	
(a) There's a need for a profitable crop and local markets	(a) *" ... because of the loss of infrastructure and the basically desperate nature of the farmers in this area, they'd be open to a crop if it pencils because they're getting so limited...I mean we're down to forage. It's the primary crop here and you know there's a limit to how much forage you can move."*
Challenges	
(a) Need payment assurance	(a) *"My expenses would have to be covered. Because it's not cheap."*
(b) Could clog tile lines	(b) *"Our cottonwood trees were actually getting down in the tile drains and filling them completely up with their roots. So we had to put in some open ditches to continue to drain the field."*
(c) Could present complications with irrigation rights	(c) *"What they do on the water rights is they'll say you get two acre feet per acre, but if you grow a crop that only takes one acre feet per acre, you didn't keep your records and they say that's your water right."*
(d) Some hayland/pasture is needed for livestock and horses	(d) *"I probably wouldn't want to plant poplar on my little bottom land. Although, wouldn't be a bad idea. It's just that I have horses, so they don't eat poplar, they eat grass."*
Questions	
(a) When would market saturation happen?	(a) *" ... a lot of these people come off with something new and bright and the earlier adopters will make a little bit of money but once it saturates so fast, that the price falls out and there's no money to be made out of it."*
(b) How many acres would an individual need?	(b) *"How many acres of trees would you estimate you would need to make a profit [as an individual]? The average farm in Lewis County is 49 acres. Most of the farm area you've got is very small acreage."*

3.2.1. Poplar and Agriculture: Local Context

Resident stakeholders expressed enjoyment of the small-town, slow-paced lifestyle and liked outdoor recreational activities, such as hiking, fishing, and boating (Figure 3). However, many participants said that the rural areas are economically depressed, suffering from poverty, downscaled industries, and unemployment. Lewis, Grays Harbor, Mason, and Wahkiakum Counties ranked in the bottom quarter of Washington State counties for per capita personal income in 2015 [30]. Participants said the agricultural sector, traditionally a primary source of livelihood in the region and part of the local character, has been hit hard with challenges (e.g., lack of infrastructure/markets, catastrophic flooding, housing development, and aging landowners). We learned from the stakeholders that a local frozen foods company processing peas and corn recently decided not to take produce from western Washington, leaving growers struggling to figure out what to do instead: *"This year we've been told the cannery isn't going to take any crops on the west side [of WA state]. Our crop growers are going, 'what are we going to grow?' I mean we can't grow the corn, the sweet corn, the peas, or the beans, which have all been harvested here before."* Participants also mentioned that other parts of our study region are increasing warehouse and housing development on former agricultural land, particularly Pierce County and along I-5.

Figure 3. A rural landscape in Lewis County in southern Washington, USA.

3.2.2. Poplar and Agriculture: Opportunities

If a poplar market emerged, it could be a useful aid for the struggling local agricultural community. Farming consultants and partners said there are not many options when it comes to selecting a profitable crop. Participants said the farmland is less than ideal, there is a lack of agricultural infrastructure and markets, and existing markets are saturated. Additionally, interviewees from or working with the agricultural community told us that many farmers are of retirement age without willing heirs to take over the family business, and young people cannot afford to buy those farms. These challenges facing agriculture are not unique to southwestern WA. An aging farm population and dependence on outside income is a national trend [31].

When asked if people would consider poplar if they could make money doing it, people indicated that they would be interested. As a participant noted, the region grew poplar before and would again if it made financial sense. Agricultural and natural resources professionals did not think it would be difficult to find enough suitable land and secure willing growers if poplar were profitable. This suggests that a clear business model would facilitate further discussion with landowners and that there is not strong opposition to growing poplar based on other factors. In contrast, Warren et al. [20] found that farmers in southwest Scotland felt short rotation woody crops did not fit with their identity, lifestyle, farming culture, and prioritization of food production. The difference in response in our study from Warren et al.'s may be a result of cultural differences, or it could be a further indicator of the *"desperate nature"* of farmers in southwestern Washington.

Given the difficulties facing agriculture our participants laid out, we see a number of ways a poplar biorefinery could present an opportunity to boost the local agriculture community. The biorefinery would create a new market and provide farmers with another cropping opportunity, even on marginal farmland. With the introduction of a potentially growing market, there could be openings for new people to get into agriculture. As a comparatively low-maintenance crop, poplar may also be appealing to older farmers.

3.2.3. Poplar and Agriculture: Challenges

As with other studies of willingness to adopt bioenergy crops [17,32,33], profit is a leading decision driver. A primary challenge would be providing reasonable assurance that poplar would be profitable. This could involve contracts between the farmer and the biorefinery [32], which a lending institute for the facility is likely to require.

Farmers with tile lines may need to switch to different draining methods for their fields. A participant from a local conservation district shared a story about how poplar trees clogged a field's tile lines on a property they worked with and alternative means of drainage were needed. Worries about tile lines could preclude some landowners from growing poplar: *"A lot of these tile line places are going to be like, we don't want to put those in, they'll break our tile lines."*

While flooding is a concern for farmland during the winter months, the region faces droughts during the summer months. Participants explained that irrigation rights are valuable property assets because no new water rights are allowed and existing rights can be lost if "beneficial use" is not demonstrated at least once every five years: *"[Department of Ecology regulators] always try to find a way to take some away. It's really hard for people to prove they've used their whole water right."* Those landowners who have irrigation water rights would need to think about how to maintain that value. The poplar trees may not require irrigation, and irrigation rights could be lost if not utilized. Irrigation can boost poplar yields, but it would be up to the landowner to determine if this is a good use of their water right and if they feel confident it would satisfy requirements for "beneficial use." A stakeholder mentioned that a water trust is an option, or property owners could choose to sell their water rights.

Not everyone in the agricultural community will be interested in poplar. Some hay and pasture is needed to support livestock and horses in the region. For example, we spoke to a retired landowner who runs a small beef operation and a resident who owns horses for personal enjoyment. In other cases, a participant explained that land has been pasture for decades because it is easy to maintain and expensive to come back to after using the land for crops: *"What happens is it's very expensive to work up a field and plant crops and then turn it back into pasture . . . a lot of farmers are letting crops stay in grass for years and years. I was a kid here, and I know pastures that have never been worked up."*

3.2.4. Poplar and Agriculture: Questions

There are concerns that need to be addressed, beyond needing assurance that poplar will be profitable. A lot of the farms in the study region are smaller parcels (<50 acres) [34]. Stakeholders wanted to know, from an individual's perspective, how much land a farmer needs to grow poplar profitably. On the other extreme, potential growers want to know at what point the market would become saturated from too much poplar farming. Southwestern WA has experiences with market saturation problems, for example the current glut of cranberries [35].

3.3. Ecosystem Services: Wastewater Management

Poplar can be a tool for wastewater management by evapotranspiring away reuse water, utilizing biosolids, or further treating wastewater for nutrients and contaminants [10]. We used the interviews and group discussions to learn from stakeholders about: (a) the current use of poplar for wastewater in the study region, (b) where participants saw opportunities and challenges for additional wastewater poplar plantations, and (c) the questions that need to be addressed. A summary of the main points and associated quotes are presented in Table 4, followed by a discussion of the results.

Table 4. Summary and quotes related to the potential for poplar to provide ecosystem services for wastewater management in southwestern Washington, from participant stakeholders in agriculture and natural resources.

Local Context	Quotes about Poplar and Wastewater
(a) Economical way to meet regulations	(a) *"It was either [the poplar plantation or] remove all of [City of Chehalis's] water during the summertime some point well north of Centralia, which would have cost way too much money."*
(b) Permits and irrigation are burdensome	(b) *"[There is] a permit for going out to the tree farm and one for going to the river. And going to the river is a lot easier to meet."*
(c) Provides other ecosystem services	(c) *"Aesthetically it's a draw … a lot of people want to take their wedding pictures out there."*
Opportunities	
(a) People like this idea	(a) *"I like the wastewater one. I think that's a terrific idea."*
(b) Treatment plant retains the reuse water rights	(b) *" … we reserve the water rights. So [if] some industry wants to potentially use reclaimed water, it's ours, we can do what we want with it."*
(c) Water stays in the immediate watershed	(c) *"We're keeping the water in our local basin."*
(d) Extracting excess nitrogen	(d) *" … the groundwater is actually improving … Before it was a farm, potentially mismanaged, nitrogen leaching through the soil."*
(e) Potential biosolids applications	(e) *" … stigma from the general public is they don't want [biosolids] to go on [food] crops … whereas with trees, nobody cares."*
(f) Two other cities expressed interested	(e,f) *"We now ship our biosolids during the wet months at a high cost."* (f) *"We may need to be looking at possibly not discharging our effluent to the [river] in a few months of the year."*
Challenges	
(a) Plants unlikely to consider it unless necessary	(a) *"I don't know about the other treatment plants being forced out [of the river]. We're kinda one of the few in western Washington."*
(b) Plant is responsible for groundwater monitoring	(b) *"I'm not going to want to give [reuse water] to farmers because the city's responsible for the ultimate disposal or reuse."*
(c) Economic concerns	(c) *"One concern might be the cost for start-up to get the farm rolling with poplar trees."*
(d) Regulations could shift to river water quantity over quality.	(d) *" … hoping sometime in the future they'll say you can put your water back in the river. At that time the plantation just becomes a bunch of trees."*
Questions	
(a) Can coppice poplar suck up enough reuse water?	(a) *"The question is would that poplar, because most of those you see grow into big trees, they don't cut them every three years, would it still work for wastewater management?"*
(b) Can overhead irrigation (rather than ground sprinklers) work?	(b) *"Irrigation is very labor intensive. [If] the trees wouldn't get that high, 30 feet at the most, and we could put some type of large irrigation pivot, it would be a lot easier to irrigate."*

3.3.1. Poplar and Wastewater: Local Context

As previously mentioned, the City of Chehalis is using a 176-acre poplar tree plantation to meet total maximum daily load (TMDL) restrictions placed by the WA Department of Ecology for the protection of river water quality (Figure 4). When the river drops below 1000 cubic feet per second, which normally happens between April and November, the reclamation facility cannot discharge to the Chehalis River. Instead, the water is used to irrigate the poplar trees.

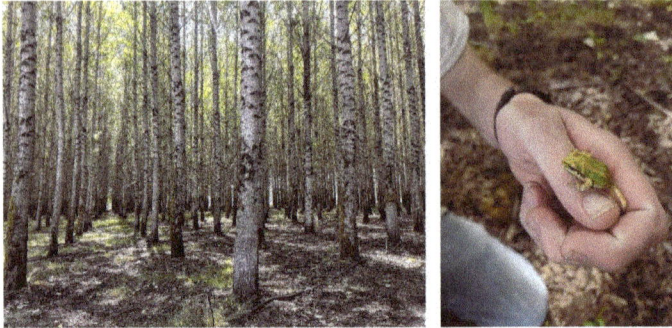

Figure 4. The City of Chehalis Poplar Plantation, where the trees are irrigated with reuse water during the dry months, has other environmental benefits like wildlife habitat.

A person closely tied to the wastewater treatment facility said that developing the poplar tree farm was the right choice for the City of Chehalis: " ... *the poplar tree plantation was the cheapest and the most beneficial option that we chose and it's still a good decision today.*" They informed us that the farm allows Chehalis to meet the TMDL restrictions without having to pump the water far from the treatment plant (which was the alternative). Also, they noted that the farm is an aesthetic draw, carbon sink, nutrient extractor, and wildlife habitat.

However, there are challenges to running the poplar plantation. The participant shared that the poplars are destined for the sawlog market but are still not big enough to sell despite being older than they thought they would need to be. The trees are now fifteen years old, pushing the plantation into a forestry rather than agricultural land-use and thus requiring different permitting. Additionally, the participant noted that permit for applying reuse water to the trees is more burdensome than the permit to discharge to the river, and the sprinkler system (Figure 5) needs constant maintenance to deal with coyote damage and clogging.

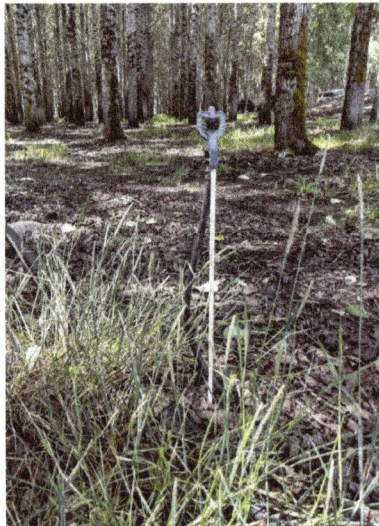

Figure 5. One of the 20,000 sprinkler heads used to irrigate the Chehalis Poplar Tree Plantation with reuse water.

3.3.2. Poplar and Wastewater: Opportunities, Challenges, and Questions

From a social perspective, the idea of combining wastewater and poplar farming was well received. However, those participants most familiar with the concept of using poplar to process reuse water questioned the ability of short rotation coppice poplar to take up enough water to get the job done. At this time, it is unknown whether a coppiced poplar farm could meet the needs of the wastewater treatment plant, but preliminary research at AHB's demonstration site in Hayden, Idaho showed similar average irrigation rates between coppiced and 20+ year-old poplar [36].

A participant noted that a potential benefit of using short rotation coppice poplar over using longer-rotation poplar is that the trees will be shorter, and overhead irrigation might be a possibility. They felt overhead irrigation could avoid the maintenance issues associated with a ground-level sprinkler system, like the coyote-chewing damage faced by the Chehalis plantation. Research would be needed to evaluate this idea. Based on conversations with outside experts, potential concerns about overhead irrigation include impacts on tree health and technical feasibility, but overhead irrigation may also facilitate a good amount of direct evaporation into the air. We should also note that some of the characteristics of the current longer-rotation plantation (e.g., aesthetics) would be different under a short rotation coppice system.

Although the Chehalis Poplar Plantation does not apply biosolids to their poplar trees, we did speak to a landowner who currently uses biosolids on his crops and would consider applying biosolids if they grew poplar. When prompted to think about biosolids, participants noted that some members of the public are adamantly opposed to applying biosolids to food crops and thought that using biosolids on a biomass crop like poplar might be more acceptable. One city told us they ship their biosolids to Eastern Washington part of the year at a high cost and would consider an alternative like operating a poplar farm on their land. There are examples of other wastewater treatment plants in the Pacific Northwest applying biosolids to poplar, like the Biocycle Farm in Eugene/Springfield, OR [37]. Another treatment plant asked whether there would be leftover poplar chips, and if so, whether they could be used in their biosolids compost operation.

A wastewater professional informed us that discharging cleaned wastewater to rivers or groundwater is easier to permit and cheaper to do than irrigating a poplar farm. Therefore, one challenge may be finding enough wastewater treatment plants interested in growing poplar to meet any significant portion of a biorefinery's poplar demand. That said, we also heard from a city that recently learned they may not be able to discharge their water into the river in the future, similar to the restrictions imposed on the City of Chehalis that lead to the establishment of the Chehalis Poplar Tree Plantation. Furthermore, although not mentioned by the participants, poplar may be able to remove chemical contaminants, like pharmaceuticals, that wastewater treatment does not capture [9,10].

3.4. Ecosystem Services: Floodplain Management

Poplars are adapted to live in riparian and flood-prone areas. Poplar may serve as flood mitigation tool by increasing landscape roughness and providing a more suitable crop for farmers in flood prone areas. An EPA review from 1999 predicted there would be an increase in hybrid poplar cultivation in degraded floodplain to supplement native hardwoods and to treat nutrient runoff [38]. A watershed management project in Minnesota proposed using hybrid poplar plantings to improve water quality, reduce excess stream flows, and provide farmers with a profitable crop [39]. In 2005, a case study explored whether a poplar riparian buffer could be economically sustainable for blueberry farmers in western Washington [40].

We used the interviews and group discussions to learn from stakeholders about: (a) local flooding issues and planning efforts, (b) where participants saw opportunities and challenges for poplars in the floodplain, and (c) the questions that need to be addressed. A summary of the main points and associated quotes are presented in Table 5, followed by a discussion of the results.

Table 5. Summary and quotes related to the potential for poplar to provide ecosystem services for floodplain management in flood-prone southwestern Washington, from participant stakeholders in agriculture and natural resources.

Local Context	Quotes about Poplar and Floodplains
(a) Floods are frequent and can be devastating	(a) *"You know it's a big deal when you shut down I-5 twice a decade. It cost hundreds of millions of dollars 'cause north-south traffic is stopped."*
(b) Restoring native riparian habitat for salmon conservation (c) Massive collaborative planning strategy underway	(b,c) *" ... to protect the habitat and to do restoration at the same time as considering these flood damage reduction actions because obviously they both have an effect on the floodplain. It's an effort to move both of those pieces of this integrated strategy forward at the same time."*
(d) Concerns among landowners about losing private property rights	(d) *"There's a lot of uncertainty or suspicion about regulations and uses of land minimizing the tax base in the community."*
Opportunities	
(a) Great if people could make money off their land while decreasing flood risk	(a) *" ... doing something that would still give people use of their land that included the floodplain I think would be pretty favorably looked at."*
(b) Could have native riparian buffers and poplar farms beyond that	(b) *" ... restoring the floodplain forest and then...we could put a poplar plantation on someone's farmland in area where water's going to be coming out of the river channel, and that would effectively slow it down."*
Challenges	
(a) Don't want non-natives near the river	(a) *" .... I think, might be a little bit of a difficult sell in the sense that you're talking about hybrid, not natural species. Most of the folks that are involved in riparian stabilization are certainly going to be a little bit leery of using non-natives."*
(b) Could be a considered a risk to property rights, if poplar farms someday become regulated as buffer	(b) *"Okay, so we make this agreement, and I put some investment into planting [poplar]. And then the rules come along and they change and then I'm just out of another investment ... it gets people a little uptight. There's a lot of private properties right advocates from this area for that very reason."*
(c) Harvest window would be limited to when the ground was dry enough for equipment	(c) *"You're not going to harvest in the winter most of the ground that's suitable [for poplar]."*
Questions	
(a) How much inundation (frequency, depth, duration) can poplar handle?	(a) *"How well do these trees deal with flood? 'Cause they could easily be flooded 6 to 8 feet deep during a good storm, and I just don't know how well they deal with that."*
(b) Would hybrid poplar crossbreed with native poplars?	(b) *"We're spending literally millions of dollars getting rid of a variety of non-native and invasive species. If it crossbreeds with the native species creating something that's different, you could very easily be creating a new, non-native invasive species along our rivers."*
(c) Would hybrid poplar drain low-flow streams in summer?	(c) *"The low flows in the summer and early fall in this basin are a huge issue. And I know that cottonwood, from what I understand at least, take a lot of water. How's that gonna affect instream flows in the summer or would it?"*

3.4.1. Poplar and Floodplains: Local Context

A challenge in the study area is how to simultaneously manage the Chehalis Basin for flood mitigation and aquatic species protection [41]. Flood events in recent decades have severely impacted the cities of Centralia and Chehalis and forced closures of Interstate 5 (the highway linking Vancouver, British Columbia, Canada, Seattle, WA and Portland, OR). At the same time, there is need for aquatic species (e.g., chinook salmon) conservation efforts. Proposed actions for the Chehalis Basin could change land-use practices in the floodplain, where much of the farmland is located (Figure 6). Balancing salmon conservation practices and agricultural land-use impacts is a long-standing issue in western Washington [42]. Conservation requires willing landowners, and stakeholders shared that landowners in the area are wary of potential threats to private property rights.

Another element of the Chehalis Basin planning efforts is a proposal to build a dam. We expected to hear the proposed dam brought up during our discussions, as the dam is generating controversy in the region [43]. Tribal opposition was a major driver of exploring alternative or concurrent strategies for protecting aquatic species and restoring native habitat [44]. Although we did occasionally hear about the dam from stakeholders (e.g., *" ... they're proposing to build a dam, and who in 2018 ever thought that would be around, but it is because of the flood problems we have on the river."*), stakeholders seemed to be focused on efforts to restore native riparian buffers. This may be an artifact of the participant sample including many people connected to habitat restoration planning.

Figure 6. Lincoln Creek (**left**), part of the Chehalis Basin, runs through an agricultural property, and floods the fields during the winter. The Conservation Reserve Easement Program (CREP) offers financial assistance to landowners for the planting of native riparian buffers, like this planting near Lincoln Creek (**right**).

3.4.2. Poplar and Floodplains: Opportunities, Challenges, and Questions

Participants shared regulatory and ecological concerns about hybrid poplar because it is non-native. Certain regulations prohibit non-native trees in riparian restoration projects. From an ecological perspective, participants questioned whether short rotation hybrid poplar could support ecosystem functions in riparian areas (food for aquatic species, habitat from downed wood, erosion control, shading) and wanted to know if the hybrid poplar would crossbreed with native populations of poplar. In addition, there were concerns about whether poplar would reduce stream flows in the summer months.

Concerns about crossbreeding may be unwarranted for two- to three-year harvest cycles because the trees can be harvested before sexual maturity. Alternatively, the trees could be genetically modified not to flower [45]. Poplar trees can reproduce vegetatively (i.e., sprout from broken branches and other tree material), so there is some risk of escape beyond plantations. Native cottonwood could be used instead of hybrid poplar, but these trees will have lower yields.

Many of the points we heard from the stakeholders in our study region were also reflected upon in the 1999 EPA review [38]. For example, farmers could plant the trees just beyond critical riparian habitat as a cost-effective way to improve the degraded floodplain, but there was fear of crossbreeding with native cottonwood and depletion of water availability. When the 1999 EPA report was written, there were few studies specifically designed to address these environmental issues. This remains largely true twenty years later.

Another challenge of producing poplar on flood-prone farmland is that the harvesting window would be limited to times when the ground was dry enough to run equipment. This concern was raised by a farmer, who had seen video of short rotation harvest elsewhere in the country happening when the leaves were off and the ground was frozen. The AHB project encountered similar challenges at its demonstrations sites in western Oregon and Washington, where muddy ground in the winter forced harvest delays. While leaves can introduce unwanted heterogeneity to the poplar chips and reduce sugar availability [46], leaves may be unavoidable if harvest must occur when the ground has dried out in the late spring. However, technology to separate leaves from the biomass is under development [47]. If the biorefinery needed a continuous flow of materials, some drier ground would need to be converted, an alternative source of biomass (like hardwood sawmill residue) could be used, or storage methods would need to be developed.

Short-rotation hybrid poplar may also serve a strategic purpose for flood mitigation: *"generally speaking, a tree crop is going to serve useful functions during flood events."* We spoke with a consultant who is thinking about how hybrid poplar might be able to provide roughness that slows down floodwaters in a reconnected floodplain. Although the participants did not emphasize the proposed dam during discussions, we believe that poplar farms may offer an alternative that is a compromise between maintaining profitable use of private land along the river and providing more riparian ecosystem services than the agricultural land use currently in place.

However, stakeholders want to know how much inundation poplar can handle. Although poplar is more tolerant of wet (and dry) conditions than many other trees and crops, stressful growing environments limit the amount of biomass that can be produced. Research is needed to understand how well poplar would be able to produce biomass in very wet ground.

Another stakeholder saw a potential risk to landowner property rights if poplar is part of a floodplain strategy—it could someday be considered part of a larger agricultural buffer and be restricted from harvest. This participant told us that forest landowners suffered when forest practice rules changed, forcing them to remove part of their land from production. Farm landowners in Lewis County are not currently required to have buffers, but regulations could change as Chehalis Basin management progresses.

Circumstances may differ across counties in the study region. One participant said that the restrictions on floodplain development are more stringent in Thurston and Pierce Counties compared to Lewis and Grays Harbor Counties. Given strict limitations in Thurston/Pierce, they thought poplar may be appealing as a land-use option, while economic development in Lewis County (specifically Centralia and Chehalis urban growth areas) may preclude poplar in those floodplains.

4. Recommendations

Participants suggested ways of framing the potential poplar-based bioeconomy that would present the industry in a positive light. One person proposed saying "cellulosic biomass" rather than starting with poplar, due to experience with past poplar market failures: *"I wouldn't run out with poplar on the top line of my banner ... [but] this is a timber county. So if you say, 'hey we're refining fuels from cellulosic materials,' they've heard that before."* Others recommended focusing on how producing poplar could be a desirable choice for private landowners and the community, rather than a top-down, mandated change for the floodplain: *" ... what we've found in other areas working in the floodplain is that people's first reaction is, 'It's my land, and I don't want anybody telling me what I can and can't do with it' ... They don't want to just give it away, but, if there's a better use for that land, that's usually [the] kind of theme that resonates with them."*

Another area where framing is important is related to poplar and the floodplain. An environmental consultant advised being clear about what part of the floodplain is being discussed. A different stakeholder observed: *"There's a difference between riparian and floodplain, especially in the lower stretch of the river. I wouldn't market [hybrid poplar] so much as for a direct buffer river habitat benefit, but if you're talking about 'hey, here's a crop of value' ... It's not like peas or wheat are sitting in there because of the environment it's in."* They felt that talking about poplar as a riparian buffer would limit the potential because of concerns over ecological restoration. However, poplar farming in the non-riparian floodplain areas may be more acceptable and may actually improve the ability for conservation and restoration by improving the overall economy.

Potential strategies for addressing some of the major challenges presented in this paper are listed in Table 6.

In addition to addressing the challenges, there are a number of remaining knowledge gaps. We recommend conducting a thorough literature review tailored to the specific questions posed by the stakeholders. Education materials could be developed from this review and tailored to the situation in southwestern Washington. We also recommend eliciting opinions on the bioeconomy and specific end products, combined with public outreach and education. Bioenergy, particularly corn ethanol, is a contentious issue [32,48], and it is important to acknowledge the role emotions and terminology play in stakeholders' acceptance of a potential bioeconomy [49,50].

Table 6. Strategies for addressing challenges to developing a poplar-based bioeconomy in southwestern WA, USA.

Challenges	Ways Forward
Damage from wildlife	(1) Test if short rotation coppice poplar experiences the same level of damage (2) Explore fencing options, particularly in flood-prone areas (3) Develop unpalatable varieties or repellents
Harvest operations	(1) Establish a framework for a harvesting cooperative (2) Develop a harvesting company business model (3) Explore chip storage or alternative feedstock options to supply the biorefinery during times when the ground is too wet for the harvester
Initial investments	(1) Develop a detailed business prospectus for potential investors and economic development stakeholders (2) Collaborate with Lewis County Economic Development Council, TransAlta, policy makers, potential investors, and private industry (3) Demonstrate a small-scale refinery using existing poplar farms (4) Explore carbon markets
Biorefinery concerns	(1) Be upfront and clear about the biorefinery's end product and what it is used for (2) Explain potential community benefits and impacts using examples of biorefineries elsewhere in the PNW or US. (e.g., biodiesel and ethanol plants)
Wastewater irrigation	(1) Establish a test site for overhead irrigation of short rotation coppice poplar to see how much recycled water could be processed by the trees and whether applying recycled water to the leaves impacts tree health

Further development toward a poplar-based bioindustry will require more community involvement. Our preliminary exploration of opportunities, barriers, and values only accounts for a subset of the natural resources and agricultural stakeholders. The next step in stakeholder participation would be reaching a wider diversity of people and evaluating the level of concrete interest in pursuing the industry (rather than theoretical feasibility). The latter will require sharing available economic information and more detailed growing guidance.

5. Conclusions

Given a convincing business model, agricultural and natural resources professionals believe it would be straightforward to find suitable land and secure willing poplar growers. Unsurprisingly, economics is the bottom line: who would make the initial investment, and would growers make money. According to participants, local agriculture is in desperate need of a boost, such as a new, lucrative crop market. The participants did not bring up concerns about short rotation poplar in relation to their identity, lifestyle, farming culture, or opinions about food production, unlike in a study mentioned earlier [20]. The region has experience growing poplar, albeit for a different industry, on longer harvest cycles, and without turning a profit.

Hybrid poplar would not be used as stream buffers, as non-natives are not desired or allowed for riparian restoration. Poplar farming could also displace some hay/pasture currently used to support livestock and horses, and could run into concerns about water scarcity, water rights, and drainage tile lines. Technical questions remain, such as how much flooding poplars can handle and how to prevent elk damage.

Overall, if economic objectives can be met, participants saw chances for win-win situations where landowners could profit on otherwise low-value agriculture/pasture land and achieve ecosystem services for wastewater or floodplain management. For example, a landscape strategy partnering native riparian buffers with adjacent poplar fields in wet bottomlands could potentially invigorate the agricultural community, protect landowners from flooding, and conserve fish habitat. However, elk may cause issues without a fence, and a fence may not work in the floodplain. Landowners would also need to be reasonably assured that their poplar plantation could not later fall under riparian conservation regulation and be lost as a private property asset.

Based on this preliminary stakeholder assessment, it seems like poplar could fit the values and meet needs of southwestern Washington, but a number of concerns and unknowns would have to be resolved. There are opportunities for poplar as an ecosystem services provider that require scientific research to verify feasibility. Next steps could include establishing local experimental poplar plantations, developing a detailed business prospectus, producing educational material about the biorefinery, and discussing with a greater number and broader diversity of stakeholders.

Author Contributions: Conceptualization, N.M.H., P.A.T., A.C. and R.G.; Formal analysis, N.M.H.; Investigation, N.M.H. and P.A.T.; Methodology, N.M.H. and P.A.T.; Validation, P.A.T.; Writing—original draft, N.M.H.; Writing—review & editing, P.A.T., A.C. and R.G.

Funding: We are grateful for funding provided by Agriculture and Food Research Initiative Competitive Grant No. 2011-68005-30407 from the U.S. Department of Agriculture National Institute of Food and Agriculture as part of the Advanced Hardwood Biofuels Northwest Project.

Acknowledgments: We would like to give a special thank you to the participants for their time, openness, and thoughtful consideration. We appreciate the help of Kirsten Harma (Chehalis Basin Partnership) and Maureen Harkcom (Lewis County Farm Bureau) for arranging our presentations. Thank you to Cat Gowan (Washington State University Extension) for contributing to research design, data collection, analysis, and editing. Our thanks to Andrew Cooke, Luke Rogers, and Jeffrey Comnick of the Natural Resources Spatial Informatics Group, Precision Forestry Cooperative, University of Washington for their poplar suitability and parcel land-use work, and for helping us map the study region. Thank you to Amy Kim (University of Washington) for review of an earlier draft. The poplar scenario graphic (Figure 1) was produced by Briana Nordaker (Washington State University Extension) for the Advanced Hardwood Biofuels Northwest project.

Conflicts of Interest: The authors declare no conflict of interest. The funder had no role in the design of the study; in the collection, analyses, or interpretation of data; in the writing of the manuscript, and in the decision to publish the results.

Appendix A. Methodological Details

The focus group was conducted during a Chehalis Basin Partnership meeting and lasted 90 min. Thirteen attendees (of an audience of 17—not counting the researchers) contributed to the focus group. Participants were a mix of representatives from government, non-profit, and private citizens, who shared a common interest: *To implement a management plan that will result in effective, economical, and equitable management of the water in the Chehalis Basin to sustain viable and healthy communities and habitat conditions necessary for native fish* (http://chehalisbasinpartnership.org/). We provided a brief presentation on the research project at the beginning of the meeting, and then facilitated an exchange of thoughts and questions based on topics similar to the interview guides. We also requested they fill out a short questionnaire about their personal opinions on their communities.

The group discussion took place during 20 min of a Lewis County Farm Bureau meeting. A brief presentation was followed by participants sharing initial reactions, concerns, and ideas. In addition to generating interesting information in-and-of itself, the meeting allowed us to connect with individuals directly connected to agriculture in the study region, including some people who had grown poplar trees in the past. For the short group discussion, we only transcribed the participants' comments, rather than the full dialog with the facilitator.

Follow-up interviews were conducted with interested participants from the two group discussions.

Appendix B. Background Information for Interviewees

This information was provided to interviewees to review prior to the interview as an email attachment or hardcopy given in-person. Figures are not captioned, as they were not captioned in the original document.

[Note: During the interviews, we discussed the different reasons for growing poplar using the term "scenarios". However, these different reasons are not mutually exclusive (as the word "scenario" may imply), and we envision a future consisting of all three applications (poplar for farm revenue, for wastewater treatment, and for floodplain management). Therefore, in this article we chose to call them "applications" instead of "scenarios". In our conversations with participants, we explained that the "scenarios" were not mutually exclusive.]

Appendix B.1. Background: Growing Hybrid Poplar for Biomass

Poplars, a group of trees that includes black cottonwood, are the fastest growing trees in temperate regions. Hybrid poplars (crossbreeds of different types of poplar) are commonly found in nature and can also be created in a nursery. These hybrid poplar trees can produce lots of biomass and are adaptable to a wide range of sites. The trees resprout after harvest and can be cut on three-year cycles as a perennial agricultural crop. Hybrid poplar chips can be turned into bioproducts.

Appendix B.2. Introduction

Models have identified Lewis County and the surrounding area as having potential for a poplar-based chemical industry. Specifically, researchers are modeling the technical and economic feasibility of a hypothetical refinery in Centralia, WA, USA supplied by poplar grown within 100 km of the site. In all scenarios, we assume 34,500 acres of land is converted (equivalent to about 25% of existing pasture land).

- Scenario 1: Farmers convert pasture land (i.e., non-cropland, not hay) or cropland into hybrid poplar farms.
- Scenario 2: Wastewater treatment plants lease land for hybrid poplar farms.
- Scenario 3: Poplar farms or buffers are developed in the floodplain.

1

Traditional ag

2

Wastewater management

3

Floodplain management

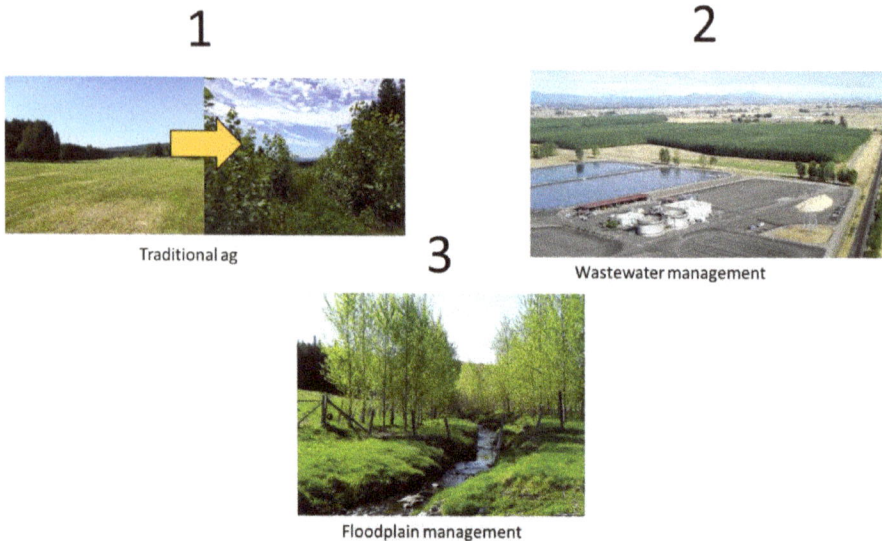

Appendix C. Interview Guide

Interviewers used this interview guide as a template to generate conversation with participants. The interviews were semi-structured, meaning that interviewers were free to explore ideas and ask additional questions based on the responses of interviewees. [See note in Appendix B about the use of the term "scenarios."]

Appendix C.1. Background about Interviewee and Community (5 min)

- Question 1: To start out, I'd like to hear a bit about yourself and your community. Where do you live? [Prompt: Town, county? rural, suburban, or urban?]
- Question 2: How long have you lived in the area?
- Question 3: What do you like about living there? [If they have been there for a long time (10+ years), ask how things have changed. If they have been there for a shorter time, ask them to compare it to where they lived before this.]
- Question 4: What is your profession? OR Where do you work? OR Who do you work for?
- Question 5: What community or communities do you consider yourself a part of? [Prompt: How about at work? How about outside of work? What is your role in those communities?]
- Question 6: What projects are going on in the area that you're aware of? [Prompt: For example, development initiatives (new business, housing development, agricultural shifts) or problems to solve (flooding, income).]

Appendix C.2. Discussion about Poplar Farms (10 min)

Now we'd like to talk about growing poplar for biomass. Let's start by reviewing the background information. [Read through the information describing how poplar is grown and the reasons to use it to make renewable chemicals.] Do you have any clarification questions?

- Question 7: What is your initial reaction to the idea of converting land into poplar farms? [Prompt: positive or negative? questions, concerns? economic, social, aesthetic?]
- Question 8: How do you feel about tree farms versus other kinds of crops?
- Question 9: When you see an open, grassy field or pasture, what do you think about?

- Question 10: Are you familiar with the poplar farm in Chehalis? If so, what do you think about as you pass by it?

Appendix C.3. Discussion about Scenarios (25 min)

Now we'd like to talk about the different scenarios we sent to you earlier. [Read through the description of the three scenarios about where and how poplar might be grown if a biorefinery were built in Lewis County, WA, USA.] Do you have any clarification questions?

- Question 11: Is there a particular scenario that you would like to discuss first?
- Question 12: What is your initial reaction to the [Traditional ag, wastewater management, or floodplain] scenario?
- Question 13: What challenges or barriers?
- Question 14: What opportunities do you see?
- Question 15: How do you think this scenario might impact your community? [Prompt: What ecosystem services could this provide? What might be the risks?]
- Question 16: Are there any modifications you would recommend for this scenario?
- Question 17: Is there anything else you would like to add about this scenario?
- Repeat Q 12–17 for each scenario.
- Question 18: Is there an alternative scenario that you think is worth exploring?
- Question 19: Is there anything else you would like to add?

Appendix C.4. Recommendations for Other Participants

- Are there any individuals or organizations you recommend we contact?
- Are there particular perspectives that you think we should try to include?

References and Note

1. Stanton, B.; Eaton, J.; Johnson, J.; Rice, D.; Schuette, B.; Moser, B. Hybrid poplar in the Pacific Northwest: The effects of market-driven management. *J. For.* **2002**, *100*, 28–33. [CrossRef]
2. Berguson, W.E.; Eaton, J.; Stanton, B. Development of hybrid poplar for commercial production in the United States: The Pacific Northwest and Minnesota experience. In *Sustainable Alternative Fuel Feedstock Opportunities, Challenges and Roadmaps for Six U.S. Regions, Proceedings of the Sustainable Feedstocks for Advanced Biofuels Workshop, Atlanta, GA, USA, 28–30 September 2010*; Braun, R., Karlen, D., Johnson, D., Eds.; Soil and Conservation Society: Ankeny, IA, USA, 2010; pp. 282–299.
3. Townsend, P.A.; Kar, S.P.; Miller, R.O. Poplar (*Populus* spp.) Trees for Biofuel Production. 6 May 2014. Available online: http://articles.extension.org/pages/70456/poplar-populus-spp-trees-for-biofuel-production (accessed on 1 August 2018).
4. Santangelo, E.; Scarfone, A.; Del Giudice, A.; Acampora, A.; Alfano, V.; Suardi, A.; Pari, L. Harvesting systems for poplar short rotation coppice. *Ind. Crop. Prod.* **2015**, *75*, 85–92. [CrossRef]
5. Licht, L.A.; Isebrands, J.G. Linking phytoremediated pollutant removal to biomass economic opportunities. *Biomass Bioenergy* **2005**, *28*, 203–218. [CrossRef]
6. Pilon-Smits, E. Phytoremediation. *Ann. Review Plant Biol.* **2015**, *56*, 15–39. [CrossRef] [PubMed]
7. Doty, S.L.; Freeman, J.L.; Cohu, C.M.; Burken, J.G.; Firrincieli, A.; Simon, A.; Kahn, Z.; Isebrands, J.G.; Lukas, J.; Blaylock, M.J. Enhanced degradation of TCE on a Superfund site using endophyte-assisted poplar tree phytoremediation. *Environ. Sci. Technol.* **2017**, *51*, 10050–10058. [CrossRef] [PubMed]
8. Kuhn, G.A.; Nuss, J. *Wastewater Management Using Hybrid Poplar*; Agroforestry Note #17, Special Applications #3; USDA-National Agroforestry Center: Lincoln, NE, USA, 2000.
9. Townsend, P.A.; Haider, N.; Boby, L.; Heavy, J.; Miller, T.; Volk, T. A roadmap for poplar and willow to provide environmental services and to build the bioeconomy. *Wash. State Univ. Ext.* in press.

10. Advanced Hardwood Biofuels Northwest. Use of Poplar Trees for Wastewater and Biosolid Utilization. Available online: https://s3.wp.wsu.edu/uploads/sites/2182/2017/09/Wastewater-Infosheet_final.pdf (accessed on 1 August 2018).

11. Johnson, J.D. Hybrid Poplar: An Overview. In *Proceedings from the Symposium on Hybrid Poplars in the Pacific Northwest: Culture, Commerce, and Capabilitym, Pasco, WA, USA, 7–9 April 1999*; Blatner., K.A., Johnson, J.D., Baumgartner, D.M., Eds.; Department of Natural Resource Sciences Cooperative Extension: Pullman, WA, USA, 1999.

12. Gordon, J.C. Poplars: Trees of the people, trees of the future. *For. Chron.* **2001**, *77*, 217–219. [CrossRef]

13. Fortier, J.; Truax, B.; Gagnon, D.; Lambert, F. Potential for hybrid poplar riparian buffers to provide ecosystem services in three watersheds with contrasting agricultural land use. *Forests* **2016**, *7*, 37. [CrossRef]

14. Merz, J.; Bandaru, V.; Hart, Q.; Parker, N.; Jenkins, B.M. Hybrid poplar based biorefinery siting web application (HP-BiSWA): an online decision support application for siting hybrid poplar based biorefineries. *Comput. Electron. Agric.* **2018**, *155*, 76–83. [CrossRef]

15. City of Chehalis: Wastewater Division. Available online: http://ci.chehalis.wa.us/publicworks/wastewater-division (accessed on 1 August 2018).

16. Lenentine, M.M. Social Perspectives on Hybrid Poplar Biofuels in the Pacific Northwest: Structuring Stakeholder Viewpoints and Analyzing Media Content. Ph.D. Dissertation, University of Washington, Seattle, WA, USA, 2017.

17. Gowan, C.H.; Kar, S.P.; Townsend, P.A. Landowners' perceptions of and interest in bioenergy crops: Exploring challenges and opportunities for growing poplar for bioenergy. *Biomass Bioenergy* **2018**, *110*, 57–62. [CrossRef]

18. Wen, Z.; Ignosh, J.; Parrish, D.; Stowe, J.; Jones, B. Identifying farmers' interest in growing switchgrass for bioenergy in Southern Virginia. *J. Ext.* **2009**, *47*, 5RIB7.

19. Villamil, M.B.; Alexander, M.; Silvis, A.H.; Gray, M.E. Producer perceptions and information needs regarding their adoption of bioenergy crops. *Renew. Sustain. Energy Rev.* **2012**, *16*, 3604–3612. [CrossRef]

20. Warren, C.R.; Burton, R.; Buchanan, O.; Birnie, R.V. Limited adoption of short rotation coppice: The role of farmers' socio-cultural identity in influencing practice. *J. Rural Stud.* **2016**, *45*, 175–183. [CrossRef]

21. Selfa, T.; Kulcsar, L.; Bain, C.; Goe, R.; Middendorf, G. Biofuels bonanza? Exploring community perceptions of the promise and perils of biofuel production. *Biomass Bioenergy* **2010**, *35*, 1379–1389. [CrossRef]

22. Cheng, J.J.; Timilsina, G.R. Status and barriers of advanced biofuel technologies: A review. *Renew. Energy* **2011**, *36*, 3541–3549. [CrossRef]

23. Chowyuk, A.; Gustafson, R.; Bura, R.; Parcel, N.; Morales-Vera, R. Utilizing Poplar-Based Ecosystem Services to Reduce Biorefinery Feedstock Costs: A Case Study in Lewis County, WA, USA. Presented at the 40th Symposium on Biotechnology for Fuels and Chemicals (SBFC), Clearwater Beach, FL, USA, 29 April–2 May 2018.

24. Rogers, L.; Cooke, A.; Comnick, J. *A Poplar Suitability and Parcel Land Use Study*; Advanced Hardwood Biofuels Northwest by the Natural Resources Spatial Informatics Group, Precision Forestry Cooperative, University of Washington: Seattle, WA, USA, 2016.

25. Weiss, R.S. *Learning from Strangers: The Art and Method of Qualitative Interview Studies*; The Free Press: New York, NY, USA, 1994; ISBN 978-0-684-82312-6.

26. Saldaña, J. *The Coding Manual for Qualitative Researchers*, 2nd ed.; SAGE Publications Ltd.: London, UK, 2013; ISBN 978-1-44624-736-5.

27. Jenkins, D. Elk Disrupt Farming in Northwest Washington Valley. Available online: http://www.capitalpress.com/Washington/20180412/elk-disrupt-farming-in-northwest-washington-valley (accessed on 12 April 2018).

28. Upreti, B.R.; van der Horst, D. National renewable energy policy and local opposition in the UK: The failed development of a biomass electricity plant. *Biomass Bioenergy* **2004**, *26*, 61–69. [CrossRef]

29. Marciano, J.A.; Lilieholm, R.J.; Teisl, M.F.; Leahy, J.E.; Neupane, B. Factors affecting public support for forest-based biorefineries: A comparison of mill towns and the general public in Maine, USA. *Energy Policy* **2014**, *75*, 301–311. [CrossRef]

30. Office of Financial Management, State of Washington. Per Capita Personal Income by County. Available online: https://www.ofm.wa.gov/washington-data-research/statewide-data/washington-trends/economic-trends/washington-and-us-capita-personal-income/capita-personal-income-county (accessed on 14 August 2018).

31. United States Department of Agriculture (USDA) National Agricultural Statistics Service (NASS). Census of Agriculture Highlights—Farm Demographics: U.S. Farmers by Gender, Age, Race, Ethnicity, and More. ACH12-3; May 2014. Available online: https://www.agcensus.usda.gov/Publications/2012/Online_Resources/Highlights/Farm_Demographics/ (accessed on 18 April 2018).

32. Youngs, H.L. The Effects of Stakeholder Values on Biofuel Feedstock choices. In *Perspectives on Biofuels: Potential Benefits and Possible Pitfalls*; Taylor, C., Lomneth, R., Wood-Black, F., Eds.; ACS Symposium Series; ACS Publications: Washington, DC, USA, 2012; Volume 1116, pp. 29–67.

33. Galik, C.S. Exploring the determinants of emerging bioenergy market participation. *Renew. Sustain. Energy Review.* **2015**, 47, 107–116. [CrossRef]

34. United States Department of Agriculture (USDA). Census of Agriculture—Washington: State and County Profiles. Available online: https://www.agcensus.usda.gov/Publications/2012/Online_Resources/County_Profiles/Washington/ (accessed on 20 June 2018).

35. Jenkins, D. USDA Orders Volume Controls on Cranberries. Available online: http://www.chinookobserver.com/co/business/20180411/usda-orders-volume-controls-on-cranberries (accessed on 11 April 2018).

36. Haider, N.; Parker, N.; Townsend, P. Potential for a hybrid poplar industry using recycled water: An environmental application of poplar in Northern Idaho. *Wash. State Univ. Ext.* in press.

37. Metropolitan Wastewater Management Commission: Biocycle Farm. Available online: http://www.ci.springfield.or.us/MWMCPartners/biocyclefarm.html (accessed on 1 August 2018).

38. Braatne, J.H. *Biological Aspects of Hybrid Poplar Cultivation on Floodplains in Western North America—A Review*; U.S. Environmental Protection Agency: Seattle, WA, USA, 1999.

39. Brooks, K.N.; Current, D.; Wyse, D. Restoring Hydrologic Function of Altered Landscapes: An Integrated Watershed Management Approach. In *Preparing for the Next Generation of Watershed Management Programmes and Projects: Water Resources for the Future, Proceedings of the International Conference, Sassari, Italy, 22–24 October 2003*; Tennyson, L., Zingari, P.C., Eds.; Food and Agriculture Organization of the United Nations: Rome, Italy, 2006; pp. 101–114.

40. Henri, C.J.; Johnson, J.D. Riparian forest buffer income opportunities: A hybrid poplar case study. *J. Soil Water Conserv.* **2005**, 60, 159–163.

41. Department of Ecology, State of Washington: Chehalis Basin Strategy. Available online: https://ecology.wa.gov/Water-Shorelines/Shoreline-coastal-management/Hazards/Floods-floodplain-planning/Chehalis-Basin-Strategy (accessed on 1 August 2018).

42. Farmers' Perceptions of Salmon Habitat Restoration Measures: Loss and Contestation; Report Presented to the Environmental Protection Agency and the Society for Applied Anthropology. Available online: http://ftp.sfaa.net/files/4913/7329/3792/breslow.pdf (accessed on 1 August 2018).

43. Department of Ecology, State of Washington. Chehalis Basin Strategy Final EIS Executive Summary. 2017. Available online: http://chehalisbasinstrategy.com/wp-content/uploads/2017/06/Chehalis-Basin-Strategy-EIS-Executive-Summary.pdf. (accessed on 20 June 2018).

44. Osowski, K. Quinault Nation Proposes New Approach to Flood Protection in Chehalis Basin. Available online: http://www.chronline.com/news/quinault-nation-proposes-new-approach-to-flood-protection-in-chehalis/article_11afd5c6-78e3-11e5-9a40-777c38545129.html (accessed on 22 October 2015).

45. Klocko, A.L.; Brunner, A.M.; Huang, J.; Meilan, R.; Lu, H.; Ma, C.; Morel, A.; Zhao, D.; Ault, K.; Dow, M.; et al. Containment of transgenic trees by suppression of LEAFY. *Nat. Biotechnol.* **2016**, 34, 918–922. [CrossRef] [PubMed]

46. Dou, C.; Marcondes, W.F.; Djaja, J.E.; Bura, R.; Gustafson, R. Can we use short rotation coppice poplar for sugar based biorefinery feedstock? Bioconversion of 2-year-old poplar grown as short rotation coppice. *Biotechnol. Biofuels* **2017**, 10, 144. [CrossRef] [PubMed]

47. Stanton, B.; O'Neill, B.; Bura, R.; Emerson, R.; Kallestad, J. The Devil is in the Details: Understanding Poplar's True Potential as an Energy Feedstock through Biomass Studies. Advanced Hardwood Biofuels Northwest Online Newsletter, Volume 6, No. 4, 2018.

48. Delshad, A.B.; Raymond, L.; Sawicki, V.; Wegener, D.T. Public attitudes toward political and technological options for biofuels. *Energy Policy* **2010**, *38*, 3414–3425. [CrossRef]
49. Cacciatore, M.A.; Scheufele, D.A.; Shaw, B.R. Labeling renewable energies: How the language surrounding biofuels can influence its public acceptance. *Energy Policy* **2012**, *51*, 673–682. [CrossRef]
50. Sleenhoff, S.; Landeweerd, L.; Osseweijer, P. Bio-basing society by including emotions. *Ecol. Econ.* **2015**, *116*, 78–83. [CrossRef]

forests

MDPI

Erratum

Erratum: Hart, N.M., et al. Stakeholder Assessment of the Feasibility of Poplar as a Biomass Feedstock and Ecosystem Services Provider in Southwestern Washington, USA. *Forests* 2018, 9, 655

Forests **Editorial Office**

MDPI AG, St. Alban-Anlage 66, 4052 Basel, Switzerland

Received: 20 November 2018; Accepted: 22 November 2018; Published: 27 November 2018

The authors have requested that the following changes be made to their paper [1].

The images published in Appendix B should be attributed as follows:

Appendix B.1. Background: Growing Hybrid Poplar for Biomass (p. 19 of 24)

Images courtesy of Advanced Hardwood Biofuels Northwest.

Appendix B.2. Introduction (p. 20 of 24)

Photo Credit 1 (Traditional ag): Advanced Hardwood Biofuels Northwest.

Photo Credit 2 (Wastewater management): ©SkyShots, Portland, OR. Image of the Biocycle Farm, Metropolitan Wastewater Management Commission in Eugene–Springfield, OR. Reproduced with permission from Dan Bigelow, Photographer/Owner, SkyShots Aerial Photography.

Photo Credit 3 (Floodplain management): Daniel Gagnon, Eastern Townships Forest Research Trust.

We apologize to the creators of these images and to the readers and editors of this journal article. The change does not affect the scientific results. The manuscript will be updated and the original will remain online on the article webpage, with a reference to this Erratum.

Reference

1. Hart, N.M.; Townsend, P.A.; Chowyuk, A.; Gustafson, R. Stakeholder Assessment of the Feasibility of Poplar as a Biomass Feedstock and Ecosystem Services Provider in Southwestern Washington, USA. *Forests* **2018**, *9*, 655. [CrossRef]

forests

MDPI

Article

Short Rotation Eucalypts: Opportunities for Biochar

Donald L. Rockwood [1,2,*], Martin F. Ellis [3], Ruliang Liu [4], Fengliang Zhao [4], Puhui Ji [4], Zhiqiang Zhu [4], Kyle W. Fabbro [2], Zhenli He [4] and Ronald D. Cave [4]

[1] Florida FGT LLC, Gainesville, FL 32635, USA; floridafgt@cox.net
[2] School of Forest Resources and Conservation, University of Florida, Gainesville, FL 32611, USA; kwfabbro@gmail.com
[3] Green Carbon Solutions (GCS), Pepper Pike, OH 44124, USA; mellis@greencarbonsolutions.com
[4] Indian River Research and Education Center (IRREC), University of Florida, Ft Pierce, FL 34945, USA; ruliang_liu@126.com (R.L.); zfl7409@163.com (F.Z.); jipuhui1983@163.com (P.J.); zhuzhiq8@163.com (Z.Z.); zhe@ufl.edu (Z.H.); rdcave@ufl.edu (R.D.C.)
* Correspondence: dlrock@ufl.edu; Tel.: +01-352-256-3474

Received: 2 February 2019; Accepted: 31 March 2019; Published: 5 April 2019

Abstract: Eucalypts can be very productive when intensively grown as short rotation woody crops (SRWC) for bioproducts. In Florida, USA, a fertilized, herbicided, and irrigated cultivar planted at 2471 trees/ha could produce over 58 green mt/ha/year in 3.7 years, and at 2071 trees/ha, its net present value (NPV) exceeded $750/ha at a 6% discount rate and stumpage price of $11.02/green mt. The same cultivar grown less intensively at three planting densities had the highest stand basal area at the highest density through 41 months, although individual tree diameter at breast height (DBH) was the smallest. In combination with an organic fertilizer, biochar improved soil properties, tree leaf nutrients, and tree growth within 11 months of application. Biochar produced from *Eucalyptus* and other species is a useful soil amendment that, especially in combination with an organic fertilizer, could improve soil physical and chemical properties and increase nutrient availability to enhance *Eucalyptus* tree nutrition and growth on sandy soils. Eucalypts produce numerous naturally occurring bioproducts and are suitable feedstocks for many other biochemically or thermochemically derived bioproducts that could enhance the value of SRWCs.

Keywords: *Eucalyptus*; short rotation woody crops; Florida; biochar; bioproducts

1. Introduction

Eucalypts are the world's most valuable and widely planted hardwoods (up to 21.7 million ha in 61 countries by 2030 [1]) and have numerous potential applications as short rotation woody crops (SRWCs) [2,3]. Several *Eucalyptus* planting stocks have promise as SRWCs in Florida [4,5], including *E. grandis* x *E. urophylla* cultivars such as EH1. After four generations of *E. grandis* genetic improvement for Florida's unique climatic and edaphic conditions starting in the 1960s and clonal testing initiated in the 1980s across a wide range of site/soil types, the University of Florida released five *G Series* cultivars in 2009 for commercial planting [4,5]. Although G1 is no longer commercially viable due to susceptibility to blue gum chalcid (*Leptocybe invasa*), G2 through G5 have shown resilience to damaging freezes, tolerance to infertile soils, exceptional stem form, improved coppicing ability, chalcid resistance, and varying degrees of windfirmness.

G Series planting density trials established on former citrus lands and phosphate mined clay settling areas in central and south Florida demonstrated maximum mean annual increments (MAI$_{max}$) as high as 75.3 to 78.2 green mt/ha/year at 4304 and 5066 trees/ha, respectively [6]. Economic analyses using current stumpage prices, high silvicultural management costs, and expected coppice yields, have shown that *G Series* cultivars can generate internal rates of return greater than 10% [6].

Most soils in Florida are sandy, with >90% of soil particles as sand, and have low nutrient and moisture holding capacities. Fertilization is necessary to sustain desired crop yield and quality. However, fertilizers are readily leached if not taken up by crop plants and consequently result in environmental pollution such as eutrophication. Applying biochar, a fine-grained, highly porous "charcoal" produced through pyrolysis (burning in a nearly oxygen-free environment) or gasification of numerous feedstocks, improves the physicochemical properties of soils, including bulk density, porosity, cation exchange capacity (CEC), and pH. It also increases soil water and nutrient holding capacities and consequently influences crop production while reducing leaching [7]. Productivity of many crops significantly increased after soils were amended with biochar [8,9]. Sandy soils are more responsive to biochar than clayey soils [10] due to their low water and nutrient holding capacities [11].

Biochar, an ancient soil-building amendment, today has wide ranging applications [12]. The International Biochar Institute (www.biochar-international.org) has identified more than 50 uses for biochar, and worldwide interest in and demand for biochar are growing quickly. Current demand estimates suggest that biochar is a billion dollar plus industry worldwide with the two largest markets being North America and Europe [13]. Depending on its particular properties, effective biochar uses include soil and crop improvement plus environmental benefits such as carbon sequestration, retention of nutrients and water, reduced leaching, and water purification, all of which are important in Florida.

Using experience in Florida, USA, we describe eucalypts' potential for maximizing SRWC productivity through site amendment and genetic improvement, document their suitability for biochar production, and assess biochar's potential for improving soil properties, tree nutrition, and eucalypts' growth.

2. Materials and Methods

2.1. EH1 Planting Density Demonstration

An 8.1 ha, intensively fertilized, herbicided, and irrigated *E. grandis* x *E. urophylla* cultivar EH1 planting density demonstration was established in May 2011 near Hobe Sound, FL, as five rows of trees on 19.8m-wide former citrus beds. Planting densities of 2071 trees/ha (3.1 × 1.2 m spacing) and 1181 trees/ha (3.1 × 2.1 m) were monitored in 21 permanent 19.8 × 12.2 m plots through harvest in December 2017. To model EH1 yield at these two densities and estimate productivity at 2471 trees/ha, 18-, 30-, 36-, 42-, 48-, 65-, and/or 81-month data were fit to stand density, dominant height and basal area development functions in *E. grandis* stand-level growth, and yield model equations used by Plessis and Kotze [14]. Yields were the same for high and low management strategies because *E. grandis* productivity under low management on citrus beds in central Florida generated similar yields to the EH1 under high management at Hobe Sound, FL [6]. Silvicultural and other forest management costs were provided by agricultural companies exploring *Eucalyptus* options in central and southern Florida. Stumpage prices were based on local *Eucalyptus* mulchwood stumpage prices reported by the same companies and a forestry consulting firm familiar with the local markets.

Maximum net present values (NPV$_{max}$) calculated for two management strategies (Table 1), two planting densities (2071 and 1181 trees/ha), two real discount rates (6% and 8%), and two stumpage prices ($11.02 and 22.05/green mt) assumed three stages (two coppice stages following the original planting) in one planting cycle. Based on young coppice in Evans Properties' EH1 plantation near Ft. Pierce, FL, expected coppice yields for stages 2 and 3 (first and second coppice, respectively) were assumed to be 90% and 80% of observed stage 1 yields, respectively. The optimum stage lengths were reported to the nearest 1/10th year; therefore, the annual interest rate was converted to an effective periodic rate.

Table 1. Low and high management activities and assumed costs for *E. grandis* x *E. urophylla* cultivar EH1 on former bedded citrus sites.

Activity	Cost
Land Preparation	$988/ha
Chemical Site Prep	$297/ha
Propagules	$0.70/tree
Planting Cost	$0.40/tree
Irrigation + Fertilization (High management only)	$1977/ha
Fertilization at Initial Establishment (Low management only)	$173/ha
Weed Control (Beginning of coppice stage)	$136/ha

2.2. EH1 Fertilizer x Planting Density Study

Cultivar EH1 was also planted in a fertilizer x planting density study in June 2015 on a former pasture at the Indian River Research and Education Center (IRREC) near Ft Pierce, FL. Five fertilizers (control, Green Edge (GE) 6–4–0 + micronutrients at 112, 224, and 336 kg of N/ha rates, and diammonium phosphate equivalent to 336 kg of N/ha) were applied as five treatment plots of three contiguous rows 3.1 m apart, for a total of 15 rows of 26 trees. Within the 26-tree rows, 5-tree row plots were systematically assigned one of three planting densities (3588, 1794, and 1196 trees/ha; 3.1 × 0.9 m, 1.8 m, and 2.7 m, respectively) such that 1794 and 1196 trees/ha were replicated twice, 3588 trees/ha once. The interior three trees of each fertilizer x planting density plot were periodically measured through November 2018. Analyses of variance and Duncan's Multiple Range Tests of fertilizer and planting density means were conducted using SAS® (SAS Institute, Cary, NC, USA).

2.3. Biochar Tests

Five test trees were used for preliminary biochar evaluations in 2017–2018. One tree in the EH1 Planting Density Demonstration, one *E. grandis* cultivar G2 in a 2012 commercial plantation near Ft Pierce, FL, one *Corymbia torelliana* tree in an adjacent progeny test, one *E. amplifolia* in a progeny test near Old Town, FL, and one *Quercus virginiana* in a nearby natural forest each provided ~23 green kg of stemwood for testing by a lab in California at different pyrolysis temperatures to determine optimal charring temperatures for the different feedstocks. Their biochar physical and chemical properties were further tested by Celignis Analytical, Ireland, to guide the processing of biochar and as a comparative benchmark.

Biochar produced in Europe by Green Carbon Solutions' (GCS') Polchar, which specializes in pyrolysis and carbonization of different feedstocks, served as a comparison for the five Florida trees. Hardwood monoculture roundwood logs only were cut to size, split, and pre-dried. Pyrolysis involved a vertical retort operating through a range of temperatures up to a maximum of ~630 °C. Post production, the biochar was sampled and tested for physical and chemical properties. Polchar's biochar was also used for the biochar–fertilizer study described below.

2.4. Biochar–Fertilizer Study

A two-row windbreak study of three *E. grandis* cultivars in one row and four *C. torelliana* progenies in an adjacent row offset 1.2 m away was established at the IRREC in July 2017 as a randomized complete block design with four complete replications of cultivars G3, G4, and G5 in 17 to 28-tree single row plots at 1.8 m within row spacing and one incomplete replication of cultivar G5 in a 13-tree single row plot. In February 2018, all four complete replications received an organic fertilizer (GE 6–4–0 + micronutrients at 336 kg of N/ha), and the two interior replications also received GCS' Polchar biochar (11.2 mt/ha) by rotovating the two treatments into the soil to a 20 cm depth between and within 1.2 m of the two rows. The incomplete replication served as a control. The cultivars in this resulting biochar–fertilizer study were measured at 5, 11, and 16 months.

To monitor soil and foliage responses, 13 trees (one in the middle of each of the 13 cultivar plots) were resampled at biochar–fertilizer treatment ages of 0, 5, and 11 months (tree ages 5, 11, and 16 months, respectively). At each time, four soil samples were collected from a 0–20 cm depth within 1.2 m around each sample tree, and recently matured foliage was taken from four representative branches around the crown of each sample tree. The collected soils were combined by tree, air dried, and ground to pass through a 2-mm sieve prior to analysis for relevant properties. The tree leaf samples were combined by tree, oven dried at 75 °C to constant weight, and powdered to pass through a 1-mm sieve prior to analysis for nutrient concentration.

Soil pH was measured using a pH/conductivity meter (AB 200, Fisher Scientific, Pittsburgh, PA, USA) at the soil to water ratio of 1:1. Electrical conductivity (EC) of soil samples was determined at the solid to water ratio of 1:2 using the pH/conductivity meter. Available soil P was determined using the method of Kuo [15]. Available metals in soil were measured by extracting the samples with Mehlich 3 (M3) solution at a solid to solution ratio of 1:10 [16]. The extracts were filtered through a 0.45-μm membrane. Subsamples of the filtrate were acidified and analyzed for the concentrations of dissolved P, K, Ca, Mg, Fe, Mn, Cu, Zn, B, and Mo using inductively coupled plasma–optical emission spectrometry (ICP–OES) (Ultima, JY Horiba Group, Edison, NJ). Portions of the plant leaf samples (0.2 g each) were digested with 6 mL of concentrated nitric acid (HNO_3)/hydrogen peroxide (H_2O_2) and diluted to 100 mL. The concentrations of P, K, Ca, Mg, Fe, Mn, Cu, Zn, B, and Mo in the digested samples were determined using ICP–OES.

Analyses of variance and Tukey–Kramer tests of cultivar tree size, soil nutrient, and tree leaf nutrient means were conducted using SAS®. Changes between soil properties and leaf nutrients from 0–5, 5–11, and 0–11 months were also analyzed.

3. Results

3.1. EH1 Planting Density Demonstration

Through 81 months, cultivar EH1 yielded more at 2071 trees/ha than at 1181 trees/ha. Maximum annual yields were directly related, and times to those peaks were inversely related, to planting density (Table 2). Annual yield at 2471 trees/ha was estimated to be over 58 green mt/ha/year in 3.7 years. At the lowest density, MAI_{max} was 44 mt/ha/year at 5.0 years. Compared to 2471 trees/ha, 2071 trees/ha had lower yields and achieved peak productivity later. However, planting density also inversely affected average tree diameter at breast height (DBH) as the higher planting density produced smaller trees (e.g., 81-month DBH = 15.4 cm at 2071 trees/ha, 19.2 cm at 1181 trees/ha), which could influence harvesting costs.

Table 2. Predicted maximum mean annual increment (MAI_{max}) and rotation age for cultivar EH1 at planting densities of 1181, 2071, and 2471 trees/ha.

Planting Density (trees/ha)	MAI_{max} (green mt/ha/year)	Rotation Age at MAI_{max} (years)
1181	44.00	5.0
2071	54.63	4.0
2471	58.98	3.7

NPVs ranged widely largely due to stumpage price (Table 3). At a stumpage price of $11.02/mt, high management intensity had negative NPVs. At $22.05/mt stumpage, all scenarios resulted in positive NPVs. Due to high establishment and planting costs, 2071 trees/ha generated higher NPVs compared to 1181 trees/ha across all scenarios. Stage lengths were always shorter with 2071 trees/ha and increasing discount rate, which for example could result in a ~15.5 year planting cycle consisting of 4.8 years from planting to first harvest, 5.1 year first coppice stage, and 5.6 year second coppice stage.

Table 3. Effects of stumpage price ($/green mt) and real discount rate on maximum net present values (NPV$_{max}$), internal rate of return (IRR), and optimum stage lengths for the first cycle of EH1 at 2071 and 1181 trees/ha with low and high management intensities and two coppice stages.

Stumpage Price ($)	Discount Rate (%)	NPV$_{max}$ ($/ha)	IRR (%)	Stage Length (years)
2071 trees/ha, Low Management Intensity ($1458/ha + $1.10/tree + $136/ha @coppice)				
11.02	6	751	8.1	5.1, 5.4, 5.8
	8	70	8.2	4.9, 5.1, 5.6
22.05	6	5365	18.1	5.1, 5.4, 5.8
	8	3986	18.6	4.8, 5.1, 5.6
2071 trees/ha, High Management Intensity ($3262/ha + $1.10/tree + $136/ha @coppice)				
11.02	6	−1053	3.6	5.1, 5.4, 5.8
	8	−1734	3.5	4.9, 5.1, 5.6
22.05	6	3561	12.2	5.1, 5.4, 5.8
	8	2182	12.5	4.8, 5.1, 5.6
1181 trees/ha, Low Management Intensity ($,458/ha + $1.10/tree + $136/ha @coppice)				
11.02	6	1377	10.3	6.3, 6.7, 7.4
	8	660	10.5	5.9, 6.3, 7.0
22.05	6	5509	19.2	6.3, 6.7, 7.4
	8	4058	19.9	5.9, 6.2, 7.0
1181 trees/ha, High Management Intensity ($3262/ha + $1.10/tree + $136/ha @coppice)				
11.02	6	−427	5.0	6.3, 6.7, 7.4
	8	−1143	4.9	5.9, 6.3, 7.0
22.05	6	3705	12.4	6.3, 6.7, 7.4
	8	2254	12.7	5.9, 6.2, 7.0

3.2. EH1 Fertilizer x Planting Density Study

Fertilizer and planting density influenced the productivity of cultivar EH1 (Table 4). While the differences among five fertilizers for 9-mo height favored the higher GE rates, subsequent differences were inconsistent due to flooding soon after planting and small plot sizes. Planting density differences were observed at all ages, with 3588 trees/ha having the tallest trees at 9 months and the largest stand basal area but smallest tree DBH at subsequent ages. For example, at 41 months, stand basal area and tree DBH at 3588 trees/ha averaged 31.9 m^2/ha and 10.3 cm, respectively, compared to 19.6 m^2/ha and 14.2 cm at 1196 trees/ha.

Table 4. Effects of three planting densities (trees/ha) and five fertilizers on EH1 tree height (m), diameter at breast height (DBH) (cm), and stand basal area (m^2/ha) at 9, 36, and 41 months at the Indian River Research and Education Center (IRREC).

Trait: Age	Planting Density	Fertilizer					Density Average
		0	GE 100	GE 200	GE 300	DAP	
Height 9-mo	3588	3.86	4.62	5.66	5.02	3.36	4.60a
	1794	2.83	2.80	3.76	3.54	2.97	3.17ab
	1196	3.81	3.35	5.03	4.41	3.65	4.02ab
	Fert. Ave.	3.43b	3.43b	4.71a	4.24ab	3.33b	3.83
DBH 36-mo	3588	9.4	9.8	9.8	10.8	6.7	9.3b
	1794	10.6	9.6	11.8	10.8	12.0	10.9ab
	1196	14.8	12.2	14.7	12.1	12.6	13.2ab
	Fert. Ave.	11.8	10.7	12.6	11.3	11.2	11.5
Basal Area 36-mo	3588	27.3	27.1	31.5	33.0	13.0	26.5a
	1794	16.1	15.0	19.9	18.1	20.7	17.8b
	1196	20.8	14.1	20.5	14.1	14.9	16.8b
	Fert. Ave.	20.4	17.1	22.5	19.5	16.5	19.2
DBH 41-mo	3588	9.9	11.1	10.8	11.7	7.7	10.3b
	1794	13.2	10.6	13.1	12.1	12.2	12.3ab
	1196	16.0	11.9	15.3	13.6	14.1	14.2a
	Fert. Ave.	13.7a	11.2b	13.2ab	12.6ab	12.0ab	12.5
Basal Area 41-mo	3588	24.7	35.5	39.9	38.9	17.6	31.9a
	1794	25.6	18.0	24.7	22.4	21.8	22.5b
	1196	24.5	14.5	22.6	17.9	18.9	19.6c
	Fert. Ave.	25.0a	20.8b	29.3a	25.4a	19.6b	23.8

* Means within Fert. or Density Averages not sharing the same letter differ at the 5% level.

3.3. Biochar Tests

Biochars from EH1, *C. torelliana*, G2, *E. amplifolia*, and *Quercus virginiana*, were relatively similar and appeared suitable for commercial biochar production (Table 5). The Cl content of G2, though, was somewhat high. Compared to Polchar biochar made from European hardwoods, which was high quality with a pH of 8.2 and electrical conductivity of 3.33 mmhos/cm, all five Florida trees appeared similar for recalcitrant carbon but higher in pH and water holding (Table 6).

Table 5. Properties of Green Carbon Solutions (GCS) biochar made from Florida *E. grandis* cultivar G2, *C. torelliana* (CT), *E. grandis* x *E. urophylla* cultivar EH1, *E. amplifolia* (EA), and *Q. virginiana* (Qv) test trees.

Property (% of Dry Weight)	G2	CT	EH1	EA	Qv
Volatile Matter	83.3	85.0	85.9	82.5	83.3
Fixed Carbon	15.7	14.4	13.7	17.0	15.5
Ash	1.00	0.54	0.37	0.50	1.15
Moisture Content	36.4	48.0	43.1	30.1	33.1
C	49.2	49.7	49.8	50.8	49.1
O	43.0	43.1	43.1	42.0	43.1
H	6.5	6.5	6.5	6.5	6.4
N	0.21	0.17	0.17	0.26	0.29
Cl	0.07	0.02	0.02	0.02	0.00
S	0.01	0.00	0.00	0.01	0.00

Table 6. Comparison of properties of biochar made from Florida *E. grandis* cultivar G2, *C. torelliana* (CT), *E. urograndis* cultivar EH1, *E. amplifolia* (EA), and *Q. virginiana* (QV) test trees with Polchar biochar.

Property	Florida Tree					Polchar Biochar
	G2	CT	EH1	EA	QV	
Recalcitrant Carbon * (%)	76.0	71.6	74.0	70.8	71.8	67.6
pH	10.6	10.4	10.5	11.1	11.9	8.2
EC (mmhos/cm)	0.57	1.76	1.56	3.88	1.14	3.33
Water Holding (mL/100 g)	75.9	78.8	79.8	69.0	68.5	43.4
Carbonate Value (%)	2.6	2.5	5.6	16.7	2.5	-

* Estimated at 80% of fixed carbon on a dry ash-free basis.

3.4. Biochar–Fertilizer Study

Soil property data from the IRREC biochar–fertilizer study suggest that biochar (BC) enhanced the nutrient properties of this inherently poor Florida soil (Table 7). GE generally increased available soil nitrogen, as indicated by increases in KCl extractable NO_3-N and NH_4-N, and GE in combination with BC further increased NH_4-N five months after GE + BC application, which may be attributed to increased NH_4-N holding capacity in the GE + BC amended soil. An increase in soil available P was significant ($p = 0.0198$) for GE + BC five months after amendment. However, both available N and P in the soil decreased 11 months after amendment, likely due to intensive uptake by the established trees. At 11 months, soil NH_4-N was significantly ($p = 0.0376$) higher with GE and GE + BC compared to the control. Replication and cultivar effects were non-significant except for beginning EC and EC from beginning to 11 months (replications, $p = 0.0147$ and 0.0368, respectively), and 11-month EC and EC from beginning to 5 months (cultivars, $p = 0.0396$ and 0.0043, respectively).

Table 7. Effects of three cultural treatments (Green Edge only (GE), Green Edge with biochar (GE + BC), and Control) on soil properties before and after treatment applications in the IRREC biochar–fertilizer study.

Treatment	Soil Property *				
	pH	EC (uS/cm)	NO_3-N (mg/kg)	NH_4-N (mg/kg)	P (mg/kg)
Before GE and GE + BC Applications					
GE	6.30 ± 0.73	46.1b ± 28.8	2.36 ± 0.51	0.80 ± 0.41	7.27 ± 3.24
GE + BC	5.32 ± 0.28	58.6a ± 23.1	4.08 ± 1.34	1.63 ± 0.74	7.26 ± 5.14
5 Months After GE and GE + BC Applications					
GE	6.25 ± 0.83	82.6 ± 28.9	3.51 ± 1.20	1.88 ± 1.07	7.84b ± 1.33
GE + BC	6.15 ± 0.69	96.1 ± 30.3	3.84 ± 1.31	2.17 ± 1.19	10.84a ± 2.42
11 Months After GE and GE + BC Applications					
GE	6.01 ± 0.55	27.1 ± 3.1	1.42 ± 0.27	0.85a ± 0.18	6.01 ± 1.09
GE + BC	5.96 ± 0.30	29.4 ± 2.5	2.15 ± 0.39	1.33a ± 0.30	6.26 ± 1.36
Control	5.44 ± ---	21.3 ± ---	0.83 ± ---	0.32b ± ---	2.01 ± ---

* Mean ± Standard Deviation; $n = 6$ for GE and GE + BC, $n = 1$ for Control; Means within a Soil Property and time not sharing the same letter are different at the 5% level.

The IRREC biochar–fertilizer study leaf nutrient data suggest that biochar also enhanced the nutrient levels of the three *E. grandis* cultivars in the study (Table 8). Application of GE generally increased the concentrations of Ca, K, Mg, P, Fe, and Mn in tree leaves, especially for K, Mg, P, Fe, and Mn, which increased 1–4 times in five months after amendment; GE + BC significantly increased ($p = 0.0161$) 5-month Zn over GE. These increases are likely due to inputs of these nutrients in GE, thus improving their availability in soil. However, Zn and Cu concentrations decreased, which may be attributed to binding of these elements to the organic components in GE, thus reducing their availability in the amended soil. Addition of BC to GE further improved tree nutrition with Ca, Mg, Zn, and Mn, and such improvement was also observed in 11 months after amendment, as BC tended to minimize leaf nutrient changes over time. However, a general decrease in leaf concentrations of Ca, K, Mg, P, and Zn also occurred, likely due to decreased availability of these nutrients in soil (Table 7) and the dilution effect of rapidly increased tree biomass. Replication and cultivar differences were detected only for initial P and Zn at 5 months (replications, $p = 0.0236$ and 0.0111, respectively) and the change in Mg from 0 to 11 months (cultivars, $p = 0.0369$).

In the IRREC biochar–fertilizer study, GE and GE + BC gradually enhanced the growth of three *E. grandis* cultivars (Table 9). Before treatment applications at age 5 months, the cultivars were 0.8 to 1.3 m tall in the plots that then received three treatments. Six months after application, the cultivars receiving GE + BC and GE only had doubled in height, approximately twice the increase with no treatment. Eleven months after application, cultivars receiving GE + BC were 1.0 and 2.8 m taller than those receiving GE only and no GE. Based on the cultivar G5 common to all three cultures, tree height and DBH were then significantly greater with GE and GE + BC.

Table 8. Effects of three cultural treatments (Green Edge only (GE), Green Edge with biochar (GE + BC), and Control) on *E. grandis* cultivars leaf nutrients (Ca, K, Mg, and P in g/kg; Zn, Cu, Fe, and Mn in mg/kg) before and after treatment applications in the IRREC biochar–fertilizer study.

Treatment	Leaf Nutrient *							
	Ca	K	Mg	P	Zn	Cu	Fe	Mn
				Before GE and GE + BC Applications				
GE	9.8 ± 4.2	10.8 ± 1.8	2.47 ± 0.52	0.83 ± 0.36	140 ± 67	22.5 ± 3.3	32.0 ± 12.6	235 ± 162
GE + BC	4.8 ± 2.5	10.7 ± 1.6	2.12 ± 0.39	0.95 ± 0.49	85 ± 50	19.2 ± 4.4	21.7 ± 5.1	191 ± 105
				5 Months After GE and GE + BC Applications				
GE	17.2 ± 2.6	20.7 ± 6.9	4.93 ± 0.84	3.78 ± 0.38	95b ± 12	14.7 ± 6.5	64.7 ± 36.6	263 ± 75
GE + BC	18.2 ± 2.9	20.4 ± 2.7	5.70 ± 1.08	3.67 ± 0.51	100a ± 12	9.8 ± 6.2	28.8 ± 19.0	317 ± 99
				11 Months After GE and GE + BC Applications				
GE	16.2 ± 1.6	4.9 ± 0.6	2.51 ± 0.29	1.50 ± 0.30	61 ± 11	19.5 ± 2.4	83.5 ± 14.1	205 ± 16
GE + BC	14.3 ± 2.6	5.7 ± 1.2	2.78 ± 0.48	1.59 ± 0.13	60 ± 11	17.5 ± 2.4	88.4 ± 14.9	234 ± 42
Control	13.1 ± --	5.3 ± --	2.64 ± ---	1.17 ± ---	74 ± --	18.1 ± --	88.6 ± ---	224 ± ---

* Mean ± Standard Deviation; $n = 6$ for GE and GE + BC, $n = 1$ for Control; Means within a Leaf Nutrient and time not sharing the same letter are different at the 5% level.

Table 9. Tree heights (m) and/or DBHs (cm) at ages 5- (before treatment applications), 11- (6 months after applications), and 16- (11 months after) months of three *E. grandis* cultivars (G3, G4, G5) receiving three treatments (Green Edge only (GE), GE with biochar (GE + BC), and Control) at the IRREC biochar–fertilizer study.

Trait: Age	Cultivar	Treatment *			All Treatments
		GE	GE + BC	Control	
Height 5-mo	G3	0.8	1.3		1.0
	G4	1.1	1.3		1.2
	G5	1.0	1.1	1.0	1.1
	All Cultivars	1.0	1.2	1.0	1.0
Height 11-mo	G3	1.6	2.5		2.1
	G4	2.3	2.7		2.5
	G5	2.2	2.3	1.4	2.1
	All Cultivars	2.0	2.5	1.4	2.2

Table 9. *Cont.*

Trait: Age	Cultivar	Treatment *			All Treatments
		GE	GE + BC	Control	
Height 16-mo	G3	2.5	4.8		3.6
	G4	5.1	5.0		5.1
	G5	4.2ab	5.1a	2.2b	4.2
	All Cultivars	4.0	5.0	2.2	4.3
DBH 16-mo	G3	1.5	4.0		2.8
	G4	4.2	4.2		4.2
	G5	3.5ab	4.4a	1.4b	3.5
	All Cultivars	3.1	4.2	1.4	3.5

* Treatment means within Cultivar G5 not sharing the same letter differ at the 5% level.

4. Discussion

Eucalypts can be very productive and economically feasible when intensively grown as SRWCs, even under our preliminary assumptions. As timber markets and forestry labor are not well established in central and southern Florida, our assumed silvicultural and other forest management costs for a start-up *Eucalyptus* operation are higher than for conventional forest plantations in the South. Stumpage prices may also change as local markets develop. Deployment of elite advanced-generation *E. grandis* families may further increase profitability of SRWCs in Florida primarily due to lower seedling costs (~$0.25/seedling versus $0.70/propagule) and economic feasibility of high-yield management regimes (>2471 trees/ha).

Even under high plantation establishment and management costs, low stumpage prices, and expected coppice yields, cultivar deployment can yield positive cash flows at real discount rates greater than 10%. Under current market conditions and management costs, low-density regimes (~1181–1483 trees/ha) are the most profitable for clonal forestry on average sites (e.g., citrus lands and flatwoods). Lower propagule costs could increase financially optimum planting densities. With proper mechanical site preparation, eucalypt plantations on clay settling areas could produce higher NPVs compared to average sites [6].

The goal of our financial analysis was to demonstrate the profitability of *Eucalyptus* plantations under moderate to high discount rates and high operational costs in central and southern Florida's developing forestry markets. Since most landowners were interested in NPV and IRR, we used NPV rather than land expectation value (LEV, also known as bare land value), even though our analysis of the EH1 planting density demonstration had unequal rotation/cycle lengths. Further background on the use of LEV for *Eucalyptus* in Florida is available (6,17).

The fertilizer and planting density differences observed in the IRREC fertilizer x planting density and biochar–fertilizer studies are consistent with previously observed influences of fertilizer and planting density on eucalypt productivity in Florida [17–20] and worldwide [21–24]. While inorganic fertilizers have been necessary for rapid growth of eucalypts on Florida's infertile sandy soils, the observed response here to a slow release organic fertilizer, and its apparently beneficial coupling with BC, is encouraging for sustainable eucalypt management. Planting density effects were evident early, with, for example, the 3588 trees/ha in the fertilizer × planting density having the tallest trees at 9 months and the largest stand basal area but smallest tree DBH at subsequent ages. Similar effects of planting density have been noted for *E. dunnii* seedlings and clones [25]. Planting density trade-offs between harvest tree size, rotation length, establishment costs, and stand productivity impact plantation economics.

While our preliminary evaluation of cultivars G2 and EH1, *C. torelliana*, *E. amplifolia*, and *Q. virginiana* suggests that all appear suitable for commercial BC production in Florida, evaluations of BCs made from various woods and other feedstocks have identified that feedstock and pyrolysis condition influence properties important for using BC as a soil amendment [26,27]. GCS' new BC production facility near Ft. Pierce, FL, scheduled to begin pilot scale testing in mid-2019, will preferably be using the cultivars G2 and EH1 and other eucalypts grown in nearby plantations. Since key objectives in BC production include minimizing the combustion of carbon, maximizing carbon content, and minimizing ash, it is imperative to ensure consistency of feedstock and the production operating environment. Known biomass characteristics, such as for the *G Series* cultivars [28], are likely to be factors in the selection of future eucalypt feedstocks.

Research on BC impact on SRWCs and forest trees in general outside Florida has generated mixed results for aspects ranging from environmental impacts to tree growth responses. When broadcast in a temperate hardwood stand in Ontario, Canada, the major short-term BC impact was an increase in limiting soil P and Ca [29]. One review of BC application in forest ecosystems found general improvements in soil physical, chemical, and microbial properties that were, however, BC-, soil-, and plant-specific [30]. A BC made from *E. marginata* decreased soil microbial carbon in a coarse soil [31], and BC added to a sandy desert soil did not significantly change soil physical properties [32]. Two BC types had different impacts on growth of young *Pinus elliottii* in subtropical China [33]. Varying doses of macadamia BC combined with two fertilizer rates had contrasting results on soil nutrients and ambiguous trends in the growth of young *E. nitens* [34]. BC did not enhance survival or growth of a *Eucalyptus* hybrid on degraded soils in southern Amazonia [35]. Compost and BC–compost mixes did not improve the performance of poplar, willow, and alder SRWCs [36]. As evidenced by presentations at a 2018 international SRWC conference [37], the biochar–fertilizer study reported here appears to be unique.

BC enhanced the soil properties of inherently poor Florida soils as well as the nutrient status of *E. grandis*, especially when applied together with organic amendments such as GE and/or chemical fertilizers. BC has a large cation exchange capacity, which facilitates retention of nutrients, particularly Ca, Mg, K, Fe, and Mn against leaching loss and thus enhances their efficient use by trees. In addition, BC has a large water holding capacity and thus improves water availability, which is especially important for Florida's sandy soils during the dry season. Due to high temperature and humidity, decomposition of organic materials in Florida's sandy soils is very rapid, and consequently these soils generally have a low organic matter content. BC can be a good organic amendment for these sandy soils, because it can stay in soil much longer than other organic materials, such as crop residues or manures.

Other potential *Eucalyptus* bioproducts may be classified as naturally occurring, generated by biochemical processes, or as the result of thermochemical processes [3,38]. Naturally occurring *Eucalyptus* bioproducts include wood products, terpenoids, phenolics, formylated phloroglucinol compounds, insecticides, repellants, antimicrobials, antifungals, and anticancers. Biorefineries such as a phosphoric lignocellulosic biorefinery [39] can produce the biochemicals lactate with parenteral and dialysis applications, succinate potentially leading to acrylic, lactic, muconic, and fumaric acids, alanine for supplements, seasonings, and antibiotics, and cellulose nanocrystals and nanofibrils for polymer nanocomposites. Sulfite paper mills and the Sulfite Pretreatment to Overcome Recalcitrant Lignin process [40] may produce jet fuel and graphene for products such as orthopedic medical implants. Thermochemical *Eucalyptus* bioproducts include biochar, syngas, and biomaterials whose carbon fiber may yield surgical implants, fabrics, filters, orthotics, chairs, beds, etc., and graphene for surgical implants, drug delivery, cancer therapy, imaging, detection of toxins, pollution, etc., graphene oxide, and batteries. These bioproducts have a broad and exciting range of applications for enhancing the value of SRWCs.

5. Conclusions

Two fast growing eucalypts adapted to Florida's climatic and edaphic conditions responded well to intensive culture in SRWC systems near Ft Pierce, Florida, USA. Plantations of the *E. grandis* x *E. urophylla* cultivar EH1 established on former citrus beds and managed at relatively low intensity were economically feasible; e.g., a planting density of 2071 trees/ha with three (original plus two coppice) 5 ± year rotations resulted in an NPV in excess of $750/ha at 6% discount rate and stumpage price of $11.02/green mt. At 3,588 trees/ha, EH1 had higher stand productivity at 41 months than at lesser densities, and an organic fertilizer generally increased its growth more than an inorganic fertilizer. The organic fertilizer combined with BC increased tree sizes of three *E. grandis G Series* cultivars on an infertile sandy soil. Given that these eucalypts were determined to be suitable for the production of BC, which in turn appears to be a useful soil amendment for their intensive culture, using BC for eucalypt plantation establishment in Florida could result in more sustainable management. High-quality feedstocks such as planted eucalypts in Florida are critical to producing consistently high-quality biochar with uniform quality and specifications.

Author Contributions: Conceptualization, D.L.R. and M.F.E.; Methodology, D.L.R., M.F.E., K.W.F., and Z.H.; Software, D.L.R., M.F.E., K.W.F., and Z.H.; Validation, D.L.R., M.F.E., K.W.F., and Z.H.; Formal Analysis, D.L.R., M.F.E., K.W.F., R.L., F.Z., P.J., Z.Z., and Z.H.; Investigation, D.L.R., M.F.E., K.W.F., R.L., F.Z., P.J., Z.Z., and Z.H.; Resources, D.L.R., M.F.E., K.W.F., Z.H.; Data Curation, D.L.R., M.F.E., K.W.F., R.L., F.Z., P.J., Z.Z., and Z.H.; Writing—Original Draft Preparation, D.L.R., M.F.E., K.W.F., R.L., F.Z., P.J., Z.Z., and Z.H.; Writing—Review and Editing, D.L.R., M.F.E., K.W.F., R.D.C., and Z.H.; Visualization, D.L.R., M.F.E., K.W.F., Z.H.; Supervision, Z.H.; Project Administration, R.D.C.; Funding Acquisition, Z.H.

Funding: This research received no external funding.

Acknowledgments: The authors gratefully acknowledge the direct and/or indirect support provided by the IRREC, GCS, GreenTechnologies, Evans Properties, US EcoGen, Becker Tree Farm, and ArborGen.

Conflicts of Interest: The authors declare no conflicts of interest.

References

1. Carle, J.; Homgren, P. Wood from planted forests: A global outlook 2005–2030. *For. Prod. J.* **2008**, *58*, 6–18.
2. Sims, R.E.H.; Hastings, A.; Schlamadinger, B.; Taylor, G.; Smith, P. Energy crops: current status and future prospects. *Glob. Chang. Biol.* **2006**, *12*, 2054–2076. [CrossRef]
3. Rockwood, D.L.; Rudie, A.W.; Ralph, S.A.; Zhu, J.; Winandy, J.E. Energy product options for *Eucalyptus* species grown as short rotation woody crops. *Int. J. Mol. Sci.* **2008**, *9*, 1361–1378. [CrossRef] [PubMed]
4. Rockwood, D.L. History and status of Eucalyptus improvement in Florida. *Int. J. For. Res.* **2012**, *2012*, 607879.
5. Rockwood, D.L.; Peter, G.F. Eucalyptus and Corymbia species for mulchwood, pulpwood, energywood, bioproducts, windbreaks, and/or phytoremediation. Florida Cooperative Extension Service Circular 1194. 2018. Available online: http://edis.ifas.ufl.edu/FR013 (accessed on 10 September 2018).
6. Fabbro, K.W.; Rockwood, D.L. Optimal management and productivity of Eucalyptus grandis on former phosphate mined and citrus lands in central and southern Florida: Influence of genetics and spacing. In *Proceedings of the 18th Biennial Southern Silvicultural Research Conference, Knoxville, TN, USA, 2–5 March 2015*, e-Gen; Tech. Rpt. SRS-212; U.S. Department of Agriculture, Forest Service, Southern Research Station: Asheville, NC, USA, 2016; pp. 510–517.
7. Ahmed, A.; Kurian, J.; Raghavan, V. Biochar influences on agricultural soils, crop production, and the environment: A review. *Environ. Rev.* **2016**, *24*, 495–502. [CrossRef]
8. Jeffery, S.; Abalos, D.; Prodana, M.; Bastos, A.C.; van Groenigen, J.W.; Hungate, B.A.; Verheijen, F. Biochar boosts tropical but not temperate crop yields. *Environ. Res. Lett.* **2017**, *12*, 053001. [CrossRef]
9. Ahmed, F.; Arthur, E.; Plauborg, F.; Razzaghi, F.; Korup, K.; Andersen, M.N. Biochar amendment of fluvio-glacial temperate sandy subsoil: Effects on maize water uptake, growth and physiology. *J. Agron. Crop Sci.* **2018**, *204*, 123–136. [CrossRef]
10. Blanco-Canqui, H. Biochar and soil physical properties. *Soil Sci. Soc. Am. J.* **2017**, *81*, 687–711. [CrossRef]
11. Bruun, E.W.; Petersen, C.T.; Hansen, E.; Holm, J.K.; Hauggaard-Nielsen, H. Biochar amendment to coarse sandy subsoil improves root growth and increases water retention. *Soil Use Manag.* **2014**, *30*, 109–118. [CrossRef]
12. Hussein, H.; Farooq, M.; Nawaz, A.; Al-Sadi, A.M.; Solaiman, Z.M.; Alghamdi, S.S.; Ammara, U.; Ok, Y.S.; Siddique, K.H. Biochar for crop production: potential benefits and risks. *J. Soils Sediments* **2017**, *17*, 685–716. [CrossRef]
13. Grand View Research. *Global Biochar Market Estimates and Forecast, 2012–2025*; Report issued; Grand View Research: San Francisco, CA, USA, 2017.

14. Du Plessis, M.; Kotze, H. Growth and yield models for *Eucalyptus grandis* grown in Swaziland. *South. For.* **2011**, *73*, 81–89. [CrossRef]

15. Kuo, K. Phosphorus. In *Methods of Soil Analysis, Part 3: Chemical Methods*; Sparks, D.L., Ed.; Soil Science Society of America: Madison, WI, USA, 1996; pp. 869–919.

16. Mehlich, A. Mehlich No. 3 soil test extractant: A modification of Mehlich No.2 extractant. *Commun. Soil Sci. Soc. Plant Anal.* **1984**, *15*, 1409–1416. [CrossRef]

17. Langholtz, M.; Carter, D.R.; Rockwood, D.L. *Assessing the Economic Feasibility of Short-Rotation Woody Crops in Florida*. Florida Cooperative Extension Service Circular 1516. 2007. Available online: http://edis.ifas.ufl.edu/FR169 (accessed on 10 September 2018).

18. Rockwood, D.L.; Dippon, D.R.; Lesney, M.S. *Woody Species for Biomass Production in Florida. Final Report 1983–1988*; ORNL/Sub/81-9050/7; Oak Ridge National Laboratory: Oak Ridge, TN, USA, 1988; 153p.

19. Rockwood, D.L.; Carter, D.R.; Stricker, JA. *Commercial Tree Crops on Phosphate Mined Lands*; FIPR Publication #03-141-225; Florida Institute of Phosphate Research: Bartow, FL, USA, 2008; 123p.

20. Zalesny, R.S., Jr.; Cunningham, M.W.; Hall, R.B.; Mirck, J.; Rockwood, D.L.; Stanturf, J.A.; Volk, T.A. Chapter 2. Woody Biomass from Short Rotation Energy Crops. In *ACS Symposium Book: Sustainable Production of Fuels, Chemicals, and Fibers from Forest Biomass*; Zhu, J., Zhang, X., Pan, X., Eds.; American Chemical Society: Washington, DC, USA, 2011; pp. 27–63.

21. Perez-Cruzado, C.; Merino, A.; Rodriguez-Soalleiro, R. A management tool for estimating bioenergy production and carbon sequestration in *Eucalyptus globulus* and *Eucalyptus nitens* grown as short rotation woody crops in north-west Spain. *Biomass Bioenergy* **2011**, *35*, 2839–2851. [CrossRef]

22. Morales, M.; Aroca, G.; Rubilar, R.; Acuna, E.; Mola-Yudego, B.; Gonzalez-Garcia, S. Cradle-to-gate life cycle assessment of *Eucalyptus globulus* short rotation plantations in Chile. *J. Clean. Prod.* **2015**, *99*, 239–249. [CrossRef]

23. Hinchee, M.; Rottman, W.; Mullinax, L.; Zhang, C.; Chang, S.; Cunningham, M.; Pearson, L.; Nehra, N. Short-rotation woody crops for bioenergy and biofuels applications. *In Vitro Cell. Dev. Biol.-Plant* **2009**, *45*, 619–629. [CrossRef] [PubMed]

24. Harper, R.J.; Sochacki, S.J.; Smettem, R.J.; Robinson, N. Bioenergy feedstock potential from short-rotation woody crops in a dryland environment. *Energy Fuels* **2010**, *24*, 225–231. [CrossRef]

25. Stape, J.L.; Binkley, D. Insights from full-rotation Nelder spacing trials with *Eucalyptus* in Sao Paulo, Brazil. *South. For.* **2010**, *72*, 91–98. [CrossRef]

26. Singh, B.; Singh, B.P.; Cowie, A.L. Characterisation and evaluation of biochars for their application as a soil amendment. *Aust. J. Soil Res.* **2010**, *48*, 516–525. [CrossRef]

27. Kloss, S.; Zehetner, F.; Dellantonio, A.; Hamid, R.; Ottner, F.; Liedtke, V.; Schwanninger, M.; Gerzabek, M.H.; Soja, G. Characterization of slow pyrolysis biochars: Effects of feedstocks and pyrolysis temperature on biochar properties. *J. Environ. Qual.* **2012**, *41*, 990–1000. [CrossRef]

28. Rockwood, D.L.; Tamang, B.; Kirst, M.; Zhu, J.Y. Field performance and bioenergy characteristics of four Eucalyptus grandis cultivars in Florida. In Proceedings of the 16th Biennial Southern Silvicultural Research Conference, Charleston, SC, USA, 15–17 February 2011; pp. 267–268.

29. Sackett, T.E.; Basiliko, N.; Noyce, G.L.; Winsborough, C.; Schurman, J.; Ikeda, C.; Thomas, S.C. Soil and greenhouse gas responses to biochar additions in a temperate hardwood forest. *GCB Bioenergy* **2015**, *7*, 1062–1074. [CrossRef]

30. Li, Y.; Hu, S.; Chen, J.; Muller, K.; Li, Y.; Fu, W.; Lin, Z.; Wang, H. Effects of biochar application in forest ecosystems on soil properties and greenhouse gas emissions: A review. *J. Soils Sediments* **2018**, *18*, 546. [CrossRef]

31. Dempster, D.; Gleeson, D.; Solaiman, Z.; Jones, D.; Murphy, D. Decreased soil microbial biomass and nitrogen mineralisation with Eucalyptus biochar addition to a coarse textured soil. *Plant Soil* **2012**, *354*, 311–324. [CrossRef]

32. Lin, Z.B.; Liu, Q.; Liu, G.; Cowie, A.L.; Bei, Q.C.; Liu, B.J.; Wang, X.J.; Ma, J.; Zhu, J.G.; Xie, Z.B. Effects of different biochars on *Pinus elliottii* growth, N use efficiency, soil N_2O and CH_4 emissions and C storage in a subtropical area of China. *Pedosphere* **2017**, *27*, 248–261. [CrossRef]

33. Mertens, J.; Germer, J.; de Araújo Filho, J.C.; Sauerborn, J. Effect of biochar, clay substrate and manure application on water availability and tree-seedling performance in a sandy soil. *Arch. Agron. Soil Sci.* **2017**, *63*, 969–983. [CrossRef]

34. Wrobel-Tobiszewska, A.; Boersma, M.; Adams, P.; Singh, B.; Franks, S.; Sargison, J.E. Biochar for eucalyptus forestry plantations. *Acta Hortic.* **2016**, *1108*, 55–62. [CrossRef]

35. De Farias, J.; Marimon, B.S.; de Carvalho Ramos Silva, L.; Petter, F.A.; Andrade, F.R.; Morandi, P.S.; Marimon-Junior, B.H. Survival and growth of native *Tachigali vulgaris* and exotic *Eucalyptus urophylla* × *Eucalyptus grandis* trees in degraded soils with biochar amendment in southern Amazonia. *For. Ecol. Manag.* **2016**, *368*, 173–182. [CrossRef]

36. Glisczynski, F.V.; Pude, R.; Amelung, W.; Sandhage-Hofmann, A. Biochar-compost substrates in short-rotation coppice: effects on soil and trees in a three-year field experiment. *J. Plant Nutr. Soil Sci.* **2016**, *179*, 574–583. [CrossRef]

37. Gardiner, E.S.; Ghezehei, S.B.; Headlee, W.L.; Richardson, J.; Soolanayakanahally, R.Y.; Stanton, B.J.; Zalesny, R.S., Jr. The 2018 Woody Crops International Conference, Rhinelander, Wisconsin, USA, 22–27 July 2018. *Forests* **2018**, *9*, 693. [CrossRef]

38. Rockwood, D.L.; Bowman, R.L. Medically related products obtainable from Eucalyptus trees. *Int. Biol. Rev.* **2017**, *1*, 1–10.

39. Castro, E.; Nieves, I.U.; Mullinnix, M.T.; Sagues, W.J.; Hoffman, R.W.; Fernández-Sandoval, M.T.; Tian, Z.; Rockwood, D.L.; Tamang, B.; Ingram, L.O. Optimization of dilute-phosphoric-acid steam pretreatment of *Eucalyptus benthamii* for biofuel production. *Appl. Energy* **2014**, *125*, 76–83. [CrossRef]

40. Wang, G.S.; Pan, X.J.; Zhu, J.Y.; Rockwood, D.L. Sulfite pretreatment to overcome recalcitrance of lignocellulose (SPORL) for robust enzymatic saccharification of hardwoods. *Biotechnol. Prog.* **2009**, *25*, 1086–1093. [CrossRef] [PubMed]

forests

MDPI

Article

Overhead Protection Increases Fuel Quality and Natural Drying of Leaf-On Woody Biomass Storage Piles

Obste Therasme, Mark H. Eisenbies and Timothy A. Volk *

Department of Forest and natural Resources Management, College of Environmental Science and Forestry, State University of New-York, Syracuse, NY 13210, USA; otherasm@syr.edu (O.T.); mheisenb@esf.edu (M.H.E.)
* Correspondence: tavolk@esf.edu; Tel.: +1-315-470-6774

Received: 19 February 2019; Accepted: 26 April 2019; Published: 1 May 2019

Abstract: Short-rotation woody crops (SRWC) have the potential to make substantial contributions to the supply of biomass feedstock for the production of biofuels and bioproducts. This study evaluated changes in the fuel quality (moisture, ash, and heating value) of stored spring harvested shrub willow (_Salix_ spp.) and hybrid poplar (_Populus_ spp.) chips with respect to pile protection treatments, location within the storage piles, and length of storage. Leaf-on willow and poplar were harvested in the spring, and wood chips and foliage with moisture content in the range of 42.1% to 49.9% (w.b.) were stored in piles for five months, from May to October 2016. Three protection treatments were randomly assigned to the piles. The control treatment had no cover (NC), so piles were exposed to direct solar radiation and rainfall. The second treatment had a canopy (C) installed above the piles to limit direct rainfall. The final treatment had a canopy plus a dome aeration system (CD) installed over the piles. Covering piles reduced and maintained the low moisture content in wood chip piles. Within 30 days of establishment, the moisture content in the core of the C pile decreased to less than 30%, and was maintained between 24%–26% until the end of the storage period. Conversely, the moisture content in the NC piles decreased in the first two months, but then increased to the original moisture content in the core (>45 cm deep) and up to 70% of the original moisture content in the shell (<45 cm deep). For all the treatments in the tested conditions, the core material dried faster than the shell material. The higher heating value (HHV) across all the treatments increased slightly from 18.31 ± 0.06 MJ/kg at harvest to 18.76 ± 0.21 MJ/kg at the end of the storage period. The lower heating value (LHV) increased by about 50% in the C and CD piles by the end of the storage period. However, in the NC piles, the LHV decreased by 3% in the core and 52% in the shell. Leaf-on SRWC biomass stored in piles created in late spring under climatic conditions in central and northern New York showed differing moisture contents when stored for over 60–90 days. Overhead protection could be used to preserve or improve the fuel quality in terms of the moisture content and heating value if more than two months of storage are required. However, the implementation of such management practice will depend on whether the end users are willing to pay a higher price for dryer biomass and biomass with a higher LHV.

Keywords: willow; poplar; wood chips; ash; LHV; HHV; storage; bioenergy

1. Introduction

To meet the goal of annually supplying a billion tons of biomass in the future, energy crops are projected to be the greatest source of biomass in the United States, with an estimated contribution in the range of 411 to 736 million tons by 2040 [1]. Short-rotation woody crops (SRWC) such as poplar (_Populus spp._) and willow (_Salix spp._) are an asset for the development of a biomass system for bioenergy and bioproducts. Willow has been in development in the United States and Canada for more than 30 years [2,3]. These crops have shown desirable characteristics such as high annual yield and the ability to regenerate by re-sprouting multiple stems after each harvest while having similarities to other woody biomass (i.e., heating value, cellulose, hemicellulose, and lignin content), as well as differences (i.e., bark to wood ratio) [4,5].

A SRWC biomass supply system includes all the processes from cultivation, harvesting, and collection, storage, and transportation to the end user, and may include some preprocessing steps depending on the conversion pathway. Management considerations may dictate that the harvested biomass material be stored prior to delivery to a conversion plant or depot [6]. Most recently, harvested willow biomass has been mixed with forest residues to produce biopower at a 60-MW facility (Black River) and power at another 22-MW facility (Lyonsdale) in northern New York (NY). Moisture content, ash content, heating value, and particle size distribution are among the key quality parameters for biopower systems, because they influence the handling and conversion systems' efficiency, and thus the cost of production. High moisture content in biomass feedstock may increase the transportation cost and affect the combustion behavior in terms of system efficiency and emissions when used in thermochemical conversion processes [7]. Moisture in biomass reduces the net calorific value, because a fraction of the heat is used to vaporize the water. Freshly harvested willow chips have a moisture content of 44 ± 2.2%, an ash content of 2.1 ± 0.59%, a lower heating value of 10.4 ± 0.52 MJ/kg, and a higher heating value of 18.6 ± 0.19 MJ/kg [8].

Several research studies have examined the effect of various factors such as the particle size, composition of the material, geometries of storage piles, geographic location, covering system, microbial communities, initial moisture, and length of storage on the fuel quality of stored wood chips [9–14]. High moisture content is an important driver for the development of fungal and bacterial communities that play a key role in reducing fuel quality and catalyzing the process of self-ignition [9]. Also, the temperature profile and moisture content in storage piles have links with weather conditions [14], which depend on the geographic location and time of year when material is stored.

To reduce the moisture content after harvest, biomass could be either stored in open-air piles [15–18] or subjected to drying measures [19,20], but both approaches have limitations. Unprotected biomass piles suffer high dry matter losses (up to 20%), while active drying requires energy consumption and incurs additional costs [15,20]. Whittaker et al. [18] investigated dry matter losses from two short rotation willow storage piles that were constructed in March and April, and found that there was a significant reduction in the moisture content in addition to an overall loss of 1.5 GJ/Mg in a six-month-old pile and 1.1 GJ/Mg in another four-month-old pile. Moisture is redistributed during the storage of wood chips in piles, resulting in a general differentiation between a wetter outside layer (shell) and a drier inside layer (core). Rates of changes in moisture in the shell and core also differ by season [21]. For instance, during late spring and the summer, the overall moisture content of the pile could decrease from 50% to 25% moisture content after several months. However, heavy rain and high air humidity could rapidly increase moisture in unprotected piles [14,22].

An alternative approach to maintain biomass quality during storage in piles is to protect the biomass from rewetting by snow or rain [22–25]. The use of gas-permeable and waterproof membranes that have several agricultural and construction applications can contribute to improved wood chip quality [25]. Covers can be deployed as a means to protect piles from precipitation and allow air flow at nominal cost with no direct energy input; however, limited information is available in the literature investigating these covers in the northeastern United States. The objective of this work was to investigate the effects of cover, storage duration, and depth in storage piles on the moisture content, ash content, and higher and lower heating value of a mixture of leaf-on willow and poplar biomass stored at the edge of the field.

2. Materials and Methods

2.1. Harvests and Storage

A mixture of approximately 50% willow and 50% poplar biomass crops were harvested in late spring starting on 26 May 2016 in a field located at Lafayette, NY (42.980° N, 76.112° W). The field was previously harvested in 2005 and 2010. Due to the layout of the plots in this field and the logistics of the commercial harvest operation, it would be challenging to harvest the poplar and willow plots separately. Leaf proportion from four samples collected from the harvested materials was less than 10% of the total mass. The site was planted in double rows in 1997 with poplar and willow cultivars including SV1 *(Salix × dasyclados)*, NM6 *(P. nigra × P. maximowiczii)*, S25 *(Salix eriocephala)*, S365 *(S. caprea × S. cinerea)*, and S301 *(S. interior)*. Harvesting was conducted with a New Holland FR9080 forage harvester equipped with a New Holland 130FB coppice header [26]. The cutting length was set at 33 mm (the largest size for its configuration). Eisenbies [27] reported the bulk density and particle size distribution of willow chips harvested with the same setting on the harvester. The harvested biomass was transported 21.3 km by dump truck to the Tully Experiment Station (College of Environmental Science and Forestry; 42.797° N, 76.120° W). A wood chip sample comprised of several scoops distributed around the dumped material was collected from each load.

The delivered chips were used to establish wood chips piles on flat and open ground at the site. The Köppen–Geiger climate class of the site is Dfb (snow, fully humid, warm summer) [28]. The piles contained 25 to 40 Mg of wood chips. A bucket loader was used to create the wood chip piles from the multiple loads. A 15-cm base of chips was maintained during pile construction to prevent soil contamination. Pile heights ranged from 2.4 to 3.5 m and the diameter of pile bases varied from 6.1 to 8.0 m.

2.2. Study Plan

To address the objectives of this study, we set up a repeated measurement experiment design. From a total of six storage piles created, we applied three protection treatments. We monitored the moisture content, ash content, and heating value in the piles for five consecutive months at two pile depth levels and two positions.

Three protection treatments were randomly assigned to the piles. Unprotected piles had no cover (NC), and were exposed to direct solar radiation and rainfall. The next group of piles had canopies (C) erected to limit direct rainfall, while allowing air to flow from one side to the other (Figure 1). The 3 × 6 m canopies were made of white powder-coated steel frames and white drawstring covers. The final group of piles had canopies plus a dome aeration system (CD). Dome aeration [24] is a method of passive aeration that is used in the aerobic biological degradation of biomass that facilitates gas (including water vapors) accumulation in the interior of the dome and flow through a pipe from the interior to the exterior of the pile. The dome structure was made of construction wood and chicken wire by creating an irregular hexahedron. A 10-cm diameter perforated, polyvinyl chloride pipe was placed at the top of the wooden structure to facilitate the air exchange between the ambient atmosphere and the core of the pile. Wood chips were mounded around the wooden structure to form a chip pile over six to eight perforated PVC pipes, followed by the erection of the canopy. Air was allowed to flow underneath the canopy between the tarp and the wood chip pile in order to maximize the flow of moisture out of the pile.

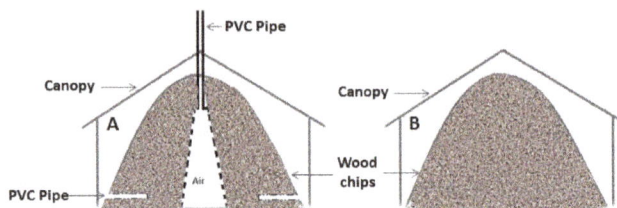

Figure 1. Cross-section of protected piles with dome (**A**) and protected piles (**B**).

2.3. Monitoring, Sampling, and Laboratory Analysis

Temperature probes were inserted at two depths in the shell (<45 cm) and core (~1/2 pile height) at two positions: two at the top and two on the sides of the piles. Precise insertion depth was achieved by attaching the probe to a threaded rod with a piece of vinyl tape and a small nut, inserting and raising a three-cm angle iron point side up to create a linear void, and inserting the probe inserted to the required depth. A detailed description of this technique and the cross-section of a pile profile showing the sample and temperature probe locations within the pile can be found in [21]. Then, the angle iron was removed, collapsing the void, and the rod was unscrewed and removed. Temperature loggers (HOBO U12-008) recorded pile temperatures automatically every 30 minutes. A weather station (HOBO U30) was installed on the site to monitor climatic conditions such as air temperature, precipitation, and relative air humidity for the first 100 days. For the remaining storage period, there was a technical issue with the power system of our installed weather station; thus, reported temperature and precipitation data for this period were gathered from PRISM (Parameter-elevation Regressions on Independent Slopes Model) climate data for Tully, NY [29]. PRISM uses elevation and measured climate data to develop climate data sets for smaller spatial grids across the United States [30].

The piles were monitored for five consecutive months; the canopies were not expected to hold up to the snow season, but freshly harvested biomass would be available by November. A minimum of six physical samples was collected monthly from the shell and the core of each pile. To have a better representation of each pile, samples were taken from both positions (the top and side) and depths (shell and core) of the pile. The shell samples were collected by using a drain spade, and the pile surface was recontoured. In order to minimize the disturbances, a custom augur was used to collect samples from the core of the pile by following the technique described by Eisenbies [21]. Then, the samples were weighed immediately to the nearest 0.1 g, and transferred to the laboratory. Moisture content was determined after drying at 60 °C until constant weight was achieved. Dried samples were ground using a Wiley mill and screened through a 0.5-mm screen before they were submitted for heating value and ash content determination. The higher heating value (HHV) was determined in accordance with ASTM (American Society for Testing and Materials) method D5865-13: The standard test method for the gross calorific value of coal and coke was completed using a Parr 6200 Oxygen bomb calorimeter [31]. The lower heating value (LHV) was calculated using the formula described by Krigstin and Wetzel [32] that takes into account the loss of energy associated with moisture and the heat of vaporization. The ash content was determined by combustion in a thermolyne muffle furnace (Model F30400) in accordance with the National Renewable Energy Laboratory NREL/TP-510-42622 method [33].

2.4. Statistical Analyses

The statistical data analysis was conducted in SAS v9.4 (Cary, NC, USA) by using the MIXED procedure. This experiment was a 3 × 2 × 2 factorial design with three protection treatments (no cover, NC; canopy, C; and canopy with a dome aeration system, CD), two depths (core and shell), and two positions (top and side). Wood chips samples were collected at regular time intervals (~four weeks); therefore, repeated measures analysis was performed. To take into account the autocorrelation resulting from the repeated measurements over time, a first-order autoregressive AR (1) covariance structure was applied. The model used includes all the main effects and the interaction terms: *(Model = Protection Depth Position Period Protection × Depth Protection × Position Protection × Period Depth×Position Depth × Period Position × Period Protection × Depth × Position Protection × Depth × Period Protection × Position × Period Depth × Position × Period Protection × Depth × Position × Period).* The differences of the least squares means were calculated for the factors that are included in the model. Significant differences were claimed for P-values less than 0.05, but the interaction tests were evaluated at a more liberal *p*-value of 0.10 [34].

3. Results

3.1. Weather Conditions and Pile Temperatures

During the first three months, the relative air humidity oscillated between 60–94% and was less than 70% for 10 consecutive days starting from 17 June (Figure 2). Daily mean temperature for the site ranged from 1.1 to 25.5 °C for the entire storage period, and the average daily mean temperature was 17 °C. From 1 June to 1 November, the site received more than 474 mm of rainfall. The months of June and July received about 115 mm of rainfall, while the month of October had 170 mm of rainfall.

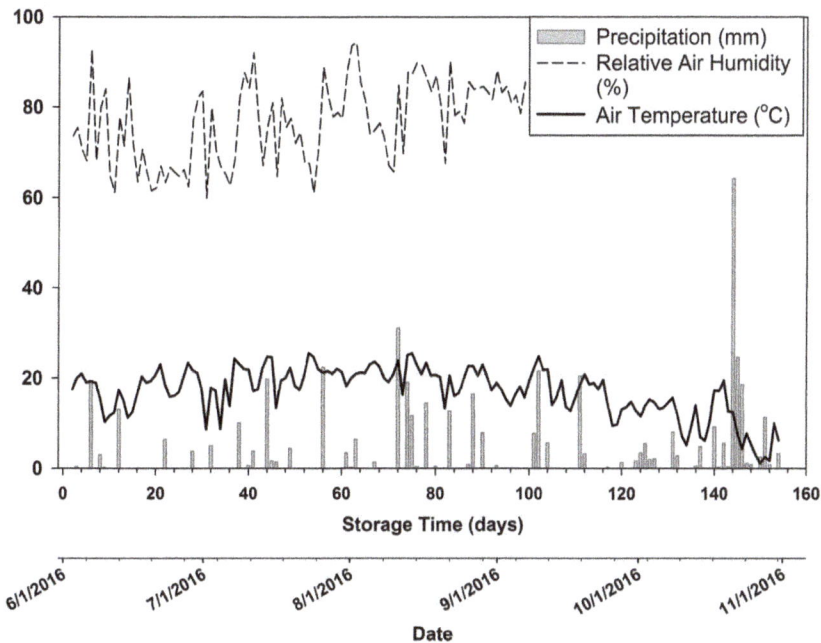

Figure 2. Daily weather patterns at the site where willow and poplar chips were stored in piles for five months (day zero is 31 May 2016).

Temperature in the piles increased rapidly, nearing 60 °C after the first week of storage and decreasing thereafter (Figure 3). The temperature in the cores of the CD piles decreased to 35 °C within 30 days of storage, while the temperature in the cores of the NC piles reached the same level after 45 days of storage. After two months, the temperatures in the shells and cores of the NC piles were generally higher than those of both the C and CD piles. The aggregate monthly temperature in the cores and the shells indicated that the shell had a lower temperature than the core for the C and CD piles. However, temperatures in the cores of the NC piles were 9.9 °C higher than those of the shell for the first month, and 1 °C higher for the second month of storage; however, during the third and fourth months, the temperatures in the shells of the NC piles were higher than those in the core.

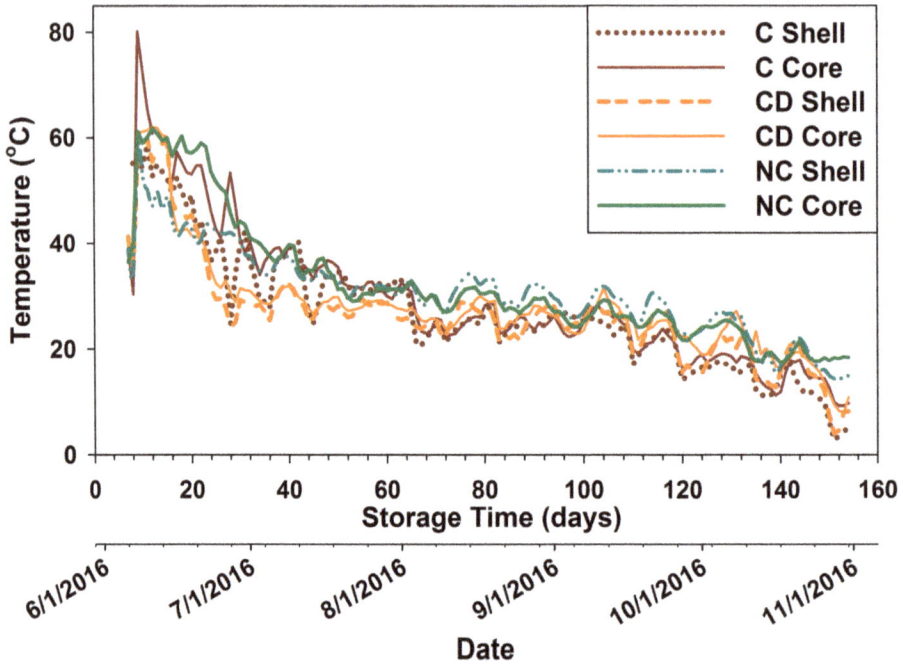

Figure 3. Daily mean temperature in the core (about $\frac{1}{2}$ depth) and shell (<45-cm depth) of C, NC, and CD piles of willow and poplar wood chips (NC—no cover on pile; C—covered pile; CD—covered pile with dome ventilation).

3.2. Moisture

The three-way interaction of depth, protection, and period was significant for moisture content ($p = 0.0002$) (Table 1). The cores of the NC piles were significantly drier than the shells after 60 days of storage. The C and CD piles had significantly drier cores within the first 90 days of storage, but beyond this period, the shells were slightly drier than the cores, with no statistical significance (Figure 4). The interaction between position and depth was significant ($p = 0.0025$).

Core samples collected from both the side and the top of the pile were not significantly different ($p = 0.7529$) to each other, but shell samples from the side were 8.8% drier than the shell samples from the top of the piles ($p < 0.0001$). The C piles were drier than the NC piles by 9% in the core and 18.6% in the shell. However, the dome feature did not result in significant moisture differences when compared to the C piles (Figure 4), suggesting that passive ventilation in piles of this size did not facilitate additional drying.

Table 1. Analyses of variance (ANOVA) for moisture, ash, higher heating value (HHV) and lower heating value (LHV) of willow and poplar chip piles that were sampled over six month period (June–October).

Source of Variation	df [a]	Moisture		Ash		df	HHV		LHV	
		F Value	p > F	F Value	p > F		F Value	p > F	F Value	p > F
Protection	2	18.05	0.0213	3.27	0.1764	2	1.12	0.4321	30.43	0.0102
Depth	1	38.85	<0.0001	12.3	0.0008	1	0.33	0.5692	18.74	<0.0001
Position	1	12.91	0.0006	0.86	0.3581	1	0.01	0.919	9.06	0.0043
Period	5	13.79	<0.0001	6.6	<0.0001	3	7.88	0.0002	24.31	<0.0001
Error (Protection)	3	-	-	-	-	3	-	-	-	-
Depth × Protection	2	3.87	0.0255	0.49	0.614	2	0.21	0.81	0.52	0.5997
Depth × Period	5	4.12	0.0025	2.29	0.0555	3	1.5	0.2276	3.48	0.0234
Depth × Position	1	9.89	0.0025	1.61	0.209	1	0.1	0.7593	6.94	0.0115
Protection × Period	10	10.96	<0.0001	1.26	0.2735	6	0.4	0.8757	16.95	<0.0001
Protection × Position	2	0.87	0.4236	1.96	0.1486	2	1.49	0.2369	0.98	0.3826
Position × Period	5	1.13	0.3518	3.12	0.0138	3	0.32	0.8104	2.02	0.1248
Protection × Position × Period	10	1.23	0.2854	0.92	0.5176	6	1.29	0.2823	1.86	0.1091
Protection × Depth × Position	2	0.56	0.5762	0.09	0.9117	2	1.05	0.3577	0.28	0.7541
Depth × Position × Period	5	1.74	0.1371	1.34	0.2576	3	0.04	0.9881	2.79	0.0511
Protection × Depth × Period	10	4.03	0.0002	0.78	0.6463	6	1.79	0.1234	3.07	0.0132
Protection × Depth × Position × Period	10	0.88	0.55	0.76	0.6647	6	0.88	0.5177	1.48	0.2064
Error	68 [b]	-	-	-	-	45	-	-	-	-
Total Error	142 [c]	-	-	-	-	95	-	-	-	-

[a] Degree of freedom; [b] df = 65 for ash; [c] df = 139 for ash.

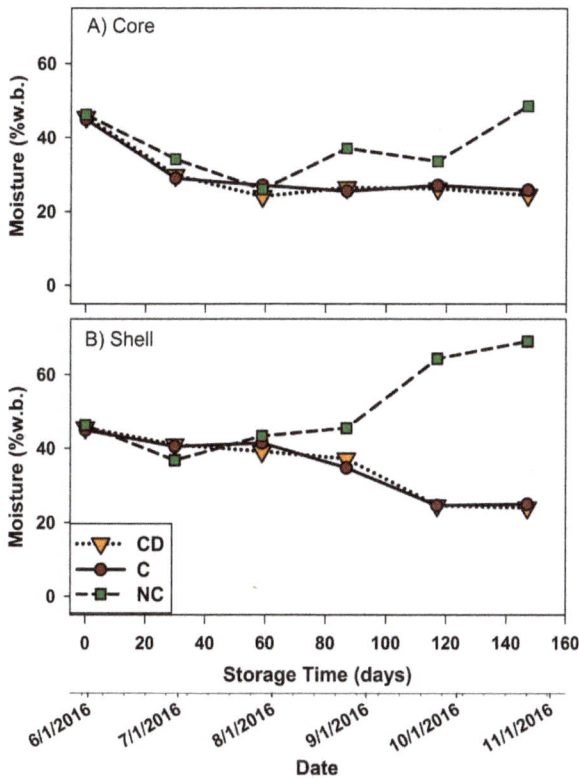

Figure 4. Variation of the moisture content of willow and poplar wood chips stored for five months in piles with three different treatments (NC—no cover on pile; C—covered pile; CD—covered pile with dome ventilation); (**A**) samples from the core; (**B**) samples from the shell.

Cover was a significant factor in decreasing and maintaining low moisture content. The moisture content in the core of both types of C piles decreased from 45% to less than 30% within 30 days of storage, and remained between 24–27% for the remainder of the storage period. Meanwhile, the biomass in the shells did not dry as quickly as that in the cores of the C piles, which could be attributed to the condensation of water vapors that migrated from the core of the piles. The most significant decrease of moisture content in the shells of the C piles occurred between 60–120 days of storage, and after the moisture content in the core had already equilibrated. Unlike the C piles, the moisture content of the NC piles initially decreased from 46% to 37% in the shell, and 26% in the core after 60 days of storage, and increased thereafter until the end of the storage period.

3.3. Higher and Lower Heating Values

The higher heating value (HHV) of stored willow chips was significantly affected by the storage time ($p = 0.0002$), but not by the factors such as cover and sample location. There was a significant linear relationship (HHV = $18.355 + 0.0031$ t, $R^2 = 0.18$, $p < 0.0001$ and $n = 96$) between the HHV and storage time across all the treatments and sample locations (Figure 5), but the change over time was small. The results suggested that the mid-term (up to 150 days) storage of leaf-on willow and poplar chips immediately after harvest in piles increased slightly from 18.31 MJ/kg to 18.76 MJ/kg.

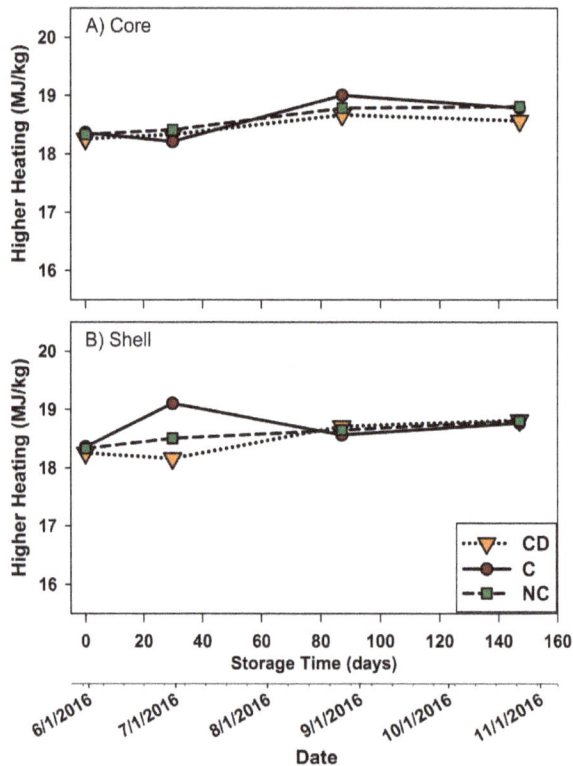

Figure 5. Variation of the higher heating value of willow and poplar wood chips stored for five months in piles with three different treatments (NC—no cover on pile; C—covered pile; CD—covered pile with dome ventilation); (**A**) samples from the core; (**B**) samples from the shell.

As for moisture content, the three-way interaction between depth, protection, and period was significant for the LHV ($p = 0.013$) (Table 1, Figure 6). Comparisons of the LHV in the shells and the cores during the storage period indicated that the LHV of the NC piles was higher in the cores than in the shells for the entire storage period. The same observation was true for the C and CD piles for the first 117 days of the trials, but by 147 days, the LHV in the shell and core were not statistically different. The LHV in the core was 4.35 MJ/kg greater than the shell for the NC piles, and the difference was less than 0.29 MJ/kg for the CD piles.

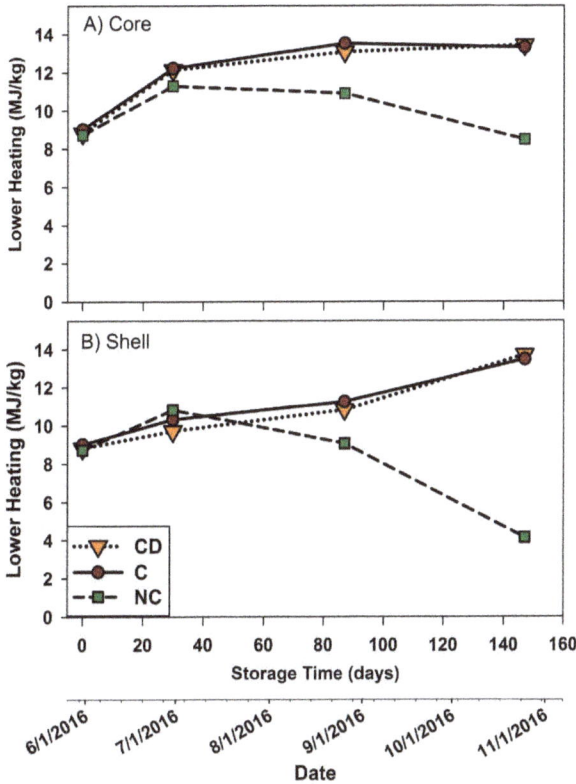

Figure 6. Variation of the lower heating value of willow and poplar wood chips stored for five months in piles with three different treatments (NC—no cover on pile; C—covered pile; CD—covered pile with dome ventilation); (**A**) samples from the core; (**B**) samples from the shell.

The interaction between depth, position, and period was significant for the LHV ($p = 0.051$). The LHV of the core samples from the side of the piles was 11.3 MJ/kg, and was not significantly different than the LHV of the core samples from the top ($p = 0.79$) and the shell samples from the side ($p = 0.23$) for the entire storage trial. However, the LHV of the shell samples from the top of the piles was 9.2 MJ/Kg, which was significantly lower than the LHV of the shell samples from the side (1.1 to 3 MJ/kg) and the core samples from the top (1.5 to 3.3 MJ/kg).

Pile protection can contribute to the enhanced fuel quality of willow and poplar chips during storage through an increase in the LHV over its initial value and providing a higher value than the NC piles. Initially, the LHV of the willow and poplar biomass chips was 8.85 MJ/kg due to the high moisture content of the freshly harvested material. Although the differences between the LHV of the C and NC piles were not significant after 30 days of storage ($p > 0.73$), cover became a significant factor resulting in a higher LHV for the rest of the trial. Furthermore, the LHV of the C and CD piles increased over time and remained higher until the end of the storage period. However, in the case of the NC piles, the LHV increased slightly during the first 30 days, and subsequently decreased to about 8.50 MJ/kg in the core and 4.15 MJ/kg in the shell at the end of the trial. These observations suggest that leaf-on willow and poplar biomass that is stored immediately after being harvested for a very short term does not require any protection to get an increase of the LHV, but protection would be necessary to maintain or improve LHV if storage was to occur for more than 60 days.

3.4. Ash

There was a significant DepthxPeriod interaction ($p = 0.055$) and a significant PositionxPeriod interaction for ash content ($p = 0.0138$) (Table 1). While there were statistical differences in ash content, mean values were all within a small range of 2.1% to 3.3% in the shell and in the core (Figure 7). The ash content of samples from the shell was more variable over time than the values in the core, and was 0.37% higher than that of the samples from the core across the entire trial.

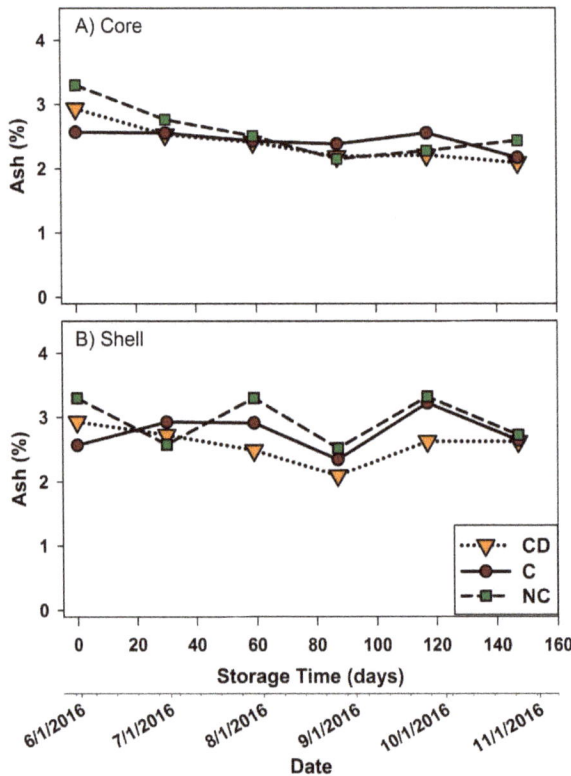

Figure 7. Variation of the ash content of willow and poplar wood chips stored for five months in piles with three different treatments (NC—no cover on pile; C—covered pile; CD—covered pile with dome ventilation); (**A**) samples from the core; (**B**) samples from the shell.

4. Discussion

Studies on wood chip storage in Europe [12,15,25] corroborate the findings of this study in regards to pile protection. Krzyżaniak [25] found that moisture content decreased from 52.5% to 20.0% over one year when biomass materials were protected with covers made of gas-permeable materials, while observing increases in moisture content for unprotected piles. The benefits associated with covering piles occur when the material is stored for longer periods of time, which will vary based on the timing of the harvest of the material and local weather conditions. In this trial, the benefits of covering the pile were most apparent after 60 days of storage.

The significant decrease in the moisture content of the NC piles during the first two months of this study could be attributed to the high temperature in the pile, low relative air humidity, and low cumulative rainfall during the first two months (Figure 2). Temperature increases in these storage piles primarily occurred due to wood respiration and microbial activities facilitated by high initial moisture content and the presence of leaves and bark in the wood chips. These conditions favor the transfer of moisture from the wood chips to the atmosphere based on the principle that moisture will migrate from the more humid (wet chips) to the less humid (dried air) environment.

Although not measured directly in this study, shoot diameter and branchiness are important parameters, as they can indirectly influence the quality of the wood chip. Previous studies reported models to predict the bark content of willow and poplar stems when the diameter is known [35,36]. Bark content decreases as the stem diameter and branchiness increase. The bark content of two three-year-old willow cultivars from two sites ranged from 14.9% to 22.6% [36]. The mean bark content of two-year-old poplar (cultivar Lux) decreased from 33.9% to 15.1% as the stem diameter increased from 1 to 9 cm [35]. Results from two experimental plots of hybrid poplar located in northern Italy demonstrated the positive correlation between nitrogen content and branchiness [37].

Increased rainfall, the lower pile temperature, and the lower air temperature thereafter were responsible for the increase in the moisture content of the NC piles (Figure 2, Table 2). Although the dome system did not show better performance in terms of moisture content than the canopy system alone, it allowed a faster dissipation of heat from the CD piles. This resulted from the higher surface to volume ratio of the CD pile, and the additional cooling by the perforated pipes installed on the side and top of the piles. The wooden structure from the center of the CD piles increased the surface area of the piles.

Table 2. Changes in moisture content from previous sampling period and cumulative rainfall over one, five, and 10 days prior to the dates when the samples were collected from poplar and willow chip piles (NC—no cover on pile; C—covered pile; CD—covered pile with dome ventilation). A positive sign indicates an increase in moisture content, and a negative sign shows a decrease in moisture content).

Storage Time (Days)	Moisture Content Differences (% Point)						Cumulative Rainfall in Days Prior to Sampling (mm)		
	NC		C		CD		1 Day	5 Days	10 Days
	Shell	Core	Shell	Core	Shell	Core			
30	−9.5	−12.1	−4.4	−16.0	−4.6	−15.8	0	3.8	10.2
59	+6.5	−8.1	+0.9	−1.9	−1.8	−5.7	0	22.4	26.8
87	+2.2	+11.1	−6.7	−1.5	−2.0	+2.4	0	12.6	27.4
117	+18.8	−3.5	−10.2	+1.5	−12.6	−0.3	0	3.2	23.6
147	+4.7	+15.0	+0.4	−1.3	−0.6	−1.8	18.6	113.4	127.6

The achievement of similar moisture content in the shells of the C and CD piles during the last two sampling periods suggested that the chip piles trend toward equilibrium moisture content. The equilibrium moisture content will vary relative to the ambient air and pile temperatures, precipitation patterns, and humidity conditions in the surroundings [32].

Maintaining low moisture content during storage is desirable to preserve feedstock quality as a fuel source. Low moisture content limits the associated dry matter loss and spontaneous combustion risk [22]. Whittaker et al. [18] reported dry matter loss in the range of 13% to 23% with the higher losses occurring in wetter samples. Other advantages are the reduction of transportation cost per ton of dry matter and better combustion efficiency. The combustion of wet biomass is less efficient; a fraction of the heat produced by the combustion process is required to evaporate the moisture in the biomass, thus reducing the lower heating value of the biomass.

Numerous studies on wood storage have reported no significant change or a decrease of the higher heating value [13,15,21,38], while others have reported a slight increase of heating value [10,16,25] at the end of the storage. Eisenbies [21] found stable HHVs in the range of 18.6 to 19.0 MJ/kg during a six-month winter storage of leaf-off willow biomass in the northeastern United States. Alternatively, a small increase of the HHV of willow biomass from 19.5 to 20.0 MJ/kg was reported in studies conducted in Poland [25]. The increase of HHV in that case, and in our study, may be associated with the degradation of the low-energy content components (e.g., leaf material) that were easily accessible to the microbial community in the storage piles. Krigstin and Wetzel [32] explained in their review that changes to the proportion of components in the biomass will affect energy content. For example, the preferential decomposition of hemicellulose will result in increased heating values, but the loss of extractives or the preferential fungal decomposition of lignin would tend to reduce heating values.

Despite the potential benefit afforded by cover in terms of increased LHV and the decrease of moisture content, the adoption of a storage pile protection system by SRWC growers in the northeastern United States would be affected by whether the value of biomass includes moisture or energy content, or is just based on as delivered mass. The current market for woody biomass in upstate NY primarily pays for wood chips based on the mass of material received, although some smaller facilities have begun to incorporate moisture content into their price. Assuming that the price of biomass is $25/Mg as received, and using the LHV of the material, the end user would pay about $2.77–2.87/MJ at the time of harvest. For NC piles, the price would decrease slightly, and then increase to $3.95/MJ at the end of the storage period (Table 3). For C piles, the price would steadily decrease over time to be $1.87/MJ at the end of the storage period.

Table 3. Biomass price and gross revenue for a storage pile with fuel purchase on a wet weight basis or an energy basis.

Protection	Storage Duration (Days)	Moisture (%)	LHV (MJ/Mg)	Biomass Price ($/MJ)	Gross Revenue ($) for a 35-Mg Pile			Biomass Price ($/Mg)	Gross Revenue ($) for a 35-Mg Pile		
					Paid $25/Mg as Received				**Paid $2.81/MJ**		
					0% DML [a]	5% DML	15% DML		0% DML	5% DML	15% DML
NC	0	46.2	8.7	2.87	875	875	875	24.52	858	858	858
NC	30	35.4	11.1	2.26	729	721	707	31.08	906	897	879
NC	59	34.6	-	-	720	705	677	-	-	-	-
NC	87	41.3	10	2.5	801	777	729	28.07	900	873	819
NC	117	48.9	-	-	921	884	810	-	-	-	-
NC	147	58.8	6.3	3.95	1141	1084	970	17.77	811	770	689
C	0	44.9	9	2.77	875	875	875	25.33	887	887	887
C	30	34.7	11.3	2.22	739	731	716	31.69	936	927	908
C	59	34.2	-	-	733	718	689	-	-	-	-
C	87	30.1	12.4	2.02	690	669	628	34.83	961	932	875
C	117	25.8	-	-	650	624	572	-	-	-	-
C	147	25.3686	13.4	1.87	646	613	549	37.64	972	924	826

[a] DML: dry matter loss. The gross revenue is estimated under three assumed scenarios: no account for dry matter loss (DML = 0), 1% monthly loss (DML = 5%), and 3% monthly dry matter loss (DML = 15).

From the SRWC grower's perspective, the value of a 35-Mg pile of chips that are purchased at $25/Mg as received (no dry matter loss is assumed) is greatest ($1141) at the end of the storage period for the NC piles, but greatest at the start of the storage period ($875) for the C piles. Factoring in different rates of dry matter loss [39,40] does not change the decision on when to sell chips. From the grower's point of view, covering storage piles does not make sense, because there is additional cost and a loss of revenue when chips are purchased on a weight as-received basis.

If end users want to stimulate growers to bring better quality biomass that has lower moisture and higher LHV, paying for chips on LHV should be considered. Using a fixed price of $2.81/MJ, the value of a 35-Mg pile for a grower is greatest at 30 days in a NC pile and at the end of the storage period for C piles. After five months, the gross value of the chips in the C piles is about 17% greater than that of the NC piles. Comparing the highest value for the NC and C piles shows that chips in the C pile after five months ($972) are worth about 7% more than the chips in the NC pile after one month ($906). When dry matter loss is factored in, the pattern is the same for the NC piles, with the highest gross value occurring at one month, but the greatest value for C piles occurs after about three months, rather than five months. There are dynamics associated with changes in the quality of biomass in storage piles, and the way that chips are purchased that will impact the supply chain dynamics of chips from SRWC.

Ash content is one of the most critical parameters for the quality testing of biomass to be used in the thermochemical conversion process. The International Standardization Organization specified three classes (A1.0, A1.5, and A3.0) for ash content in the ISO 17225 standard for solid biofuels [41]. Overall, for all the samples collected during this storage trial, 77% complied with class A3.0. However, 95% of the samples had ash content less than 3.5%, which is lower than many herbaceous biomass sources [42]. Throughout the entire trial, the ash content of the chips was consistent, and there was little variation. Consistency in the ash content with a limited amount of variation in biomass supply over time is important for end users [42]. If end users require lower ash values than are present in SRWC from this kind of harvesting and storage system, then pretreatments such as hot water extraction, washing, or air classification could be employed [43,44], or different types of woody material could be blended.

5. Conclusions

This study characterizes the changes in biomass quality (moisture, ash, and heating values) during the storage of SRWC chips over a five-month period with respect to overhead protection and sample location. Storage piles constructed in late spring went through a natural drying process during the first 60 days. For a longer storage time, covering piles impacted the changes in moisture content. The LHV varied based largely on moisture content, because the HHV variation was small for the duration of the study. Ash content was fairly low and consistent. A covering system is desirable and perhaps necessary to maintain the fuel quality (moisture and LHV) of biomass stored for more than two months, such as wood chip piles in New York state or other regions with similar climatic conditions. However, how end users value wood chips will influence how storage piles are managed at the edge of the field.

Author Contributions: Conceptualization, O.T., M.H.E., and T.A.V.; formal analysis, O.T. and M.H.E.; investigation, O.T.; data curation, O.T.; writing—original draft preparation, O.T.; writing—review and editing, M.H.E., O.T. and T.A.V.; visualization, O.T.; project administration and funding acquisition, T.A.V.

Funding: Funding to complete this research was provided by the US Department of Energy Bioenergy Technologies Office under award number DE- EE0002992, the New York State Energy Research and Development Authority (NYSERDA) Award 30713, and the Agriculture and Food Research Initiative Competitive Grant No. 2012-68005-19703 from the USDA National Institute of Food and Agriculture.

Acknowledgments: The authors would like to thank Karl Hallen for his contribution in the construction of the storage piles, Qian Wang, Dana J. Carris, Eleanor C. Clark, Allison Rutherford and Vanessa Gravenstine for their contribution in samples collection and processing. We thank Dr. Stephen V. Stehman, Distinguished Teaching Professor at SUNY ESF, for reviewing the sampling design and analysis. The comments from three anonymous reviewers contribute to the improvement of this manuscript.

Conflicts of Interest: The authors declare no conflict of interest.

References

1. Brandt, C.C.; Davis, M.R.; Davison, B.; Eaton, L.M.; Efroymson, R.A.; Hilliard, M.R.; Kline, K.; Langholtz, M.H.; Myers, A.; Sokhansanj, S.; et al. *2016 Billion-Ton Report: Advancing Domestic Resources for a Thriving Bioeconomy, Volume 1: Economic Availability of Feedstocks*; Oak Ridge National Lab. (ORNL): Oak Ridge, TN, USA, 2016.

2. Amichev, B.Y.; Hangs, R.D.; Konecsni, S.M.; Stadnyk, C.N.; Volk, T.A.; Bélanger, N.; Vujanovic, V.; Schoenau, J.J.; Moukoumi, J.; Van Rees, K.C. Willow Short-Rotation Production Systems in Canada and Northern United States: A Review. *Soil Sci. Soc. J.* **2014**, *78*, 78. [CrossRef]

3. Volk, T.; Abrahamson, L.; Nowak, C.; Smart, L.; Tharakan, P.; White, E. The development of short-rotation willow in the northeastern United States for bioenergy and bioproducts, agroforestry and phytoremediation. *Biomass Bioenergy* **2006**, *30*, 715–727. [CrossRef]

4. Volk, T.A.; Abrahamson, L.; Buchholz, T.; Caputo, J.; Eisenbies, M. Development and Deployment of Willow Biomass Crops. In *Cellulosic Energy Cropping Systems*; Wiley: Hoboken, NJ, USA, 2014; pp. 201–217.
5. Volk, T.A.; Eisenbies, M.H.; Heavey, J.P. Recent developments in shrub willow crops in the U.S. for bioenergy, bioproducts and bioremediation. In Proceedings of the Biomass and energy crops V, Brussels, Belgium, 20–22 October 2015; Association of Applied Biologists: Wellesbourne, UK, 2015; pp. 1–10.
6. Lamers, P.; Tan, E.C.D.; Searcy, E.M.; Scarlata, C.J.; Cafferty, K.G.; Jacobson, J.J. Strategic supply system design—A holistic evaluation of operational and production cost for a biorefinery supply chain: Strategic biorefinery feedstock supply system design. *Biofuels Bioprod. Biorefining* **2015**, *9*, 648–660. [CrossRef]
7. Yang, Y.; Sharifi, V.; Swithenbank, J. Effect of air flow rate and fuel moisture on the burning behaviours of biomass and simulated municipal solid wastes in packed beds. *Fuel* **2004**, *83*, 1553–1562. [CrossRef]
8. Eisenbies, M.H.; Volk, T.A.; Posselius, J.; Foster, C.; Shi, S.; Karapetyan, S. Evaluation of a Single-Pass, Cut and Chip Harvest System on Commercial-Scale, Short-Rotation Shrub Willow Biomass Crops. *BioEnergy Res.* **2014**, *7*, 1506–1518. [CrossRef]
9. Noll, M.; Jirjis, R. Microbial communities in large-scale wood piles and their effects on wood quality and the environment. *Appl. Microbiol. Biotechnol.* **2012**, *95*, 551–563. [CrossRef] [PubMed]
10. Brand, M.A.; De Muñiz, G.I.B.; Quirino, W.F.; Brito, J.O. Storage as a tool to improve wood fuel quality. *Biomass Bioenergy* **2011**, *35*, 2581–2588. [CrossRef]
11. Filbakk, T.; Høibø, O.A.; Dibdiakova, J.; Nurmi, J. Modelling moisture content and dry matter loss during storage of logging residues for energy. *Scand. J.* **2011**, *26*, 267–277. [CrossRef]
12. Pari, L.; Civitarese, V.; Del Giudice, A.; Assirelli, A.; Spinelli, R.; Santangelo, E. Influence of chipping device and storage method on the quality of SRC poplar biomass. *Biomass Bioenergy* **2013**, *51*, 169–176. [CrossRef]
13. Pari, L.; Brambilla, M.; Bisaglia, C.; Del Giudice, A.; Croce, S.; Salerno, M.; Gallucci, F. Poplar wood chip storage: Effect of particle size and breathable covering on drying dynamics and biofuel quality. *Biomass Bioenergy* **2015**, *81*, 282–287. [CrossRef]
14. Bedane, A.H.; Afzal, M.T.; Sokhansanj, S. Simulation of temperature and moisture changes during storage of woody biomass owing to weather variability. *Biomass Bioenergy* **2011**, *35*, 3147–3151. [CrossRef]
15. Lenz, H.; Idler, C.; Hartung, E.; Pecenka, R. Open-air storage of fine and coarse wood chips of poplar from short rotation coppice in covered piles. *Biomass Bioenergy* **2015**, *83*, 269–277. [CrossRef]
16. Barontini, M.; Scarfone, A.; Spinelli, R.; Gallucci, F.; Santangelo, E.; Acampora, A.; Jirjis, R.; Civitarese, V.; Pari, L. Storage dynamics and fuel quality of poplar chips. *Biomass Bioenergy* **2014**, *62*, 17–25. [CrossRef]
17. Hofmann, N.; Mendel, T.; Schulmeyer, F.; Kuptz, D.; Borchert, H.; Hartmann, H. Drying effects and dry matter losses during seasonal storage of spruce wood chips under practical conditions. *Biomass Bioenergy* **2017**, *111*, 196–205. [CrossRef]
18. Whittaker, C.; Yates, N.E.; Powers, S.J.; Misselbrook, T.; Shield, I. Dry Matter Losses and Greenhouse Gas Emissions from Outside Storage of Short Rotation Coppice Willow Chip. *BioEnergy Res.* **2016**, *9*, 288–302. [CrossRef]
19. Gigler, J.; Van Loon, W.; Vissers, M.; Bot, G. Forced convective drying of willow chips. *Biomass Bioenergy* **2000**, *19*, 259–270. [CrossRef]
20. Shahrukh, H.; Oyedun, A.O.; Kumar, A.; Ghiasi, B.; Kumar, L.; Sokhansanj, S. Net energy ratio for the production of steam pretreated biomass-based pellets. *Biomass Bioenergy* **2015**, *80*, 286–297. [CrossRef]
21. Eisenbies, M.H.; Volk, T.A.; Patel, A. Changes in feedstock quality in willow chip piles created in winter from a commercial scale harvest. *Biomass Bioenergy* **2016**, *86*, 180–190. [CrossRef]
22. Shinners, K.J.; Wepner, A.D.; Muck, R.E.; Weimer, P.J. Aerobic and Anaerobic Storage of Single-pass, Chopped Corn Stover. *BioEnergy Res.* **2011**, *4*, 61–75. [CrossRef]
23. Shah, A.; Darr, M.J.; Webster, K.; Hoffman, C. Outdoor Storage Characteristics of Single-Pass Large Square Corn Stover Bales in Iowa. *Energies* **2011**, *4*, 1687–1695. [CrossRef]
24. Trois, C.; Polster, A. Effective pine bark composting with the Dome Aeration Technology. *Waste Manag.* **2007**, *27*, 96–105. [CrossRef] [PubMed]
25. Krzyżaniak, M.; Stolarski, M.J.; Niksa, D.; Tworkowski, J.; Szczukowski, S. Effect of storage methods on willow chips quality. *Biomass Bioenergy* **2016**, *92*, 61–69. [CrossRef]
26. Eisenbies, M.H.; Volk, T.A.; Posselius, J.; Shi, S.; Patel, A. Quality and Variability of Commercial-Scale Short Rotation Willow Biomass Harvested Using a Single-Pass Cut-and-Chip Forage Harvester. *BioEnergy Res.* **2014**, *8*, 546–559. [CrossRef]

27. Eisenbies, M.H.; Volk, T.A.; Therasme, O.; Hallen, K. Three bulk density measurement methods provide different results for commercial scale harvests of willow biomass chips. *Biomass Bioenergy* **2019**, *124*, 64–73. [CrossRef]

28. Kottek, M.; Grieser, J.; Beck, C.; Rudolf, B.; Rubel, F. World Map of the Köppen-Geiger climate classification updated. *Meteorol. Z.* **2006**, *15*, 259–263. [CrossRef]

29. PRISM Climate Group, Oregon State University. Available online: http://prism.oregonstate.edu (accessed on 27 December 2018).

30. Daly, C.; Halbleib, M.; Smith, J.I.; Gibson, W.P.; Doggett, M.K.; Taylor, G.H.; Curtis, J.; Pasteris, P.P. Physiographically sensitive mapping of climatological temperature and precipitation across the conterminous United States. *Int. J. Clim.* **2008**, *28*, 2031–2064. [CrossRef]

31. *ASTM Standard Test Method for Gross Calorific Value of Coal and Coke*; ASTM standard D5865; American Society for Testing and Materials: West Conshohocken, PA, USA, 2013.

32. Krigstin, S.; Wetzel, S. A review of mechanisms responsible for changes to stored woody biomass fuels. *Fuel* **2016**, *175*, 75–86. [CrossRef]

33. Sluiter, A.; Hames, B.; Scarlata, C.; Sluiter, J.; Templeton, D. *Determination of Ash in Biomass*; National Renewable Energy Laboratory (NREL): Golden, CO, USA, 2008; p. 8.

34. Stehman, S.V.; Meredith, M.P. Practical analysis of factorial experiments in forestry. *Can. J. For. Res.* **1995**, *25*, 446–461. [CrossRef]

35. Guidi, W.; Piccioni, E.; Ginanni, M.; Bonari, E.; Nissim, W.G. Bark content estimation in poplar (*Populus deltoides* L.) short-rotation coppice in Central Italy. *Biomass Bioenergy* **2008**, *32*, 518–524. [CrossRef]

36. Eich, S.; Volk, T.A.; Eisenbies, M.H. Bark Content of Two Shrub Willow Cultivars Grown at Two Sites and Relationships with Centroid Bark Content and Stem Diameter. *BioEnergy Res.* **2015**, *8*, 1661–1670. [CrossRef]

37. Paris, P.; Mareschi, L.; Sabatti, M.; Tosi, L.; Scarascia-Mugnozza, G. Nitrogen removal and its determinants in hybrid Populus clones for bioenergy plantations after two biennial rotations in two temperate sites in northern Italy. *iFor. Biogeosci. For.* **2015**, *8*, 668–676. [CrossRef]

38. Jirjis, R. Effects of particle size and pile height on storage and fuel quality of comminuted Salix viminalis. *Biomass Bioenergy* **2005**, *28*, 193–201. [CrossRef]

39. Whittaker, C.; Yates, N.E.; Powers, S.J.; Misselbrook, T.; Shield, I. Dry matter losses and quality changes during short rotation coppice willow storage in chip or rod form. *Biomass Bioenergy* **2018**, *112*, 29–36. [CrossRef] [PubMed]

40. Wihersaari, M. Evaluation of greenhouse gas emission risks from storage of wood residue. *Biomass Bioenergy* **2005**, *28*, 444–453. [CrossRef]

41. *ISO 17225-4 Solid Biofuels—Fuel Specifications and Classes*; ISO: Geneva, Switzerland, 2014.

42. Kenney, K.L.; A Smith, W.; Gresham, G.L.; Westover, T.L. Understanding biomass feedstock variability. *Biofuels* **2013**, *4*, 111–127. [CrossRef]

43. Therasme, O.; Volk, T.A.; Cabrera, A.M.; Eisenbies, M.H.; Amidon, T.E. Hot Water Extraction Improves the Characteristics of Willow and Sugar Maple Biomass with Different Amount of Bark. *Front. Energy Res.* **2018**, *6*, 93. [CrossRef]

44. Williams, C.L.; Emerson, R.M.; Hernandez, S.; Klinger, J.L.; Fillerup, E.P.; Thomas, B.J. Preprocessing and Hybrid Biochemical/Thermochemical Conversion of Short Rotation Woody Coppice for Biofuels. *Front. Energy Res.* **2018**, *6*, 74. [CrossRef]

forests

MDPI

Article

Productivity, Growth Patterns, and Cellulosic Pulp Properties of Hybrid Aspen Clones

Marzena Niemczyk [1,*], Piotr Przybysz [2,3], Kamila Przybysz [3], Marek Karwański [4], Adam Kaliszewski [5], Tomasz Wojda [1] and Mirko Liesebach [6]

[1] Department of Silviculture and Forest Tree Genetics, Forest Research Institute, Braci Leśnej 3, Sękocin Stary, 05-090 Raszyn, Poland; T.Wojda@ibles.waw.pl

[2] The Faculty of Wood Technology, Warsaw University of Life Sciences, Nowoursynowska 159, 02-776 Warszawa, Poland; piotrprzybysz@interia.pl

[3] Natural Fibers Advanced Technologies, 42A Błękitna str., 93-322 Lódź, Poland; kamila.przybysz@interia.pl

[4] The Faculty of Applied Informatics and Mathematics, Warsaw University of Life Sciences, Nowoursynowska 159, 02-776 Warszawa, Poland; marek_karwanski@sggw.pl

[5] Department of Forest Resources Management, Forest Research Institute, Braci Leśnej 3, Sękocin Stary, 05-090 Raszyn, Poland; A.Kaliszewski@ibles.waw.pl

[6] Thünen Institute of Forest Genetics, Sieker Landstr. 2, D-22927 Großhansdorf, Germany; mirko.liesebach@thuenen.de

* Correspondence: M.Niemczyk@ibles.waw.pl; Tel.: +48-22-7150-681

Received: 9 April 2019; Accepted: 22 May 2019; Published: 24 May 2019

Abstract: Research Highlights: This research provides a firm basis for understanding the improved aspen hybrid performance that aims at facilitating optimal clone selection for industrial application. Background and Objectives: Rapid growth and wood properties make aspen (*Populus tremula* L.) suitable for the production of pulp and paper. We assessed the potential of tree improvement through hybridization to enhance aspen productivity in northern Poland, and investigated the effects of *Populus tremula* hybridization with *Populus tremuloides* Michaux and *Populus alba* L. on the growth and cellulosic pulp properties for papermaking purposes. Materials and Methods: A common garden trial was utilized that included 15 hybrid aspen clones of *P. tremula* × *P. tremuloides*, four of *P. tremula* × *P. alba*, and one, previously tested *P. tremula* clone. Clones of *P. tremula*, plus trees from wild populations, were used as a reference. Tree height and diameter at breast height (DBH) were measured after growing seasons four through seven. At seven years of age, the three clones representing all species combinations were harvested, and their cellulosic pulp properties and paper sheet characteristics were assessed. Results: The clones from wild populations exhibited the poorest growth. In contrast, the clone 'Wä 13' (*P. tremula* × *P. tremuloides*) demonstrated the highest DBH, height, volume production, and mean annual increment (MAI) (25.4 m^3 ha^{-1} year^{-1}). The MAI ratio calculated for interspecific crosses ranged from 1.35- to 1.42-fold, higher than that for the *P. tremula*. Chemical properties of pulp, fiber morphology, and the physical properties of paper sheets were more desirable for interspecific hybrid clones than those for the pure *P. tremula* clone. Conclusions: The results indicated that plantations of hybrid aspen may constitute an important additional source of wood for pulp and paper products in Poland. Our findings further suggested that the standard rotation of these trees may be reduced from 40 to 20 years, increasing overall biomass yield and enhancing atmospheric carbon sequestration.

Keywords: *Populus tremula*; heterosis; growth rate; cellulosic pulp; paper properties

---

1. Introduction

Populus tremula L. (known as common, European, or Eurasian aspen) has the largest native range of any species from the *Populus* genus, being one of the most widely-distributed trees globally [1].

This species occurs from 40° to 70° N latitude [2] on the Eurasian continent, across a wide variety of soils, elevations, and climatic conditions. In Poland, common aspen is considered as the only forest species of the *Populus* genus. Being an early successional species, *P. tremula* L. plays an important role in the first generation of a forest on previously non-forested areas, or following considerable forest disturbances. Tree stands with common aspen as the dominant tree species cover only 0.8% of all forest area in Poland; however, owing to its high ecological value, *P. tremula* constitutes a respected supplementary (co-occurring) tree species. In addition, because of its relatively short life span, *P. tremula* provides numerous birds and mammals with habitats and food, thereby contributing to the enhancement of forest biodiversity. Aspen's advantage over all other native tree species of Poland manifests however in large part through its rapid growth, especially in the juvenile phase. Through its favorable wood properties, low lignin, and high carbohydrate content along with fibers of small diameter and thin wall, *P. tremula* comprises a suitable and desirable tree species for the production of pulp and paper [3]. These qualities appear to be particularly significant in consideration of the projections showing that paper production is expected to increase worldwide from the current value of 400 million Mg annually, to 700–900 million Mg by 2050 [4]. Thus, plantations of fast-growing aspen may represent a promising source of wood for satisfying the increasing demand for wood-based products, providing the opportunity to reduce the timber harvest from natural forests. Moreover, the latter possibility provides added incentive to improve both the growth rates and wood properties of fast-growing species such as aspen, because of the limited wood resources that have been increasingly exploited for papermaking, as well as other cellulose-based products and timber purposes.

Taking into account the considerable genetic diversity in natural populations of aspen and its predisposition to hybridization, the most promising results of aspen improvement might be achieved through breeding, inter-specific hybridization, and cloning [5–9]. Hybrid aspens, in particular the offspring of geographically distant species (e.g., *P. tremula* with *P. tremuloides* Michaux), demonstrate superior performance over the average of both their genetically distinct parents [10]. This phenomenon, known as heterosis, or hybrid vigor, constitutes a multi-genetic complex trait, and can be extrapolated as the sum of multiple physiological and phenotypic traits, including the magnitude and rate of growth, flowering time, yield, and resistance to biotic and abiotic environmental factors [11,12]. However, despite the ability of numerous studies to achieve desirable increases in the growth of F_1 hybrids of *Populus* spp. including aspen [13,14], the genetic mechanism underlying heterosis remains incompletely understood.

The hybrid vigor phenomena in aspen breeding began to be widely utilized in Europe in the early 20th century [10,15–18]. At the end of the 1950s, *P. tremula* × *P. tremuloides* hybrids were already produced in almost all European countries [17], as well as in the USA and Canada [8,19]. In Poland, the first studies on poplar hybridization within the *Populus* section began in 1953 [10]. The main objective was to cross aspen (*P. tremula*) with native white poplar (*P. alba* L.) or trembling (American) aspen (*P. tremuloides*). Concurrently, an independent breeding program for aspen was also initiated in the Polish forestry. Although the goals of this program were never fully realized, over 50 trees were phenotypically selected, the majority of which originated from the Białowieża Forest. Owing to their high qualitative and quantitative value, these selected *P. tremula* genotypes were used as the mother trees in the hybridization program for control crosses with *P. tremuloides* or *P. alba* as the paternal parents.

The primary factor that supports the improvement and testing of a fast-growing species as a feedstock source for papermaking, as well as renewable energy, is based on the restrictions placed on the harvesting of natural forests, stemming from ecological needs and social expectations concerning the use of forests. Nevertheless, fast-growing aspen and its hybrids have attracted relatively little research attention in Poland.

The aims of this study were therefore to (1) evaluate the productivity of aspen and its hybrids in the environmental conditions of Poland, and (2) investigate the effects of *P. tremula* with *P. tremuloides* and *P. alba* hybridization on the growth and wood quality for papermaking purposes, in the context of further genotypes selection and developing recommendations for industrial application.

2. Materials and Methods

2.1. Experimental Location, Soil, and Climatic Condition

The experimental area of the study was located in northern Poland (54°4′26″ N, 20°30′4″ E) in the proximity of Lidzbark Warmiński. Average annual temperature is 8.0 °C with an annual precipitation of 683 mm. The experiment was initiated in April 2011 on post-agricultural land. According to the World Reference Base for Soil Resources (WRB) the main soil type was Luvisol. The soil texture was determined as sandy clay with a pH in H_2O of 4.7–5.0 (acidic), and the C:N ratio of 8.7–9.3 in the top 20 cm. The soil preparation was done by plowing, and the saplings were planted in holes, created by an earth auger powered by a tractor.

2.2. Planting Material

Planting material consisted of 15 hybrid aspen clones of *P. tremula* × *P. tremuloides* (TA × TE), four hybrid clones of *P. tremula* × *P. alba* (TA × A), and one clone of *P. tremula* (TA × TA), which performance was tested previously in Germany. Additionally, a mixture of 30 clones of *P. tremula* (TA) plus trees from wild populations in the Białowieża Forest was used as a reference (Table 1). Hybrid clones had been crossed and selected in Poland and Germany. The German genotypes were propagated by tissue culture. Polish clones used in the experiment were crossed and selected at the Forest Research Institute, and were propagated vegetatively from root cuttings. Prior to planting, all saplings were maintained for one year in the forest nursery.

Table 1. Source and parentage of the clones used in the study.

Clone	Cross	Taxon Code	Background	Source
IBL 264/2/2	*Populus tremula* 19 (Anin) × *P. tremuloides* 84 (Sweden)	TA × TE	Selected clone within a progeny crossed at the Forest Research Institute	Poland
IBL 55/8	*Populus tremula* 38 (Białowieża) × *P. alba var. nivea* 31 (Sadłowice)	TA × A	Selected clone within a progeny crossed at the Forest Research Institute	
IBL 91/78	*Populus tremula* 19 (Anin) × *P. alba* 30 (Grodzisk Maz.)	TA × A	Selected clone within a progeny crossed at the Forest Research Institute	
IBL 91/2	*Populus tremula* 19 (Anin) × *P. alba* 30 (Grodzisk Maz.)	TA × A	Selected clone within a progeny crossed at the Forest Research Institute	
Białowieża (Clone mixture = reference)	*Populus tremula*	TA	Vegetative progeny of 30 plus trees) originated in wild populations in Białowieża Forest	
CA-2-75	*P. tremula* × *P. alba*	TA × A		Germany
Kh 83	*P. tremula* (Wedesbüttel 3) × *P. tremuloides* (North Wisconsin Clone 13)	TA × TE	Selected clone within a progeny crossed at the Thünen Institute	
ESCH 5	*P. tremula* (Brauna 11) × *P. tremuloides* (New Hampshire Turesson 141)	TA × TE	Selected clone within a progeny crossed at the Thünen Institute at the site Escherode	

Table 1. *Cont.*

Clone	Cross	Taxon Code	Background	Source
ESCH 8	*P. tremula* (Brauna 11) × *P. tremuloides* (New Hampshire Turesson 141)	TA × TE	Selected clone within a progeny crossed at the Thünen Institute at the site Escherode	
L 176 (Se 3)	*P. tremula* (Brauna 11) × *P. tremuloides* (New Hampshire Turesson 141)	TA × TE	Selected clone within a progeny crossed at the Thünen Institute at the site Seedorf	
Se 1	*P. tremula* (Brauna 11) × *P. tremuloides* (New Hampshire Turesson 141)	TA × TE	Selected clone within a progeny crossed at the Thünen Institute at the site Seedorf	
W 3	*P. tremula* (Wedesbüttel 3)	TA	Selected clone in the stand Wedesbüttel (which was originated Tapiau)	
174/10	*P. tremula* (Wedesbüttel 18) × *P. tremuloides* (North Wisconsin Clone 13)	TA × TE	Selected clone within a progeny crossed at the Thünen Institute	
Wä 1	*P. tremula* (Großdubrau 1) × *P. tremuloides* (Ontario Maple)	TA × TE	Selected clone within a progeny crossed at the Thünen Institute at the site Wächtersbach	
L 191 (Wä 14)	*P. tremula* (Großdubrau 1) × *P. tremuloides* (Ontario Maple)	TA × TE	Selected clone within a progeny crossed at the Thünen Institute at the site Wächtersbach	
Astria	*P. tremula* × *P. tremuloides*	TA × TE	A triploid clone which is on the market as forest reproductive material of the category "tested"	
KH 73	*P. tremula* (Wedesbüttel 3) × *P. tremuloides* (North Wisconsin Clone 13)	TA × TE	Selected clone within a progeny crossed at the Thünen Institute at the site Klausheide	
Wä 13	*P. tremula* (Großdubrau 1) × *P. tremuloides* (Ontario Maple)	TA × TE	Selected clone within a progeny crossed at the Thünen Institute at the site Wächtersbach	
Se 4	*P. tremula* (Brauna 11) × *P. tremuloides* (New Hampshire Turesson 141)	TA × TE	Selected clone within a progeny crossed at the Thünen Institute at the site Seedorf	
Ihlendieksweg (Ihl 174/59)	*P. tremula* (Wedesbüttel 18) × *P. tremuloides* (North Wisconsin Clone 13)	TA × TE	Selected clone within a progeny crossed at the Thünen Institute at the site Schmalenbeck	
164A	*P. tremula* (Wedesbüttel 5) × *P. tremuloides* (T 13-58)	TA × TE	Selected clone within a progeny crossed at the Thünen Institute	

2.3. Study Design

The study layout comprised a randomized complete block design with four replicates. Each block was divided into 21 plots equal to the number of the tested group of clones. A total of 25 saplings of a given clone were planted within each plot with a spacing of 2.5 × 3.0 m, resulting in a planting density of 1333 saplings ha^{-1}. A bordering row was planted around the experimental area. The area was fenced to prevent browsing by wild animals. During the first three years, the plantation was weeded mechanically once annually. No irrigation or fertilization was applied to the tested area.

2.4. Measurement of Tree Characteristics

The diameter at breast height (DBH; measured at a height of 1.3 m) of all trees was measured after four, five, six, and seven growing seasons on the plantation. The basal area for each tree was calculated, based on its DBH. Height was recorded for 20% of trees systematically in each plot and year. The height curve was constructed separately for each clone in a given block and year, according to the following function [20]:

$$h = \left(\frac{DBH}{\alpha + \beta \times DBH}\right)^2 + 1.3 \tag{1}$$

where *h* represents tree height (m), *DBH* is the diameter at breast height (cm), and α, β are the fitted coefficients.

The estimated coefficients (α, β) of the regression function for each clone in each block were used to estimate the height of trees from the entire range of DBH, which was utilized in the volume equation. Tree volume was calculated based on the function developed by Wróblewski and Zasada [21] based on the volume table of Orłow (elaborated by Czuraj [22]):

$$V = 0.0000529644 \times DBH^{1.882362} \times h \tag{2}$$

where *V* represents an individual tree stem volume, *h* is tree height, and *DBH* is breast height diameter.

Based on individual tree volumes, we calculated tree volume on an area basis and the mean annual increment (MAI) for each clone at a given age. The survival rate was assessed based on the number of living trees.

2.5. Properties of Cellulosic Pulp

At the age of seven years, three clones representing all species combination types and characterized by the good performance ('Wä 13' TA × TE, 'IBL 91/78' TA × A, and 'W 3' TA) were selected to evaluate the properties of cellulosic pulp. From each selected clone in a given replication, three trees (total number 9 trees per clone), representing average parameters, were felled. The 'average trees' were identified according to the basal area of trees. From harvested trees, 70 cm long samples were taken at the middle of every 2 m section, beginning from the stem base to the tree top. The total volume of the samples represented approximately 50% of the volume of the whole stem. All samples were numbered and transferred to the laboratory immediately after collection. The wood samples were manually debarked and deprived of knots, and subsequently sawn into chips using an electric Milwaukee MD 304 saw (Milwaukee Electric Tool Corporation, Hilden, Germany).

Poplar cellulosic pulp was prepared by the Kraft method [23,24] from air-dried woodchips (25 mm × 15 mm × 8 mm). Dry weight (d.w.) of all materials was determined prior to pulping. Pulping processes were conducted in a 0.015 m^3 stainless steel reactor with regulated temperature (PD-114, Danex, Rosko, Poland) with agitation (3 swings min^{-1}, swinging angle of 60°). Suspensions of woodchips (1000 g d.w. of wood chips in 0.004 m^3 of alkaline sulfate solution) were heated for 120 min to achieve the temperature of 165 °C, and incubated at this temperature for a further 120 min. Then, the temperature was decreased to ambient temperature (22 ± 1 °C) using a jacket with cold tap water, and the insoluble residue was separated by filtration within the reactor, washed with approximately 0.050 m^3 of demineralized water, and incubated overnight (12 h) in demineralized water to remove the residues of the alkali-soluble fractions.

The obtained fibrous biomass was disintegrated using a laboratory JAC SHPD28D propeller pulp disintegrator (Danex, Rosko, Poland) at 10,000 rpm and screened using a PS-114 membrane screener (Danex, Rosko, Poland) (0.2 mm gap). Fibers were collected, dried for 48 h in ambient conditions (21 ± 1 °C), then weighed. Analysis of the chemical composition of cellulosic pulps included a quantification of extractives, lignin, cellulose, hemicelluloses, and ash. The lignin mass fraction was determined by a gravimetric method in compliance with the TAPPI T222 standard after the removal of extractives (TAPPI T204). The holocellulose mass fraction was determined in accordance with the TAPPI T249 standard. Cellulose was quantified as alpha cellulose (TAPPI T203). The hemicelluloses mass fraction was calculated as the difference between the holocellulose and cellulose mass fractions.

Ash content was determined by a gravimetric method in compliance with TAPPI T211. All these assays were performed in triplicate for each raw material.

2.6. Properties of Paper Sheets

Sheets of paper were produced under laboratory conditions from rewetted pulp samples (22.5 g d.w. samples were soaked in water for at least 8 h) that were subjected to disintegration using the laboratory JAC SHPD28D propeller pulp disintegrator at 20,000 rpm, according to the ISO 5263-1 (2004) standard. The disintegrated pulps were concentrated to a dry weight mass fraction of 10% and refined in a JAC 12DPFIX PFI mill (Danex, Rosko, Poland) under standard conditions (ISO 5264-2 (2011)). The ultimate standard freeness of the pulp was approximately 30° Schopper-Riegler. Water retention values were determined according to the ISO 23714 (2014) standard. Dimensions of fibers were measured using a Morfi Compact Black Edition apparatus (Techpap, Saint Martin d'Hères, France) according to ISO 16065-2 (2014).

Paper sheets (grammage of approximately $80 \, g \, m^{-2}$) were produced from the refined and unrefined pulps, using a Rapid-Koethen class apparatus (Danex, Rosko, Poland) according to ISO 5269-2 (2004). Mechanical properties of paper were determined only for the sheets with a grammage of $80 \pm 1 \, g \, m^{-2}$.

The sheets were stored for 24 h at the relative moisture mass fraction of $50 \pm 2\%$ and temperature of $23 \pm 1 \, °C$ (ISO 187 (1990)), prior to the determination of mechanical properties such as tensile index, stretch, tensile energy absorption (ISO 1924-2 (2008)), and tear index (ISO 2758 (2014)). Other parameters measured included bulk (ISO 534 (2011)), brightness (TAPPI T452), and opacity (TAPPI T519).

2.7. Data Analysis

To assess differences in biometric characteristics and productivity (volume and MAI) between clones and the interaction between clones over time, generalized linear mixed models with a log link function and normal-distributed errors were used (the GLIMMIX procedure). The choice of the optimal model was based on the Akaike Information Criterion. The model was expressed using the following equation:

$$log(E(Y_i)) = \mu + b + C_j + T_k + (CT)_{jk} + E(B_l) \tag{3}$$

where μ represents the general mean, b is the mean of random effect, C_j is the j-clone effect, T_k is the k-year effect, $(CT)_{jk}$ represents the kj-interaction between clones and time, and B_l is the l-random block effect. The log link function represented the multiplicative form of the explanatory variables. The type III tests of fixed effects were used to determinate the significance of specified effects. The size of these effects was calculated as means ratios according to the base level achieved for TA clones from wild populations (reference) in Białowieża Forest. The same generalized linear mixed models and subsequent procedures were implemented for comparisons between groups of different species crosses (clones were grouped in accordance with their parental species). In this case, the pure species TA (= clone mixture from wild populations in the Białowieża Forest and clone 'W 3') was used as a reference.

The papermaking traits between species crosses were compared using Analysis of Variance (ANOVA) with the PROC GLM procedure. The model assumed:

$$Y_i = \mu + S_j + \varepsilon_i \tag{4}$$

where Y_i represents a dependent variable, μ is a general mean, S_j is a j-clone effect, and ε_i is a random component from the normal distribution. Such model choice was dictated by the fact that data were collected on an annual basis (after seven years of tree growth), with the results relating to optimized processes (without considering the block effect). The *post-hoc* comparisons between clones/species combination types were performed using the Tukey HSD test. Moreover, the SAS CORR procedure was used to determine the correlations between papermaking traits for clones. All statistical analyses were performed with SAS/STAT (rel. 14.3) statistical package (SAS Institute Inc., Cary, NC, USA).

3. Results

3.1. Biometric Characteristics and Productivity of Clones

Survival rate for all clones was generally high (Figure 1) and stable for most clones over time. 'Kh 73' (TA × TE) constituted the only clone for which the recorded survival rate was lower than 90%.

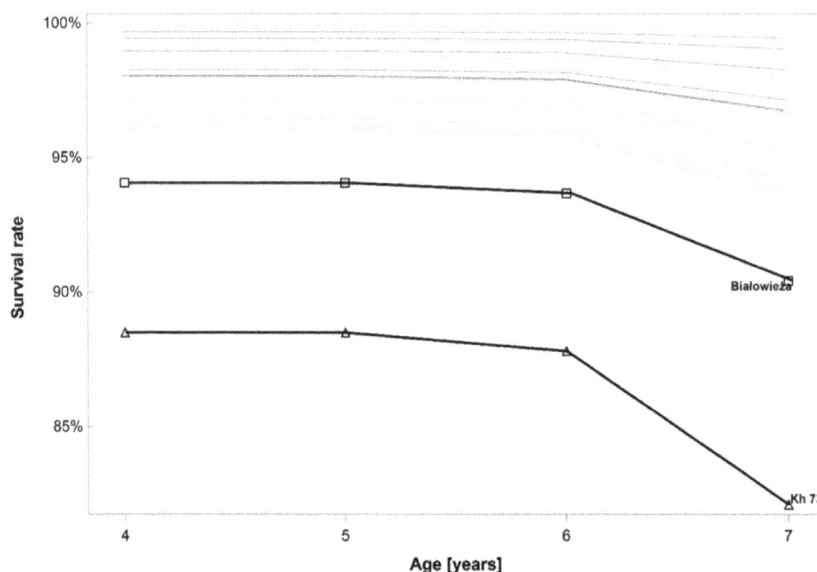

Figure 1. Trend lines in survival rate for clones over time. The bold line with square symbols indicates the trend for a mixture of 30 clones of *P. tremula* (TA) plus trees from wild populations in the Białowieża Forest (Białowieża = reference), and the bold line with triangle symbols for the clone with the lowest survival rate.

Overall, biometric characteristics differed noticeably among clones and, for most traits, significant clone-by-year interaction was observed (Table 2). The average diameters of the various clones showed statistically significant differences. Clone 'Wä 13' (TA × TE) had the largest mean DBH values in each researched year (14.34 cm at the age of seven years) (Figure 2). Of all analyzed clones, the reference clones exhibited the lowest DBH (8.70 cm at the age of seven years). Mean DBH values were significantly greater for all researched clones in comparison to the reference, which was reflected in the results of our Tukey test (Table 3). The results obtained from all clones in different years are shown in Figure 2.

Table 2. Results of the generalized linear mixed model (GLIMMIX) analysis of growth traits of aspen clones: Type III tests of fixed effects.

Effect	DF	DBH	H	V	MAI
		F Value (*p*-Value)			
Clone	20	221.61 (<0.0001)	359.61 (<0.0001)	150.27 (<0.0001)	266.33 (<0.0001)
Age	3	2599.52 (<0.0001)	7119.56 (<0.0001)	2156.28 (<0.0001)	1157.05 (<0.0001)
Clone × Age	60	1.03 (0.4105)	6.37 (<.0001)	1.47 (0.0105)	1.70 (0.0006)

Abbreviations: DF—degrees of freedom; DBH—diameter at breast height; H—Height; V—Volume; MAI—Mean annual increment.

Similarly, clone 'Wä 13' produced significantly taller trees (arithmetic mean from 10.35 m at the age of four years to 13.20 m at the age of seven years), than those of all other analyzed clones.

The reference exhibited the lowest mean heights in each consecutive year, achieving from 6.17 m at the age of four years to 10.21 m at the age of seven years.

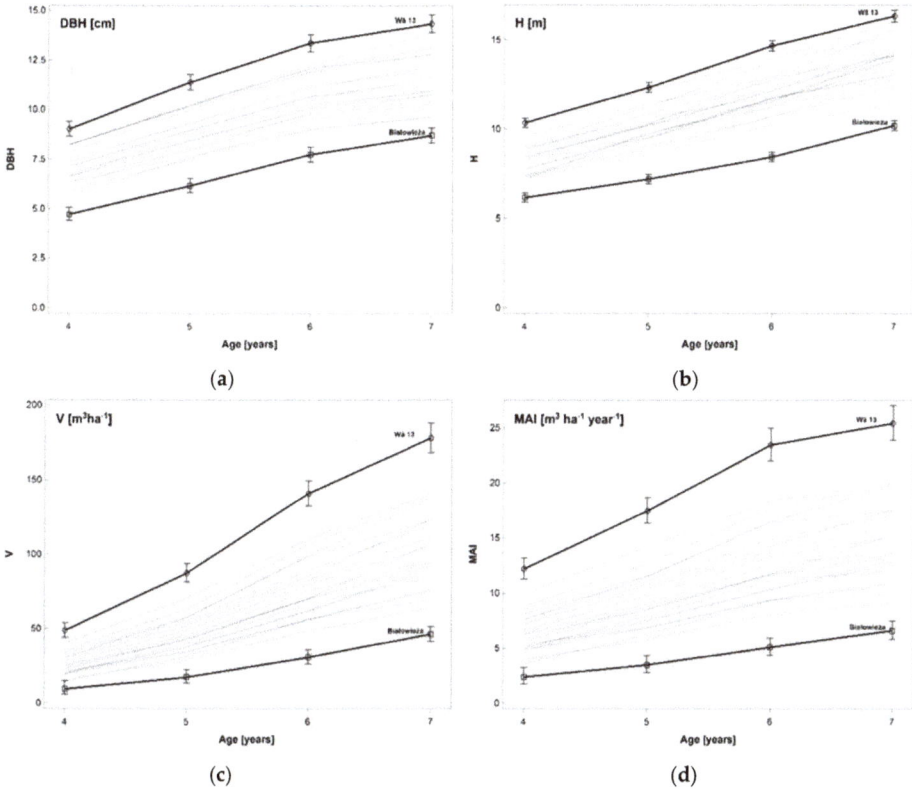

Figure 2. Trend lines of (**a**) mean diameter at breast height (DBH), (**b**) height (H), (**c**) stem volume (V) and (**d**) mean annual increment (MAI)) for clones over time. The bold line with square symbols indicates a trend for a mixture of 30 clones of *P. tremula* L. (TA) plus trees from wild populations in the Białowieża Forest (Białowieża = reference), and the bold line with circle symbols for the best performing clone. Vertical bars denote standard errors.

Volume calculated on an area basis indicated large differences between clones. The mean volume ranged from 46 m^3 ha^{-1} for the reference to 178 m^3 ha^{-1} for the best performing hybrid aspen clone, 'Wä 13' (TA × TE), at the age of seven years (Figure 2). The mean annual increment of 'Wä 13' ranged from 12.19 m^3 ha^{-1} at the age of four years to 25.42 m^3 ha^{-1} at the age of seven years. In contrast, the MAI produced by the reference at the same age amounted to 2.37 and 6.60 m^3, respectively. During the studied period, clones showed an increasing MAI that progressed over each analysis year (Figure 2). The mean estimated stem volume of all hybrid clones was greater than that of the wild species. In case of the best performing clone 'Wä 13', the mean estimated stem volume ratio was 4.6-fold greater than the reference (Table S1).

3.2. Biometric Characteristics and Productivity for Species Combination

To reduce the number of comparisons, we grouped together clones with the same species parentage. The differences between taxa were statistically significant ($\alpha = 0.05$) for all the researched traits (Table 3). The reference (TA) exhibited inferior biometric parameters in comparison to the adequate traits of crosses TA × TE and TA × A (Figure 3). The inferiority of TA was further reflected in the Tukey test results and ratio values (Table S2). The MAI ratio calculated for inter-specific crosses ranged from 1.35- to 1.42-fold higher than the reference (Table S2). The absolute values for taxa over time are presented in Figure 3.

Table 3. Results of the generalized linear mixed model (GLIMMIX) analysis of growth traits of taxa: type III tests of fixed effects.

Effect		DBH	H	V	MAI
	DF		F Value (*p*-Value)		
Taxon	2	156.05 (<0.0001)	274.46 (<0.0001)	41.74 (<0.0001)	73.97 (<0.0001)
Age	3	817.24 (<0.0001)	1593.65 (<0.0001)	584.37 (<0.0001)	309.17 (<0.0001)
Taxon × Age	6	0.79 (0.5775)	3.66 (0.0013)	0.20 (0.9769)	0.21 (0.9730)

Abbreviations: DF—degrees of freedom; DBH—Diameter at breast height; H—Height; V—Volume; MAI—Mean annual increment.

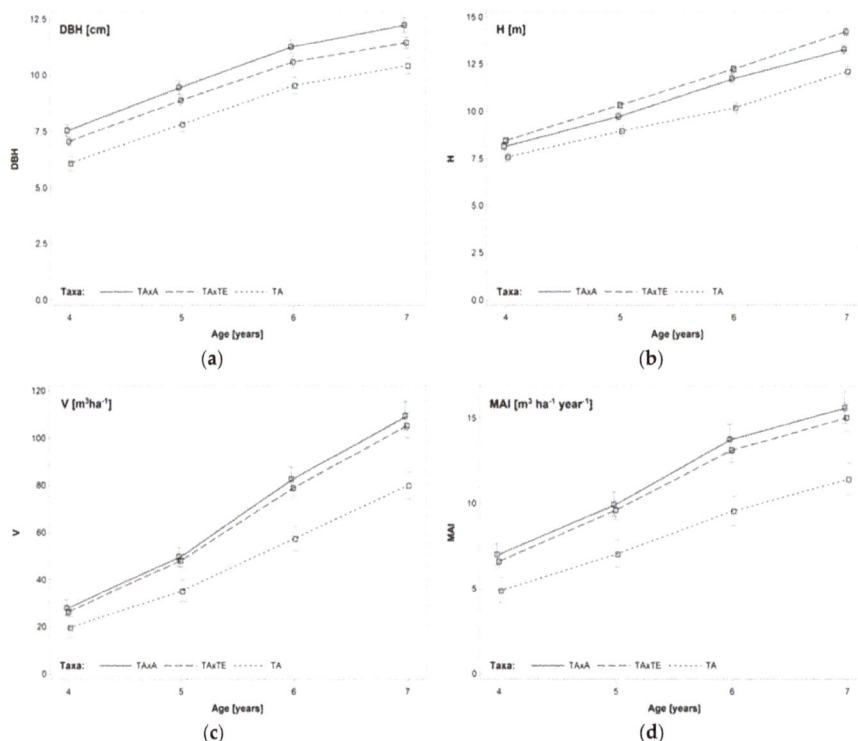

Figure 3. Trend lines of (**a**) mean diameter at breast height (DBH), (**b**) height (H), (**c**) stem volume (V ha^{-1}) and (**d**) mean annual increment (MAI)) for taxa over time. Vertical bars denote standard errors. Abbreviations: TA: *P. tremula* L.; TA × A: *P. tremula* × *P. alba*; TA × TE: *P. tremula* × *P. tremuloides*.

3.3. Properties of Cellulosic Pulp and Paper

The main component of the Kraft pulps was cellulose (Table 4). The cellulose mass fraction ranged from 87% d.w. for 'W 3' (TA) to >92% d.w. for the 'Wä 13' (TA × TE). The lignin mass fraction was also diversified between different clones, being significantly lower in 'Wä 13' (TA × TE) pulp in comparison to that in pulps of other clones. The screened yield for pulps ranged between 41.6 and 46.8%. The Kraft pulping results and chemical composition of cellulosic pulps are presented in Table 4.

Table 4. Kraft cellulosic pulps and chemical composition of different clones ±SE, and results of Tukey's honestly significant difference test. The same lowercase letters indicate statistically homogenous groups at α = 0.05 in Tukey test.

Clone	Taxon	Kappa Number [-]	Screened Yield [%]	Cellulose [%, d.w]	Hemicellulose [% d.w.]	Lignin [%, d.w]	Extractives [%, d.w]	Minerals [%, d.w]
Wä 13	TA × TE	19.2 ± 0.3 a	46.8	92.4 ± 0.3 a	3.4 ± 0.6 a	3.3 ± 0.2 a	0.2 ± 0.1 a	0.6 ± 0.2 a
IBL 91/78	TA × A	18.3 ± 0.1 b	41.6	88.1 ± 0.9 b	5.1 ± 1.2 b	5.6 ± 0.8 b	0.5 ± 0.1 b	0.7 ± 0.3 a
W 3	TA	18.0 ± 0.1 b	43.7	86.7 ± 0.6 c	6.1 ± 1.3 b	6.0 ± 0.8 b	0.5 ± 0.1 b	0.7 ± 0.2 a

Abbreviations: SE: standard error; TA: *P. tremula*; TA × A: *P. tremula* × *P. alba*; TA × TE: *P. tremula* × *P. tremuloides*.

With regard to refined pulp properties, the greatest weighted fiber length among the researched clones was obtained for 'Wä 13' (TA × TE) followed by 'IBL 91/78' (TA × A). Overall, these crosses were characterized by significantly more desirable pulp characteristics in comparison to those of 'W 3' (TA) (Table 5).

Table 5. Characteristics of the refined pulps (freeness 30° SR) means ± SE and results of Tukey's honestly significant difference test. The same lowercase letters indicate statistically homogenous groups at α = 0.05 in Tukey test.

Clone	Taxon	Mean Arithmetic Fiber Length [μm]	Mean Weighted Fiber Length [μm]	Mean Fiber Width [μm]	Coarseness [mg m⁻¹]	Fine Content [% in Length]
Wä 13	TA × TE	619.5 ± 1.7 a	712.8 ± 3.0 a	24.6 ± 0.2 a	0.0836 ± 0.001 a	11.9 ± 0.4 a
IBL 91/78	TA × A	622.3 ± 1.0 a	697.5 ± 2.1 b	24.4 ± 0.2 a	0.087 ± 0.001 a	11.2 ± 0.3 a
W 3	TA	585 ± 1.8 b	672.3 ± 1.9 c	22.7 ± 0.1 b	0.076 ± 0.001 b	13.4 ± 0.5 b

Abbreviations: SR: Schopper-Riegler; SE: standard error; TA × TA: *P. tremula*; TA × A: *P. tremula* × *P. alba*; TA × TE: *P. tremula* × *P. tremuloides*.

The refined pulps were used to produce sheets of paper under laboratory conditions. Static tensile properties, i.e., braking length and strain, show very well in plane tensile properties of the paper produced from all the tested clones. However, due to low weighted average fiber length and relatively high fines content, the tear index for the 'W 3' TA was significantly lower than that for other hybrid aspen clones. The brightness (at 36%) and CIE L*a*b* color of paper were similar for researched clones. The opacity for all clones was very high, reaching 100% (data not shown). The remaining physical properties of sheets did not differ notably among clones (Table 6).

Table 6. Physical properties of paper sheets for researched clones, means ± SE and results of Tukey's honestly significant difference test. The same lowercase letters indicate statistically homogenous groups at α = 0.05 in Tukey test.

Clone	Taxon	Apparent Density [g cm⁻³]	Breaking Length [m]	Strain [%]	Breaking Energy/TEA [J]	Tear Resistance [mN]	Tensile Index [N m g⁻¹]	Tear Index [mN m² g⁻¹]	WRV [%] Beaten
Wä 13	TA × TE	0.79 ± 0.10 a	11033 ± 367 a	2.6 ± 0.1 a	0.20 ± 0.01 a	227 ± 11 a	102.7 ± 3.4 a	2.8 ± 0.1 a	2.155 ± 0.005 a
IBL 91/78	TA × A	0.81 ± 0.01 b	11467 ± 260 a	2.7 ± 0.2 a	0.22 ± 0.01 a	220 ± 18 a	106.8 ± 2.4 a	2.8 ± 0.2 a	2.107 ± 0.008 b
W 3	TA	0.82 ± 0.10 b	11483 ± 328 a	2.4 ± 0.2 a	0.20 ± 0.02 a	189 ± 15 b	106.9 ± 3.1 a	2.4 ± 0.2 b	2.088 ± 0.017 b

Abbreviations: SE: standard error; TA × TE: *P. tremula* × *P. tremuloides*; TA × A: *P. tremula* × *P. alba*; TA: *P. tremula*.

3.4. Correlations

The chemical composition of the pulp showed the highest correlations among the papermaking properties. Cellulose mass fraction was highly negatively correlated with other carbohydrates and increased with the increase in Kappa number. The inverse correlation was found with regard to lignin mass fraction and Kappa number. The chemical composition of the pulp positively influenced fiber length and width with regard to the cellulose mass fraction and negatively with regard to the lignin mass fraction. Each fiber characteristic was positively correlated with each other. Among physical properties of the sheets, the tear index was strongly correlated with fiber properties. Table S3 summarizes the correlation coefficients between the studied traits along with their significance at the level of $\alpha = 0.05$.

4. Discussion

Aspen is among the most widely used poplars for papermaking purposes in the Northern Hemisphere. Because hybrid aspens may benefit from heterosis, the majority of practically oriented hybridization programs are aimed at using this phenomenon for the improvement of growth and wood quality parameters. This is also confirmed by the present study, which showed that during a seven-year long experiment, all hybrid aspen clones were characterized by superior growth and productivity in comparison with those of wild species.

The MAI of the stem wood of the best performing clone 'Wä 13' (TA × TE) was at the highest expected level for climatic conditions of the Baltic Region, which encompasses central and northern European countries. Numerous studies have reported that the MAI of hybrid aspen in the southern part of the Nordic region and the Baltic area may reach up to 20 m^3 ha^{-1} $year^{-1}$ during rotation periods of 20–25 years [25–31]. The highest MAI measured to date, 25.8 m^3 ha^{-1} $year^{-1}$, was recorded for a 23 year old hybrid aspen (TA × TE) plantation in southern Sweden [32]. For hybrid aspen trials established in north-western Germany, the modeled MAI reached its maximum at the age of 24 years [33]. In the present study, the majority of aspen hybrid clones examined at the age between four and seven years demonstrated a high MAI, with that for the best clone amounting to 25 m^3 ha^{-1} $year^{-1}$ at the age of seven years, which progressed in each consecutive year of the analysis. Considering that the MAI for aspen plantations is not expected to culminate before the age of 20 years [34], we can anticipate further improvement of this MAI in subsequent years.

In Poland, a rotation period of native stands of *P. tremula* typically comprises 40 years, which in the present case can be assigned to vegetatively propagated offspring of wild populations from the Białowieża Forest. Taking into account that in the present study inter-specific crosses resulted in 1.4-fold greater mean annual increment, as compared to the clones of trees from wild populations of TA, and ranging from 1.6- to 4.6-fold enhancement for particular hybrid clones, in comparison to the reference, our data suggest that the rotation age for selected hybrid aspen clones might be reduced to 20 years. Similar conclusions were provided by Li et al. [8], who postulated that the expected rotation for aspen clones in the north-eastern part of the United States might also be limited to 20 years. This remains consistent with the considered periods of 30 years in the Swedish short-rotation program [35] along with the estimated period for reaching merchantable volumes of American aspen crosses in Alberta, Canada [5].

The reduced rotation periods for superior hybrid aspen clones may complement traditional, multifunctional forest management with the long-rotation periods, widely implemented in Central Europe. Shortened production cycles may serve as a partial remedy for the loss of stability of forest ecosystems in conditions of rapidly progressing climate change. Over the period 1951–2008, the mean temperature in Poland increased by 0.24 °C per 10 years [36]. Climate change not only involves directional changes in the mean values of climatic variables (e.g., global warming) but the frequency and magnitude of various extreme climatic events [37,38]. In many regions of Europe, extreme weather events (heat waves, heavy precipitation, droughts, heavy storms) are likely to intensify in the coming decades [39,40]. In the case of hurricanes/storms, which in recent years frequently cause serious damages in Polish and in European forestry [41], it was proved that a greater stand age and a taller stand height increase stand susceptibility [41–43].

Replacement of such high-risk stands, and associated economic losses, can be made by introduction of stands with shortened production cycles, which have also the potential to significantly contribute to both increasing biomass supply and capturing carbon from the atmosphere [44–47].

Understanding the growth increment curve of various tree species in the juvenile period holds exceptional importance for silviculture [48], determining e.g., the choice of spacing and term of first thinning. Among fast-growing species, aspen is considered to achieve the highest annual increment, albeit somewhat later than poplars from the *Aigeiros* or *Tacamahaca* sections or species of the *Salix* genus [49]. However, the stem volume produced by hybrid aspen in northern Poland at the age of seven years did not reflect inferior productivity in comparison to that of hybrid poplars at the same age, growing in the same soil and climatic condition [50]. As highlighted by Liesebach et al. [51], considerable growth increment is produced by aspen from the age of 6 years onward with the differences occurring at a very early stage of tree growth. This may explain the comparable results of productivity achieved by species crosses from the *Populus* section in the present study with those of the different crosses among the *Aigeiros* or *Tacamahaca* sections.

Taking into account differences in productivity among various parental species and combinations of species within the *Populus* section, we revealed that inter-specific hybrids of TA × TE and TA × A performed significantly better than vegetative progenies of TA. Notably, however, on the clone level, statistically significant differences between the selected TA clone 'W 3', and other clones of inter-specific hybrid aspen crosses, in terms of growth parameters, were not found. This finding suggests that selection and testing genotypes in the area of the future utilization may result in good performing clones, with adaptive ability to specific climatic and soil conditions. Furthermore, it seems reasonable to add the clone 'W 3' to further stages of the tree improvement program, thereby enriching the native population of *P. tremula*, and providing new genotypes for future mating strategies.

Hybrid aspen clones were characterized by more desirable properties related to papermaking traits than those of the clone of the pure species TA. Moreover, the chemical composition of the cellulosic pulp was diversified among the clones. It is notable that the cellulose mass fraction was greater in the pulps of both hybrid aspen clones than that in the pulp of the pure species, indicating the enhanced desirability of the former. This result impacts yield and Kappa number, as reflected in the positive correlation with the cellulose mass fraction and a negative correlation with the lignin mass fraction. As indicated by Hart et al. [5], the further consequences of particular carbohydrate concentrations can be revealed in ethanol biofuel applications. Therefore, the obtained differences in chemical composition (yield) between clones may represent important factors when considering the utilization of the biomass of a particular clone for paper and biofuel production at the industrial scale.

The present results regarding the absolute values of screened yield and Kappa number are comparable with the findings of several previous studies. It has to be mentioned however, that a much higher yield of hardwood pulps can be achieved by trees older than those we tested in our study [52]. When considering yields and characteristics of cellulosic pulps obtained from multiple softwood and hardwood species, along with fast growing grasses in Poland [24], the yield of cellulosic pulp from aspen clones obtained in the present study confirms their high suitability for pulp production. In comparison, the only alternatives for hybrid aspen pulp in Poland appear to comprise pulps obtained from silver birch (*Betula pendula* Roth.) and the hybrid poplar *P. maximowiczii* × *P. trichocarpa* 'NE-42' (syn. 'Hybrid 275').

Fiber properties of hybrid aspen clones were enhanced and more desirable, compared to those of pure species fibers, albeit they remained typical for the *Populus* section [45]. According to Francis et al. [53], 31 year old aspens produced much longer fibers than those of 7.5 year old poplars. Consequently, fiber and strength properties are expected to increase with age. This observation highlights the need for a reassessment of pulp and paper properties at the end of the rotation period in the experimental plantation in northern Poland. The differences in fiber morphology between aspen crosses resulted in different physical properties of paper sheets, which is consistent with the results of a previous study by Gurnagul et al. [54].

Generally, sheet properties were affected by fiber morphology in manners that could be predicted from the principles of paper physics [54,55]. In particular, tearing resistance, as the most important strength parameter, was directly correlated with fiber length.

5. Conclusions

Overall, the high evaluated productivity identified in the present study clearly indicated that short rotation plantation of hybrid aspen might be considered as an important additional source of woody biomass for pulp and paper products in Poland. Such plantations may complement traditional, multifunctional forest management, in which conservation approaches play an increasing role. The reduced rotation periods for superior hybrid aspens would likely contribute to an increase in economic benefits, and partly mitigate the uncertainty accompanying the long-rotation periods applied in traditional silviculture. In addition, this aspect appears to be particularly important with regard to rapid climate change and its impact on forests.

Furthermore, significant improvement of hybrid aspen traits related to growth, pulp, and paper properties was observed in the present study, which dominated over those of pure species. The effect of hybrid vigor manifested as 1.4-fold greater MAI compared to that of wild species, with particular hybrid clones exhibiting a 1.6- to 4.6-fold enhancement. The superiority of hybrid aspen also translated into papermaking properties, as the chemical properties of the pulp, yield, fiber morphology, and physical properties of the final paper sheets were more desirable for *P. tremula* hybrids with both *P. tremuloides* and *P. alba*, than those for pure *P. tremula*. Together, our findings confirmed that a good knowledge of the maximal growth parameters that could be achieved with particular site conditions; species, or hybrid selection, combined with the concise relationship between chemistry, fiber morphology, and sheet properties, could facilitate the optimal clone choice for each specific region for industry purposes.

Supplementary Materials: The following are available online at http://www.mdpi.com/1999-4907/10/5/450/s1, Table S1. Least squares means and ratios for growth traits by clone from the GLIMMIX. Standard errors (SE) are shown in brackets. Reference ratio for trees from the wild population (Białowieża) = 1. Tukey's honestly significant difference test, α = 0.05 was applied to test statistically significant differences between clones. The same lowercase letters accompanying LS-means indicate statistically homogenous groups of clones. Abbreviations: DBH: Diameter at breast height; H: Height; V: Stem volume; MAI: Mean annual increment; SE: Standard error; TA: *P. tremula*; TA × A: *P. tremula* × *P. alba*; TA × TE: *P. tremula* × *P. tremuloides*; TA: *P. tremula* a mixture of 30 clones of plus trees from wild populations in Białowieża (reference). Table S2. Least squares means and ratios for growth traits by taxon from the GLIMMIX. Standard errors (SE) are shown in brackets. Reference ratio for TA = 1. Tukey's honestly significant difference test, α = 0.05 was applied to test statistically significant differences between taxa. The same lowercase letters accompanying LS-means indicate statistically homogenous groups of taxa. Abbreviations: DBH: Diameter at breast height; H: Height; V: Stem volume; MAI: Mean annual increment; TA: *P. tremula*; TA × A: *P. tremula* × *P. alba*; TA × TE: *P. tremula* × *P. tremuloides*. Table S3. Pearson's correlations coefficient between papermaking traits. Values statistically significant at the level α = 0.05 are shown in bold. The green tones indicate positive correlation and the red tones negative relationship.

Author Contributions: Conceptualization, M.N.; methodology, M.N., P.P., K.P.; Investigation M.N., T.W., P.P. and K.P.; resources, M.N., T.W., M.L.; formal analysis M.N. and M.K.; Writing—Original draft preparation, M.N.; Writing—Review and editing, M.N., P.P, K.P., A.K., M.K., T.W. and M.L.; funding acquisition, M.N.

Funding: This research was funded by the Directorate General of the State Forests (Poland), grant number 500 425.

Acknowledgments: We thank our colleagues from the Forest Research Institute: Jan Matras for valuable comments to the manuscript as well as Szymon Krajewski and Władysław Kantorowicz, for providing help during field data collection. We also thank two anonymous reviewers for their significant impact in the improvement of our paper.

Conflicts of Interest: The authors declare no conflict of interest. The funders had no role in the design of the study; in the collection, analyses, or interpretation of data; in the writing of the manuscript, or in the decision to publish the results.

References

1. Worrell, R. European aspen (*Populus tremula* L.): A review with particular reference to Scotland I. Distribution, ecology and genetic variation. *Forestry* **1995**, *68*, 93–105. [CrossRef]
2. Richardson, J.; Isebrands, J.G.; Ball, J.B. Ecology and Physiology of Poplars and Willows. In *Poplars and Willows: Trees for Society and the Environment*; Isebrands, J.G., Richardson, J., Eds.; FAO: Rome, Italy, 2014; pp. 92–123.

3. Yu, Q.; Pulkkinen, P.; Rautio, M.; Haapanen, M.; Alén, R.; Stener, L.G.; Beuker, E.; Tigerstedt, P.M.A. Genetic control of wood physicochemical properties, growth, and phenology in hybrid aspen clones. *Can. J. For. Res.* **2001**, *31*, 1348–1356. [CrossRef]

4. Bajpai, P. *Pulp and Paper Industry: Energy Conservation*; Elsevier: Amsterdam, The Netherlands, 2016.

5. Hart, J.F.; de Araujo, F.; Thomas, B.R.; Mansfield, S.D. Wood quality and growth characterization across intra- and inter-specific hybrid aspen clones. *Forests* **2013**, *4*, 786–807. [CrossRef]

6. Einspahr, D.W.; Winton, L.L. Genetics of quaking aspen. *Aspen Bibliogr. Paper* **1976**, *5026*, 1–23.

7. Einspahr, D.W. Production and utilization of triploid hybrid aspen (*Populus tremuloides, Populus tremula*). *Iowa State J. Resour.* **1984**, *58*, 401–409.

8. Li, B.; Wyckoff, G.W.; Einspahr, D.W. Hybrid aspen performance and genetic gains. *North J. Appl. For.* **1993**, *10*, 117–122.

9. Melchior, G.H. Breeding of aspen and hybrid aspen and their importance for practical use. *Allg. For. Jagdztg.* **1985**, *156*, 112–122.

10. Zajączkowski, K. *Hodowla Lasu: Plantacje Drzew Szybko Rosnących*; PWRiL: Warsaw, Poland, 2013.

11. Baranwal, V.K.; Mikkilineni, V.; Zehr, U.B.; Tyagi, A.K.; Kapoor, S. Heterosis: Emerging ideas about hybrid vigour. *J. Exp. Bot.* **2012**, *63*, 6309–6314. [CrossRef] [PubMed]

12. Lippman, Z.B.; Zamir, D. Heterosis: Revisiting the magic. *Trends Genet.* **2007**, *23*, 60–66.

13. Li, B.; Wu, R. Genetic causes of heterosis in juvenile aspen: A quantitative comparison across intra- and inter-specific hybrids. *Theor. Appl. Genet.* **1996**, *93*, 380–391. [CrossRef]

14. Li, B.; Howe, G.T.; Wu, R. Developmental factors responsible for heterosis in aspen hybrids (*Populus tremuloides* × *P. tremula*). *Tree Physiol.* **1998**, *18*, 29–36. [CrossRef] [PubMed]

15. Wettstein-Westerheim, W. Die Kreuzungsmethode und die Beschreibung von F1-Bastarden bei *Populus*. *Zeitschrift für Züchtung* **1933**, *18*, 597–626.

16. Chmielewski, W. Mieszańce Topolowe Hodowli Instytutu Badawczego Leśnictwa i ich Wstępna Ocena. Ph.D. Thesis, Forest Research Institute, Warsaw, Poland, 1966; p. 131.

17. Jobling, J. Poplars for Wood Production and Amenity. In *Forestry Commission Bulletin 92*; Forestry Commission: London, UK, 1990.

18. Liesebach, M.; Schneck, V. Züchtung, Zulassungen, Vermehrung. In *Agrarholz—Schnellwachsende Bäume in der Landschaft*; Böhm, C., Veste, M., Eds.; Springer: Berlin, Germany, 2018; pp. 119–145.

19. Dickmann, D.I.; Stuart, K.W. *The Culture of Poplars in Eastern North America*; Michigan State University: East Lansing, MI, USA, 1983; p. 168.

20. Näslund, M. Skogsförsöksanstaltens gallringsförsök i tallskog. *Meddelanden från Statens Skogsförsöksanstalt* **1936**, *29*, 1–169.

21. Wróblewski, L.; Zasada, M. Wzory do określania miąższości grubizny dla modrzewia, osiki, grabu, topoli i lipy. *Sylwan* **2001**, *11*, 71–79.

22. Czuraj, M. *Tablice miąższości Kłód Odziomkowych i Drzew Stojących*; PWRiL: Warsaw, Poland, 1991.

23. Modrzejewski, K.; Olszewski, J.; Rutkowski, J. *Analysis in Papermaking Industry*; Editorial Office of the Lodz University of Technology: Łódź, Poland, 1969; pp. 60–89, 206–250.

24. Przybysz, K.; Malachowska, E.; Martyniak, D.; Iłowska, J.; Kalinowska, H.; Przybysz, P. Yield of pulp, dimensional properties of fibers, and properties of paper produced from fast growing trees and grasses. *BioResources* **2018**, *13*, 1372–1387. [CrossRef]

25. Johnsson, H. Hybridaspens ungdomsutveckling och ett försök till framtidsprognos. *Svenska Skogsvårdsföreningens Tidskrift* **1953**, *51*, 73–96.

26. Jakobsen, B. Hybrid aspen (*Populus tremula* L. × *Populus tremuloides* Michx.). *Forstlige Forsøgsvæsen Danmark* **1976**, *34*, 317–338.

27. Rytter, L. Nutrient content in stems of hybrid aspen as affected by tree age and tree size, and nutrient removal with harvest. *Biomass Bioenergy* **2002**, *23*, 13–25. [CrossRef]

28. Karačić, A.; Verwijst, T.; Weih, M. Above-ground woody biomass production of short-rotation *Populus* plantations on agricultural land in Sweden. *Scand. J. For. Res.* **2003**, *18*, 427–437. [CrossRef]

29. Rytter, L.; Stener, L.G. Productivity and thinning effects in hybrid aspen (*Populus tremula* × *P. tremuloides* Michx.) stands in southern Sweden. *Forestry* **2005**, *78*, 285–295. [CrossRef]

30. Rytter, L. A management regime for hybrid aspen stands combining conventional forestry techniques with early biomass harvests to exploit their rapid early growth. *For. Ecol. Manag.* **2006**, *236*, 422–426. [CrossRef]

31. Tullus, A.; Rytter, L.; Tullus, T.; Weih, M.; Tullus, H. Short-rotation forestry with hybrid aspen (*Populus tremula* L. × *P. tremuloides* Michx.) in Northern Europe. *Scand. J. For. Res.* **2012**, *27*, 10–29. [CrossRef]

32. Johnsson, H. Das potential der hybrid aspen produktion (*Populus tremula* × *tremuloides*) in Südschweden. *Holzzucht* **1976**, *30*, 19–22.

33. Poker, J. Ökologische und ökonomische Beurteilung unterschiedlicher Produktionsprogramme für Pappeln in Nordwestdeutschland. Ph.D. Thesis, University of Hamburg, Hamburg, Germany, 1984; p. 134.

34. Perala, D.A. Stand equations for estimating aerial biomass, net productivity, and stem survival of young aspen suckers on good sites. *Can. J. For. Res.* **1973**, *3*, 288–292. [CrossRef]

35. Christersson, L.; Sennerby-Forsse, L. The Swedish programme for intensive short-rotation forests. *Biomass Bioenergy* **1994**, *6*, 145–149. [CrossRef]

36. Biernacik, D.; Filipiak, J.; Miętus, M.; Wójcik, R. Zmienność Warunków Klimatycznych w Polsce po Roku 1951. Rezultaty Projektu Klimat; In *Klimat Polski na tle Klimatu Europy: Zmiany i Konsekwencje*; Bednorz, E., Kolendowicz, L., Eds.; Bogucki Wydawnictwo Naukowe: Poznań, Poland, 2010; pp. 9–21.

37. Van de Pol, M.; Ens, B.J.; Heg, D.; Brouwer, L.; Krol, J.; Maier, M.; Exo, K.M.; Oosterbeek, K.; Lok, T.; Eising, C.M.; et al. Do changes in the frequency, magnitude and timing of extreme climatic events threaten the population viability of coastal birds? *J. Appl. Ecol.* **2010**, *47*, 720–730. [CrossRef]

38. Brang, P.; Spathelf, P.; Larsen, J.B.; Bauhus, J.; Bončína, A.; Chauvin, C.; Drössler, L.; García-Güemes, C.; Heiri, C.; Kerr, G.; et al. Suitability of close-to-nature silviculture for adapting temperate European forests to climate change. *Forestry* **2014**, *87*, 492–503. [CrossRef]

39. Dmyterko, E.; Mionskowski, M.; Bruchwald, A. Risk of the wind damage to the forests in Poland on the basis of a stand damage risk model. *Sylwan* **2015**, *159*, 361–371.

40. Beniston, M.; Stephenson, D.B.; Christensen, O.B.; Ferro, C.A.T.; Frei, C.; Goyette, S.; Halsnaes, K.; Holt, T.; Jylhä, K.; Koffi, B.; et al. Future extreme events in European climate: An exploration of regional climate model projections. *Clim. Chang.* **2007**, *81*, 71–95. [CrossRef]

41. Kjellström, E.; Nikulin, G.; Hansson, U.; Strandberg, G.; Ullerstig, A. 21st century changes in the European climate: Uncertainties derived from an ensemble of regional climate model simulations. *Tellus Ser. A Dyn. Meteorol. Oceanogr.* **2011**, *63*, 24–40. [CrossRef]

42. Spiecker, H. Silvicultural management in maintaining biodiversity and resistance of forests in Europe—Temperate zone. *J. Environ. Manag.* **2003**, *67*, 55–65. [CrossRef]

43. Mayer, P.; Brang, P.; Dobbertin, M.; Hallenbarter, D.; Renaud, J.P.; Walthert, L.; Zimmermann, S. Forest storm damage is more frequent on acidic soils. *Ann. For. Sci.* **2005**, *62*, 303–311. [CrossRef]

44. Updegraff, K.; Baughman, M.J.; Taff, S.J. Environmental benefits of cropland conversion to hybrid poplar: Economic and policy considerations. *Biomass Bioenergy* **2004**, *27*, 411–428. [CrossRef]

45. Arevalo, C.B.; Bhatti, J.S.; Chang, S.X.; Sidders, D. Land use change effects on ecosystem carbon balance: From agricultural to hybrid poplar plantation. *Agricult. Ecosyst. Environ.* **2011**, *141*, 342–349. [CrossRef]

46. Rytter, R.M. The potential of willow and poplar plantations as carbon sinks in Sweden. *Biomass Bioenergy* **2012**, *36*, 86–95. [CrossRef]

47. Rytter, L.; Rytter, R.M. Growth and carbon capture of grey alder (*Alnus incana* (L.) Moench.). *For. Ecol. Manag.* **2016**, *373*, 56–65. [CrossRef]

48. Jaworski, A. *Hodowla Lasu: Sposoby Zagospodarowania, Odnowienie Lasu, Przebudowa i Przemiana Drzewostanów*; PWRiL: Warsaw, Poland, 2011.

49. Muhs, H.J. Growth Potential of Aspen Grown under Traditional Forest Management and Some Aspects of Using Aspen as an Alternative Crop in Agriculture. In *Elaboration of a Problem-Oriented, Strategic Decision-Matrix for Research Project in Agriculture, Short Agricultural Surpluses*; Schliephake, D., Krämer, P., Eds.; Dechema: Frankfurt/Main, Germany, 1986; pp. 94–102.

50. Niemczyk, M.; Kaliszewski, A.; Jewiarz, M.; Wróbel, M.; Mudryk, K. Productivity and biomass characteristics of selected poplar (*Populus* spp.) cultivars under the climatic conditions of northern Poland. *Biomass Bioenergy* **2018**, *111*, 46–51. [CrossRef]

51. Liesebach, M.; von Wuehlish, G.; Muhs, H.J. Aspen for short-rotation coppice plantations on agricultural sites in Germany: Effects of spacing and rotation time on growth and biomass production of aspen progenies. *For. Ecol. Manag.* **1999**, *121*, 25–39. [CrossRef]

52. Thykesson, M.; Sjöberg, L.A.; Ahlgren, P. Paper properties of grass and straw pulps. *Ind. Crops Prod.* **1998**, *7*, 351–362. [CrossRef]

53. Francis, R.C.; Hanna, R.B.; Shin, S.J.; Brown, A.F.; Riemenschneider, D.E. Papermaking characteristics of three *Populus* clones grown in the north-central United States. *Biomass Bioenergy* **2006**, *30*, 803–808. [CrossRef]

54. Gurnagul, N.; Page, D.H.; Seth, R.S. Dry sheet properties of Canadian hardwood kraft pulps. *J. Pulp Pap. Sci.* **1990**, *16*, 36–41.

55. Niskanen, K. *Mechanic of Paper Products*; De Gruyter: Berlin, Germany, 2012.

forests

MDPI

Article

The Economics of Rapid Multiplication of Hybrid Poplar Biomass Varieties

Brian J. Stanton *, Kathy Haiby, Carlos Gantz, Jesus Espinoza and Richard A. Shuren

GreenWood Resources 1500 SW 1st Avenue, Portland, OR 97201, USA; kathy.haiby@gwrglobal.com (K.H.);
carlos.gantz@gwrglobal.com (C.G.); jesus.espinoza@gwrglobal.com (J.E.); rshuren@comcast.net (R.A.S.)
* Correspondence: brian.stanton@gwrglobal.com; Tel.: +1-971-533-7037

Received: 20 April 2019; Accepted: 21 May 2019; Published: 23 May 2019

Abstract: Background: Poplar (*Populus* spp.) hybridization is key to advancing biomass yields and conversion efficiency. Once superior varieties are selected, there is a lag in commercial use while they are multiplied to scale. Objective: The purpose of this study was to assess the influence of gains in biomass yield and quality on investment in rapid propagation techniques that speed the time to commercial deployment. Material and Methods: A factorial experiment of propagation method and hybrid variety was conducted to quantify the scale-up rate of in vitro and greenhouse clonal multiplication. These data were used in modeling the internal rate of return (IRR) on investment into rapid propagation as a function of genetic gains in biomass yield and quality and compared to a base case that assumed the standard method of supplying operational varieties in commercial quantities from nurseries as hardwood cuttings, capable of yields of 16.5 Mg ha^{-1} year^{-1}. Results: Analysis of variance in macro-cutting yield showed that propagation method and varietal effects as well as their interaction were highly significant, with hedge propagation exceeding serial propagation in macro-cutting productivity by a factor of nearly 1.8. The *Populus deltoides* × *P. maximowiczii* and the *Populus trichocarpa* × *P. maximowiczii* varieties greatly exceeded the multiplication rate of the *P.* × *generosa* varieties due to their exceptional response to repeated hedging required to initiate multiple tracks of serial propagation. Analyses of investment into rapid propagation to introduce new material into plantation establishment followed by a 20-year rotation of six coppice harvests showed that gains in biomass yield and quality are warranted for a commitment to rapid propagation systems. The base case analysis was generally favored at yields up to 18 Mg^{-1} year^{-1} dependent on pricing. The rapid multiplication analysis proved superior to the base case analysis at the two highest yield levels (27.0 and 31.5 Mg ha^{-1} year^{-1},) at all price levels and at yields of 22.5 Mg^{-1} year^{-1}, dependent on price and farm location. Conclusion: Rapid multiplication is a reliable method to move improved plant material directly into operations when valued appropriately in the marketplace.

Keywords: hybrid poplar; genetic improvement; clonal propagation; biofuels; renewable energy

1. Introduction

The profitability of hybrid poplar biomass production and its biochemical conversion to transportation fuels is dependent on genetically improved interspecific varieties. Both farm and refinery economics are influenced by genetic gains in biomass yield [1,2] and biomass quality, the latter manifested in the ease of cell wall deconstruction and sugar release during pretreatment and hydrolysis [3–5]. The improvement cycle is not trivial, encompassing up to seven years to complete the process of hybridization, clonal propagation, and yield testing through the first coppice cycle and laboratory analyses of biomass composition and conversion efficiency. Additionally, once superior varieties have been identified, it remains for them to be multiplied to levels for widespread initial varietal deployment. Considering the density of bioenergy plantations approximates 3500–4500 stems per hectare (ha) and the sizable acreage that needs to be cultivated to produce the tonnage required by refinery operations, the time to fully expand the supply of planting stock of newly-selected varieties impedes the expeditious delivery of gains in yield and conversion efficiency. While traditional nursery propagation may take seven years and 150 ha to introduce a new varietal into a moderately sized plantation (i.e., 10,000 ha), a combination of laboratory and greenhouse propagation has demonstrated a condensed delivery timeline [6]. An argument for such rapid multiplication propagation techniques as indispensable to the initial propagation of new hybrid varieties for bioenergy farms has been made since the beginnings of clonal forestry [7].

Commercial biomass farms employing interspecific poplar hybrids between sections Aigeiros and Tacamahaca are normally planted as clonal stands using inexpensive one-year-old hardwood cuttings that establish in the field by formation of adventitious roots that elongate from initials formed on nursery stock the previous growing season during shoot development. Although measurable genetic variance in vegetative propagation has been reported, field rooting of Aigeiros × Tacamahaca hybrid taxa using hardwood cuttings is a generally reliable propagation method [8]. However, for the initial scaling of clonal selections, varieties are rapidly multiplied in greenhouses using succulent cuttings rooted under mist propagation in soil or hydroponically to produce containerized planting stock [9,10]. Alternatively, in vitro micropropagation systems of exceedingly greater capacity are available, albeit costly [11,12]. Recent micropropagation research—producing plantlets in liquid-phase bioreactors for ex vitro greenhouse rooting—may ultimately prove cost-effective for industrial clonal propagation [13], but as currently practiced, micropropagation is prohibitively expensive for commercial planting stock quantities [14]. Until the affordability of in vitro systems is proven, the rapid multiplication technique as developed in Finland for hybrid aspen (*Populus* × *wettsteinii* Hämet-Ahti) may be the most cost-effective avenue for scaling new selections [15]. This approach utilizes in vitro micro-cuttings that are produced in laboratories, rooted in greenhouses, and subsequently expanded by hedge propagation of macro-cuttings to generate multiple serial propagation tracks. In vitro propagation takes place in a laboratory using culture media and glass vessels [16]. Hedge propagation is the continuous harvesting of re-sprouting plants for re-planting under greenhouse conditions. Serial propagation begins with greenhouse propagation of cuttings that become primary ramets, from which cuttings are collected that become secondary ramets that lead to tertiary ramets and so on.

The economics of this combined approach has not been studied for the Aigeiros × Tacamahaca taxa, although generally recognized to be more expensive than conventional nursery propagation. A factorial study was therefore conducted to assess the economics of a rapid multiplication method based on the propagation of micro- and macro-cuttings for Aigeiros × Tacamahaca hybrid varieties patterned on the Finnish *P.* × *wettsteinii* model. For this paper, micro-cuttings are shoots produced through multiple rounds of in vitro propagation from explants containing a shoot-tip or nodal meristem. Conversely, shoots produced using ex vitro greenhouse rooting techniques for containerized planting stock are referred to as macro-cuttings in this study. The intent of the investigation was to: (1) Quantify the efficiency of greenhouse hedge and serial macro-cutting multiplication initiated with in vitro micro-cuttings for several varieties of diverse taxa and (2) determine the profitability of greenhouse rapid multiplication as a function of genetic gains in biomass yield and biomass quality using the internal rate of return as a standard metric of economic performance [17].

2. Materials and Methods

2.1. Selection of Plant Material

Twenty-two centimeter (cm) hardwood cuttings of four experimental varieties were collected in January 2013 from a clonal bank maintained by GreenWood Resources at its Tree Improvement Center, Westport, Oregon, USA. The varieties were chosen to provide a contrast of nursery growth rates and four distinct hardwood cutting production categories. A further consideration in the choice was the assemblage of four distinct taxa, *Populus* × *generosa* (Henry) and its reciprocal and hybrids formed from separate crosses between *Populus deltoides* (Bartram ex Marsh.) and *Populusmaximowiczii* (Henry) and *Populus balsamifera* subsp. *trichocarpa* (Torr. and Gray) (Table 1). The hardwood cuttings were used to grow a single stock plant for each variety in 7.6 cubic decimeters (dm^3) pots in a greenhouse to provide nodal cuttings, with which in vitro propagation was initiated. Before entering dormancy in the fall, the stock plants were sheared during the 2013 growing season to encourage branching. Stock plants were forced in January 2014 and grown for three months to an average height of 92 cm when succulent cuttings were collected from axillary and terminal shoots. The shoots were trimmed to a terminal bud and two nodal buds, disinfected with a fungicidal application, refrigerated, and shipped to a contract micropropagation laboratory for in vitro production of micro-cuttings for greenhouse rooting trials. Between 47 and 87 succulent shoots approximately 10 to 14 cm in length were provided in March–July 2014 for culture establishment and shoot proliferation.

Table 1. Experimental varieties, taxa, and hardwood nursery production metrics.

Varietal Identity	Taxon [1]	Sprouts Stool^{-1}	DBH [2] (mm)	Height [2] (dm)	Cutting Utilization [3]	Production Category
790-99-28596	T × D	3.8	17	31	0.47	Moderate growth; low usage
846-00-30120	D × M	1.7	26	40	0.29	Excessive growth; low usage
854-00-30517	T × M	2.3	21	42	0.72	Excessive growth; high usage
893-01-31899	D × T	1.5	18	33	0.56	Moderate growth and usage

[1] Taxa coded as: D × T and T × D (*Populus deltoides* × *P. balsamifera* subsp. *trichocarpa* and reciprocal, aka *Populus* × *generosa*), T × M (*Populus balsamifera* subsp. *trichocarpa* × *P. maximowiczii*), D × M (*Populus deltoides* × *P. maximowiczii*). [2] Breast-height diameter and height of the largest sprout per stool following the fourth coppice year. [3] Proportion of sprout length without sylleptic branching from which 24 cm cuttings bearing axillary buds required for quality hardwood cuttings.

Five hundred micro-cuttings were returned to the greenhouse and transplanted into 144-cell trays (23 cm^3 capacity) filled with a commercial soil mix in the spring of 2015 and grown under mist-propagation using artificial lighting to extend daylength to 18 h. Temperatures were set at 23 °C for acclimation and root initiation. The total number of surviving ramets was recorded for each variety after two months, at which time they ranged from five to 10 cm in height. The tallest 50 rooted plants of each variety were then transplanted into plastic containers (163 cm^3 capacity) filled with the same soil mix to serve as a hedge bank. These were arranged in racks by variety at a density of 37 cm^2 per plant. The plants were grown through 2015, allowed to go dormant, and then forced in January 2016. Succulent 20–22 mm (mm) macro-cuttings were collected in March 2016 from the hedges and used in establishing one track of serial propagation that was cycled through the quinary ramet stage (Figure 1). The 50-ramet hedges were harvested five additional times in propagating additional primary ramets throughout all four seasons.

2.2. Greenhouse Rooting Trials

Succulent macro-cuttings were dipped in a commercial rooting hormone (1000 ppm indole acetic acid and 500 ppm naphthalene acetic acid) during hedge and serial propagation before sticking in to 200 cell trays (20 cm^3 capacity). Trays were grown under an 18 h day length at 23 °C and misted every 20 min for 10 s for the first week of acclimation. Mist settings were changed to a 30 min frequency of 10 s duration for the second week. Thereafter, the plants were misted every 60 min for 20 s. The schedule of hedge propagation cycles was not coincident with the schedule of the serial propagation due to the time to regrow the hedges following the first harvest to initiate serial propagation (Table 2). Thereafter, the cycles of hedge propagation lagged the serial propagation cycles by approximately six weeks. Succulent macro-cuttings were harvested after six to eight weeks dependent upon the season.

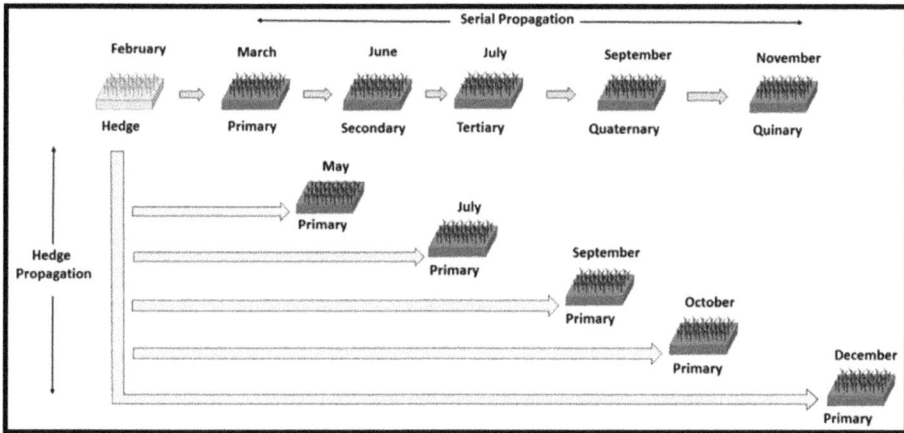

Figure 1. Schematic showing five cycles of greenhouse hedge and serial propagation.

Table 2. Greenhouse schedule of five season cycles of serial and hedge propagation.

Propagation Cycle	Serial Propagation		Hedge Propagation	
(Seasons)	Initiation	Harvest	Initiation	Harvest
1 (Early spring–midsummer)	March 25	June 2	May 27	July 18
2 (Late spring–summer)	June 2	July 25	July 18	September 9
3 (Late summer–early autumn)	July 25	September 19	September 9	October 21
4 (Autumn)	September 19	November 9	October 21	December 12
5 (Late autumn–midwinter)	November 9	December 30	December 12	February 13

2.3. Data Collection and Analysis

Survival was recorded as the percentage of ramets surviving each cycle of hedge or serial propagation; yield of macro-cuttings was specified in terms of the number of cuttings harvested per surviving ramet per cycle. Data were analyzed as a 2 × 4 factorial analysis of variance (i.e., two levels of propagation method, hedge versus serial, and four levels of hybrid variety (Equation 1). Both factors were considered fixed effects in the following analytical model:

$$Y_{ijk} = \alpha + \beta_i + \beta_j + \beta_{ij} + \varepsilon_{k(ij)}, \tag{1}$$

where Y_{ijk} is ramet survival or macro-cutting yield for the *i*th variety using the *j*th propagation method during the *k*th propagation cycle, β_i is the fixed effect of the *i*th variety, β_j is the fixed effect of the *j*th method of propagation, β_{ij} is the interaction of variety and propagation method, and $\varepsilon_{k(ij)}$ is experimental error. (The five cycles of propagation were nested within propagation method as replications since the timing of the serial and hedge propagation cycles were not coincident.)

Varietal and interaction effects were investigated using three single-degree of freedom a priori orthogonal contrasts based on the excessive and moderate levels of hardwood cutting production of the varieties in field nurseries (Table 1). These are: (1) Comparison of the *Populus × generosa* varieties (i.e., 790-99-28596 versus 893-01-31899,) (2) comparison of the *P. maximowiczii* interspecific varieties (i.e., 846-00-30120 versus 854-00-30517,) and (3) the varietal contrast between the two main taxonomic groups (i.e., contrast 1 versus contrast 2).

2.4. Economic Analysis

Survival and yield data from the greenhouse rooting trials were used in deriving the cost of producing containerized macro-cuttings. The internal rate of return on investment into improved planting stock produced by this system was then modeled as a function of increases in (1) yield and (2) pricing for feedstock that undergoes hydrolysis more cost-effectively. Research costs in breeding the improved varietals were not considered in the analysis; only the higher propagation and planting costs of containerized macro-cuttings were accounted for. (This approach assumes research and development costs are borne by a timber investment management organization that offers improved plant material exclusively to its investors at a price to cover propagation costs as part of its investment thesis.) Analyses were based on the average varietal propagation costs.

The analyses utilized a cost-of-production model developed by the USDA-NIFA AFRI CAP program, Advanced Hardwood Biofuels Northwest (AHB) for hybrid poplar as a biomass investment opportunity in renewable energy feedstock [18]. The model details the range of agronomic production activities and the costs of land leases, equipment, labor, chemicals and fuel needed to grow the crop and the expense of biomass harvesting and transportation. Inputs are integrated with yields and anticipated biomass pricing in deriving internal rates of return specific to poplar production regions in Washington, Oregon, Idaho, and Northern California, where four large-scale (i.e., 20–40 ha) hybrid poplar biomass AHB demonstration farms were managed from 2012 to 2018 [19]. The demonstration farms were situated on lands previously managed for grain, pasture, hay or row crops. The Oregon and Washington farms were unirrigated, while the Idaho and California farms were irrigated. Lease rates and production costs varied among farms, being highest in California due to the cost of land and irrigation and lowest in Washington where biomass cropping followed pasture (Table 3). The economic analyses supposed a medium-sized biorefinery supported by a 10,117 ha biomass farm managed for a 20-year rotation comprising a two-year establishment cycle followed by six three-year coppice cycles netting seven harvests. The planting density was set at 3,588 trees ha^{-1}. Plantation composition was set at five varieties each stocking 674 ha per each of the two years of planting before coppice regeneration was initiated. Based on the greenhouse propagation trials, the required number of rooted cuttings for the each of the first- and second-year's establishment was 2,424,035 per variety, after which, regeneration was transitioned to coppicing every three years.

Table 3. Land and biomass productions costs for Advanced Hardwood Biofuels Northwest (AHB) poplar demonstration farms throughout a 20-year rotation of seven harvest cycles.

Farm Region	Lease Rate (USD ha^{-1} year^{-1}) [1]	Production Cost (USD Dry Mg^{-1} Rotation^{-1}) [2]
Washington	$111	$77.70
Oregon	$383	$92.38
Idaho	$175	$99.46
California	$865	$146.28

[1] http://quickstats.nass.usda.gov/. [2] Biomass production costs are based on a yield of 16.5 Mg ha^{-1} year^{-1}.

Internal rates of return were modeled across six levels of yield (9.0, 13.5, 18.0, 22.5, 27.0, 31.5 dry Mg ha^{-1} year^{-1} averaged over one 20-year rotation) and eight levels of market pricing ($66, $77, $88, $99, $110, $121, $132, and $145 dry Mg^{-1} biomass) using land, management, and harvesting costs specific to each AHB region (Table 3). The modeled yields up to 22.5 Mg ha^{-1} year^{-1} are consistent with AHB farm inventories and with growth rates projected for the AHB regions using the 3PG growth model (Physiological Processes Predicting Growth) [20]. Inclusion of yields of 27.0 and 31.5 Mg ha^{-1} year^{-1} in the analyses represents future genetic improvements. Likewise, modeled market prices ranging up to \$110 Mg^{-1} are consistent with supply simulations in the 2016 Billion Ton Study (BTS) and the Bioenergy KDF (Knowledge Discovery Framework) supplementary database [21]. This price range is in line with prices modeled for hardwoods managed as short rotation bioenergy crops elsewhere in the US [22,23]. Inclusion of market prices between \$121 and \$145 Mg^{-1} in the analysis reflects future premiums for varietals capable of highly effective hydrolysis. Modeled results of investment into rapid multiplication were compared to a base case that assumed the standard method of supplying operational varieties in commercial quantities from nurseries as unrooted hardwood cuttings capable of yields of 16.5 Mg ha^{-1} year^{-1} based on prevailing yield estimates for the Pacific Northwest [2,24]. Prices modeled for the base case were not limited to the BTS range (\$88–\$110 Mg^{-1}), although it was realized that commercial poplar varieties have been developed for fiber and veneer markets, in which superior lignin chemistry that enables effective hydrolysis has not been an emphasis, and premium market pricing was not likely. The price of rooted planting stock, including packaging was gauged at approximately 4× the price of commercial hardwood cuttings in the base case. Likewise, the cost of cold storage for rooted cuttings, transportation of the rooted cuttings to the field, and the cost of planting the rooted cuttings was modeled at of 2×, 2×, and 6× of the respective cost for commercial hardwood cuttings.

3. Results

3.1. Laboratory

Micro-cuttings were successfully produced for all four varieties during in vitro propagation. The varieties were sub-cultured every two weeks during the shoot proliferation stage; between three and five in vitro cycles were required before enough high-quality micro-cuttings of each variety were available for the greenhouse rooting trials. Marked varietal differences were observed in the consistency of in vitro growth rates (Table 4). Variety 790-99-28596 was rated as a consistent grower, while varieties 854-00-30517 and 893-01-31899 exhibited sporadic growth. The slowest growth pattern was typical of variety 846-00-30120 that also showed the lowest multiplication rate. Multiplication rates were otherwise not too discrepant for the remaining three varieties, varying by just 5.3%.

Table 4. Varietal performance during in vitro propagation on two-week sub-culturing schedules.

Variety	Taxon	Growth Pattern	Multiplication Factor [1]
790-99-28596	T × D	Consistent	2.025
846-00-30120	D × M	Slow	1.680
854-00-30517	T × M	Sporadic	2.075
893-01-31899	D × T	Sporadic	2.133

[1] Number of quality micro-cuttings produced per transplant averaged over three-to-five sub-culture cycles following high grading.

3.2. Greenhouse

In vitro micro-cuttings were rooted nearly completely in the greenhouse for all varieties but one. Survival rates were recorded as 99.5% (790-99-28596), 87.3% (846-00-30120), and 100% (854-00-30517 and 893-01-31899). Considerable variation in greenhouse macro-cutting production was observed (Table 5). The number of macro-cuttings harvested during hedge propagation (2.992 cuttings plant^{-1}) exceeded the amount harvested during serial propagation (1.669 cuttings plant^{-1}) by 79% when averaged over the four varieties (Table 6). Likewise, substantial varietal variation in macro-cutting yield was evidenced when averaged over the two propagation methods. Variety 854-00-30517 exhibited the highest mean macro-cutting yield (3.034 cuttings plant^{-1}), with varieties 790-99-28596 and 893-01-31899 showing the lowest yield, 1.959 and 1.754 cuttings plant^{-1}, respectively. Variety 846-00-30120 was intermediate in its production at 2.573 cuttings plant^{-1} (Table 6). Correspondingly, the main effects of the variety and propagation method were both deemed highly significant sources of variation in macro-cutting yield in the analysis of variance (Table 5). The orthogonal contrasts suggested further that there were no significant yield differences for varieties within either the *P.* × *generosa* taxon or the *P. maximowiczii* interspecific taxa, but a large and significant effect associated with the third orthogonal contrast between the *P.* × *generosa* varieties (790-99-28596 and 893-01-31899) and the *P. maximowiczii* interspecific varieties (846-00-30120 and 854-00-30517).

Table 5. Analysis of variance of the yield of macro-cutting yield.

Source of Variation	Degrees of Freedom	Sum of Squares	Mean Square	F Ratio
Variety	3	10.2443	3.4148	18.52 *
Within *P.* × *generosa*	1	0.2110	0.2110	1.1444
Within *P. maximowiczii* hybrids	1	1.0635	1.0635	5.7673
Between *P.* × *generosa* and *P. maximowiczii*	1	8.9698	8.9698	48.6429 *
Propagation Method	1	17.5001	17.5001	94.89 *
Variety-by-Propagation Method	3	8.7568	2.9189	15.83 *
Within *P.* × *generosa*	1	0.0127	0.0127	0.0688
Within *P. maximowiczii* hybrids	1	0.0007	0.0007	0.0038
Between *P.* × *generosa* and *P. maximowiczii*	1	8.7434	8.7434	47.4156 *
Error	32	5.9018	0.1844	
Total	39	42.4030		

* Significant at the 0.0001 probability level.

Table 6. Macro-cutting survival and yield (cuttings plant^{-1}) during greenhouse propagation [1].

Variety	Taxon	Serial Propagation		Hedge Propagation		Mean	
		Survival (%)	Yield	Survival (%)	Yield	Survival (%)	Yield
790-99-28596	T × D	98.1	1.791	95.9	2.128	97.1	1.959
846-00-30120	D × M	97.5	1.438	97.4	3.708	95.9	2.573
854-00-30517	T × M	94.1	1.911	96.2	4.157	97.9	3.034
893-01-31899	D × T	95.0	1.535	95.1	1.973	94.9	1.754

[1] Standard errors of varietal, propagation, and interaction means for yield are 0.134, 0.096, and 0.192 cuttings plant^{-1}, respectively. Standard errors of survival means are 1.03, 0.73, and 1.46%, respectively.

Very little varietal variation was recorded in survival during the macro-cutting propagation trials, averaging 96.6% for both serial and hedge propagation, with a standard deviation among varieties of 1.92% (serial propagation) and 1.23% (hedge propagation) (Table 6). Consequently, analysis of variance did not show any significant main or interaction effects of survival (Supplementary Table S1).

The first order interaction of variety and propagation method was also highly significant (Table 5). Orthogonal contrast revealed that the interaction again resulted mainly from the much stronger response of the two *P. maximowiczii* hybrids to hedging in comparison to the *P.* × *generosa* varieties (Figure 2).

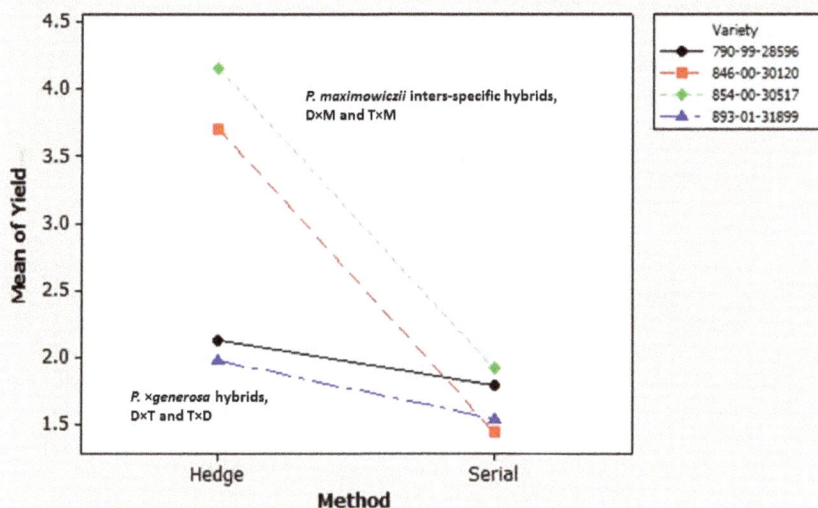

Figure 2. Varietal variation in the response of macro-cutting yield to propagation method.

3.3. Economic Analysis

Using the survival and yield data from the greenhouse rooting trials averaged over varieties (Table 6), the production of the 2.4 million rooted cuttings was projected in terms of five cycles of hedge and serial macro-cutting propagation beginning in the spring and concluding in the winter following the delivery of the micro-cuttings. To produce this quantity of macro-cuttings, it was estimated that 29,000 in vitro micro-cuttings were needed based on the mean varietal yield of macro-cuttings during greenhouse propagation.

Investment returns under the rapid multiplication scenario for each of the four demonstration regions are detailed in Table 7 as a function of biomass yield and market pricing. The profile of internal rates of return (IRRs) at the two non-irrigated farms in Oregon and Washington was superior to the irrigated Idaho and California farms. The low least rate at the Idaho farm partially offset the cost of irrigation, with better returns compared to California. The percentage of positive IRRs recorded for the 48 scenarios of yield and market prices at each demonstration farm was 75% for Washington (36 positive returns), 58% for Oregon (28 positive returns), 50% for Idaho (24 positive returns), and 25% (12 positive returns) for California. Averaged over yields and market prices, the mean for the positive IRRs for Washington, Oregon, Idaho, and California was 15.91%, 13.89%, 13.35%, and 6.89%, respectively. The maximum return in Washington 37.90% and 31.96% in Oregon. Maximum returns for the other two regions were 29.42% (Idaho) and 16.73% (California). Returns estimated for Washington were the most variable, with a standard deviation of 10.30%, followed by Oregon (8.60%) Idaho (7.71%), and California (5.35%).

Both the mean and the number of positive IRRs (n) for the rapid multiplication scenario increased steadily within the boundaries of the BTS market prices ($66–$110 Mg^{-1}) when results were averaged by yield and compiled for all farms. The progression was 3.29% ($n = 2$) at $66 Mg^{-1}, 5.76% ($n = 6$) at $77 Mg^{-1}, 9.02% ($n = 9$) at $88 Mg^{-1}, 10.14% ($n = 13$) at $99 Mg^{-1}, and 12.63% ($n = 15$) at $110 Mg^{-1}. Positive values of IRR recorded above the BTS price range ($121–145 Mg^{-1}) were 15.37% ($n = 16$) at $121 Mg^{-1}, 16.01% ($n = 19$) at $132 Mg^{-1}, and 18.55% ($n = 20$) at $145 Mg^{-1}. Likewise, the number of positive IRRs and their mean value increased with yield when averaged across market prices being 2.09% ($n = 2$), 6.10% ($n = 10$), 9.87% ($n = 16$), 13.41% ($n = 20$), 15.23% ($n = 25$), and 18.23% ($n = 27$), respectively, for the 9.0, 13.5, 18.0, 22.5, 27.0, and 31.5 Mg ha^{-1} year^{-1} yield levels. These IRRs compare to an upper bound of IRRs for hybrid poplar bioenergy programs in the northeastern US of 14.6% at prevailing stumpage prices of $11 Mg^{-1} and not accounting for harvest and transportation costs [25].

Returns from investment into rapid multiplication frequently surpassed those of the base case analysis within the range of the BTS prices across the four farm regions. Base case analyses (yield set to 16.5 Mg ha^{-1} year^{-1}) showed no positive IRRs at the $66 and $77 Mg^{-1} pricing levels at any farm, and only one positive return was recorded at the $88 Mg^{-1} price level (8.81%, Washington) (Table 7). Two positive IRRs were recorded for the base case at the $99 market price, averaging 9.87% at Washington and Oregon; these increased to a mean IRR of 12.31% at the $110 price level averaged across Washington, Oregon, and Idaho. The base case IRRs at the $99 and $110 Mg^{-1} price points exceeded those observed for rapid multiplication at Oregon and Idaho, when newly released varieties produced no more than 18.0 Mg ha^{-1} year^{-1}. The base case analysis for Washington and Oregon was marginally better than the rapid multiplication analysis at yields of 22.5 Mg ha^{-1} year^{-1} across a price range of $110–$145 Mg^{-1} (Table 7). However, when the comparison was made for new varieties characterized by yields at the two uppermost levels (27.0 and 31.5 Mg ha^{-1} year^{-1},) the rapid multiplication scenario was uniformly superior to the base case analysis both within and above the price BTS range at Washington, Oregon, and Idaho farms. No positive IRRs were returned for the base case analysis at California.

Table 7. Internal rates of return (%) as a function of yield and market pricing [1].

Yield (Mg ha^{-1} year^{-1})				Price (USD Mg^{-1})					(+) IRRs	
Washington	$66	$77	$88	$99	$110	$121	$132	$145	Mean	N
9.0							0.54	3.63	2.09	2
13.5				0.57	4.48	7.56	10.19	12.98	7.16	5
18.0			3.57	8.05	11.64	14.76	17.57	20.68	12.71	6
22.5		3.84	9.41	13.70	17.37	20.67	23.71	27.11	16.54	7
27.0	0.91	8.78	14.1	18.50	22.33	25.85	29.12	32.81	19.05	8
31.5	5.66	12.80	18.20	22.70	26.74	30.47	33.95	37.90	23.55	8
Base case (16.5)			8.81	14.9	19.96	24.42	28.51	33.09	21.62	6
Oregon	$66	$77	$88	$99	$110	$121	$132	$145	Mean	N
9.0									0.00	0
13.5						2.31	5.44	8.57	5.44	3
18.0				1.95	6.26	9.74	12.76	16.00	9.34	5
22.5			2.83	7.89	11.92	15.41	18.56	22.03	13.11	6
27.0		1.15	7.64	12.50	16.57	20.21	23.55	27.28	15.55	7
31.5		5.41	11.5	16.40	20.60	24.43	27.98	31.96	19.75	7
Base case (16.5)				4.80	10.69	15.45	19.63	24.17	14.95	5
Idaho	$66	$77	$88	$99	$110	$121	$132	$145	Mean	N
9.0	.								0.00	0
13.5						2.79	6.12	4.46	2
18.0					3.54	7.22	10.33	13.61	8.68	4
22.5				5.25	9.45	13.00	16.16	19.61	12.69	5
27.0			4.96	10.00	14.14	17.79	21.11	24.8	15.47	6
31.5		2.53	8.99	13.94	18.17	21.98	25.49	29.42	17.22	7
Base case (16.5)					6.27	11.21	15.40	19.88	13.19	4
California	$66	$77	$88	$99	$110	$121	$132	$145	Mean	N
9.0									0.00	0
13.5									0.00	0
18.0								0.28	0.28	1
22.5							3.28	7.11	5.20	2
27.0					0.76	5.08	8.64	12.33	6.70	4
31.5				0.44	5.47	9.51	13.00	16.73	9.03	5
Base case (16.5)									0.00	0

[1] Empty cells signify negative internal rates of return.

4. Discussion

The present study looked at returns that accrue over a 20-year rotation when rapid multiplication is used to produce planting stock of new varieties directly for operations, compared with the coincident use of less expensive hardwood cuttings to initiate planting with standard varieties that are available in commercial quantities. Thus, this comparison focused on the last component of the entire tree improvement cycle when newly-released varieties are propagated en masse as containerized rooted cuttings to originate a bioenergy planting that later transitions to serial coppice regeneration. An implicit assumption is that if the route of rapid propagation of new varieties is chosen, that process takes place at the same time that land is acquired and prepared, meaning that the new varieties are available at the corresponding time that standard varieties are purchased from a commercial nursery. Returns were estimated under both options over the course of a 20-year rotation. Research costs in developing new hybrids were not factored into the analysis, reasoning that a timber investment management organization produces proprietary varieties as leverage in raising capital into its bioenergy investment funds.

Investments into hybrid poplar bioenergy operations at all four AHB regions with current commercial varieties were previously shown to generate insufficient returns to attract private-sector capital and indicated the need for marked increases in biomass pricing and yields [19]. Other AHB analyses have suggested that the adoption of poplar feedstock operations is also impacted by the opportunity cost and the demand for competing crop commodities [26]. Despite the influence of biomass yields, selling prices and alternative crop options, dedicated energy crops are nonetheless expected to provide a significant component of biorefinery feedstock supply in the AHB region at a benchmark biofuel price of $19.6 GJ^{-1} [27].

Rapid multiplication was clearly superior to the base case at the two highest rates of yield irrespective of market pricing and region. For investments into rapid multiplication to make sense at the Washington, Oregon, and Idaho locations within the BTS range, yields of 22.5 Mg ha^{-1} year^{-1} and above would be needed for the most part. Conversely, a decision to forego investment into rapid multiplication in favor of the base case analysis of current commercial varieties would be generally expected at yields of 18–22.5 Mg^{-1} year^{-1} at all but the California location; this is true especially at market prices above the BTS range. However, it is doubtful whether current commercial hybrid poplar varieties that have been bred and selected for markets other than bioenergy would have the preferred biomass chemistry that would incentivize the market to offer a premium represented by prices above the BTS range. The regional influence of the study was noteworthy: The most favorable investment analysis, in terms of the overall mean and number of positive IRRs, was noted for the two non-irrigated farms, Washington and Oregon; of the two, the former is pasture land with the lowest lease rate in the study, while the Oregon farm was situated on cropland that commanded a higher lease rate. The two dryland farms requiring irrigation-Idaho and California-had lower overall IRRs and counts of positive returns. The California farm with the highest lease rate in the study exhibited the least favorable profile of returns under both the rapid multiplication and the base case. Returns at California under the base case were uniformly negative, and appreciable returns for the most part under the rapid multiplication case were not seen until yields of 27 and 31.5 Mg^{-1} year^{-1} at market prices above the BTS range.

The argument for rapid multiplication is that the exploitation of gains in biomass yield and chemistry requires a swift and seamless introduction of those gains into operations that regenerate by frequent coppicing to justify the expense of propagation [7]. However, are the genetic gains in yield and feedstock quality presented in this study realistic as requisites to poplar bioenergy farms investment? They may be. Poplar growth and yield simulations support the yields modeled in this study up to 22.5 to 27.0 Mg ha^{-1} year^{-1} [25,28,29], and preliminary estimates from bioenergy plantation density trials of 67 hybrid poplar varieties support theoretical yields of 35 Mg ha^{-1} year^{-1} for the most productive varieties [30]. Achieving yields above this level through ongoing hybridization is conceivable based on projections of 90% biomass gains for hybrid poplar when using reciprocal, intraspecific recurrent breeding in advance of first generation interspecific hybridization and within-family clonal selection [31].

Similarly, to accomplish improvements in biomass quality, gains in cell wall deconstruction, sugar release and biomass hydrolyzability may be anticipated using genomic selection of naturally-occurring mutant alleles in the lignin biosynthetic pathway [32–34], as well as lignin transformation [35,36], with the expectation that low-lignin biomass may significantly reduce the operating cost of mechanical and chemical pre-treatment. To illustrate, research at Oak Ridge National Laboratory's Bioenergy Research Center demonstrated a 15% increase in sugar yield associated with the biochemical conversion of low-lignin poplar varieties [37]. AHB research confirms this with reports of increases of 19% in sugar yield leading to 10% increases in biorefinery revenue that accompany individual hybrid selections [38]. However, it should be recognized that achieving these gains in large-scale plantations where soil quality and water supply vary at the landscape scale will be challenging.

Although gains in both biomass quantity and quality are likely, genetic improvement in the former is preferred from the perspective of an investor into biomass farms: Farm management is far better positioned to exploit increases in yield rather than bargaining the price at which the biomass is sold in to the energy markets. Increased yield directly provides incremental revenue, while an increase in price could be difficult to negotiate to the investor's favor, recognizing that biorefineries may not be able to adjust their processes to specific biomass characteristic-the superiority of cell wall chemistry notwithstanding-when managing a diversity of cellulosic feedstock sources.

Beyond the genetic gains that strengthen the attractiveness of rapid multiplication, its appeal may be improved per se by exploiting differences in the varietal performance during both the in vitro and greenhouse propagation phases [39–41]. For instance, the statistically-significant yield differences in macro-cutting propagation noted can be exploited to optimize the system, as recognized elsewhere [42]. This is especially true of the *P. maximowiczii* interspecific varieties and their strong response to hedge propagation that increased cutting yield by approximately twofold. The superior greenhouse performance of the *P. maximowiczii* hybrids mirrored their exceptional height growth and extremely low rates of sylleptic branching that maximizes nursery production of hardwood cuttings. Beyond these manipulations, the process can be shortened considerably using test trees from final stage yield trials to collect succulent macro-cuttings during spring shoot extension which are then used to initiate greenhouse propagation circumventing the in vitro production of micro-cuttings with cost and time savings.

Finally, there are other valuable features of the rapid multiplication system associated with the use of containerized rooted planting stock that reduce the risk of plantation failures. First, vagaries of the weather-frost injury and drought-are much more easily mitigated when planting stock is grown in controlled greenhouse environments in comparison to field nurseries. Second, greenhouse production of newly-deployed varieties allows for far greater flexibility than field nurseries when turning over and initiating new varieties in to scale-up operations when, for example, such varieties must be removed from production due to a loss of pest resistance or a change in market preference. Third, rooted cuttings greatly reduce the risk of planting failures compared to the risk incurred using unrooted hardwood cuttings when an unfavorable environment-protracted dry or cold periods-is encountered during the time of establishment.

5. Conclusions

1. No variety failed to respond to in vitro micro-cutting propagation or greenhouse macro-cutting propagation.

2. The *Populus deltoides* × *P. maximowiczii* and the *Populus trichocarpa* × *P. maximowiczii* varieties greatly exceeded the multiplication rate of the *P.* × *generosa* varieties under greenhouse propagation. This was largely due to their exceptional response to repeated hedging required to initiate multiple tracks of serial propagation. It mirrored the stronger performance of the *P. maximowiczii* taxon under traditional stoolbed culture.

3. Rapid multiplication was clearly superior to the base case at the two highest rates of yield irrespective of market pricing and region. For investments into rapid multiplication to make sense at

the Washington, Oregon, and Idaho locations within the BTS range, yields of 22.5 Mg ha^{-1} year^{-1} and above would be needed for the most part.

4. Conversely, a decision to forego investment into rapid multiplication in favor of the base case analysis of current commercial varieties would be generally expected at yields of 18.0–22.5 Mg^{-1} year^{-1} at all but the California location; this is true especially at market prices above the BTS range. However, it is doubtful whether current commercial hybrid poplar varieties for which preferred biomass chemistry has not been an improvement emphasis would justify market prices above the BTS range.

Supplementary Materials: The following are available online at http://www.mdpi.com/1999-4907/10/5/446/s1, Table S1: Analysis of variance in survival.

Author Contributions: Conceptualization, B.J.S. and C.G.; Data curation, R.A.S.; Investigation, B.J.S.; Methodology, K.H., C.G. and J.E.; Project administration, B.J.S.; Resources, J.E.; Supervision, B.J.S. and R.A.S.; Validation, R.A.S.; Writing—original draft, B.J.S.; Writing—review and editing, R.A.S.

Funding: This research was funded by the United States Department of Agriculture, National Institute for Food and Agriculture, Grant no. 2011-68005-30407, Agriculture and Food Research Initiative.

Acknowledgments: The authors gratefully acknowledge the support of the United States Department of Agriculture National Institute of Food and Agriculture (USDA-NIFA). The work reported here was conducted under Agriculture and Food Research Initiative (AFRI) Competitive Grant Number 2011-68005-30407.

Conflicts of Interest: The authors declare no conflict of interest. The funders had no role in the design of the study; in the collection, analyses, or interpretation of data; in the writing of the manuscript, or in the decision to publish the results.

References

1. Stanton, B.J. Clonal variation in basal area growth patterns during stand development in hybrid poplar. *Can. J. For. Res.* **2001**, *31*, 2059–2066. [CrossRef]
2. Volk, T.A.; Berguson, B.; Daly, C.; Halbleib, M.D.; Miller, R.; Rials, T.G.; Abrahamson, L.P.; Buchman, D.; Buford, M.; Cunningham, M.W.; et al. Poplar and shrub willow energy crops in the United States: Field trial results from the multiyear regional feedstock partnership and yield potential maps based on the PRISM-ELM model. *Glob. Chang. Biol. Bioenergy* **2018**, *10*, 735–751. [CrossRef]
3. Dou, C.; Marcondes, W.F.; Djaja, J.E.; Bura, R.; Gustafson, R. Can we use short rotation coppice poplar for sugar based biorefinery feedstock? Bioconversion of 2-year-old poplar grown as short rotation coppice. *Biotechnol. Biofuels* **2017**, *10*, 144. [CrossRef] [PubMed]
4. Xie, M.; Wellington, M.; Bryan, A.C.; Yee, K.L.; Guo, H.-B.; Zhang, J.; Tschaplinski, T.; Singan, V.R.; Lindquist, E.; Payyavula, R.S.; et al. A 5-enolpyrearuvylshikimate 3-phosphate synthase functions as a transcriptional repressor in *Populus*. *Plant Cell* **2018**, *30*, 1645–1660. [CrossRef] [PubMed]
5. Abramson, M.; Shoseyov, O.; Shani, Z. Plant cell wall reconstruction toward improved lignocellulosic production and processability. *Plant Sci.* **2010**, *178*, 61–72. [CrossRef]
6. Bhojwani, S.S. Micropropagation method for a hybrid willow (Salix matsudana × alba NZ-1002). *N. Z. J. Bot.* **1980**, *18*, 209–214. [CrossRef]
7. Thorpe, T.A. Biotechnological applications of tissue culture to forest tree improvement. *Biotechnol. Adv.* **1983**, *1*, 263–278. [CrossRef]
8. Zalesny, R.S., Jr.; Riemenschneider, D.E.; Hall, R.B. Early rooting of dormant hardwood cuttings of *Populus*: Analysis of quantitative genetics and genotype × environment interactions. *Can. J. For. Res.* **2005**, *35*, 918–929. [CrossRef]
9. Randall, W.K.; Miller, A.J.E. Mist propagation recommended for expanding cottonwood clones rapidly. *Tree Plant. Notes* **1971**, *22*, 9–13.
10. Phipps, H.M.; Hansen, E.A.; Tolsted, D.N. Rooting greenwood tip cuttings of several *Populus* clones hydroponically (hydroponic rooting of *Populus* cuttings). *Can. J. For. Res.* **1979**, *10*, 107–110. [CrossRef]
11. Jiang, C.; Liu, Z.; Zheng, Q. Direct regeneration of plants derived from in vitro cultured shoot tips and leaves of poplar (*Populus* × euramericana 'Neva'). *J. Life Sci.* **2015**, *9*, 366–372.
12. Thakur, A.K.; Saraswat, A.; Srivastava, D.K. In vitro plant regeneration through direct organogenesis in *Populus* deltoides clone G48 from petiole explants. *J. Plant Biochem. Biotechnol.* **2012**, *21*, 23–29. [CrossRef]

13. Arencibia, A.D.; Gomez, A.; Poblete, M.; Vergara, C. High-performance micropropagation of dendroenergetic poplar hybrids in photomixotrophic Temporary Immersion Bioreactors (TIBs). *Ind. Crops Prod.* **2017**, *96*, 102–109. [CrossRef]

14. Louis, K.A.; Eils, L.E. Application of tissue culture systems for commercial plant production. In *Micropropagation, Genetic Engineering and Molecular Biology of Populus*; Klopfenstein, N.B., Chun, Y.W., Kim, M.-S., Ahuja, M.R., Eds.; USDA U.S. Forest Service Rocky Mountain Research Station General Technical Report RM-GTR-297; U.S. Forest Service: Fort Collins, CO, USA, 1997; Volume 30, pp. 236–240.

15. Haapala, T.; Pakkanen, A.; Pulkkinen, P. Variation in survival and growth of cuttings into clonal propagation methods for hybrid aspen. *For. Ecol. Manag.* **2004**, *193*, 345–354. [CrossRef]

16. Ahuja, M.R.; Libby, W.J. Glossary. In *Clonal Forestry I Genetics and Biotechnology*; Ahuja, M.R., Libby, W.J., Eds.; Springer: Berlin/Heidelberg, Germany, 1993; pp. 255–265.

17. Thomson, T.A.; Lester, D.T.; Martin, J.A.; Foster, G.S. Using economic and decision making concepts to evaluate and design a corporate tree improvement program. *Silvae Genet* **1989**, *38*, 21–27.

18. Shuren, R.A.; Busby, G.; Stanton, B.J. The Biomass Production Calculator: A Decision Tool for Hybrid Poplar Feedstock Producers and Investors. In Proceedings of the Woody Crops International Conference, Rhinelander, WI, USA, 22–27 July 2018.

19. Chudy, R.P.; Busby, G.M.; Binkley, C.S.; Stanton, B.J. The economics of dedicated hybrid poplar biomass plantations in the western U.S. *Biomass Bioenergy* **2019**, *124*, 114–124. [CrossRef]

20. Hart, Q.J.; Tittmann, P.W.; Bandaru, V.; Jenkins, B.M. Modeling poplar growth as a short rotation woody crop for biofuels in the Pacific Northwest. *Biomass Bioenergy* **2015**, *79*, 12–27. [CrossRef]

21. U.S. Department of Energy. *Billion-Ton Report: Advancing Domestic Resources for a Thriving Bioeconomy, Volume 1 Economic Availability of Feedstocks*; Langholtz, M.H., Stokes, B.J., Eaton, L.M., Eds.; ORNL/TM-2016/160; Oak Ridge National Laboratory: Oak Ridge, TN, USA, 2016; p. 448.

22. Frank, J.R.; Brown, T.R.; Volk, T.A.; Heavey, J.P.; Malmsheimer, R.W. A stochastic techno-economic analysis of shrub willow production using EcoWillow 3.0S. *Biofuel Bioprod. Biorefin.* **2018**, *12*, 846–856. [CrossRef]

23. Munn, I.A.; Hussain, A.; Grebner, D.L.; Grado, S.C.; Measells, M.K. Mississippi Private Landowner Willingness for Diverting Land to Growing Short Rotation Woody Crops for Bioenergy Enterprises. *For. Sci.* **2018**, *64*, 471–479. [CrossRef]

24. Clifton-Brown, J.; Harfouche, A.; Casler, M.D.; Jones, H.D.; Macalpine, W.J.; Murphy-Bokern, D.; Smart, L.B.; Adler, A.; Ashman, C.; Awty-Carroll, D.; et al. Breeding progress and preparedness for mass-scale deployment of perennial lignocellulosic biomass crops switchgrass, miscanthus, willow, and poplar. *Glob. Chang. Biol. Bioenergy* **2018**, *11*, 118–151. [CrossRef]

25. Stanturf, J.A.; Young, T.M.; Perdue, J.H.; Dougherty, D.; Pigott, M.; Guo, Z.; Huang, X. Potential Profitability Zones for *Populus* spp. Biomass Plantings in the Eastern United States. *For. Sci.* **2017**, *63*, 586–595. [CrossRef]

26. Bandaru, V.; Parker, N.; Hart, Q.; Jenner, M.; Yeo, B.; Crawford, J.; Li, Y.; Titman, P.; Rogers, L.; Kaffka, S.; et al. Economic sustainability modeling provides decision support for assessing hybrid poplar-based biofuel development in California. *Calif. Agric.* **2015**, *69*, 171–176. [CrossRef]

27. Parker, N.; Titman, P.; Hart, Q.; Nelson, R.; Skog, K.; Schmidt, A.; Gray, E.; Jenkins, B. Development of a biorefinery optimized biofuel supply curve for the western United States. *Biomass Bioenergy* **2010**, *34*, 1597–1607. [CrossRef]

28. Deckmyn, G.; Laureysens, I.; Garcia, J.; Muys, B.; Ceulemans, R. Poplar growth and yield in short rotation coppice: Model simulations using the process model SECRETS. *Biomass Bioenergy* **2004**, *26*, 221–227. [CrossRef]

29. Wang, D.; LeBauer, D.; Dietze, M. Predicting yields of short-rotation hybrid poplar (*Populus* spp.) for the United States through model–data synthesis. *Ecol. Appl.* **2013**, *23*, 944–958. [CrossRef]

30. Yaneza, M.A.; Zamudio, F.; Espinoza, S.; Ivkovic, M.; Guerra, F.; Espinosa, C.; Baettig, R.M. Genetic variation and growth stability of hybrid poplars in high-density short-rotation coppice stands in central Chile. *Biomass Bioenergy* **2018**, *120*, 81–90.

31. Berguson, W.E.; McMahon, B.E.; Riemenschneider, D.E. Additive and Non-Additive Genetic Variances for Tree Growth in Several Hybrid Poplar Populations and Implications Regarding Breeding Strategy. *Silvae Genet.* **2017**, *66*, 33–39. [CrossRef]

32. Studer, M.H.; DeMartini, J.D.; Davis, M.F.; Sykes, R.W.; Davison, B.; Keller, M.; Tuskan, G.A.; Wyman, C.E. Lignin content in natural *Populus* variants affects sugar release. *Proc. Natl. Acad. Sci. USA* **2011**, *108*, 6300–6305. [CrossRef]

33. Vanholme, B.; Cesarino, I.; Goeminne, G.; Kim, H.; Marroni, F.; Acker, R.; Vanholme, R.; Morreel, K.; Ivens, B.; Pinosio, S.; Morgante, M.; et al. Breeding with rare defective alleles (BRDA): A natural *Populus* nigra HCT mutant with modified lignin as a case study. *New Phytol.* **2013**, *198*, 765–776. [CrossRef] [PubMed]

34. Davison, B.H.; Drescher, S.R.; Tuskan, G.A.; Davis, M.F.; Nghiem, N.P. Variation of S/G Ratio and Lignin Content in a *Populus* Family Influences the Release of Xylose by Dilute Acid Hydrolysis. *Appl. Biochem. Biotechnol.* **2006**, *130*, 427–435. [CrossRef]

35. Chen, F.; Dixon, R.A. Lignin modification improves fermentable sugar yields for biofuel production. *Nat. Biotechnol.* **2007**, *25*, 759–761. [CrossRef] [PubMed]

36. Mansfield, S.D.; Kang, K.-Y.; Chapple, C. Designed for deconstruction—Poplar trees altered in cell wall lignification improve the efficacy of bioethanol production. *New Phytol.* **2012**, *194*, 91–101. [CrossRef] [PubMed]

37. Bhagia, S.; Muchero, W.; Kumar, R.; Tuskan, G.; Wyman, C. Natural genetic variability reduces recalcitrance in poplar. *Biotechnol. Biofuels* **2016**, *9*, 106. [CrossRef] [PubMed]

38. Dou, C.; Gustafson, R.; Bura, R. Bridging the gap between feedstock growers and users: The study of a coppice poplar-based biorefinery. *Biotechnol. Biofuels* **2018**, *11*, 77. [CrossRef]

39. Noel, N.; Leple, J.-C.; Pilate, G. Optimization of in vitro micropropagation and regeneration for *Populus* × interamericana and *Populus* × euramericana hybrids (*P. deltoides*, *P. trichocarpa*, and *P. nigra*). *Plant Cell Rep* **2002**, *20*, 1150–1155.

40. Rutledge, C.B.; Douglas, G.C. Culture of meristem tips and micropropagation of 12 commercial clones of poplar in vitro. *Physiol. Plant.* **1988**, *72*, 367–373. [CrossRef]

41. Sellmar, J.C.; McCown, B.H.; Haissig, B.E. Shoot culture dynamics of six *Populus* clones. *Tree Physiol.* **1989**, *5*, 219–227. [CrossRef]

42. Stenvall, N.; Haapala, T.; Pulkkinen, P. Effect of genotype, age and treatment of stock plants on propagation of hybrid aspen (*Populus* tremula × *Populus* tremuloides) by root cuttings. *Scand. J. For. Res.* **2004**, *19*, 303–311. [CrossRef]

![forests logo] *forests*

MDPI

Article

Adaptability of *Populus* to Physiography and Growing Conditions in the Southeastern USA

Solomon B. Ghezehei [1,*], Elizabeth G. Nichols [1], Christopher A. Maier [2] and Dennis W. Hazel [1]

[1] Department of Forestry & Environmental Resources, NC State University, Raleigh, NC 27695, USA; egnichol@ncsu.edu (E.G.N.); hazeld@ncsu.edu (D.W.H.)

[2] Southern Research Station, USDA Forest Service, Research Triangle Park, NC 27709, USA; cmaier@fs.fed.us

* Correspondence: sbghezeh@ncsu.edu; Tel.: +1-919-513-1371

Received: 22 December 2018; Accepted: 30 January 2019; Published: 2 February 2019

Abstract: *Populus* species have a high productivity potential as short-rotation woody crops, provided that site-suitable varieties are planted. The Coastal Plain, the Piedmont, and the Blue Ridge Mountains make up a significant part of the eastern and southeastern USA, and an insight into poplar productivity and adaptability will be valuable for the successful implementation of large-scale poplar stands in these regions. The objectives of this study were to examine the green wood biomass (hereafter biomass), biomass allocation, and wood properties of poplars in relation to growing conditions, physiography, and topography. The biomass of 4-year-old poplars was estimated using an equation derived through destructive sampling. Biomass-based clonal rankings were compared across the various site conditions (fertility, irrigation, land marginality, soil preparation, and topography) and the three physiographic provinces. Although not all clonal differences in biomass were significant, growing conditions, physiography, and soil preparation affected the clonal rankings and the significance of the clonal differences. Biomass changes due to physiography and land conditions were more structured at the genomic-group level. A higher-altitude physiography led to greater biomass increases in *Populus trichocarpa* × *Populus deltoids* (TD) clones than in *P. deltoids* × *P. deltoids* (DD) clones and vice versa. Favorable soil quality or management generally led to greater biomass of DD clones than of TD and *P. deltoids* × *Populus maximowiczii* (DM) clones. Weather-related variables were not clearly correlated with biomass, while land aspect was a significant influence on the biomass of genomic groups and clones. The site significantly affected wood density, moisture content, and carbon and nitrogen concentrations, while the clonal effects on wood composition and the clonal and site effects on biomass allocation were insignificant. Although clones showing greater biomass responses to growing conditions generally belonged to the same genomic group, clone-level selection could produce greater biomass gains than selection at the genomic-group level.

Keywords: biomass allocation; land marginality; *Populus*; physiographic provinces; soil preparation; topography; wood biomass productivity; wood composition and properties

1. Introduction

According to the 2005 Billion-Ton Study, the 2011 update, and the 2016 report, in the United States, woody feedstocks are expected to make a great contribution to the sustainable biomass supply for the bioeconomy [1]. However, with only 2% of the 2017 total energy consumption in the United States coming from wood and wood waste [2] and with the contribution of woody feedstocks to biofuels currently being non-existent for practical purposes, achieving the bioeconomy target will require a great enhancement in the productivity of woody crops [1]. In the southeastern United States, there is significant potential for conventional forestry to contribute to the bio-based economy mainly in the form of wood wastes and logging residues. However, a much greater potential lies in purpose-grown and fast-growing woody species in 'whole-tree' forms [1], also known as short-rotation woody crops (SRWCs).

The production of SRWCs is mainly restricted to marginal lands, including underutilized, reclaimed, and contaminated lands or lands that are poorly fertile for food production or convectional forestry. The marginality of lands varies greatly, and in the eastern and southeastern United States, the variations may be further emphasized by the variable and even contrasting physiographic features and growing conditions. Economic feasibility is key to the adoption of SRWCs and requires high productivity, suitable establishment, and high feedstock prices. It is important to assess and match SRWC species and varieties to particular site and growing conditions so as to maximize productivity. This has additional significance in the southeastern United States due to the standing of the region as the leading wood pellet exporter to Europe.

Poplars (*Populus* spp.) have great productivity potential as SRWCs nationally, provided that suitable clones are planted, and are one of the key non-coppice woody energy crop types identified in the Billion-Ton report [1]. Wood from poplars can be used for a number of applications, including the production of bioenergy, pulp, veneer, plywood, and timber [3,4]. Poplars can be grown in shorter rotations for bioenergy or pulp, or for up to 20 years when targeting high-value products [4,5].

Populus clones have been subjected to extensive studies in the United States since the 1920s and specifically in the southeast since the 1960s [6]. The adaptation of poplars to growing conditions is greatly dictated by, among other things, clone varieties and the quality of sites [4,7]. Moreover, previous studies involving multiple poplar clones and genotypes have shown great variations in clonal productivity [4,6,8–11]. Some of these studies have shown that genotype–environment interactions can affect poplar productivity [9,11]. The early results of a study comparing several *Populus* clones, native hardwoods, and eucalyptus species in the southeastern United States showed that *Populus* had superiority in growth and survival compared to other species [12]. Another early growth study indicated that although poplars can be highly productive, clonal and genotypic suitability to site conditions is a decisive consideration for its success as an SRWC [11].

An insight into poplar productivity and clonal and genotypic adaptability to growing conditions (site quality, land marginality, topography, and soil management) and physiography is crucial for identifying the best poplar varieties to grow under a particular set of conditions and for maximizing the efficacy of poplars as SRWCs. In the southeastern United States, such a study is greatly significant due to the physiographic variations in the region, which includes the Blue Ridge Mountains, the Piedmont, and the Coastal Plains, and the various growing conditions that can be considered for growing SRWCs. This information regarding clonal and genotypic adaptability to growing conditions and physiography can be applicable to eastern and southeastern states such as Pennsylvania, Maryland, Virginia, Tennessee, North Carolina, South Carolina, and Georgia, which have the mountain–Piedmont–coastal physiographic variations.

The objectives of this study were to examine how physiography, topography, and growing conditions in the southeastern USA can affect the adaptability and green wood biomass (hereafter biomass) of poplars, and whether poplar wood properties and composition vary due to position on tree stem, clone effect, and physiography. The following hypotheses were formulated:

1. The woody biomass proportions of stem and branches, the allometry between dimensions (height and stem diameter), and the woody biomass of poplars are affected by sites and clones.
2. The biomass productivity rankings of poplar clones are affected by physiography.
3. The biomass productivity rankings and adaptability of poplars (within a physiographic region) are affected by growing conditions including the marginality of lands, soil preparation, topographic positions (upslope versus downslope), and aspects of the land.
4. The density, moisture content, and carbon and nitrogen concentrations of the wood of poplars are affected by site, clones, and position on trees (basal, breast height, top).

2. Materials and Methods

2.1. Description of the Study Sites

The study sites were located in North Carolina, and represented three physiographic provinces present in the eastern and southeastern USA, namely the Coastal Plains, the Piedmont, and the Blue Ridge Mountains (Table 1). The main physiographic feature of this study was altitude. Furthermore, four weather-related site variables were studied: Growing degree days (GDD) using 10 °C as a base temperature, total amount of precipitation and irrigation, Penman–Montheith reference evapotranspiration (ET_0), and photosynthetically active radiation (PAR). GDD and ET_0 increased as the altitudes of the physiographic regions decreased. The totals of GDD, precipitation and irrigation, ET_0, and PAR at the sites for the 4-year study are presented in Figure 1.

Table 1. Details of the locations, physiography, and growing conditions of the sites where stands used for the biomass sampling and growth monitoring of poplar trees were located.

Site; Altitude (m, above Sea Level)	Latitude; Longitude	Physiography	Precipitation; Irrigation (mm year^{-1})	Soil Texture	Planted
Salisbury (SB); 215	35.6974; −80.6219	Piedmont	1118	Loam	2014
Mills River (MR); 630	35.4272; −82.5589	Lower southern Blue Ridge Mountains	1261	Loam	2014
Laurel Springs (LS); 975	36.4023; −81.2971	Upper southern Blue Ridge Mountains	1244	Sandy clay loam	2014
Williamsdale (WD); 26	34.7641; −78.0983		1400	Loam	2013
Gibson (GB); 76	34.7672; −76.5962	Coastal Plain	1300; 1102	Loamy sand	2013
Tidewater (TW); 5	35.8555; −76.6508		1200; 550	Loam	2014

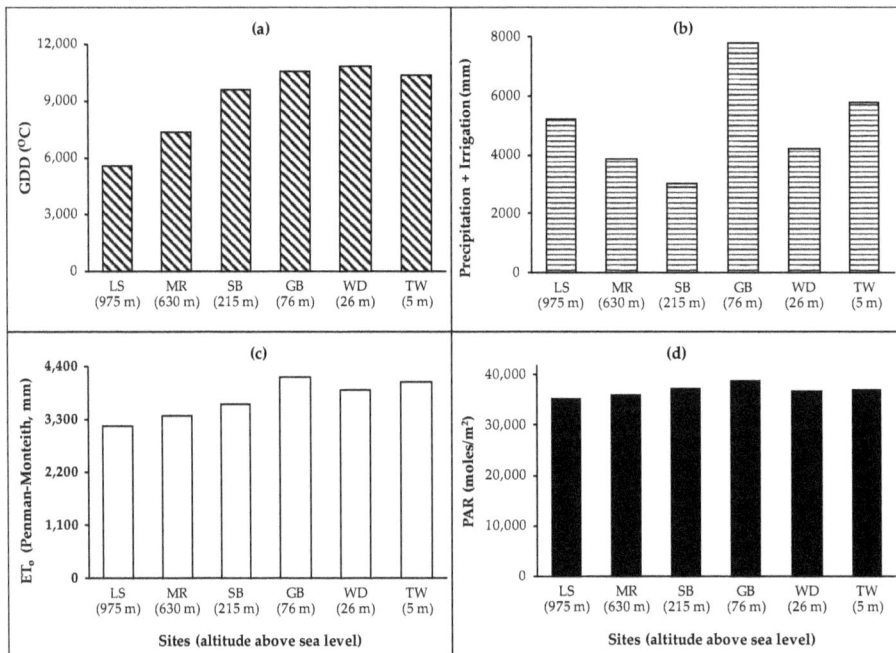

Figure 1. Totals of (**a**) growing degree days (GDD) based on a base temperature of 10 °C, (**b**) precipitation and irrigation, (**c**) Penman–Montheith reference evapotranspiration, and (**d**) photosynthetically active radiation (PAR) at the sites during the 4-year study.

The clones, genomic groups, and experimental designs used for the current study are presented in Table 2. Site preparation entailed weed control using mowers and post-emergent (Gly Star®Pro, Albaugh, INC., Ankeny, IO 50021, USA) and pre-emergent (Pendulum®3.3 EC Herbicide, BASF Corporation, Research Triangle Park, NC 27709, USA) herbicides, and subsoiling using a V-Ripper. Trees were planted in early spring as 30-cm-long cuttings. Prior to planting, the cuttings were immersed in water for 48 hours. Post-establishment weed control at all study sites entailed banding as needed with Gly Star®Pro along tree rows and mowing as needed between tree rows.

Table 2. Details of the experimental design, clones, and genomic grouping used for the study of the effects of physiography and growing conditions on poplar woody productivity, biomass allocation, and wood properties and composition. At Salisbury, three separate trials were used: SBC (randomized block design or RBD) with a 2500-tree/ha density, and SBS2 × 1 (CRD) and SBS2 × 2 (CRD) with 5000-tree/ha and 2500-tree/ha densities, respectively. Two separate LS trials were used: LS2 × 2 with a 2500-tree/ha density and LS1 × 1 with a 10,000-tree/ha density. WDD and WDSS denote disking and subsoiling treatments of soil preparation used at Williamsdale, respectively. Gibson (GB) and Williamsdale (WD) had the same clones and design details.

Study Effects and Stands (Trees per ha)		Genomic Groups (Clones)	Design
Physiography	MR (lower Blue Ridge Mountains, 2500); SBC (Piedmont, 2500)	**DD** (140, 177, 210, 373, 379); **TD** (185, 187, 188, 229, 302, 339, 342, 5077)	RBD
	TW (Coastal, 5000); SBS2 × 1 (Piedmont, 5000)	**DD** (140, 312, 356); **TD** (187, 188, 302, 342)	
	LS2 × 2 (upper Blue Ridge Mountains, 2500), SBS2 × 2 (Piedmont, 2500); WD (coastal, 1495)	**DD** (140, 176, 356); **TD** (185, 187, 188, 229)	
Growing Conditions	Land marginality (coastal, 1495): WD vs GB	**DD** (140, 176, 356, 373); **DM** (230); **TD** (185, 187, 188, 229, 339)	CRD
	Soil preparation (coastal, 1495): WDD, WDSS	**DD** (140, 176, 356, 373); **DM** (230); **TD** (185, 187, 188, 229, 339)	
	Land topography and aspect (upper Blue Ridge Mountains, 10,000): LS1 × 1 (upper Blue Ridge Mountains)	**DD** (176, 210, 312, 356, 373, 379, 419, 426, 443); **TD** (185, 187, 188, 229, 339, 342); **DM** (230); **DN** (DN-34, OP-367)	

RBD: Randomized block design; CRD: Cluster randomized design; D: *Populus deltoids*; T: *Populus trichocarpa*; M: *Populus maximowiczii*; N: *Populus nigra*.

2.2. Data Collection

Tree height and stem diameter at breast height (DBH), which is 1.3 m above the soil surface, of all trees at the study sites were measured at the end of fourth year of growth. Whole-tree destructive sampling was carried out in March and April 2016 on 3-year-old trees representing small, medium, and large trees of clones 140, 187, and 188 from stands at Salisbury (SB) in the Piedmont, Mills River (MR) in the lower southern Blue Ridge Mountains, and Laurel Springs (LS) in the upper southern Blue Ridge Mountains. These three clones were selected for destructive sampling because of their availability at almost all stands located in the above three physiographic regions, and the total number of sampled trees, 30, included 10 trees from each clone. The classification of the trees as small, medium, and large were clone- and site-specific. That is, at each site and for each clone, three size classes were formed based on the tree height measurements at the end of 2015, namely, the bottom third (small), the middle third (medium), and the top third (large). From these clone- and site-specific classes, one random tree per clone per class was harvested at LS and MR (a total of nine trees at each site); whereas four trees were randomly sampled per clone at SB, including one small, two medium, and one large tree (a total of 12 trees). Prior to cutting down the trees, stem diameters at breast height (DBHs) and total heights were measured. The total fresh biomass of the trees (wood and foliage) was determined using a scale mounted on a tall sling, ensuring that the samples being measured were fully suspended. After the total tree biomasses were determined, all leaves were removed, and the leafless samples were weighed to determine the total fresh wood biomasses. The tree stem biomass was determined by weighing tree stems after removing branches and twigs. Cross-sectional fresh wood samples (5 cm high) were collected from the base, breast height, and two-thirds of tree height positions, and their weights were recorded. The samples were placed in air-tight containers and kept in a cool and dry place for further processing in the laboratory. The cross-sectional wood samples were oven-dried at 60 °C until constant masses were obtained. The samples were then sent to the Forest Genetics and Biological Laboratory of the US Forest Service—Southern Research Station located in the Research Triangle Park in North Carolina for the determination of wood density, wood moisture, and carbon and nitrogen contents (%).

2.3. Data Analyses

A generalized linear model (GLM) analysis using Proc GLM of SAS [13] ($\alpha = 0.05$) was applied for analyzing woody biomass allocation to stems and branches and the elemental composition (carbon and nitrogen), moisture content, and density of the wood (Hypotheses 1 and 4). The GLM was applied due to its flexibility in handling error distribution models with both normal and non-normal distributions. Where there were significant interaction effects, Proc Slice (SAS) was applied to further examine if all clones had interaction effects, and to identify clones showing interactions and those that do not show interaction effects. Clone-specific, site-specific, and generalized (all-data) allometric equations were derived by plotting volume indices (height multiplied by DBH squared) of the sampled trees against the ratio of the destructively sampled fresh woody biomass and mean wood density obtained from the cross-sectional wood samples. An analysis of covariance (ANCOVA) was applied using Proc REG (SAS) to examine if the allometric equations were significantly different (Hypothesis 1). ANCOVA was used because it enables correlation analysis between independent and dependent variables while meeting the assumptions of linearity between the variables, normality and independence of error terms, and homogeneity of error variances. Based on the ANCOVA results, the generalized wood volume equation was used to estimate the green wood biomass of standing trees using the average wood density and the heights and DBHs of standing 4-year-old trees. The wood biomass estimates were used to examine and compare the effects of physiography, land, and growing conditions on poplar productivity and adaptability.

Biomass analyses entailed ranking the biomasses of clones at individual sites and examining how the rankings would change with changes in physiography and growing conditions. With the ranking-based assessment, it was possible to concentrate on the site and physiographic variables in question while avoiding errors that could arise from potential differences in confounding factors between sites. For each study, an additional comparison of percentage changes in the green wood biomass of common clones at the paired sites was added to examine productivity differences between the sites. For analyzing physiography effects on poplars, a non-parametric significance analysis known as the Kruskal–Wallis test was applied using the SAS procedure Proc NPAR1WAYS to the clonal comparisons or ranks of wood productivity as the normality conditions of some of the studies were not met. For analyzing the effects of growing conditions on poplars, the Proc GLM (SAS procedure) was applied and where interaction effects existed, while the Proc SLICE (SAS) was applied to check if the interaction effects were present at all levels of the interacting factors/treatments. The effects of land topographic positions and aspects on poplar wood biomass were examined using LS1 $\times$ 1 (Table 2), which included three plots with mean slopes of 14%–14.7% and upslope and downslope topographic positions on each plot. The aspects of the plots were southerly aspect or south-facing slope (N to S), westerly aspect or west-facing slope (N47E to S47W), and southwesterly aspect or southwest-facing slope (N65E to S65W). The Proc GLM (SAS) was applied for analyzing the effects and, where there were interaction effects, the Proc SLICE (SAS) was applied to examine the presence of interactions at various levels of the interacting treatments/factors.

3. Results

3.1. Woody Biomass

Based on the biomass sampling, woody biomass allocation to branches and stems was not significantly affected by clone, site (physiography), and clone–site interaction (Figure 2). Clonal ($p = 0.3148$), site ($p = 0.2998$), and clone–site interaction ($p = 0.4334$) effects on the correlation between volume and volume index ($H \times DBH^2$) were insignificant. Hence, the following generalized equations were developed:

$$\text{Volume (m}^3) = H \times DBH^2 \times 0.99185 + 0.00188 \qquad (R^2 = 0.94) \qquad (1)$$

$$\text{Biomass (t)} = \text{volume} \times \text{wood density} \times 0.001 \qquad (2)$$

$$\text{Green wood biomass (kg)} = 408.31 \times H \times DBH^2 - 0.2883 \qquad (R^2 = 0.98) \qquad (3)$$

$$\text{Green wood biomass (kg)} = 2656.7 \times DBH^2 - 0.2923 \qquad (R^2 = 0.98) \qquad (4)$$

where volume and biomass denote the volume and the biomass of the green wood of poplars, the unit of H and DBH is meter (m), and wood density (kg/m^3) is the mean density of wood samples from all tree positions of all trees sampled.

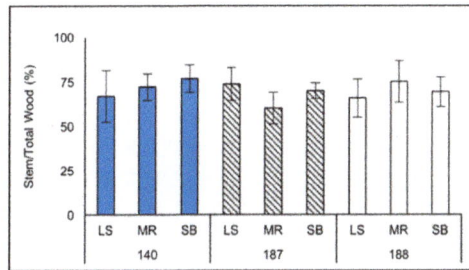

Figure 2. Biomass-based percentages of stem wood (from the total tree wood) for three sampled clones (140, 187, and 188) sampled from stands located in the upper Blue Ridge Mountains (LS), the lower Blue Ridge Mountains (MR), and the Piedmont (SB) of North Carolina.

Green wood biomasses of all studied clones (Sections 2.3, 3.2 and 3.3) were estimated using Equations (1) and (2); direct estimations of poplar woody biomass using Equation (3) or Equation (4) were identical to biomass estimates obtained from Equations (1) and (2).

3.2. The Effects of Physiography on Adaptability and Biomass Productivity

Many of the studied sites showed an absence of significant clonal differences (at $p > 0.05$). However, differences in physiography led to some changes in the clonal rankings and significance of clonal differences. There were significant clonal differences in biomass at MR ($p = 0.018$) but not at SBC ($p = 0.1981$), and there were differences in clonal standings between the sites (Figure 3a). The biomass of all studied clones was much higher at SBC than it was at MR. DD clones were affected by site changes to a much greater extent (approximately 1570% to 5000% increases) than TD clones (Figure 3b). Noting that only the GDD during the growth period was considerably higher at SBC than it was at MR, it can be inferred that the biomass increases of the clones could mainly be attributed to better soil fertility at SBC.

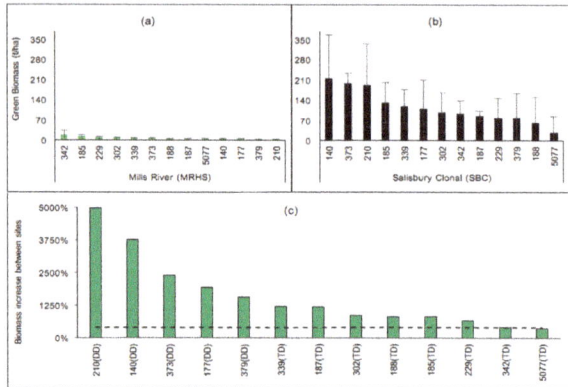

Figure 3. Four-year-old green wood biomass (tonne/ha) of non-irrigated poplars located in (**a**) the lower Blue Ridge Mountains (MR) and (**b**) the Piedmont (SBC), and (**c**) increases (%) of clonal biomass at SBC relative to those at MR.

A comparison of poplars in the Piedmont (SBS2 × 1) vs the Coastal Plain (Tidewater, (TW)) with the same planting density (5000 trees/ha) showed that clonal biomass differences were significant at TW ($p < 0.0001$) but not at SBS2 × 1 ($p = 0.0534$). Regardless of the physiographic differences between the Piedmont (SBS2 × 1) and the Coastal Plains (TW), and although the TW stand was irrigated, both SBS2 × 1 and TW stands had common high-performing (140 and 356) and low-ranked clones (302 and 188) (Figure 4a). The TD clones showed biomass improvements at SBS2 × 1 compared to those at TW, while DD clones showed increases or decreases to a lesser extent at SBS2 × 1 compared to those at TW (Figure 4b). It should be noted that TW had considerably higher GDD and precipitation than SBS2 × 1. It is also worth noting that the biomass increases of clone 140 were outstanding at lower-precipitation (irrigated) sites.

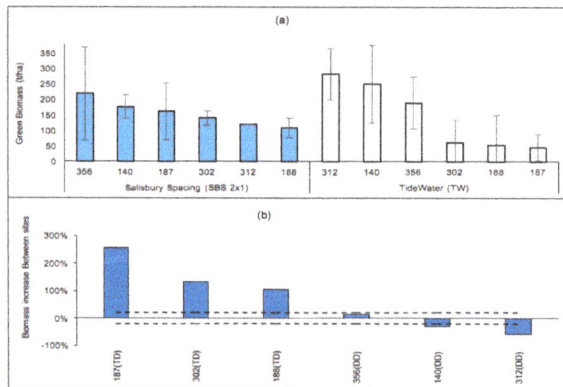

Figure 4. (**a**) Four-year-old green wood biomass (t/ha) of poplar stands located in the Piedmont (SBS2 × 1, non-irrigated) and in the Coastal Plains (TW, irrigated) and (**b**) changes (%) in clonal wood biomasses at SBS2 × 1 relative to those at TW.

The clonal biomass differences were insignificant at LS2 × 2 ($p = 0.1243$), SBS2 × 2 ($p = 0.2451$), and Williamsdale (WD) ($p = 0.1193$). The LS2 × 2 and SBS2 × 2 sites had common high-biomass clones (229,185) and a low-producing clone (187). Although TD clones were the highest producers at both sites, the clonal rankings changed between the sites (Figure 5a), with DD clones showing greater biomass increases at the Piedmont site (SBS2 × 2) than at the upper Mountain site (LS2 × 2), and TD clones showing variations within 20% (Figure 5b). Most clones had similar productivities in LS2 × 2 and SBS2 × 2, and it can be inferred that the overall fertility of these sites may be similar. Irrespective of the possible fertility similarities of the sites, DD clones in general and clone 140 in particular showed greater increases in wood biomass at the sites where the GDD was higher even when the precipitation was lower. Some variations in clonal rankings also occurred between WD and SBS2 × 2 (Figure 5a). Overall, tree growth was similar at the sites, while the per-hectare productivity was higher at SBS2 × 2 due to the higher planting density. However, the selected weather-related variables (GDD, precipitation, and PAR) were higher at WD than those at SBS2 × 2. The greatest growth increase (75%–118%) at SBS2 × 2 compared to that at WD was achieved by TD clones, while DD clones had a much lower proportional increase (9%–45%). The clonal rankings were more definite at LS2 × 2 than at WD, and there were changes in the rankings between the sites (Figure 5a). The higher precipitation at LS2 × 2 may have contributed to the higher overall productivity at the site compared to that at WD. All clones that had higher productivity at LS2 × 2 were TD, and all DD clones had greater biomass at WD (Figure 5b), which had a higher 4-year total GDD and lower precipitation.

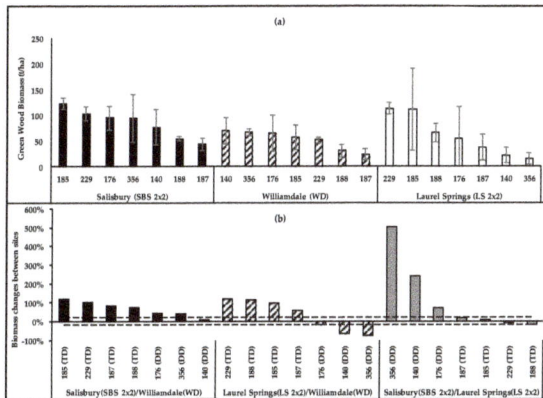

Figure 5. (a) Four-year-old green wood biomass (t/ha) of non-irrigated poplars located in the Piedmont (Salisbury, SBS2 × 2) and the upper Blue Ridge Mountains (Laurel Springs, LS2 × 2) and the Coastal (WD) Plain; (b) clonal biomass changes (%) at SBS2 × 2 and LS2 × 2 relative to those at WD, and at SBS2 × 2 relative to those at LS2 × 2.

3.3. The Effects of Growing Conditions on Adaptability and Biomass Productivity

Differences in the growing conditions led to some changes in the clonal rankings and significance of clonal differences of wood biomass productivity (at $p > 0.05$). In the Coastal Plain, clonal differences in biomass were significant at Gibson (GB) ($p = 0.03$) but not at WD ($p = 0.0534$). The biomass of all studied clones was much higher at the non-irrigated agricultural land at WD than at the irrigated marginal land at GB, with the exception of clones 187 and 339. The clonal ranking of biomass differed between the sites (Figure 6a), which could be due to fertility differences between the sites rather than any of the weather variables considered since none of these variables were higher at WD than they were at GB. The greatest growth responses (500%–655% increases) associated with changes in sites were experienced by all clones belonging to the DD genomic group with the TD clones showing 156% or less biomass increases at WD than at GB (Figure 6b).

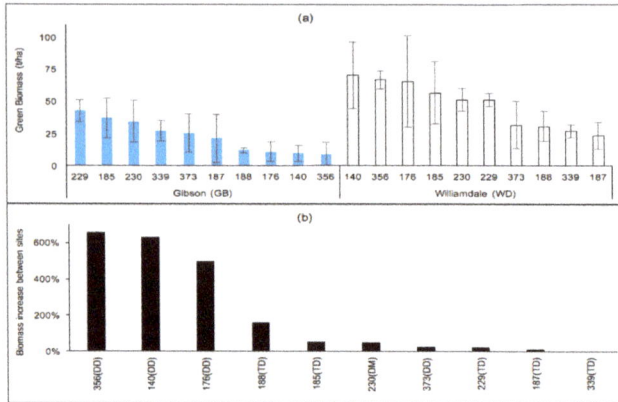

Figure 6. (**a**) Four-year-old green wood biomass (t/ha) of poplars located in the Coastal Plains under irrigated (GB) and non-irrigated (WD) conditions; (**b**) clonal biomass increases (%) at WD relative to those at GB.

There were significant clonal differences in biomass in subsoiled plots (WDSS, $p = 0.0091$) but not in disked plots (WDD, $p = 0.3245$), as shown in Figure 7a. Overall, the average biomass was higher in WDSS (52 tonne/ha) than in WDD (43 t/ha). Most clones produced significantly greater biomass (>20%) in WDSS than in WDD, while the reverse was true for clone 187 (Figure 7b). With the exception of clone 339, biomass increases in WDSS compared to WDD were greater for DD clones than for TD clones.

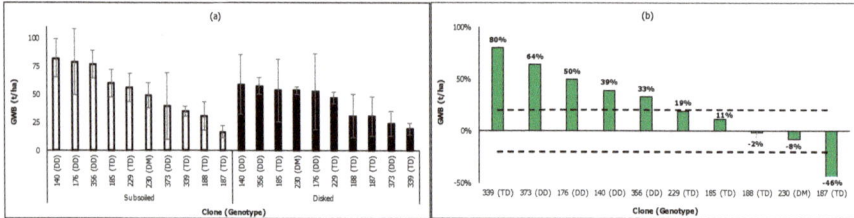

Figure 7. (**a**) Four-year-old green wood biomass (t/ha) of non-irrigated poplar stands in the coastal region (WD) using two soil preparation methods; (**b**) increases (%) in clonal wood biomass at subsoiled plots (WDSS) relative to those at disked plots (WDD).

The results of the analysis of the effects of land topography and aspect on mean wood productivity of poplars (Figure 8) at clonal and genomic-group levels are presented in Table 3. The significance of the clone–aspect–position interaction effects was present for both upslope and downslope land positions and all land aspects, but not all clones showed the interaction effect. Clones 176, 419, 426, and 443, all of which belong to the DD genomic group, and clone DN-34 (DN) did not show interaction effects with the positions and aspects of the land. The genomic group–aspect interaction effect occurred in the southerly (south-facing) and the westerly (west-facing) plots but not in the southwest (SW)-facing plot, and was seen in TD and DM clones but not in DD and DN clones. The aspect–position interaction effect occurred in the SW-facing plot only while the interaction effect was present for both topographic positions.

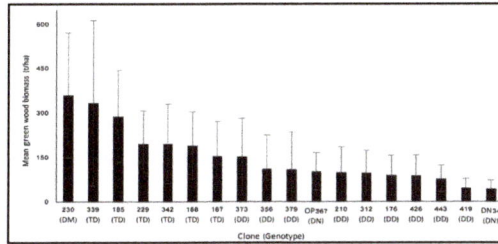

Figure 8. Mean green wood biomass (t/ha) of high-density (10,000 trees/ha) and non-irrigated poplars after 4 years of growth at a study site in the upper Blue Ridge Mountains.

Table 3. Results of statistical analyses of the effects of topographic position (upslope vs. downslope) and aspect of the land as well as possible interactions with clones and genomic groups on the wood productivity of poplars.

Clone			Genomic Group		
Effect	*F* Value	*p*	Effect	*F* Value	*p*
Clone	52.2	<0.0001	Genomic group	146.2	<0.0001
Topographic position	1.4	0.2325	Topographic position	2.8	0.0906
Aspect	70.7	<0.0001	Aspect	65.1	<0.0001
Clone–position interaction	2.7	0.0002	Genomic group–position interaction	0.8	0.5137
Clone–aspect interaction	15.2	<0.0001	Genomic group–aspect interaction	26.3	<0.0001
Aspect–position interaction	1.3	0.2754	Position–aspect interaction	4.8	0.0086
Clone–aspect–position interaction	3.2	<0.0001	Genomic group–aspect–position interaction	1.7	0.1083

3.4. Wood Properties and Composition

3.4.1. Wood Density

Figure 9 presents the wood density of poplar clones based on samples obtained from various positions on the tree and from three sampling sites. The wood density was significantly affected by clone ($p < 0.0001$), site ($p < 0.0001$), position on trees ($p = 0.0016$), and clone–site interactions ($p = 0.0356$). The results of Proc Slice (SAS) show that the clone–site interaction effects were present for the three clones (140, 187, and 188; $p \leq 0.0001$), and at SB ($p < 0.0001$) and LS ($p = 0.001$) and but not at MR ($p = 0.2573$).

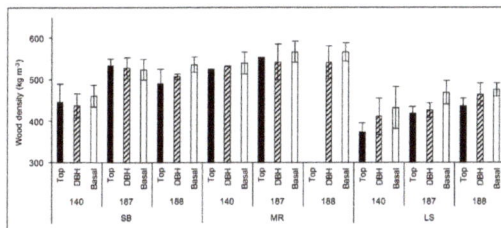

Figure 9. Wood densities (kg m^{-3}) at various tree positions of three poplar clones (140, 187, and 188) from stands located in the Piedmont (SB), the upper Blue Ridge Mountains (LS), and the lower Blue Ridge Mountains (MR).

3.4.2. Wood Composition

The nitrogen content of the wood (%) was significantly affected by site ($p = 0.0105$) and position on trees ($p < 0.0001$), but clonal differences were not significant ($p = 0.2963$) as shown in Figure 10a. The wood carbon content (%), on the other hand, was significantly affected by site only

($p = 0.0063$). The wood moisture content was significantly affected by site ($p < 0.0001$), position on trees ($p < 0.0001$), and interaction effects including clone–site ($p = 0.0002$), site–position ($p = 0.0098$), clone–position ($p = 0.0009$), and clone–site–position ($p = 0.0002$) interactions. However, the clonal effect was insignificant in characterizing the variability in moisture content ($p = 0.2819$).

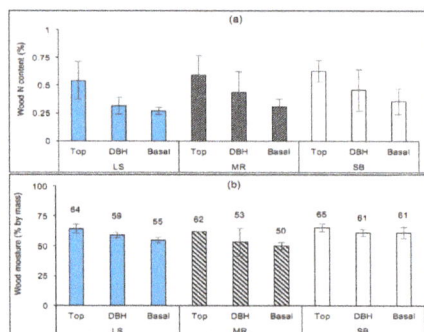

Figure 10. (**a**) Nitrogen content (N, %) and (**b**) moisture content by mass (%) of poplar wood at various tree positions for three clones (140, 187, and 188) sampled from stands located in the Piedmont (SB), the upper southern Blue Ridge Mountains (LS), and the lower southern Blue Ridge Mountains (MR).

Based on the results of Proc Slice (SAS), the clone–position interaction effects were present for the three clones sampled (140, 187, and 188; $p \leq 0.0100$), especially at DBH ($p = 0.0003$) but not at the basal position ($p = 0.3846$). The site–position interaction effects were significantly present at the three sites (LS, MR, and SB; $p \leq 0.0026$) and at the basal tree position ($p \leq 0.0100$), but were marginally significant for DBH position ($p = 0.0582$) and insignificant at the tree-top position ($p = 0.5426$). Clone–site interaction effects were significant for all sites ($p \leq 0.0114$) and for clones 140 and 187 ($p < 0.0001$) but not for clone 188 ($p = 0.1847$). In other words, the wood moisture content of clone 188 did not significantly vary among the sampled sites, and hence had no contribution to the clone–site interaction effects was reported. The clone–site–position interaction effects were present for all sites, positions (on tree), and clones.

4. Discussion

Several global studies have corroborated that the biomass productivity of poplars is greatly affected by clones and parentage [9,14–18]. Nevertheless, clonal rankings of wood volume after 1 year of growth in the upper southern Blue Ridge Mountains (LS) and the Piedmont (SB) were considerably different [11] from the 4-year woody productivity rankings of the current study. In another study, the height-based clonal ranks of 1-year-old poplar trees were not adequate indicators of rankings at medium and later stages of rotations [19]. For poplars, clonal survival after 1 year generally indicates a successful root development; yet, some clones suffer subsequent high mortality resulting from a lack of adaptability to growing conditions [6]. This leads to an overall decline in productivity on a per-area basis. The findings of the current study were in agreement with the results of Fortier et al. [4], indicating that poplar clonal differences in stem biomass were significant and that site had a greater effect on productivity.

A remarkable observation of the current study was regarding the response of genomic groups to changes in growing conditions. Improved growing conditions within the same physiography and physiographic differences between the Blue Ridge Mountains and the Piedmont prompted greater productivity increases from DD clones than from other genomic groups. Conversely, physiographic differences between the Coastal Plains and the Blue Ridge Mountains or the Piedmont and between the Piedmont and the Blue Ridge Mountains led to greater wood productivity increases from TD clones than from DD clones. Variations in productivity performances across different growing conditions for clones belonging to particular genomic groups were less definite, which generally agreed with the findings of Verlinden et al. [18], who stated that closely related genotypes are expected to have similar biomass productivities. Many studies have shown the strong effects of site on poplar growth and biomass productivity [9,14,16,17,20–23]. One study demonstrated strong effects of soil conditions and fertility on poplar biomass productivity [17], while another study emphasized that elevation and fertility of sites have stronger effects on poplar biomass productivity than clones [23]. Our result of biomass productivity performance of pure versus mixed hybrid poplars was contrary to the results of Reference [24], in which pure poplar species produced greater biomass than mixed poplar species when growing conditions were non-conducive. Benetka et al. [25,26] found that mixed poplar species produced greater biomass than pure poplar species under sub-optimal site conditions, while differences in biomass between pure and mixed poplar species under conducive conditions were not significant. Moreover, other studies reported contradicting results, with one study reporting no clear biomass-based advantage of pure or mixed hybrid poplars [16], while others showed the superiority of interspecific hybrid poplars compared to pure hybrids on the basis of biomass productivity [27–29]. Although differences in water use have been observed among clones [30], poplars vary in water use efficiency (WUE) at the genotype level [30,31], and the more distinctive productivity differences at the genomic-group level in the current study could possibly be WUE-related.

The current findings of woody biomass allocation (where clone, site, and interaction effects were insignificant) were in contrast with the results of a multi-clonal study of 6-year-old poplars [4]. Percentage allocations of wood biomass to stem and branches are likely to vary with different phases of tree growth. Wood biomass allocation can be used to identify less-branching clones preferable for high-end wood products, possibly due to the presence of fewer irregularities and lesser pruning required [4], and to identify suitable silvicultural practices in line with the purpose of the stands.

Although allometric equations were developed using destructively obtained data, the use of the equations enabled accurate, straightforward, and non-destructive estimations of the growth, biomass, and carbon storage of trees and forests using easily measurable variables [32]. The use of generalized equations to estimate green wood biomass in all current sites, including sites not used as data sources for developing the equations, is supportable by studies that reported most allometric relationships developed for the same species but different sites to be similar [33] and supported the use of equations developed for a species in a different place if site-specific or generalized equations for the species are not present [34]. In the current study, the addition of height as a predictor variable for estimating biomass did not improve the accuracy of the estimations. Hence, green woody biomass can be reliably estimated using only the DBH (Equation (4)), which is an important result since DBH is easy to measure. However, the DBH-only equation used in the work of Zalesny et al. [9] and references therein (wood biomass = $6.16 - 2.23\,\text{DBH} + 0.353\,\text{DBH}^2$) provided biomass predictions that deviated significantly ($p = 0.0035$, $t_{0.05} = 3.20$) from both the actual biomass sampled (from SB, MR, and LS) and the estimates of the allometric equations developed in the current study.

With the ranking-based analysis approach employed in this study, poplar productivity and adaptability to growing conditions was assessed as an outcome inclusive of variables that change with sites and physiography. This approach captured real-world planting conditions, including existing land marginality types and the variability and interactions of growth factors across growing and physiographic conditions. It was possible to concentrate on the site and physiographic variables in question with the paired-site approach. Nonetheless, some paired sites likely had different confounding factors, which makes isolating the effects of a particular growing condition on clones challenging. The poplars clones ranked as the highest biomass producers after 3 years are the clones that can be expected to have the highest biomass-based ranking after 10 growth years [6]. Biomass-based genotype rankings of poplars are also expected to be maintained beyond the first rotation [18]. Hence, our findings can be used to identify the best-adapting and highest-producing poplar clones and genomic groups for the duration of rotations under the studied and similar conditions. However, the woody biomass estimates presented should be interpreted bearing in mind that there is an inherent bias of overestimation associated with using trial-scale productivity to estimate productivity at a production scale [9].

Wood characteristics affect the processing and use of poplar wood. Wood density significantly differed among sampled sites of the current study, while Headlee et al. [10] reported significant wood density differences among sites at year 20 but not at year 10. Headlee et al. [10] also reported a poplar wood density of 267–495 kg/m^3 from their study and of 270–500 kg/m^3 from other studies (3- to 22-year-old trees). A mean wood density range of 300–390 g/cm^3 was also reported for North American natural forests including hybrid poplars [3]. In the current study, wood density, based on 80 samples from 30 trees representing three clones and three physiographic regions, had a wider range of 345–578 kg/m^3 (mean: 492 ± 58 kg/m^3). Poplar wood is relatively weak in many strength criteria (tensile, compressive, shearing, etc.) but has high transverse strength and elasticity. Still, regional differences in these strength attributes can be significant [3]. Since a positive correlation exists between the wood density and the wood strength [35], poplar wood density differences between the current and the above studies are likely to have implications on wood strength. The processing and the use of poplar wood are also affected by properties related to wood water content. Due to its high water content, poplar wood is convenient for specialized cutting and shrinks by as much as 12% (by volume) when drying, although its shrinking pattern makes the wood prone to deformations [3].

There were three pest attacks by cottonwood leaf beetles (*Chrysomela scripta* F.) during the current study, early in the first and the second years of growth at the Piedmont site (SB) and early in the third year at one of the coastal sites (WD). Early detection of the insects and leaf damages and prompt treatment of the affected stands with SEVIN®Insect Killer Concentrate (GardenTech®, Bayer CropScience LP, Research Triangle Park, NC 27709, USA) allowed leafing of the trees in time for summer. Leaf rust (caused by *Melampsora medusae*) often occurred toward the end of the growing seasons at all sites. Although a detailed study on poplar clonal vulnerability to the rust infections was not conducted, there seemed to be somewhat visible differences in the extent of the rust among some trees. However, there have not been any definite indications that poplar susceptibility or resistance to pests and diseases is affected by growing condition or physiography.

Based on the current and cited [6,36] studies, poplars have a great potential for the production of sustainable woody feedstocks, and their productivity and feasibility can be greatly enhanced provided that site-suitable genotypes and clones are selected. Optimum stand management is crucial for the feasibility of the species as an SRWC [6], and should include robust monitoring to prevent and minimize disease and pest damages. Finally, the economic feasibility of the clones studied cannot be realized under the current market conditions of smaller-diameter woody feedstocks and feedstock prices. However, optimally established and managed stands of site-suitable clones growing on productive sites could have positive economic returns at longer rotations (>7 years) and lower feedstock prices ($30/ton) or at higher feedstock prices (>$36/ton) and shorter rotations (5 years).

5. Conclusions

The green wood productivity, wood properties and composition, and wood biomass allocation of poplars were examined in relation to site conditions (fertility, irrigation, and soil preparation), physiography, and land topography (position and aspect). Although not all clonal differences in the wood biomass were significant, the clonal rankings and the significance of clonal differences were largely affected by growing conditions and physiography. Changes in poplar wood productivity due to physiography and growing conditions were more structured at the genomic-group level than at the clonal level. A higher-altitude physiography led to greater wood biomass increases in TD clones than in DD clones (compared to a lower-altitude physiography) and vice versa. Within a physiographic province, favorable growing conditions largely led to greater wood biomass productivity of DD clones than of TD and DM clones. No clear correlations were observed between poplar productivity and regional adaptability and weather-related variables (GDD, precipitation amount, reference evapotranspiration, and PAR). The aspect of the land affected the wood productivity of poplar genomic groups and clones significantly. The site significantly affected the wood properties and composition studied, while clonal effects on wood composition and clonal and site effects on wood biomass allocation and biomass–dimension allometric relationships were not significant. Although clones that showed greater biomass responses to growing conditions generally belonged to the same genomic group, clone-level selection could produce greater wood biomass gains than selections made at the genomic-group level.

Author Contributions: Conceptualization, methodology & investigation: S.B.G., E.G.N. and D.W.H.; Data curation, formal analysis, validation & original draft preparation: S.B.G.; Resources (wood analyses): C.A.M.; Funding acquisition: E.G.N. and D.W.H.; Supervision & project administration: D.W.H., E.G.N. and S.B.G.; Reviewing & editing: S.B.G., E.G.N., D.W.H. and C.A.M.

Funding: This research was funded by the North Carolina Department of Agriculture and Consumer Services, Bioenergy Research Initiatives (NCDA&CS-BRI), grant numbers: G40100278914RSD and G40100278314RSD.

Acknowledgments: We would like to thank the NCDA&CS-BRI (formerly the Biofuels Center of North Carolina) for funding and Arborgen, LLC, and GreenWood Resources for their technical support and for supplying trees. We are also grateful to the NCDA&CS research stations at the study sites and the Town of Gibson, North Carolina for their cooperation and assistance at the study sites, and to Karen Sarsony from the US Forest Service, Southern Research Station for her analysis of the wood samples used for the current study.

Conflicts of Interest: The authors declare no conflict of interest.

References

1. U.S. Department of Energy. 2016 Billion-Ton Report: Advancing Domestic Resources for a Thriving Bioeconomy. In *Economic Availability of Feedstocks*; Langholtz, M.H., Stokes, B.J., Eaton, L.M., Eds.; ORNL/TM-2016/160; Oak Ridge National Laboratory: Oak Ridge, TN, USA, 2016; Volume 1, 448p. [CrossRef]

2. U.S. Energy Information Administration (EIA), Independent Statistics and Analysis. Wood and Wood Waste: Energy Explained, Your Guide to Understanding Energy. 2018. Available online: https://www.eia.gov/energyexplained/index.php?page=biomass_wood (accessed on 18 August 2018).

3. Balatinecz, J.J.; Kretschmann, D.E.; Leclercq, A. Achievements in the utilization of poplar wood—Guideposts for the future. *For. Chron.* **2001**, *77*, 265–269. [CrossRef]

4. Fortier, J.; Gagnon, D.; Truax, B.; Lambert, F. Biomass and volume yield after 6 years in multiclonal hybrid poplar riparian buffer strips. *Biomass Bioenergy* **2010**, *34*, 1028–1040. [CrossRef]

5. Berthelot, A.; Ranger, J.; Gelhaye, D. Nutrient uptake and immobilization in a short-rotation coppice stand of hybrid poplars in North-West France. *For. Ecol. Manag.* **2000**, *128*, 167–179. [CrossRef]

6. Kaczmarek, D.J.; Coyle, D.R.; Coleman, M.D. Survival and growth of a range of *Populus* clones in central South Carolina USA through age ten: Do early assessments reflect longer-term survival and growth trends? *Biomass Bioenergy* **2013**, *49*, 260–272. [CrossRef]

7. Stanturf, J.A.; van Oosten, C.; Netzer, D.A.; Coleman, M.D.; Portwood, C.J. Ecology and silviculture of poplar plantations. In *Poplar Culture in North America*; Dickmann, D.I., Isebrands, J.G., Eckenwalder, J.E., Richardson, J., Eds.; NRC Research Press, National Research Council of Canada: Ottawa, ON, Canada, 2001; Part A, Chapter 5; pp. 153–206.

8. Lo, M.H.; Abrahamson, L.P. Principal component analysis to evaluate the relative performance of nine-year-old hybrid poplar clones. *Biomass Bioenergy* **1996**, *10*, 1–6. [CrossRef]

9. Zalesny, R.S.; Richard, B.H.; Zalesny, J.A.; McMahon, B.G.; Berguson, W.E.; Stanosz, G.R. Biomass and genotype×gnvironment interactions of *Populus* energy crops in the Midwestern United States. *BioEnergy Res.* **2009**, *2*, 106–122. [CrossRef]

10. Headlee, W.L.; Zalesny, R.S., Jr.; Hall, R.B.; Bauer, E.O.; Bender, B.; Birr, B.A.; Miller, R.O.; Randal, J.A.; Wiese, A.H. Specific gravity of hybrid poplars in the North-Central Region, USA: Within-tree variability and site × genotype effects. *Forests* **2013**, *4*, 251–269. [CrossRef]

11. Ghezehei, S.B.; Nichols, E.G.; Hazel, D.W. Early clonal survival and growth of poplars grown on North Carolina Piedmont and Mountain marginal lands. *BioEnergy Res.* **2016**, *9*, 548–558. [CrossRef]

12. Shifflett, S.D.; Hazel, D.W.; Frederick, D.J.; Nichols, E.G. Species trials of short rotation woody crops on two wastewater application sites in North Carolina, USA. *BioEnergy Res.* **2014**, *7*, 157–173. [CrossRef]

13. SAS 9.4 for Windows x64 Based Systems. Available online: https://ualberta.onthehub.com/WebStore/OfferingDetails.aspx?o=56e487ad-057e-e311-93f9-b8ca3a5db7a1 (accessed on 30 January 2019).

14. Pliura, A.; Zhang, S.Y.; MacKay, J.; Bousque, J. Genotypic variation in wood density and growth traits of poplar hybrids at four clonal trials. *For. Ecol. Manag.* **2007**, *238*, 92–106. [CrossRef]

15. Guo, X.; Zhang, X. Performance of 14 hybrid poplar clones grown in Beijing, China. *Biomass Bioenergy* **2010**, *34*, 906–911. [CrossRef]

16. Nielsen, U.B.; Madsen, P.; Hansen, J.K.; Nord-Larsen, T.; Nielsen, A.T. Production potential of 36 poplar clones grown at medium length rotation in Denmark. *Biomass Bioenergy* **2014**, *64*, 99–109. [CrossRef]

17. Pliura, A.; Suchockas, V.; Sarsekova, D.; Gudynaite, V. Genotypic variation and heritability of growth and adaptive traits, and adaptation of young poplar hybrids at northern margins of natural distribution of *Populus nigra* in Europe. *Biomass Bioenergy* **2014**, *70*, 513–529. [CrossRef]

18. Verlinden, M.S.; Broeckx, L.S.; Ceulemans, R. First vs. second rotation of a poplar short rotation coppice: Above-ground biomass productivity and shoot dynamics. *Biomass Bioenergy* **2015**, *73*, 174–185. [CrossRef]

19. Brown, K.R.; Beall, F.D.; Hogan, G.D. Establishment-year height growth in hybrid poplars; Relations with longer-term growth. *New For.* **1996**, *12*, 175–184.

20. Riemenschneider, D.E.; Berguson, W.E.; Dickmann, D.I.; Hall, R.B.; Isebrands, J.G.; Mohn, C.A.; Stanosz, G.R.; Tuskan, G.A. Poplar breeding and testing strategies in the North-Central U.S.: Demonstration of potential yield and consideration of future research needs. *For. Chron.* **2001**, *77*, 245–253. [CrossRef]

21. Pliura, A.; Zhang, S.Y.; Bousquet, J.; MacKay, J. Age trends in variation of wood density and its intra-ring components of young poplar hybrid crosses. *Ann. For. Sci.* **2006**, *63*, 673–685. [CrossRef]

22. Dillen, S.Y.; Marron, N.; Bastien, C.; Ricciotti, L.; Salani, F.; Sabatti, M.; Pinel, M.P.C.; Rae, A.M.; Taylor, G.; Ceulemans, R. Effects of environment and progeny on biomass estimations of five hybrid poplar families grown at three contrasting sites across Europe. *For. Ecol. Manag.* **2007**, *252*, 12–23. [CrossRef]

23. Traux, B.; Cagnon, D.; Fortier, J.; Lambert, F. Yield in 8-year-old hybrid poplar plantations on abandoned farmland along climatic and soil fertility gradients. *For. Ecol. Manag.* **2012**, *3*, 228–239. [CrossRef]

24. Dillen, S.; Djomo, S.N.; Al Afas, N.; Vanbeveren, S.; Ceulemans, R. Biomass yield and energy balance of a short-rotation poplar coppice with multiple clones on degraded land during 16 years. *Biomass Bioenergy* **2013**, *56*, 157–165. [CrossRef]

25. Benetka, V.; Bartáková, I.; Mottl, J. Productivity of *Populus nigra* L. ssp. nigra under short-rotation culture in marginal areas. *Biomass Bioenergy* **2002**, *23*, 327–336. [CrossRef]

26. Benetka, V.; Novotná, K.; Štochlová, P. Biomass production of *Populus nigra* L. clones grown in short rotation coppice systems in three different environments over four rotations. *iForest* **2014**, *7*, 233–239. [CrossRef]

27. Heilman, P.E.; Ekuan, G.; Fogle, D. Above- and below-ground biomass and fine roots of 4-year-old hybrids of *Populus trichocarpa* × *Populus deltoides* and parental species in short-rotation culture. *Can. J. For. Res.* **1994**, *24*, 1186–1192. [CrossRef]

28. Ceulemans, R.; Deraedt, W. Production physiology and growth potential of poplars under short-rotation forestry culture. *For. Ecol. Manag.* **1999**, *121*, 9–23. [CrossRef]

29. Verlinden, M.S.; Broeckx, L.S.; Van den Bulcke, J.; Van Acker, J.; Ceulemans, R. Comparative study of biomass determinants of 12 poplar (Populus) genotypes in a high-density short-rotation culture. *For. Ecol. Manag.* **2013**, *307*, 101–111. [CrossRef]

30. Jones, T.; McIvor, I.; McManus, M. Drought tolerance and water-use efficiency of five hybrid poplar clones. In *Integrated Nutrient and Water Management for Sustainable Farming*; Currie, L.D., Singh, R., Eds.; Occasional Report No. 29; Fertilizer and Lime Research Centre, Massey University: Palmerston North, New Zealand, 2016; pp. 1–13. Available online: http://flrc.massey.ac.nz/publications.html (accessed on 24 August 2018).

31. Rancourt, G.T.; Éthier, G.; Pepin, S. Greater efficiency of water use in poplar clones having a delayed response of mesophyll conductance to drought. *Tree Physiol.* **2015**, *35*, 172–184. [CrossRef]

32. Ghezehei, S.B.; Annandale, J.G.; Everson, C.S. Shoot allometry of *Jatropha curcas*. *South. For.* **2009**, *71*, 279–286. [CrossRef]

33. Tritton, L.M.; Hornbeck, J.W. Biomass estimation for Northeastern forests. *Ecol. Soc. Am. Bull.* **1981**, *62*, 106–107.

34. Pastor, J.; Abet, J.D.; Melillo, J. Biomass prediction using generalized allometric regressions for some northeast tree species. *For. Ecol. Manag.* **1984**, *7*, 265–274. [CrossRef]

35. WoodProdcuts.fi. Strength Properties of Wood. Wood Products. 13 November 2013. Available online: https://www.woodproducts.fi/content/wood-a-material-1 (accessed on 9 August 2018).

36. Zalesny, R.S.; Deahn, D.M.; Coyle, D.R.; Headlee, W.L. An approach for siting poplar energy production systems to increase productivity and associated ecosystem services. *For. Ecol. Manag.* **2012**, *284*, 45–58. [CrossRef]

forests

MDPI

Article

Quality Testing of Short Rotation Coppice Willow Cuttings

Katrin Heinsoo * and Kadri Tali

Institute of Agricultural and Environmental Sciences, Estonian University of Life Sciences, Kreutzwaldi 5D, 51006 Tartu, Estonia; kadri.tali@emu.ee
* Correspondence: katrin.heinsoo@emu.ee; Tel.: +372-529-5325; Fax: +372-731-3037

Received: 29 April 2018; Accepted: 20 June 2018; Published: 23 June 2018

Abstract: The production and feasibility of Short Rotation Coppice depend on cutting early performance. The shoot and root biomass production of *Salix* cuttings in hydroponic conditions was studied. The amount of sprouted biomass after four weeks of growth depended on cutting the diameter, but the original position of the cutting along the rod or number of visible buds was not in correlation with biomass produced. Application of mineral fertilizer or soil originating from the willow plantation did not increase the total production. On the contrary, the addition of soil tended to decrease biomass production and we assumed this was a result of a shortage of light. Under the influence of fertilization, plants allocated greater biomass to roots. Comparison of different clones revealed that those with *S. dasyclados* genes tended to allocate less biomass to roots and the poorest-performing clone in our experiment, also had the lowest wood production in the plantation. The number of visible buds on the cutting was also clone-specific.

Keywords: clone selection; cutting size; cutting sprouting; production; short rotation coppice; willow

---

1. Introduction

High quality and good physiological condition of cutting material are key aspects of successful and economically feasible short rotation coppice (SRC) plantations [1]. In addition to an insufficient availability of nutrients, unequal plant growth and dieback of cuttings are found to be the main factors resulting in a decline of production during consecutive SRC harvests [2]. These kinds of problems are most severe in *Salix* L. SRC where very high planting densities are usually suggested [3]. In addition to economic considerations, SRC plantations with variable plant densities and large numbers of gaps in the SRC vegetation cover can cause ecological problems if the SRC is used as a phytoremediation site or in terms of vegetation filter [4].

It is especially important to ensure fast, equal sprouting of cuttings with consecutive rapid shoot growth occur during the first growing season of SRC, since this is the period where the largest seed bank of weed species in the plantation soil exists and competition between sprouting cuttings and weeds can be expected [5–7]. Therefore, a lot of attention has been paid to the study of critical factors that may impact the growth potential of cuttings [8]. Besides the selection of planting material (highly productive genotypes and vigorous hybrid cultivars), the quality of cutting material survival has been found to be influenced by pre-planting storage conditions [1], the quality of soil preparation [9,10] and planting techniques [11]. There are also a number of publications that demonstrate the benefits of horizontal planting instead of the more typical vertical planting method and include discussion of the reasonable length of the cutting [12–14]. Moreover, the impact of cutting diameter has been found to be an important predictor of plant growth and hence when mechanically planting using long rods, the effectiveness of placing the thinnest top of one rod adjacent to the thickest bottom of another rod is considered to be arguable [8]. The potential for decreasing the typical cutting length from between

18 and 25 cm to between 1 and 5 cm (micro-cuttings) in order to minimize plantation costs has also been studied [15].

Another approach to enhancing the economic feasibility and sustainability of modern agriculture including SRC management is to promote the distribution of soil microorganisms (fungi or bacteria) that support nutrient uptake from growing substrate and thus enable a reduction in the use of mineral fertilizers [16]. The rhizosphere microbiome is considered to be dependent on root exudates and to be species-specific. Some studies have revealed that *Salix* can take advantage of specific fungal assemblages [17,18]. It has been demonstrated that planting of *Salix* in former permanent grassland areas resulted with differences in the soil microbial community [19]. Moreover, there is strong evidence that *Salix* also has some endophytes in terms of symbionts for nitrogen uptake [20], hence, the inoculation of planting material with specific microorganisms is suggested [3].

There is information available about the physiological responses of different willow genotypes to variable water and nutrient conditions in pot experiments (e.g., [21]). During the first growing season in some genotypes, even a short one-week period of water stress can lead to a significant decrease in plant biomass and height [22]. Supplementation of nutrients in terms of mineral fertilizer or various organic residues from human activities (e.g., sewage sludge, wastewater, biogas digestate, etc.) may increase the biomass increment of young plants significantly [23,24]. Differences in biomass production can be observed only in cases where large differences exist between the "control" growing substrate and artificially created altered growing conditions [4]. The maximal fertilization, however, must be carefully considered, since young cuttings cannot withstand such large doses of additional nutrients or utilize residues in the same way that mature SRC plants do [23]. The duration of such greenhouse or pot-experiments however, is usually limited by the size of a growing area or difficulties in maintaining homogenous artificial growing conditions.

There are greenhouse data available on the leaf phenology of several *Salix* genotypes that can be correlated with the wood yield of particular clones in SRC [25,26], little less is known about cutting biomass production in the absence of additional nutrients or growth substrates. Willow is a species that is easy to propagate vegetatively by chopping dormant stems into cuttings. New roots on the cutting sprout from latent root primordia that have been formed during stem development and are ready to grow as soon as the cutting is placed in a humid environment [27]. However, studies have shown that *Salix* species exhibit great variability in bud structure and therefore coppicing ability [28]. Hence the rapid visual evaluation of bud number can be a useful indicator for predicting the early stage biomass production of cuttings.

While hydroponic experiments have been undertaken to study plant tolerance to a range of chemicals or to gather information about the synthesis of root exudates [29,30], information from hydroponic experiments for selecting energy crops, or even SRC planting material is very limited. However, this can add valuable information about particular clone preferences and shortages. Therefore, we decided to undertake a repeated experiment with the main aim of gaining an understanding of whether a quick, cheap, and simple to use method can be reasonably employed to obtain additional knowledge on those *Salix* genotypes commonly grown in SRC.

The experiment was designed to test the following hypotheses:

- cutting genotype and diameter is a predictor of early stage biomass
- application of fertilizer and/or microorganisms originating from mature SRC promote the early stage biomass production of cuttings
- there are significant differences in the sprouting speed and early stage biomass production between different willow clones
- allocation of photosynthesis products between shoots and roots varies by clones and/or by applied substances
- early stage biomass production of a cutting can be predicted by counting the number of visible buds on the cutting.

2. Materials and Methods

The shoots of the willow clones studied, were harvested from the Kambja vegetation filter, South-Estonia (58°14′57″ N, 26°42′39″ E) on 15 April 2009 and 15 April 2010. The harvesting time was selected to avoid snow cover in SRC and start the experiment before the beginning of the growing period or any leaf bud burst in Estonian climatic conditions. In the first year, shoots of clones "Tora" and 78,183 (according to the Swedish clone numbering system) were studied. In the second year, material from clones 81,090 and "Gudrun" was added to the above clones in the study. Hence the clone selection represented both clones of *Salix viminalis* L. (78,183), *S. dasyclados* Wimm. (81,090) and the relatively new commercial hybrids of *S. dasyclados* "Helga" burjatica × *S. viminalis* (clone "Gudrun") and *S. schwerinii* E. Wolf × *S. viminalis* (clone "Tora"). In the field the tallest (1.6–2.5 m) one-year-old, straight shoots from inside the plantation were selected to harvest the rods and cut at 5–10 cm height above the ground. Within a few hours the collected material was transported to the laboratory for further experiment preparation the following day.

The experiment (between 16 April and 14 May both in 2009 and 2010) was carried out in a laboratory of the Estonian University of Life Sciences at a room temperature of between 18–23 °C. The experiment room had large windows facing in a southerly direction and the natural light on windowsills was considered to be equal for all studied materials. At the start of the experiment the rods were divided into cuttings of 20 ± 0.1 cm in length as long as their diameter was more than 7 mm and their location along the rod in terms of position from the ground (1–8), number of visible buds on the bark and diameters of the apical cutting part were recorded and the cutting was marked using a waterproof pencil. In total, 120 cuttings per clone in 2009 and 60 cuttings in 2010 were made. After this procedure, 15 cuttings originating from the same clone were placed in a similar orientation vertically into a specially designed aquarium-like glass box of 8 × 40 × 25 cm and placed on the windowsill to get the maximum natural light available. The open glass boxes with different clones were located on the windowsills alternately and filled to a 15 cm height with water which originated from an open well near to the Kambja SRC that met all the criteria for drinking water. Since the evaporation from the boxes was small and no significant algae intervention occurred, we did not add any water or disturb the glass boxes in any other way during the experiment. The cuttings were placed along the longer edge (40 cm) of the box and hence the distance between cuttings was more than 2 cm to avoid any competition for space between young plants during the experiment (Figure S1).

In 2009, every second glass box received 150 g of soil collected from the upper 10 cm of the Kambja vegetation filter control plots (sandy loam soil with pH 5.5) or/and 10 g of typical solid NPK (nitrogen, phosphorus and potassium) fertilizer (17.0:3.1:11.6) was added to the ca 4.8 L of water in the boxes, thus creating four different treatment options (control, fertilizer, soil, and fertilizer + soil) (Table 1, Figure S2). The application of soil decreased water transparency significantly. In 2010, only the addition of mineral fertilizer was repeated. Since the space with homogeneous natural light conditions was limited, this change allowed us to keep the number of cuttings in one treatment constant ($n = 30$ replicated into two boxes) but to increase the number of clones studied in 2010.

Table 1. Experiment design included four treatments and two clones in 2009 and two treatments and four clones in 2010. The sample size in each treatment combination was 15. The numbers and text in bold are further used in the paper as abbreviations to indicate various factors.

Year	Treatments	Harvesting	Clones
2009	Control Fertilizer Soil Soil + Fertilizer	2 weeks from start 4 weeks from start	78,183 Tora
2010	Control Fertilizer	2 weeks from start 4 weeks from start	78,183 81,090 Gudrun Tora

Two weeks after the start of the experiment (30.04 in both years), each second cutting from each box/treatment/replication box was harvested. The cutting was taken out of the box and all sprouted shoots and roots were carefully removed from the cutting with the help of scissors or a razor blade. All harvested biomass was separately collected into paper envelopes and dried for at least 48 h at 85 °C to constant weight. After four weeks of the experiment, the same procedure was carried out with the remainder of the cuttings.

For statistical analyses the software package SAS System v. 9.4 (SAS Institute Inc., Cary, NC, USA) was used. In all cases, General Linear Models were exploited, and the significance of all factors was tested with SAS GLM (generalized linear model). The regressions between cutting diameter and biomass production were found with Linear Regression Model, the differences between average values of various datasets were detected with ANOVA or MANOVA procedures Ryan-Einot-Gabriel-Welsh Multiple Range Test (REGW) and the covariation of different factor combinations analyzed with the Least Square Means test (LSM). In all cases the confidence level of tests was set to 0.05.

From previous studies we assumed that there is a strong linear correlation between the cutting diameter and its biomass [8]. In our experiment the studied shoots were younger and the range of diameters in each data series was variable. Therefore, we found it useful to normalize the biomass data over the diameter values and used the equation of the GLM linear regression fit for all collected data (Equation (1)):

$$\text{Biomass production} = a \times \text{diameter (year, harvest, treatment, clone, replication)} - \text{Intercept} \quad (1)$$

where biomass production is the measured dry biomass (g) of a particular cutting and diameter is the measured upper diameter of the same cutting in the start date of the experiment.

The slope coefficient from this equation (a), further labelled as "relative biomass production", was calculated separately for each cutting data and used as the dependent variable instead of the actual biomass production data to enable a more detailed statistical analysis on a range of experiment factors. In practice this slope indicates the influence of cutting diameter (mm) on the production of new biomass (g) from the cutting.

3. Results

The general model, that included all experiment data, revealed that the total new biomass production of cuttings was significantly dependent on study year, on treatment, on clone and on cutting diameter, but the replication did not add any significant information to the model (Table 2). Hence, further normalization was necessary as the range of cutting diameters under investigation varied by clones (e.g., Figure 1). The intercept value of the general linear regression model ($R^2 = 0.83$ $p < 0.001$) Equation (1) was −0.270173.

Table 2. The results of the multiple ANOVA analysis ($R^2 = 0.71$; $p < 0.0001$) of the pooled data with biomass (g) as independent variable. The factors in bold are found to be statistically significant.

Factor	Degrees of Freedom	F Value	$p > F$
diameter	1	51.29	<0.0001
year	1	47.57	<0.0001
clone	3	3.39	0.0181
harvest	1	776.13	<0.0001
treatment	3	8.70	<0.0001
position	7	0.96	0.4796
replication	1	2.97	0.0855
clone$^\times$treatment	5	2.15	0.0584

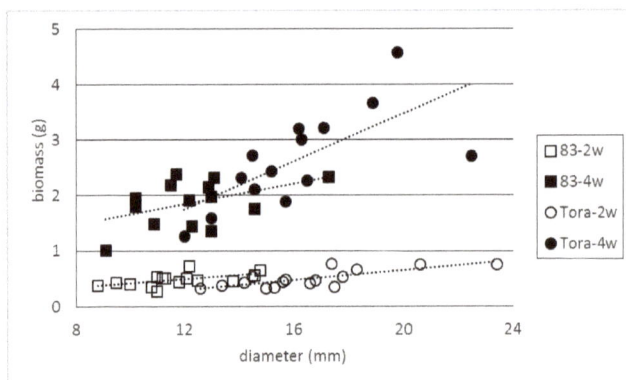

Figure 1. Relationship between cutting diameter and biomass production in fertilizer treatment after two or four weeks in 2009.

The 2009 data show that the relative biomass production was significantly ($p < 0.05$) affected by harvesting date, treatment and clone. The other factors studied (the replication and the cutting position) were found to be insignificant and removed from further analyses. The detailed analysis was carried out separately for both harvesting dates (Table 3). In both cases the treatment and clone were significant factors having an impact on relative biomass production. The interaction between these two factors was found to be insignificant in most cases and hence we assumed that cuttings from both clones reacted to the treatment in a similar way.

Table 3. The impact of clone and treatment on relative biomass (g mm^{-1}) in 2009.

Factor	2 Weeks			4 Weeks		
	DF	F Value	$p > F$	DF	F Value	$p > F$
Clone	1	74.91	<0.0001	1	14.74	0.0002
Treatment	3	22.90	<0.0001	3	17.42	<0.0001
clone×treatment	3	2.37	0.0743	3	4.52	0.0049

The statistical analyses of pooled data from both clones revealed that the application of different materials to the water boxes did not have any positive effect on the biomass production of the cuttings after four weeks of treatment when compared to the control dataset in 2009 (Figure 2). In contrast, soil (S) supply decreased the production alone and in combination with fertilizer (S+F) both when the biomass production was measured after two weeks or after four weeks from the start of the experiment. At the end of the experiment a negative impact of the fertilizer application was also detected on clone 83, but this pattern was not found to be statistically significant, when the clone factor was excluded from the analysis.

In order to better understand the impact of the treatments on cutting biomass production we also studied the ratio between shoot and root biomass production. At the end of the experiment the cuttings which were treated with fertilizer alone or in combination with soil tended to have a smaller shoot to root ratio than cuttings from other treatments (Figure 3).

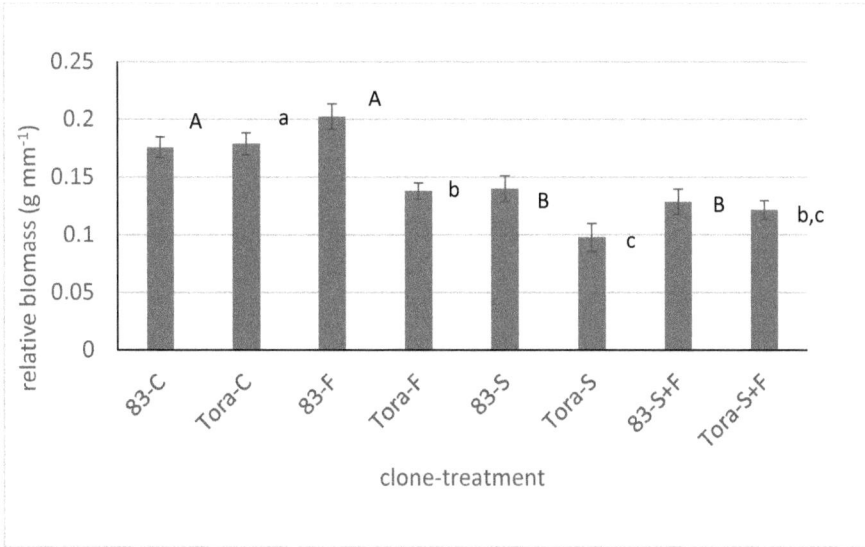

Figure 2. The average relative biomass production, at four-week harvest, by cuttings originating from different clones and treatments (C—control; F—fertilizer, S—soil; S+F—soil and fertilizer) in 2009. The bars represent SE (n = 14–15). The particular clone values with distinct letters (capitals for clone 83 and lowercase for clone Tora) differ from each other significantly (REGW test).

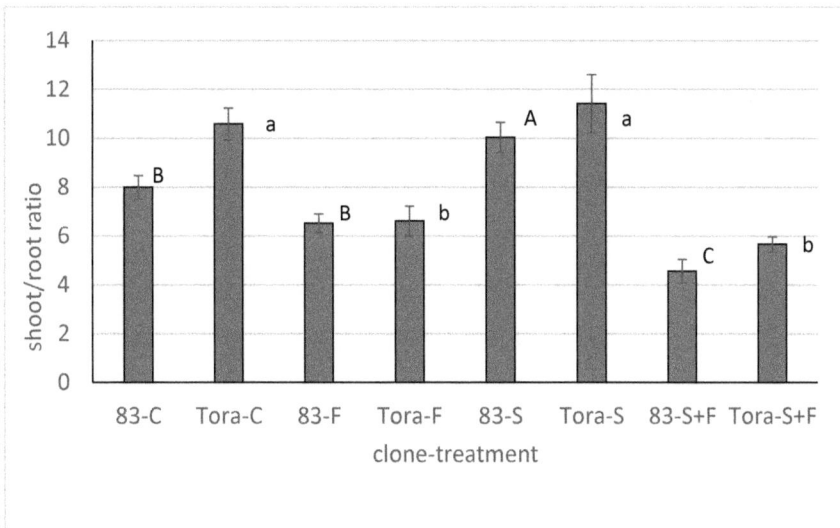

Figure 3. The average ratio between cuttings' shoot and root biomass productions of various clones and treatments at the end of the experiment in 2009 (C—control; F—fertilizer, S—soil; S+F—soil and fertilizer). The bars represent SE (n = 14–15). The particular clone values with distinct letters (capitals for clone 83 and lowercase for clone Tora) differ from each other significantly (REGW test).

The variability of relative biomass production by clones was studied in detail on the basis of the data from 2010. In this year the covariation of clone and treatment factors were insignificant both in two-week harvest and four-week harvest data (Table 4). These results enabled pooling of the data excluding the fertilization factor during this year and to increase the sample number for further analysis. We found strong evidence that the relative biomass production was clone-dependent. Biomass production for all the clones studied was approximately the same during the first two weeks and the final two weeks of the experiment. Clone 90 had significantly lower relative biomass production per cutting diameter than the rest of the clones studied for both the two-week and four-week harvest dates (Figure 4).

Table 4. The impact of clone and treatment on relative biomass (g mm^{-1}) in 2010.

Factor	2 Weeks			4 Weeks		
	DF	F Value	$p > F$	DF	F Value	$p > F$
clone	3	20.97	<0.0001	3	5.87	0.0009
treatment	1	15.59	0.0001	1	71.15	<0.0001
clone$^\times$treatment	3	0.89	0.4487	3	0.14	0.9349

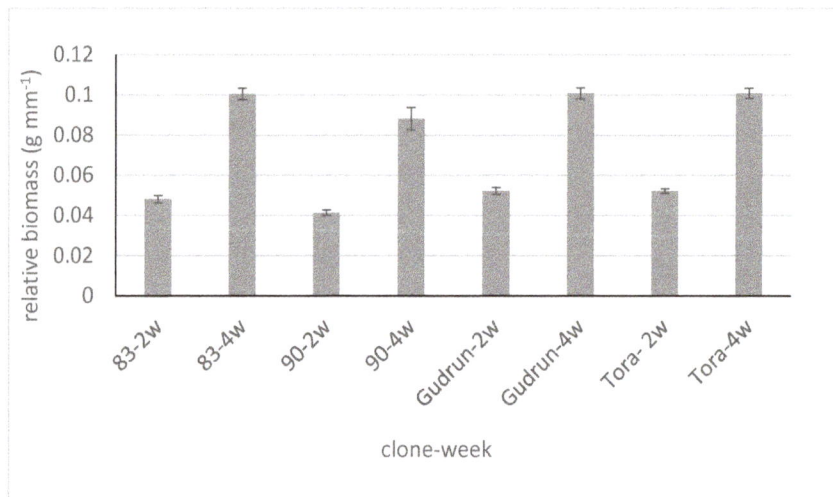

Figure 4. The average relative biomass production by cuttings originating from different clones during two or four weeks in 2010. The bars represent SE (*n* = 30). The values of particular harvest time with distinct letters differ from each other significantly (REGW test).

The lower relative biomass growth of clone 90 was accompanied by the smaller biomass of sprouted roots: The shoot/root biomass ratio of this clone was significantly larger than that of other clones (Figure 5). Clone Gudrun also had a larger shoot/root production compared with clones 83 and Tora.

Analysis of the 2010 data also shows that the number of visible buds on cuttings was an unreliable parameter for predicting absolute biomass production when harvested after the first four weeks. No significant relationship was found between the number of visible buds and cutting biomass production by clones ($R^2 < 0.15$ in all cases) either (Figure 6). However, the General Linear Model revealed that both the cutting diameter and clone were significant factors for predicting the number of buds on the cutting ($p < 0.0001$ and $p < 0.05$, respectively), but the model reliability was rather low ($R^2 = 0.29$). Inclusion of data about the origin of the cutting from the donor rod did not improve the model neither was this factor significant. The detailed analysis demonstrated that the pattern between cutting diameter and number of visible buds varied by clones (Figure 7). For clone 83 there was a tendency to positive correlation ($R^2 = 0.10$) between cutting diameter and the number of buds. At the same time clone Tora tended to have a similar number of buds regardless of the cutting diameter. In many cases no visible buds were detected on cuttings and it was especially difficult to detect buds in clone 90, where 28 cuttings out of 60 were marked with 0 visible buds. The number of cuttings from clone Tora with no visible buds was tenfold smaller than that of clone 90 in our experiment. Hence, the average numbers per cutting were 7.7; 6.0; 5.0 and 1.8 for Tora, 83, Gudrun and 90, respectively.

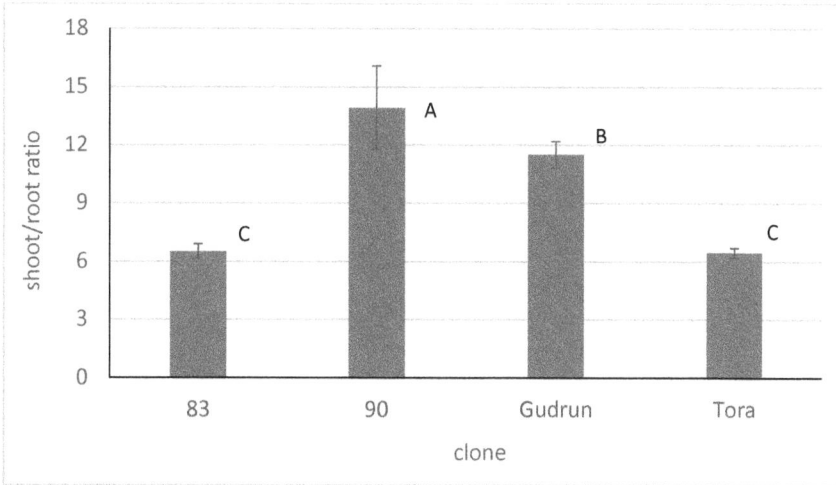

Figure 5. The average ratio between shoot and root biomass for cuttings of various clones at four-week harvest in 2010. The bars represent SE ($n = 30$). The values with distinct letters differ from each other significantly (REGW test).

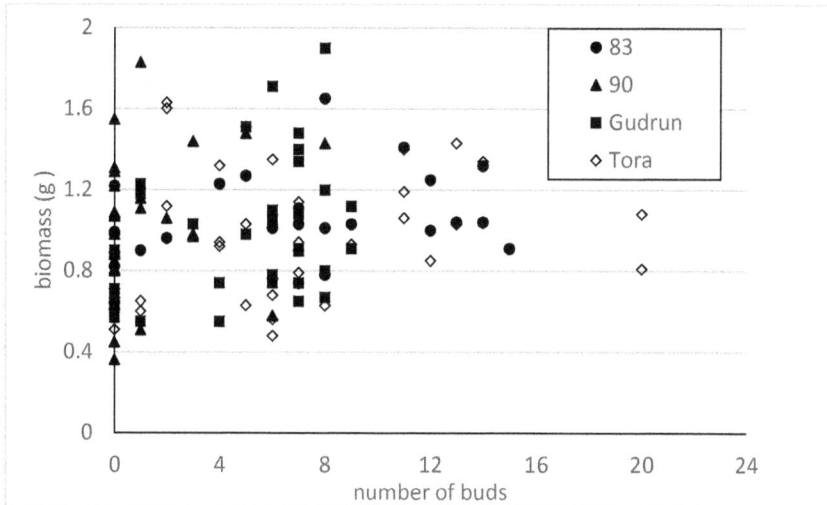

Figure 6. Relationship between the number of visible buds on the cuttings and cutting total biomass production for different clones in 2010.

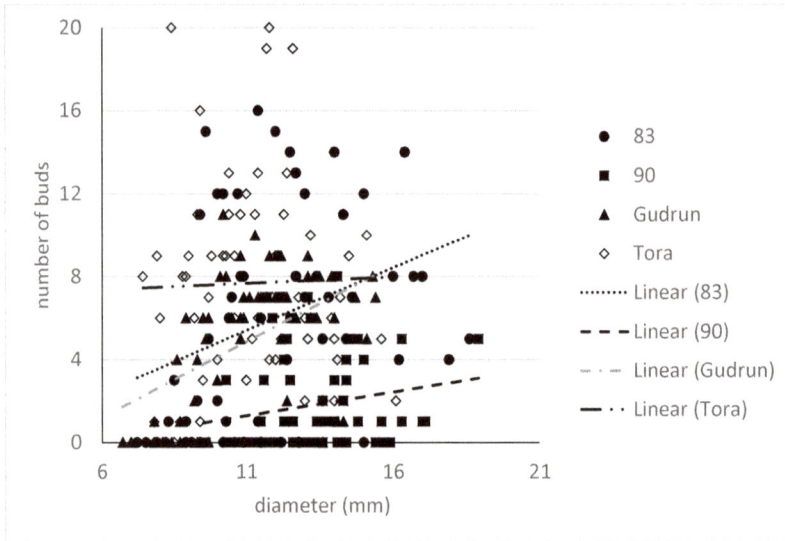

Figure 7. Relationship between the cutting diameter and number of visible buds on the cuttings of various clones in 2010.

4. Discussion

Our lab-scale experiment with *Salix* cuttings revealed that there are some patterns in the biomass production and allocation between roots and shoots that can be linked with studied factors. In general, hydroponic conditions differ from the natural growing environment by both unlimited water supply and more anaerobic conditions of rhizosphere. However, to our knowledge the studied *Salix* species in nature prefer wet growing conditions. Therefore we assume that both their roots and potential symbiotic microorganisms behaved as usual in our laboratory conditions. In this case our study results do not contain only purely new scientific information but can be extrapolated to the practical SRC establishment as well.

The strong linear regression between cutting diameter and the biomass of newly formed roots and shoots detected in these experiments has also been reported by other scientists [8,31]. This phenomenon is most probably caused by the larger volume of products from photosynthesis stored in cuttings of a greater diameter, thus larger volumes of energy were available for new biomass growth. The intercept reciprocal value of the regression model (−0.270173 in our case) indicates the minimal hypothetical value of the cutting diameter that is still able to produce new biomass. The crucial numeric value of this parameter for each clone that marks the lower diameter limit for vigorous cuttings which are capable of sprouting can be calculated by dividing this constant with the relative biomass production value for particular data series. These critical diameters are much lower than the cutting diameters that are currently used for practical SRC establishment in order to avoid cutting debarking or breaking. However, this information can contain useful additional information for planning manual planting, operating with micro-cuttings or assuming the reasonable rod length for horizontal planting of *Salix* shoots [14].

Differences in the available cutting diameter range for various clones was the reason for using the normalized biomass production number in our analysis instead of absolute values. This correction evened out the differences in the cutting biomass production of clones 93 and Tora in 2009. Moreover, this analysis gave a negative result to our hypothesis, that the original position of cuttings on the rod may have an impact on the biomass production of cuttings of the same diameter. Cuttings with a given diameter from apical parts of rods are reported to have a larger biomass production than cuttings originating from basal parts, but this result was not confirmed by our study with other *Salix* clones.

The lack of statistical relevance for replication/box number in our general model revealed that both boxes with particular clone/treatment cuttings were located in equal environmental conditions in the laboratory. Differences in both absolute and relative biomass production by study years are most probably caused by differences in the available direct sunlight by years that could also have a slight impact on the water temperature. After four weeks of experiment in 2010 both Tora and 83 control series produced significantly less biomass per cutting diameter than in 2009. According to the Estonian Weather Service database (I.Saaremäe, personal communication) in the closest weather station to our experiment location there was 15% less direct sunshine in April 2010 than in the previous year. Hence, we conclude that light conditions during sprouting are important and the results obtained in different weather conditions cannot be compared in detail.

We did not detect any leaf discoloration or any other symptom of nutrient deficit during the experiment even in the control treatment without any nutrient supply. Application of soil originating from the local SRC had no positive effect on the relative biomass production of cuttings both alone and in combination with fertilizer. There can be two explanations for such a result. First, we cannot be sure if we got sufficient useful microorganisms from the SRC soil to promote *Salix* growth or if they were able to survive in anaerobic hydroponic conditions. Second, it is possible that in the sprouting stage no support from microorganisms is available or required. Even more interesting however, is that the harvested biomass of soil-treated cuttings was significantly lower both at two and four weeks from the experiment start and with the ratio of shoots and roots biomass remaining similar with the control series. Since all growing conditions, with the exclusion of water transparency, remained the same, we have to assume that cutting bark is sensitive to light capture. As this phenomenon appeared throughout the experiment, light availability does not only trigger cuttings' sprouting but most probably also has a positive impact on biomass production without changing the resource allocation between roots and shoots. The phenomenon of carbohydrate refixation by plant's green parts, other than leaves, is reported to occur in *Populus tremula* L. [32]. This hypothesis is also supported by the report by S. Edelfeldt et al that shoots emerge from the ground more quickly from shallower horizontally planted cuttings than those from deeper planted [12], as in this case some differences in the light availability can also be expected.

For practical SRC establishment it is usually recommended that cuttings are planted in the ground with no more than 1–2 cm of soil on top to avoid drought impacting cuttings before root formation [33]. However, if light availability is important to promoting the speed of cutting biomass production in both above and below ground parts, it is perhaps useful to cover the upper part with a material which could avoid evaporation yet allow light penetration to promote fast sprouting of cuttings followed by rapid growth and thus avoid competition with the weeds.

Application of the fertilizer in 2009 did not have any significant impact on cuttings' relative biomass production and changes were noticeable with only one clone. Repeating the experiment with the same fertilizer a year later did not confirm any significant differences either and therefore we conclude that additional nutrient availability is not important for cutting biomass production at this stage of growth or the fertilizer concentration in our experiments was too low to have any impact on the cutting productivity. However, in 2009 we found that newly formed biomass in treatments with fertilizer tended to be allocated towards the roots. This indicates that the cuttings were sensitive to nutrients in the hydroponic conditions and allocated their resources to get access to them. Such a plant strategy may be disadvantageous for practical SRC establishment into fertile soil. In this case the young *Salix* plants that allocate more of their resources to root production and less to shoot production suffer more from competition for light with weeds that have an opposite strategy.

It has been claimed that some of the new commercialized *Salix* clones have an earlier bud burst than others [8,25]. Our experiment in 2010 which included four *Salix* clones did not reveal any differences in cutting sprouting speed that could be detected by the biomass production during the first two weeks of the growing period. Biomass production during the final two weeks of the experiment was almost the same as that of the first two weeks, indicating similar growing speeds for all studied clones in the early growing stage. Clone 90 (*S. dasyclados*), had the lowest relative biomass production at the two-week stage and its cuttings also performed the poorest at the end of the experiment. This result concurs with our data from the cuttings original SRC where clone 90 also had the lowest biomass production per plant during the first rotation cycle [34]. This confirms the statement from our Swedish colleagues that pot-experiment results enable the prediction of performance of the plants in the field [35]. Clone 90 also had a significantly higher shoot/root production ratio than the other clones. Higher shoot production compared to root production than in the rest of the clones studied, was also detected in clone Gudrun. This hybrid clone also has some *S dasyclados* genes from its ancestors and therefore species-specific strategy can be speculated. The probability of gaining an advantage in practice from this strategy as well as from any other pattern described in the current experiment depends on the abiotic conditions in SRC. The most crucial factor here is a sufficient water supply that allows the cuttings to use their full potential and suppress the other stress factors. Moreover, the biomass production of cuttings is not a linear function in time and the first produced grams of biomass cannot be multiplied over the SRC lifetime to predict the wood yield.

Visible number of buds on a cutting was in correlation with the cutting diameter and clone but the statistical model only explained a relatively small proportion of the variability. There was strong evidence that the number of buds is larger on thicker cuttings. However, on clone Tora the number of visible buds was similar despite the cutting diameter. Although quite a number of cuttings were marked with no buds, they performed well in our biomass production experiment. The number of such cuttings was especially large among clone 90 cuttings. Hence, we must conclude that it is difficult to accurately measure the bud number on shoots in the dormant stage and that perhaps bud counting should take place after the cuttings have been in hydroponic conditions for some time and the vigorous buds are better detectable. However, such a technique would not be suitable as an express method in the field. Moreover, we are not sure that the counting of buds enables us to predict total biomass production since the relationship between the counted visible number of buds and absolute biomass production without any normalization with cutting diameter does not demonstrate any promising tendencies. Additionally, horizontal planting density must be based on other cutting characters than the number of visible buds on the rod. Most probably, during vegetative propagation the availability of other plant tissues including latent root primordia is more important for sprout formation than that of buds which are primarily grown to form new leaves in the future.

5. Conclusions

The *Salix* cutting diameter is a useful parameter for prediction of plant growth in a hydroponic experiment, while origin of the cutting from the donor rod and number of visible buds on the cutting are not. Neither application of SRC soil or fertilizer into boxes with liquid promote biomass production. Lower production from treatments with added soil indicates the importance of light for biomass production even at this early stage. The clone with the lowest early stage biomass production (90) in these experiments also produced the lowest wood yield in the SRC from where the cuttings originated. According to the results of this study the clones with *S. dasyclados* genes tend to invest less in roots and fertilizer-treated cuttings allocate more to root growth. The visible number of buds on cuttings is clone-specific and in general cutting diameter does not predict the number of buds on the cutting.

Supplementary Materials: The following are available online at http://www.mdpi.com/1999-4907/9/7/378/s1, Figure S1: Photo about experiment in three weeks after establishment in 2009, Figure S2: Illustrative materials about experiment design.

Author Contributions: Conceptualization, K.H.; Methodology, K.H.; Software, K.H.; Validation, K.H and K.T; Formal Analysis, K.H.; Investigation ,K.H.; Resources, K.H. and K.T.; Data Curation, K.H.; Writing—Original Draft Preparation, K.H.; Writing—Review & Editing, K.T.; Visualization, K.H.; Supervision, K.H.; Project Administration, K.T.; Funding Acquisition, K.H. and K.T.

Funding: The study was partly supported by Grant No. 9375 of the Estonian Science Foundation and by Grants PUT1463 and IUT21-1 of the Estonian Research Council.

Conflicts of Interest: The authors declare no conflicts of interest.

References

1. Sennerby-Forsse, L. Hägnförsök med kemisk repellent mot älgbetning. *Vaxtskyddsnotiser* **1982**, *45*, 165–169.
2. Verwijst, T. Stool mortality and development of a competitive hierarchy in a *Salix* viminalis coppice system. *Biomass Bioenergy* **1996**, *10*, 245–250. [CrossRef]
3. De Maria, S.; Rivelli, A.R.; Kuffner, M.; Sessitsch, A.; Wenzel, W.W.; Gorfer, M.; Strauss, J.; Puschenreiter, M. Interactions between accumulation of trace elements and macronutrients in *Salix caprea* after inoculation with rhizosphere microorganisms. *Chemosphere* **2011**, *84*, 1256–1261. [CrossRef] [PubMed]
4. Heinsoo, K.; Dimitriou, I. Growth Performance of Willow Clones in Short Rotation Coppice after Sewage Sludge Application. *Balt. For.* **2014**, *20*, 70–77.
5. Albertsson, J.; Hansson, D.; Bertholdsson, N.O.; Ahman, I. Site-related set-back by weeds on the establishment of 12 biomass willow clones. *Weed Res.* **2014**, *54*, 398–407. [CrossRef]
6. Albertsson, J.; Verwijst, T.; Hansson, D.; Bertholdsson, N.O.; Åhman, I. Effects of competition between short-rotation willow and weeds on performance of different clones and associated weed flora during the first harvest cycle. *Biomass Bioenergy* **2014**, *70*, 364–372. [CrossRef]
7. Edelfeldt, S.; Lundkvist, A.; Forkman, J.; Verwijst, T. Establishment and early growth of willow at different levels of weed competition and nitrogen fertilization. *Bioenergy Res.* **2016**, *9*, 763–772. [CrossRef]
8. Verwijst, T.; Lundkvist, A.; Edelfeldt, S.; Forkman, J.; Nordh, N.-E. Effects of Clone and Cutting Traits on Shoot Emergence and Early Growth of Willow. *Biomass Bioenergy* **2012**, *37*, 257–264. [CrossRef]
9. Watts, C.W.; Clark, L.J.; Chamen, W.; Whitmore, A.P. Adverse effects of simulated harvesting of short-rotation willow and poplar coppice on vertical pressures and rut depths. *Soil Tillage Res.* **2005**, *84*, 192–199. [CrossRef]
10. Arevalo, C.B.M.; Drew, A.P.; Volk, T.A. The effect of common Dutch white clover (*Trifolium repens* L.), as a green manure, on biomass production, allometric growth and foliar nitrogen of two willow clones. *Biomass Bioenergy* **2005**, *29*, 22–31. [CrossRef]
11. Edelfeldt, S.; Verwijst, T.; Lundkvist, A.; Forkman, J. Effects of mechanical planting on establishment and early growth of willow. *Biomass Bioenergy* **2013**, *55*, 234–242. [CrossRef]
12. Edelfeldt, S.; Lundkvist, A.; Forkman, J.; Verwijst, T. Effects of Cutting Length, Orientation and Planting Depth on Early Willow Shoot Establishment. *Bioenergy Res.* **2015**, *8*, 796–806. [CrossRef]
13. McCracken, A.R.; Moore, J.P.; Walsh, L.R.; Lynch, M. Effect of planting vertical/horizontal willow (*Salix* spp.) cuttings on establishment and yield. *Biomass Bioenergy* **2010**, *34*, 1764–1769. [CrossRef]
14. Lowthe-Thomas, S.C.; Slater, F.M.; Randerson, P.F. Reducing the establishment costs of short rotation willow coppice (SRC)—A trial of a novel layflat planting system at an upland site in mid-Wales. *Biomass Bioenergy* **2010**, *34*, 677–686. [CrossRef]
15. Guidi Nissim, W.; Labrecque, M. Planting microcuttings: An innovative method for establishing a willow vegetation cover. *Ecol. Eng.* **2016**, *91*, 472–476. [CrossRef]
16. Bakker, M.G.; Manter, D.K.; Sheflin, A.M.; Weir, T.; Vivanco, J.M. Harnessing the rhizosphere microbiome through plant breeding and agricultural management. *Plant Soil* **2012**, *360*, 1–13. [CrossRef]
17. Corredor, A.H.; Van Rees, K.; Vujanovic, V. Changes in root-associated fungal assemblages within newly established clonal biomass plantations of *Salix* spp. *For. Ecol. Manag.* **2012**, *282*, 105–114. [CrossRef]
18. Hrynkiewicz, K.; Toljander, Y.K.; Baum, C.; Fransson, P.M.A.; Taylor, A.F.S.; Weih, M. Correspondence of ectomycorrhizal diversity and colonisation of willows (*Salix* spp.) grown in short rotation coppice on arable sites and adjacent natural stands. *Mycorrhiza* **2012**, *22*, 603–613. [CrossRef] [PubMed]
19. Truu, M.; Truu, J.; Heinsoo, K. Changes in soil microbial community under willow coppice: The effect of irrigation with secondary-treated municipal wastewater. *Ecol. Eng.* **2009**, *35*, 1011–1020. [CrossRef]
20. Doty, S.L.; Oakley, B.; Xin, G.; Kang, J.W.; Singleton, G.; Khan, Z.; Vajzovic, A.; Staley, J.T. Diazotrophic endophytes of native black cottonwood and willow. *Symbiosis* **2009**, *47*, 23–33. [CrossRef]
21. Weih, M.; Bonosi, L.; Ghelardini, L.; Ronnberg-Wastljung, A.C. Optimizing nitrogen economy under drought: Increased leaf nitrogen is an acclimation to water stress in willow (*Salix* spp.). *Ann. Bot.* **2011**, *108*, 347–1353. [CrossRef] [PubMed]
22. Bonosi, L.; Ghelardini, L.; Weih, M. Growth responses of 15 *Salix* genotypes to temporary water stress are different from the responses to permanent water shortage. *Trees-Struct. Funct.* **2010**, *24*, 843–854. [CrossRef]

23. Holm, B.; Heinsoo, K. Biogas digestate suitability for the fertilisation of young *Salix* plants. *Balt. For.* **2014**, *20*, 263–271.

24. Dimitriou, I.; Aronsson, P. Wastewater and sewage sludge application to willows and poplars grown in lysimeters—Plant response and treatment efficiency. *Biomass Bioenergy* **2011**, *35*, 161–170. [CrossRef]

25. Weih, M. Genetic and environmental variation in spring and autumn phenology of biomass willows (*Salix* spp.): Effects on shoot growth and nitrogen economy. *Tree Physiol.* **2009**, *29*, 1479–1490. [CrossRef] [PubMed]

26. Brereton, N.J.B.; Pitre, F.E.; Shield, I.; Hanley, S.J.; Ray, M.J.; Murphy, R.J.; Karp, A. Insights into nitrogen allocation and recycling from nitrogen elemental analysis and ^{15}N isotope labelling in 14 genotypes of willow. *Tree Physiol.* **2014**, *34*, 1252–1262. [CrossRef] [PubMed]

27. Sennerby-Forsse, L.; Zsuffa, L. Bud structure and resprouting in coppiced stools of *Salix viminalis* L., *S. eriocephala* Michx., and *S. amygdaloides* Anders. *Trees* **1995**, *9*, 224–234. [CrossRef]

28. Pijut, P.M.; Woeste, K.E.; Michler, C.H. Promotion of Adventitious Root Formation of Difficult-to-Root Hardwood Tree Species. *Hortic. Rev.* **2011**, *38*, 213–251. [CrossRef]

29. Mleczek, M.; Rutkowski, P.; Rissmann, I.; Kaczmarek, Z.; Golinski, P.; Szentner, K.; Strażyńska, K.; Stachowiak, A. Biomass productivity and phytoremediation potential of *Salix alba* and *Salix viminalis*. *Biomass Bioenergy* **2010**, *34*, 1410–1418. [CrossRef]

30. Magdziak, Z.; Kozlowska, M.; Kaczmarek, Z.; Mleczek, M.; Chadzinikolau, T.; Drzewiecka, K.; Golinski, P. Influence of Ca/Mg ratio on phytoextraction properties of *Salix viminalis*. II. Secretion of low molecular weight organic acids to the rhizosphere. *Ecotoxicol. Environ. Saf.* **2011**, *74*, 33–40. [CrossRef] [PubMed]

31. Vigl, F.; Rewald, B. Size matters?—The diverging influence of cutting length on growth and allometry of two Salicaceae clones. *Biomass Bioenergy* **2014**, *60*, 130–136. [CrossRef]

32. Wittmann, Z.; Aschan, G.; Pfanz, H. Leaf and twig photosynthesis of young beech *(Fagus sylvatica)* and aspen *(Populus tremula)* trees grown under different light regime. *Basic Appl. Ecol.* **2001**, *2*, 145–154. [CrossRef]

33. Sennerby-Forsse, L.; Johansson, H. *Energiskog—Handbok i Praktisk Odling*; Sveriges Lantbruksuniversitet Speciella Skrifter: Uppsala, Sweden; 1989; Volume 38, pp. 1–40.

34. Aasamaa, K.; Heinsoo, K.; Holm, B. Biomass production, water use and photosynthesis of *Salix* clones grown in a wastewater purification system. *Biomass Bioenergy* **2010**, *34*, 897–905. [CrossRef]

35. Weih, M.; Nordh, N.E. Determinants of biomass production in hybrid willows and prediction of field performance from pot studies. *Tree Physiol.* **2005**, *25*, 1197–1206. [CrossRef] [PubMed]

forests

Article

The Performance of Five Willow Cultivars under Different Pedoclimatic Conditions and Rotation Cycles

Werther Guidi Nissim [1,2], Benoit Lafleur [1,3] and Michel Labrecque [1,*]

[1]   Institut de Recherche en Biologie Végétale, Université de Montréal and Montreal Botanical Garden, Montréal, QC H1X 2B2, Canada; werther.guidinissim@unifi.it (W.G.N.); benoit.lafleur@uqat.ca (B.L.)
[2]   Department of Agrifood Production and Environmental Science—University of Florence, Firenze 50144, Italy
[3]   Forest Research Institute and Industrial Chair NSERC-UQAT-UQAM in Management of Sustainable Forest, Université du Québec en Abitibi-Témiscamingue, Rouyn-Noranda, QC J9X 5E4, Canada
*   Correspondence: michel.labrecque@umontreal.ca; Tel.: +1-514-872-1862

Received: 1 May 2018; Accepted: 11 June 2018; Published: 13 June 2018

Abstract: A plant's genotype, their environment, and the interaction between them influence its growth and development. In this study, we investigated the effect of these factors on the growth and biomass yield of willows in short-rotation coppice (SRC) under different harvesting cycles (i.e., two- vs. three-year rotations) in Quebec (Canada). Five of the commercial willow cultivars most common in Quebec, (i.e., *Salix × dasyclados* Wimm. 'SV1', *Salix viminalis* L. '5027', *Salix miyabeana* Seeman 'SX61', 'SX64' and 'SX67') were grown in five sites with different pedoclimatic conditions. Yield not only varied significantly according to site and cultivar, but a significant interaction between rotation and site was also detected. Cultivar '5027' showed significantly lower annual biomass yield in both two-year (average 10.8 t ha^{-1} year^{-1}) and three-year rotation (average 11.2 t ha^{-1} year^{-1}) compared to other cultivars (15.2 t ha^{-1} year^{-1} and 14.6 t ha^{-1} year^{-1} in two- and three-year rotation, respectively). Biomass yield also varied significantly with rotation cycle, but the extent of the response depended upon the site. While in some sites the average productivity of all cultivars remained fairly constant under different rotations (i.e., 17.4 vs. 16 t ha^{-1} year^{-1} in two- and three-year rotation, respectively), in other cases, biomass yield was higher in the two- than in the three-year rotation or vice versa. Evidence suggests that soil physico-chemical properties are better predictors of willow SRC plantation performance than climate variables.

Keywords: *Salix*; SRC; biomass; willow cultivars; environmental conditions

1. Introduction

Woody biomass is a renewable resource with multiple applications, and can be used as feedstock for pulp and paper production as well as by the energy or biofuel industry [1]. Short-rotation coppice (SRC) is a well-established plant production technique used to manage a broad variety of woody species, mostly for bioenergy purposes. This cropping system was defined in the late 1980s as a silvicultural approach based on short clear-felling cycles, which uses intensive cultural techniques with genetically superior planting material and often relies on coppice regeneration [2]. Since then, it has been applied to a number of species, including poplars [3], willows [4], eucalyptus [5], black locust [6], alders [7] and *Leucaena* spp. [8]. In the early years, SRC plantations were designed to be managed much like any other agricultural crop, by methods including tilling, fertilization, weed and pest control, and irrigation [9]. However, concerns were raised about the need to reduce its footprint on the environment while maintaining economic profitability [10]. This has led to management models that rely on less external input, often by recycling different types of waste, including ash [11], sewage sludge [12,13], pig slurry [14] and wastewaters [15–17]. The productivity of SRC plantations relies heavily on the cultivar's genotypic characteristics, habitat-related factors and cultivation techniques. Along with poplar, willow SRC represents a very popular woody crop in temperate regions of North America and Eurasia. In fact, willow SRCs have been shown to be one of the best-suited biomass crops for many Canadian regions, because under optimal conditions they can achieve very high biomass yields (on average 21 t ha^{-1} year^{-1} of dry biomass for the most productive cultivar) [18], although yields are highly dependent on genotypic characteristics, soil fertility, climate, and crop management [19]. A recent study conducted to determine which set of soil, climatic conditions, and cultivars are responsible for greater willow SRC yields in eastern Canada showed that both geographic location and cultivar play a significant role in determining annual yields. In particular, biomass yield was positively correlated to some climatic (i.e., average annual temperature, total annual precipitation, average growing season temperature, average growing season precipitation, and degree days) and soil (pH, extractable P) factors and negatively affected by others (e.g., soil clay concentration) [20]. However, this study was conducted over a single rotation. Furthermore, other studies report higher average annual biomass yield in four-year rotation willow plantations compared to those in a three-year cycle [18]. In this context, it appears necessary to assess the response of different willow cultivars grown over several rotations and under different rotation lengths. Because it is an early successional species, willow shows fast juvenile growth and rapid adjustments in leaf area and shoot morphology in response to environmental conditions and management practices [21–23]. This means that in dense plantations, the maximum mean annual increment (MAI) in biomass yield is reached at an early stage. However, available information about this aspect is still inconclusive. Some studies have reported the highest MAI in willow grown in Sweden at a density of 20.000 plants ha^{-1} and managed with 4-year rotation harvesting [24]. Therefore, in order to identify optimal rotation length in willow SRC, both cultivar characteristics and environmental parameters must be considered along with the principles determining the dynamics of a population under inter-plant competition.

The aim of the current study was: (i) to assess the response of several willow cultivars in SRC to different rotations; and (ii) to highlight which set of environmental parameters play a major role in determining the yield of different willow cultivars.

2. Materials and Methods

2.1. Experimental Sites

This study was carried out over a three-year period (2015–2016–2017) at five sites in southern Quebec, Canada, located along a climatic gradient. Prior to plantation establishment, each site had distinct soil properties (Table 1).

Table 1. Soil Physical and Chemical Characteristics at the Time of Establishment for Five short-rotation coppice (SRC)-Willow Plantations.

Site	pH	OM %	P	K	Ca	Mg	Sand	Silt	Clay	Texture	Soil Type
				(kg ha^{-1})				%			
Beloeil	7.3	4.2	31	729	7570	2490	10	40	50	Silty clay	Gleysols
Boisbriand	6.9	4	21	166	4935	1120	31	33	36	Clayey loam	Brunisols
La Morandière	5.6	2.9	20	410	2796	1261	1	36	63	Clay	Luvisols
La Pocatière	6.1	5.6	101	475	5889	1138	20	31	49	Clay	Gleysols
St-Siméon	5.2	2.1	50	155	1662	72	60	20	20	Sandy loam	Brunisols

Soil texture was determined by granulometric analysis; P, K, Ca, and Mg were extracted by Melich-3 digestion and determined using inductively coupled plasma mass spectrophotometry (ICP-MS); OM organic matter [20].

The Beloeil (45°35′24″ N–73°13′48″ W) and Boisbriand (45°35′24″ N–73°13′48″ W) sites showed the highest pH values (7.1 on average), as well as high organic matter (OM) (4.1%) and Ca, Mg, K concentrations, whereas the La Morandière (48°40′12″ N–77°36′00″ W) and St-Siméon (48°04′12″ N–65°33′36″ W) sites showed the lowest pH values (5.4 on average) and OM concentration

(2.5%), as well as lower Ca, Mg and K concentrations. The La Pocatière site (47°21′36″ N–70°01′48″ W) showed intermediate properties and, compared to the other sites, had the highest OM (5.6%) and P concentrations. Sites also differed in soil texture; the La Morandiere site is clay Luvisol, whereas the St-Siméon site is a sandy loam Brunisol. More details on soil properties are reported in Lafleur et al. [20].

2.2. Willow Varieties and Experimental Setup

The experimental sites were all located on former agricultural farmland. In the fall of 2010, the soil was prepared by ploughing, and disc harrowing was performed the following spring prior to planting. One-year-old dormant cuttings (0.20–0.25 m) were planted in the spring of 2011. Plant density at establishment was 18,500 cuttings ha^{-1} (1.80 m between rows and 0.30 m within each row). Five of the willow commercial cultivars most common in Quebec were selected, including 'SV1' (*Salix* × *dasyclados* Wimm.), '5027' (*Salix viminalis* L.), 'SX61', 'SX64' and 'SX67' (*Salix miyabeana* Seeman). The five willow cultivars were arranged in a complete randomized block experimental design with four replicates for each cultivar (total 20 plots per site, 100 plants per plot per each cultivar). Throughout the trial, weeding was performed manually in willow rows, and mechanically between rows. No chemical control of weeds, pests or diseases was performed during the trial. All plots in all sites were harvested (cut back) at the end of the first growing season. Since high mortality was recorded at the La Morandière site in 2011, the site was replanted in 2012. The following spring (2012), mineral fertilizer was applied, supplying 100 kg ha^{-1} of N, 100 kg ha^{-1} of P$_2$O$_5$. While N was supplied to all sites, only those showing less than 100 kg ha^{-1} of P prior to planting received P$_2$O$_5$ fertilizer (e.g., the La Pocatière site received none). Plants were harvested from all sites in 2014 when stems were three (S3) and roots four (R4) years old respectively (except for the La Morandière site, were roots were three years old). Data related to the first rotation are described in Lafleur et al. [20]. After the first harvest, each site was split into biannual (S2R6) and triennial (S3R7) rotation cycles with blocks 1 and 2 harvested in 2016 (biannual) and blocks 3 and 4 harvested in 2017 (triennial). Plants were harvested at the end of the growing season, that is, in November.

2.3. Measurement and Sampling

At the end of the growing season of each rotation (2016 and 2017) the height and diameter (0.2 m above ground) of the main stem were measured on nine randomly selected plants per block for each species in each site. Plants were randomly selected within the two central rows in order to avoid edge effect. The growth rates were expressed as annual diameter and height increments by dividing the value measured by the age of the stems. In addition, after leaf drop, the same plants were harvested and weighed in the field using an electronic scale. To evaluate dry matter of willow aboveground biomass, the whole green stem samples collected from the field were oven-dried at 80 °C (to constant mass) before being reweighed. Annualized biomass yield was calculated by dividing plantation density as re-evaluated at the end of each growing season by the age of the plants.

2.4. Meteorological Conditions

The climatic conditions at different experimental sites are shown in Figure 1. According to the nearest weather station for each site, from 2015 to 2017, La Morandière was the site recording the lowest average annual temperature (i.e., 1.3 °C) and the lowest rainfall (579 mm), whereas Beloeil showed the highest average annual temperature (i.e., 7.3 °C) and St-Siméon the highest annual rainfall (1143 mm). During the growing season (i.e., May to September), average temperatures varied between 13.5 °C (St-Siméon) and 18.8 °C (Boisbriand), whereas total precipitation varied between 340 mm (La Pocatière) and 490 mm (St-Siméon).

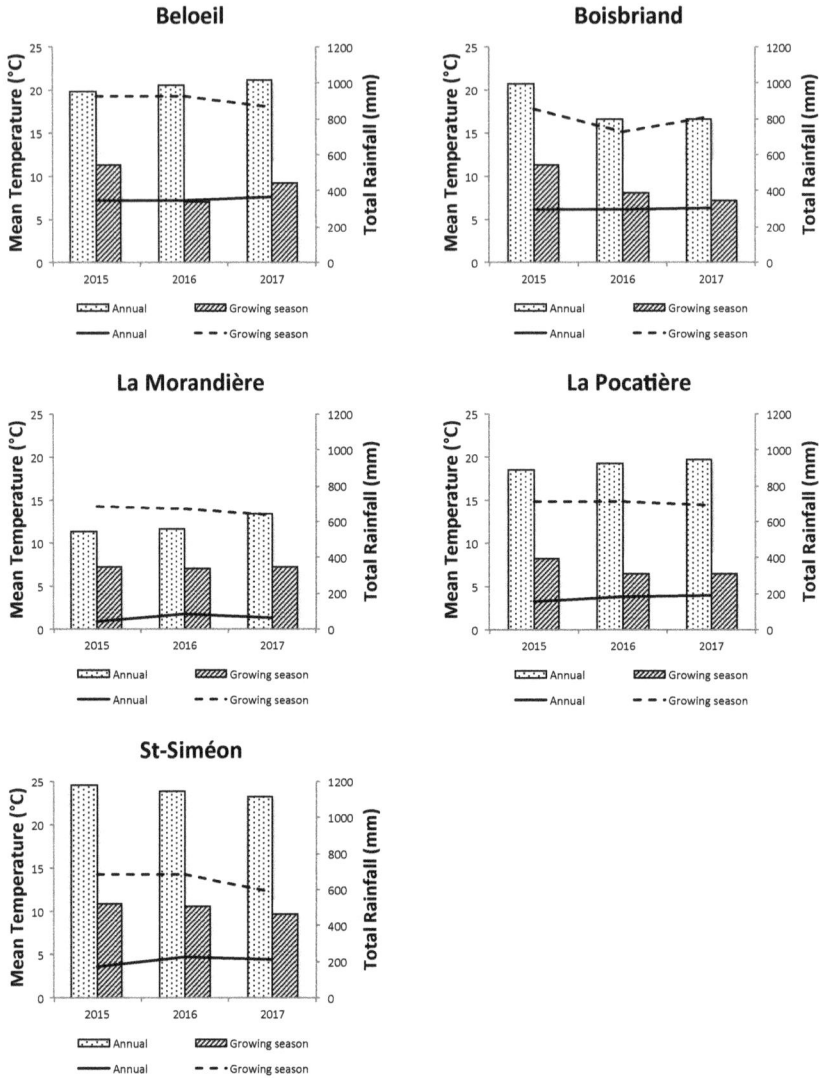

Figure 1. The mean temperature and rainfall on an annual basis and during the growing season (May–September 2015–2017) at different experimental sites. Bars represent annual precipitation and lines temperature.

2.5. Data Analysis

Within each rotation cycle, the differences in plant growth and biomass yield of the five cultivars in the five locations were assessed through two-way ANOVA tests followed by multiple comparisons of means according to Tukey's Honestly Significant Difference (HSD). A 5% significance level was adopted for identifying significant treatment effects (i.e., cultivar, site, cultivar × site). Subsequently, a regression tree approach was used to predict, independently for the two- and three-year rotations, plant height and diameter annual increment and annual yield at the site level from climate and soil variables. Regression trees function by partitioning a dataset into increasingly homogenous subsets.

This approach was selected because it is non-parametric, can account for non-linear relationships between variables, and tends to be robust in regard to errors in both the independent and dependent variables [25]. Regression trees were carried out using JMP 10.0 (SAS, 2012).

3. Results

3.1. Plant Growth and Biomass Yield

The growth of willow plants varied according to cultivar, site and the interaction between these two factors. For the two-year rotation cycle, the diameter growth rate was significantly influenced by both the cultivar ($p < 0.0001$) and the site ($p < 0.0001$) (Figure 2). In particular, cultivars of the *S. miyabeana* group (i.e., 'SX61', 'SX64' and 'SX67') showed the highest diameter growth rates (average 14.1 mm year^{-1}) compared to 'SV1' and '5027' (10.6 mm year^{-1}). The highest diameter growth rates (average 14.4 mm year^{-1}) were recorded at two specific sites (i.e., La Pocatière and Beloeil) for all cultivars, whereas for cultivar *S. viminalis* '5027', the most productive site was in La Morandière, with an average growth rate of 12.0 mm year^{-1}. On the other hand, for the three-year rotation, ANOVAs showed significant differences only among cultivars, whereas no significant difference was found among the sites (Figure 2). In particular, we found that although diameter growth was reduced compared to the two-year rotation, at all sites the *S. miyabeana* cultivars showed higher growth rates (average 8.0 mm year^{-1}) compared to '5027' (average 6.1 mm year^{-1}) and 'SV1', which showed intermediate values (7.0 mm year^{-1}).

Height increment rate of the willow stems submitted to a two-year rotation was influenced by site ($p < 0.0001$), cultivar ($p < 0.0001$) and the interaction between them (Figure 3). Thus, the tallest willow cultivars were 'SX61', 'SV1', 'SX67' and 'SX64' at Beloeil (average 243 cm year^{-1}), 'SX61' and 'SX64' at Boisbriand (average 230 cm year^{-1}), 'SX64' at La Morandière (average 162 cm year^{-1}), 'SX61' at La Pocatière (average 255 cm year^{-1}), and 'SX61' and 'SX67' at St-Siméon (average 235 cm year^{-1}). On the other hand, except at La Morandière (122 cm year^{-1}), cultivar '5027' showed the poorest performance (average 154 cm year^{-1}). In the three-year rotation, the rate of height increment of willow stems varied according to cultivar ($p < 0.0001$), and, to a lesser extent, site ($p = 0.0332$), although the average increment values were lower than in the two-year rotation. In this case, 'SX61' was the best performing cultivar in all sites (average 137 cm year^{-1}) and '5027' the worst performing cultivar (average 92 cm year^{-1}). For all cultivars, the La Pocatière site (average 141 cm year^{-1}) recorded significant higher stem height increments than at La Morandière.

With respect to biomass yield, we observed significant differences according to site ($p < 0.0001$) and cultivar ($p < 0.0001$), as well as a significant interaction between rotation and site ($p < 0.0001$). In particular, in all locations, cultivar '5027' showed a significantly lower annual biomass yield in both two-year (average 10.8 t ha^{-1} year^{-1}) and three-year rotations (average 11.2 t ha^{-1} year^{-1}), compared with the other cultivars, which on average showed a biomass yield ranging from 15.2 t ha^{-1} year^{-1} to 14.6 t ha^{-1} year^{-1} in two- and three-year rotations respectively. (Figure 4). The most productive site was La Pocatière, with 30.9 t ha^{-1} year^{-1} and 21 t ha^{-1} year^{-1} in the two- and three-year rotations respectively, whereas the least productive site was La Morandière, where average biomass yield ranged between 4.6 t ha^{-1} year^{-1} and 5.5 t ha^{-1} year^{-1} in the two- and three-year rotations respectively. Intriguingly, biomass yield also varied significantly with rotation cycle, but the extent of the response was site-dependent (Table 2). Thus, although at Beloeil the average biomass yield of all cultivars remained fairly constant under different rotations (i.e., 17.4 vs. 16 t ha^{-1} year^{-1} in two- and three-year rotations respectively), in other cases, biomass yield was higher in two-year than in three-year rotation (La Pocatière and St-Siméon) or the opposite, higher in three-year than in two-year rotation (Boisbriand and La Morandière).

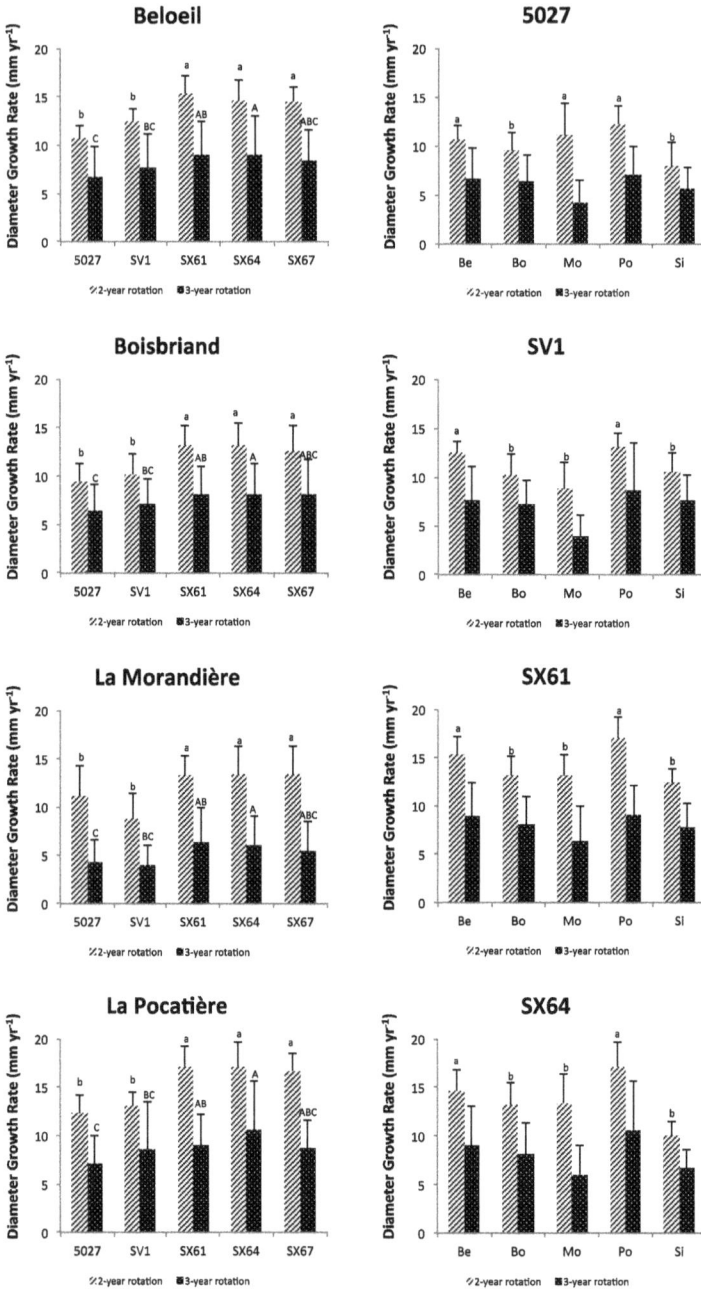

Figure 2. *Cont.*

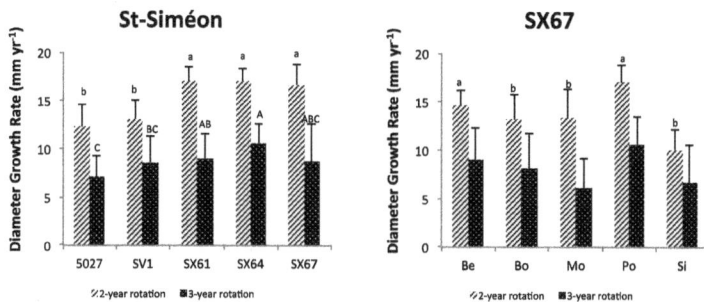

Figure 2. Stem diameter growth rates of willow stems under different rotation cycles. Values are means (*n* = 9; standard deviation on the top of each bar) for each site (Be = Beloeil; Bo = Boisbriand; Mo = La Morandière; Po = La Pocatière, Si = St-Siméon) and cultivar. A 5% significance level was adopted for identifying significant treatment effects according to Tukey's HDS test. Lowercase letters were used to highlight significant difference in two-year rotation. Capital letters identify significant differences in the three-year rotation.

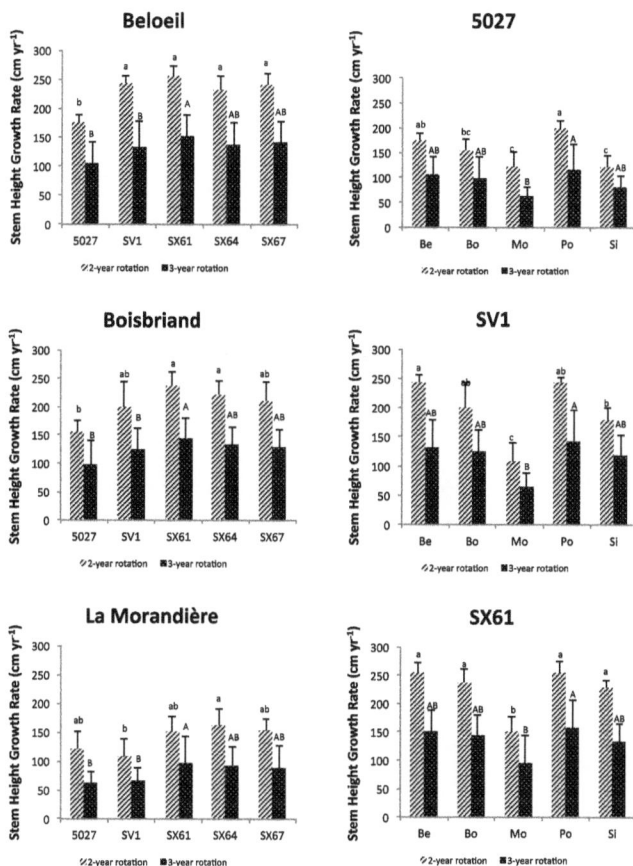

Figure 3. *Cont.*

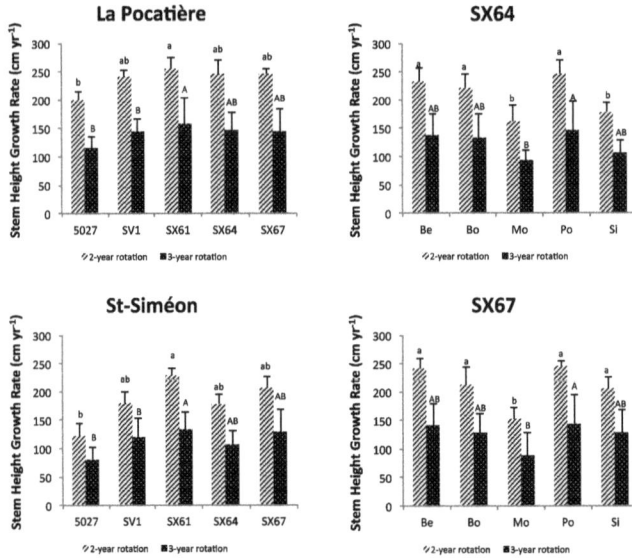

Figure 3. Stem height growth rates of willow stems under different rotation cycles. Values are means (*n* = 9; SD on the top of each bar) for each site (Be = Beloeil; Bo = Boisbriand; Mo = La Morandière; Po = La Pocatière, Si = St-Siméon) and cultivar. A 5% significance level was adopted for identifying significant treatment effects according to Tukey's HDS test. Lowercase letters were used to highlight significant difference in two-year rotation Capital letters identify significant differences in the three-year rotation.

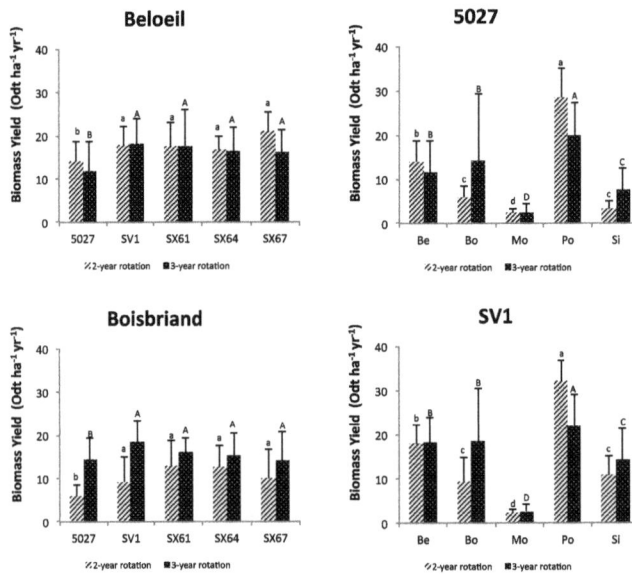

Figure 4. *Cont.*

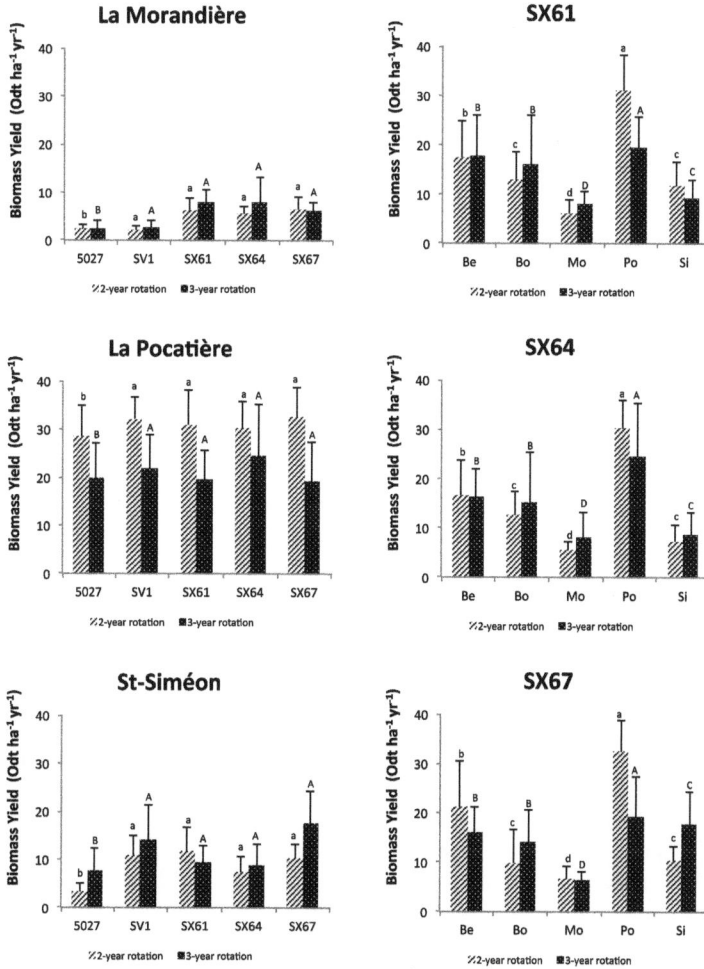

Figure 4. Annual biomass yield of willow plants under different rotation cycles. Values are means (*n* = 9; SD on the top of each bar) for each site (Be = Beloeil; Bo = Boisbriand; Mo = La Morandière; Po = La Pocatière, Si = St-Siméon) and cultivar. A 5% significance level was adopted for identifying significant treatment effects according to Tukey's HDS test. Lowercase letters were used to highlight significant difference in two-year rotation. Capital letters identify significant differences in the three-year rotation.

Table 2. Results of ANOVA tests describing the statistical significance of site, cultivar, and rotation on willow biomass yield.

Source	Nparm	DF	DFDen	F Ratio	Prob. > F
Site	4	4	50	99.0831	<0.0001
Cultivar	4	4	50	7.8034	<0.0001
Site × Cultivar	16	16	50	1.752	0.0668
Rotation	1	1	50	0.0422	0.838
Site × Rotation	4	4	50	9.9017	<0.0001
Cultivar × Rotation	4	4	50	0.3704	0.8286
Site × Cultivar × Rotation	16	16	50	0.6305	0.8437

	Least Square Mean			
Site	Rotation			
	Two-year		Three-year	
Beloeil	16.729656	A	15.294729	A
Boisbriand	9.270192	B	14.119425	A
La Morandière	4.1636872	B	4.8795649	A
La Pocatière	30.626091	A	20.151016	B
La Pocatière	10.681216	A	7.98445	B

The table highlights the interaction between site and rotation and their impact on biomass yield. Different letters between columns indicate significant differences ($p \leq 0.05$) according to Tukey's HSD test.

3.2. Relating Plant Height and Diameter Annual Increment and Annual Yield to Climate and Soil Variables

Regression tree analyses conducted on the two-year rotation showed that along the pedoclimatic gradient, soil variables were more accurate than climate variables at predicting height and diameter annual increment and annual biomass yield than climate variables (Figure 5). With respect to both height and diameter annual increment (Figure 5a,b), soil pH and OM content best explained differences observed among sites, with sites with better annual increment having higher soil pH and OM content. Soil extractible P was the best predictor of annual biomass yield, and the next best predictor was mean annual temperature (Figure 5c). More specifically, higher extractible P and mean annual temperature were conducive to greater annual biomass yield. With respect to these predictors, the La Pocatière, Beloeil and Boisbriand sites showed the best performance. For the three-year rotation, regression tree analyses showed that soil extractible P and clay content were the best predictors of height annual increment (Figure 6a), whereas soil exchangeable Ca, extractible P and pH were the best predictors of annual biomass yield (Figure 6c). More specifically, high soil P, Ca, pH and clay content were associated with sites with better performance. With respect to diameter increment, three-year mean annual temperature, followed by soil sand content, were the best predictors; high mean annual temperature and low sand content were related to better growth (Figure 6b). Taken together, these results strongly suggest that soil variables exerted a stronger effect than climate variables on plant growth for the three-year rotation. As was the case for the two-year rotation, the La Pocatière, Beloeil and Boisbriand sites showed the best performance with respect to these predictors. Soil physico-chemical properties therefore appeared better significant predictors of willow SRC plantation performance.

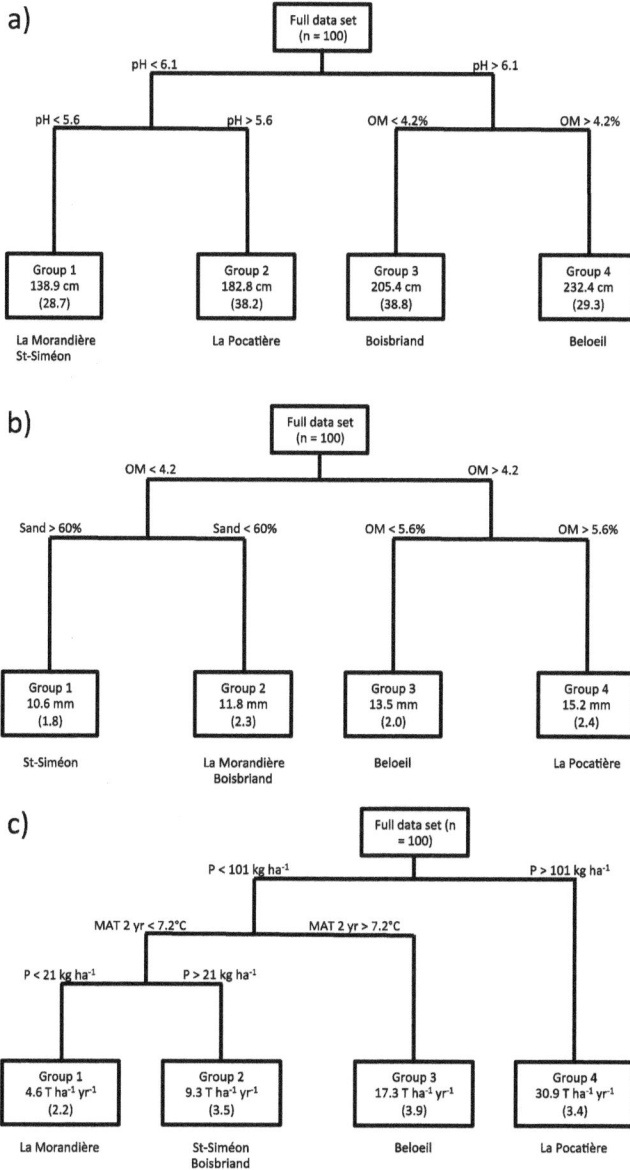

Figure 5. Regression tree model for stem height (**a**), stem diameter (**b**), and biomass yield (**c**) in the two-year rotation willow SRC. MAT, mean annual temperature; OM, soil organic matter; P, soil extractible phosphorus; pH, soil pH; Sand, soil sand content. Splitting values are indicated along the branches, and boxes indicate terminal node mean (±standard error).

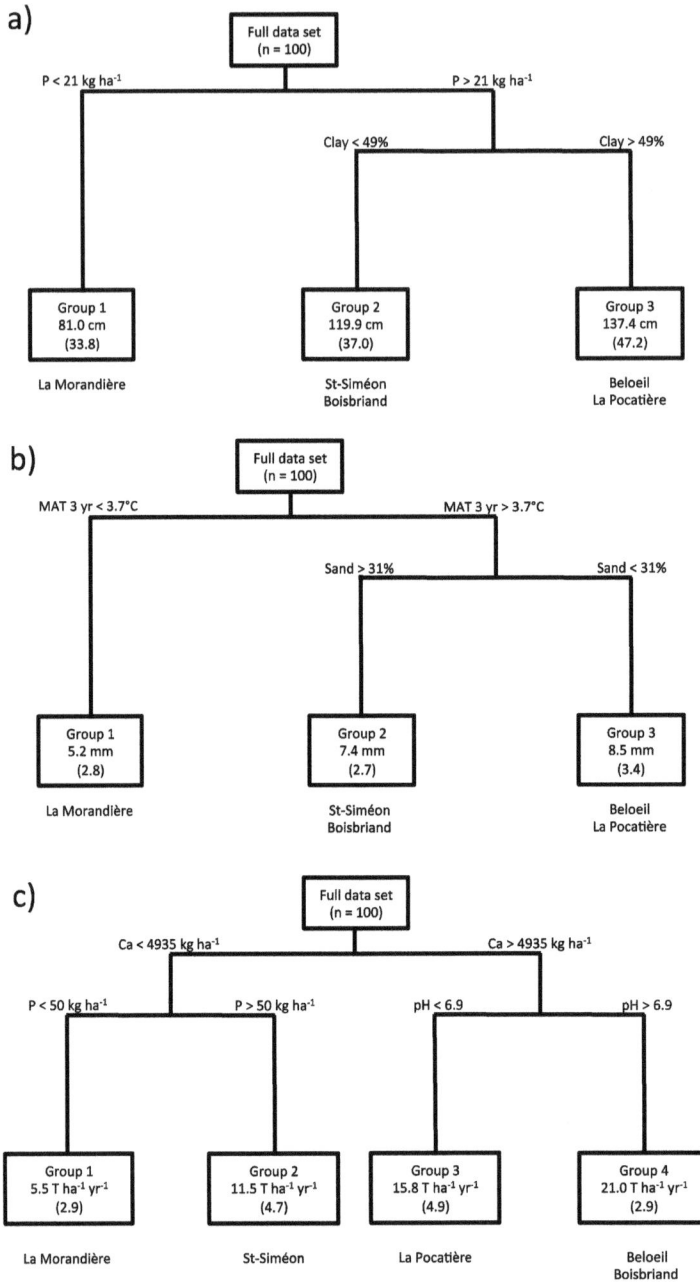

Figure 6. Regression tree model for stem height (**a**), stem diameter (**b**), and biomass yield (**c**) in the three-year rotation willow SRC. Ca, soil exchangeable calcium; Clay, soil clay content; MAT, mean annual temperature; P, soil extractible phosphorus; pH, soil pH; Sand, soil sand content. Splitting values are indicated along the branches, and boxes indicate terminal node mean (± standard error).

4. Discussion

Willow SRC plantations have been studied for a number of years in many temperate and boreal regions of the world. Previous studies carried out on willows in short-rotation coppice showed that the climate of south-eastern Canada offers favourable growing conditions [26,27]. In the current study, the annualized diametric growth of willow was significantly affected by both cultivar and site in the two-year rotation and only by the cultivar in the three-year rotation, with the best performing cultivars all belonging to the same *S. miyabeana* cultivar group. With respect to height increment, we found that site played a significant role (at least for some cultivars) in determining performance. In any case, since diameter and height play a major role in determining the total woody biomass of the plant at the end of each growing season, biomass of the willow plants varied significantly according to site ($p < 0.0001$) and cultivar ($p < 0.0001$). Strong genetic, site, and genetic by site interactions have previously been observed in other experiments carried out in willow SRC [28]. In some cases, site quality differences were most likely driven by nutrient concentration, physical traits, and water holding capacity of the soil [29], whereas in others, differences were mainly influenced by climatic variables, such as growing degree-days [30] or total rainfall during the growing season [31]. Likewise, plant genetics has been shown to play an important role in determining the biomass yield of willow SRC, both under optimal [32] and harsh [33] growing conditions. In our study, the lowest biomass yield was shown by cultivar '5027', which is consistent with previous results published for Canada [34], and reveals once again the poor performance of *S. viminalis* cultivars in such regions. Even though this result is surprising, considering that the most productive willow cultivars in other suitable regions do originate from *S. viminalis* [35], its poor performance in our study is likely due to its high sensitivity to pests and diseases, including the willow leaf beetle (*Plagiodera versicolora* Laicharteg.) and potato leafhopper (*Empoasca fabae* Harris). On the other hand, our results confirm the high performance of *S. miyabeana* cultivars also reported in other studies [36]. This species offers a wide range of genetic material for farmers to select from for probable successful willow growth in SRC for commercial purposes in Quebec; breeders, too, can select from among new high-performing commercial varieties. Intriguingly, we also found a significant interaction between rotation length and site ($p < 0.0001$). This finding could have a very strong impact on the management of willow SRC plantations. In fact, if the ultimate goal of willow SRC is to maximize annual yield, our study shows that in some cases a two-year rotation would be preferable to a three-year cycle. In general, willow SRC plantations show higher annual biomass yield in longer, rather than shorter, rotation cycles [24] and specifically triennial harvesting has been shown to provide higher annual biomass production than biennial harvesting [37]. On the basis of these observations, the explanation for this differentiated response should be sought elsewhere. Actually, the differences in the two rotations observed in the current study were evident in both very suitable (i.e., Boisbriand and La Pocatière) and harsher (i.e., La Morandière and St-Siméon) growing sites. In this case, the only factor that seemed to differentiate these groups of sites was P soil concentration, which was significantly higher (average 75.5 kg ha^{-1}) in those where the highest biomass yield was recorded in the two-year rotation (i.e., La Pocatière and St-Siméon) and lower (average 20 kg ha^{-1}) where the highest biomass yield was recorded for most cultivars in the three-year rotation (i.e., Boisbriand and La Morandière). The fact that the only site where biomass yield was rather stable showed an intermediate P soil concentration (average 30 kg ha^{-1}) could support this observation. In fact, some researchers have shown that the concentration of nutrients in the soil (including fertilization to enhance this pool) affects the mean annual increment (MAI) of some short-rotation forestry species, including poplar and sycamore [38].

The fact remains that the annual yield values reported for most of the cultivars on the La Pocatière site are surprisingly high. Yet the climatic conditions of this site are far from the most favourable among these experimental sites. The average annual degree days (>5 °C) in La Pocatière are 1582, compared to 2029 in Boisbriand and 2122 in Beloeil. Soil characteristics, in particular the high percentage of organic matter, certainly contribute to higher yield. Very high yield was also reported by Héneault-Éthier [39] for willow, *S. miyabeana* 'SX64', grown in riparian buffer strips on organic rich soils in southern Quebec.

When data concerning cultivars were pooled, some soil (i.e., extractible P concentration) and meteorological (i.e., mean annual temperature) factors in the two-year rotation and soil exchangeable Ca, extractible P and pH in the three-year rotation were the best predictors of annual biomass yield. This information is partially consistent with the results of a recent published study in which silt content, soil organic matter, pH, exchangeable Ca and Mg, and total N and Zn were significantly and positively related to aboveground yield of willow in SRC [23]. In fact, in acid soils, soluble inorganic P is fixed by aluminum and iron, thereby reducing its availability for the crop [40]. This would explain the strong dependence of willow biomass yield and pH and Ca.

5. Conclusions

Taken together, our findings show that most commercial willow cultivars can be grown in short rotation coppice and achieve high biomass yields under most of Quebec's pedoclimatic conditions. However, care should be taken to both avoid choosing poor performing genetic material (e.g., *S. viminalis* '5027') and match the right cultivar with the right environmental conditions. In particular, since there is evidence that soil variables (i.e., pH, soil extractible P and Ca) have a stronger impact than climate variables on willow growth and productivity, the site on which the crop will be grown should be selected with care, and site soil should be modified to decrease physico-chemical imbalances (e.g., liming) that prevent good crop development.

Author Contributions: M.L. conceived and designed the experiment; M.L. and B.L. supervised the experiments and the samplings; B.L. and W.G.N. analyzed the data; W.G.N. wrote the first draft of the paper; B.L. and M.L. provided extensive proof-reading and suggested edits.

Acknowledgments: This study was supported by CÉROM (Seed Research Center) and funded by the RPBQ (Quebec Bio-industrial Crop Network) of the Quebec Ministry of Agriculture, Fisheries and Food.

Conflicts of Interest: The authors declare no conflict of interest.

References

1. Hinchee, M.; Rottmann, W.; Mullinax, L.; Zhang, C.; Chang, S.; Cunningham, M.; Pearson, L.; Nehra, N. Short-rotation woody crops for bioenergy and biofuels applications. *In Vitro Cell. Dev. Biol. Plant* **2009**, *45*, 619–629. [CrossRef] [PubMed]
2. Drew, A.P.; Zsuffa, L.; Mitchell, C.P. Terminology relating to woody plant biomass and its production. *Biomass* **1987**, *12*, 79–82. [CrossRef]
3. Auclair, D.; Bouvarel, L. Intensive or extensive cultivation of short rotation hybrid poplar coppice on forest land. *Bioresour. Technol.* **1992**, *42*, 53–59. [CrossRef]
4. Kenney, W.A.; Sennerby-Forsse, L.; Layton, P. A review of biomass quality research relevant to the use of poplar and willow for energy conversion. *Biomass* **1990**, *21*, 163–188. [CrossRef]
5. Sims, R.E.H.; Senelwa, K.; Maiava, T.; Bullock, T. Eucalyptus species for biomass energy in Nex Zealand-I: Growth screening trials at first harvest. *Biomass Bioenergy* **1999**, *16*, 199–205. [CrossRef]
6. Grünewald, H.; Böhm, C.; Quinkenstein, A.; Grundmann, P.; Jörg Eberts, J.; von Wühlisch, G. *Robinia pseudoacacia* L.: A Lesser Known Tree Species for Biomass Production. *Bioenergy Res.* **2009**, *2*, 123–133. [CrossRef]
7. Rytter, L.; Šlapokas, T.; Granhall, U. Woody biomass and litter production of fertilized grey alder plantations on a low-humified peat bog. *For. Ecol. Manag.* **1989**, *28*, 161–176. [CrossRef]
8. Lopez, F.; Garcia, M.M.; Ynez, R.; Tapias, R.; Fernàndez, M.; Diaz, M.J. Leucaena species valoration for biomass and paper production in 1 and 2 year harvest. *Bioresour. Technol.* **2008**, *99*, 4846–4853. [CrossRef] [PubMed]
9. Mitchell, C.P. Ecophysiology of short rotation forest crops. *Biomass Bioenergy* **1992**, *2*, 25–37. [CrossRef]
10. Dimitriou, I.; Baum, C.; Baum, S.; Busch, G.; Schulz, U.; Köhn, J.; Lamersdorf, N.; Leinweber, P.; Aronsson, P.; Weih, M.; et al. The impact of Short Rotation Coppice (SRC) cultivation on the environment. *vTI Agric. For. Res.* **2009**, *59*, 159–162.

11. Park, B.B.; Yanai, R.D.; Sahm, J.M.; Ballard, B.D.; Abrahamson, L.P. Wood Ash Effects on Soil Solution and Nutrient Budgets in A Willow Bioenergy Plantation. *Water Air Soil Pollut.* **2004**, *159*, 209–224. [CrossRef]

12. Labrecque, M.; Teodorescu, T.I.; Daigle, S. Biomass productivity and wood energy of salix species after 2 years growth in SRIC fertilized with wastewater sludge. *Biomass Bioenergy* **1997**, *12*, 409–417. [CrossRef]

13. Guidi Nissim, W.; Cincinelli, A.; Martellini, T.; Alvisi, L.; Palm, E.; Mancuso, S.; Azzarello, E. Phytoremediation of sewage sludge contaminated by trace elements and organic compounds. *Environ. Res.* **2018**, *164*, 356–366. [CrossRef] [PubMed]

14. Cavanagh, A.; Gasser, M.O.; Labrecque, M. Pig slurry as fertilizer on willow plantation. *Biomass Bioenergy* **2011**, *35*, 4165–4173. [CrossRef]

15. Guidi Nissim, W.; Voicu, A.; Labrecque, M. Willow short-rotation coppice for treatment of polluted groundwater. *Ecol. Eng.* **2014**, *62*, 102–114. [CrossRef]

16. Dimitriou, I.; Aronsson, P. Wastewater and sewage sludge application to willows and poplars grown in lysimeters—Plant response and treatment efficiency. *Biomass Bioenergy* **2011**, *35*, 161–170. [CrossRef]

17. Börjesson, P.; Berndes, G. The prospects for willow plantations for wastewater treatment in Sweden. *Biomass Bioenergy* **2006**, *30*, 428–438. [CrossRef]

18. Guidi Nissim, W.; Pitre, F.E.; Teodorescu, T.I.; Labrecque, M. Long-term biomass productivity of willow bioenergy plantations maintained in southern Quebec, Canada. *Biomass Bioenergy* **2013**, *56*, 361–369. [CrossRef]

19. Amichev, B.Y.; Hangs, R.D.; Konecsni, S.M.; Stadnyk, C.N.; Volk, T.A.; Bélanger, N.; Vujanovic, V.; Schoenau, J.J.; Moukoumi, J.; Van Rees, K.C.J. Willow Short-Rotation Production Systems in Canada and Northern United States: A Review. *Soil Sci. Soc. Am. J.* **2014**, *78*, S168–S181. [CrossRef]

20. Lafleur, B.; Lalonde, O.; Labrecque, M. First-Rotation Performance of Five Short-Rotation Willow Cultivars on Different Soil Types and Along a Large Climate Gradient. *BioEnergy Res.* **2017**, *10*, 158–166. [CrossRef]

21. Grime, J.P. *Plant Strategies and Vegetation Processes*; John Wiley & Sons, Ltd.: Chichester, UK; New York, NY, USA; Brisbane, Australia; Toronto, ON, Cadana, 1979.

22. Fontana, M.; Labrecque, M.; Messier, C.; Courchesne, F.; Bélanger, N. Quantifying the effects of soil and climate on aboveground biomass production of *Salix miyabeana* SX67 in Quebec. *New For.* **2017**, *48*, 817–835. [CrossRef]

23. Fontana, M.; Labrecque, M.; Messier, C.; Bélanger, N. Permanent site characteristics exert a larger influence than atmospheric conditions on leaf mass, foliar nutrients and ultimately aboveground biomass productivity of *Salix miyabeana* 'SX67'. *For. Ecol. Manag* **2018**, in press. [CrossRef]

24. Willebrand, E.; Ledin, S.; Verwijst, T. Willow coppice systems in short rotation forestry; effects of plant spacing, rotation length and clonal disposition on biomass production. *Biomass Bioenergy* **1993**, *4*, 323–331. [CrossRef]

25. Breiman, L.; Friedman, J.H.; Olshen, R.A.; Stone, C.J. *Classification and Regression Trees*; Chapman and Hall/CRC: Boca Raton, FL, USA, 1998.

26. Labrecque, M.; Teodorescu, T.I. Field performances and biomass production of 12 willow and poplar clones in short-rotation coppice in southern Quebec (Canada). *Biomass Bioenergy* **2005**, *29*, 1–9. [CrossRef]

27. Guidi, W.; Pitre, F.E.; Labrecque, M. Short-Rotation Coppice of Willows for the Production of Biomass in Eastern Canada. In *Biomass Now—Sustainable Growth and Use*; Matovic, M.D., Ed.; InTech: London, UK, 2013; Chapter 17; pp. 421–448.

28. Larsen, S.; Jørgensen, U.; Lærke, P.E. Willow Yield Is Highly Dependent on Clone and Site. *BioEnergy Res.* **2014**, *7*, 1280–1292. [CrossRef]

29. Mosseler, A.; Major, J.E.; Labrecque, M. Genetic by environment interactions of two North American Salix species assessed for coppice yield and components of growth on three sites of varying quality. *Trees* **2014**, *28*, 1401–1411. [CrossRef]

30. Fabio, E.S.; Volk, T.A.; Miller, R.O.; Serapiglia, M.J.; Gauch, H.G.; Van Rees, K.C.J.; Hangs, R.D.; Amichev, B.Y.; Kuzovkina, Y.A.; Labrecque, M.; et al. Genotype × environment interaction analysis of North American shrub willow yield trials confirms superior performance of triploid hybrids. *GCB Bioenergy* **2016**, *9*, 445–459. [CrossRef]

31. Aylott, M.J.; Casella, E.; Tubby, I.; Street, N.R.; Smith, P.; Taylor, G. Yield and spatial supply of bioenergy poplar and willow short, rotation coppice in the UK. *New Phytol.* **2008**, *178*, 358–370. [CrossRef] [PubMed]

32. Wilkinson, J.M.; Evans, E.J.; Bilsborrow, P.E.; Wright, C.; Hewison, W.O.; Pilbeam, D.J. Yield of willow cultivars at different planting densities in a commercial short rotation coppice in the north of England. *Biomass Bioenergy* **2007**, *31*, 469–474. [CrossRef]

33. Grenier, V.; Pitre, F.E.; Guidi Nissim, W.; Labrecque, M. Genotypic differences explain most of the response of willow cultivars to petroleum-contaminated soil. *Trees* **2015**, *29*, 871–881. [CrossRef]

34. Mosseler, A.; Major, J.E. Coppice growth responses of two North American willows in acidic clay soils on coal mine overburden. *Can. J. Plant Sci.* **2014**, *94*, 1269–1279. [CrossRef]

35. Stolarski, M.; Szczukowski, S.; Tworkowski, J.; Klasa, A. Productivity of seven clones of willow coppice in annual and quadrennial cutting cycles. *Biomass Bioenergy* **2008**, *32*, 1227–1234. [CrossRef]

36. Thevathasan, N.; Coleman, B.; Zabek, L.; Ward, T.; Gordon, A. Agroforestry in Canada and its Role in Farming Systems. In *Temperate Agroforestery Systems*, 2nd ed.; Gordon, A., Newman, S.M., Coleman, B., Eds.; CABI: London, UK, 2018; Chapter 2; pp. 7–49.

37. Kopp, R.F.; Abrahamson, L.P.; White, E.H.; Burns, K.F.; Nowak, C.A. Cutting cycle and spacing effects on biomass production by a willow clone in New York. *Biomass Bioenergy* **1997**, *12*, 313–319. [CrossRef]

38. Coyle, D.R.; Aubrey, D.P.; Coleman, M.D. Growth responses of narrow or broad site adapted tree species to a range of resource availability treatments after a full harvest rotation. *For. Ecol. Manag.* **2016**, *362*, 107–119. [CrossRef]

39. Hénault-Ethier, L.; Gomes, M.P.; Lucotte, M.; Smedbol, É.; Maccario, S.; Lepage, L.; Juneau, P.; Labrecque, M. High yields of riparian buffer strips planted with *Salix miyabena* 'SX64' along field crops in Québec, Canada. *Biomass Bioenergy* **2017**, *105*, 219–229. [CrossRef]

40. Adnan, A.; Mavinic, D.S.; Koch, F.A. Pilot-scale study of phosphorus recovery through struvite crystallization-examining to process feasibility. *J. Environ. Eng. Sci.* **2003**, *2*, 315–324. [CrossRef]

![forests logo] *forests*

MDPI

Article

Planting Density and Site Effects on Stem Dimensions, Stand Productivity, Biomass Partitioning, Carbon Stocks and Soil Nutrient Supply in Hybrid Poplar Plantations

Benoit Truax [1,*], Julien Fortier [1], Daniel Gagnon [1,2] and France Lambert [1]

[1]   Fiducie de Recherche sur la Forêt des Cantons-de-l'Est/Eastern Townships Forest Research Trust,
      1 rue Principale, Saint-Benoît-du-Lac, QC J0B 2M0, Canada; fortier.ju@gmail.com (J.F.);
      daniel.gagnon@uregina.ca (D.G.); france.lambert@frfce.qc.ca (F.L.)
[2]   Department of Biology, University of Regina, 3737 Wascana Parkway, Regina, SK S4S 0A2, Canada
[*]   Correspondence: btruax@frfce.qc.ca; Tel.: +1-819-821-8377

Received: 17 April 2018; Accepted: 22 May 2018; Published: 25 May 2018

Abstract: In this study, planting density and site effects on hybrid poplar productivity and stem dimensions were evaluated on the mid-term and longer term (8 and 14 years) in southern Québec (Canada). We also evaluated the effects of planting density and site on biomass accumulation and carbon stocks in different plantation compartments, on biomass partitioning at the stand-level, on soil carbon stocks and on soil nutrient supply rate after 14 years. The experimental design consisted of three replicate poplar stands located along a site fertility gradient. Each stand contained six planting densities (ranging from 494 to 1975 trees/ha) and a single genotype (*Populus canadensis* × *P. maximowiczii* hybrid). Planting density had a large effect on stem dimensions, but a minor effect on stand volume, aboveground woody biomass production, and aboveground biomass carbon stocks. Site selection and tree survival were more important factors affecting these variables. At all sites, and independent of planting density, mean annual volume increments were also higher after 14 vs. 8 years. On fertile sites, strong correlations between area per tree at planting and biomass partitioning, carbon allocation belowground, soil nutrient supply rate and soil carbon stocks were observed. Aboveground, higher competition for light with increasing planting density resulted in an increase in the stem to branch ratio. Belowground, higher competition for soil resources with increasing planting density reduced soil macronutrient availability (except for potassium), which likely stimulated carbon allocation belowground and carbon accumulation in the soil. Over the longer-term, higher density plantations of poplars could provide greater benefits in terms of carbon storage belowground (soil and roots).

Keywords: spacing trial; stocking level; carbon sequestration; allocation; ecosystem services; abandoned farmland; *Populus canadensis* × *Populus maximowiczii*; southern Québec (Eastern Canada); short-rotation woody crops; intraspecific competition

1. Introduction

In temperate ecosystems, fast-growing poplars and their hybrids (*Populus* × spp.) are among the most widely used species for afforestation and agroforestry [1–3]. Poplars are planted for the production of bioenergy, pulp and solid wood products, but also for the provision of other ecosystem services (atmospheric carbon dioxide sequestration, refuge for forest biodiversity, nurse crop for valuable hardwoods, fodder for livestock, soil stabilization, wind protection, soil phytoremediation, flood control) [1,2,4–6]. Thus, poplar afforestation has the potential to address both local and global environmental problems, while providing an opportunity to intensify biomass or wood production.

To reduce competition with food crops, abandoned farmland are increasingly used for poplar afforestation [7,8]. However, marginal agricultural land are often characterized by sub-optimal growth conditions [9], and appropriate management decisions and silvicultural practices are needed to maximize stand productivity. These include intensive site preparation, vegetation management, irrigation, fertilization and the selection of genotypes (clones) adapted to local environmental conditions [10]. Manipulation of tree spacing or planting density (i.e., the number of trees planted per hectare) can increase fast-growing plantation productivity by ensuring rapid and complete site occupancy [11,12]. Maximum site productivity, which generally occurs when mean annual biomass increment peaks in the rotation, is a function of the stocking level [12]. Hence, the selection of planting density has important economic consequences because it influences not only the rotation length, but also plantation establishment costs, weed control strategies, and the dimensions of stems at harvest [12]. Also, ecosystem services such as carbon (C) sequestration in tree biomass is directly linked to stand biomass productivity and may thus be affected by planting density.

Studies involving fast-growing species (*Populus*, *Eucalyptus*, *Pinus*) have shown strong positive effects of wider tree spacing on individual stem diameter, volume and quality [13–17]. For example, DeBell et al. [16] observed that diameter at breast height (DBH) of two hybrid poplar clones ranged 3.2–4.5 cm in the 0.5 × 0.5 m spacing, 6.0–7.7 cm in the 1 × 1 m spacing and 10.8–13.5 m in the 2 × 2 m spacing after 7 years. However, conflicting evidence of linkage between planting density and biomass productivity have been reported in both short-rotation coppices and more widely spaced poplar plantations. In India, 9 year-old *P. deltoides* plantations showed major increase in total biomass production (from 72 to 250 t/ha), with increasing planting density (from 278 to 2500 trees/ha) [18]. Similar observations were reported in the Eastern Canadian boreal zone after 6 growing seasons (3.2 t/ha at 400 trees/ha, 4.9 t/ha at 1111 trees/ha and 29.6 t/ha at 10,000 trees/ha) [19]. In China, C stocks in the total biomass of 10 year-old hybrid poplars increased significantly (from 58 to 72 t C/ha) with increasing planting density (from 500 to 1111 trees/ha) [20]. In England, biomass yields were generally higher in the 10,000 stems/ha plantations compared to 2500 stems/ha plantations, but the significance of the planting density effect varied between sites, clones and cutting cycles [21]. Spacing effect on poplar biomass yield was also found to vary with plantation design (inter-planted design vs. short-rotation coppice alone) [22].

On the other hand, several studies suggest that planting density has only minor and/or temporary effects on the yield of planted poplars. Although biomass production was found to increase exponentially with planting density after the first growing season, planting density (ranging 6666–33,333 cuttings/ha) was no longer a significant factor affecting productivity after 3 years across 12 sites in Spain [23]. In the Pacific Northwest region of the United States, little effect of planting density (2500, 10,000 and 40,000 cuttings/ha) was reported on the biomass productivity after 7 years, despite a trend towards higher productivity in narrower spacings earlier in the rotation [16]. Longer-term studies from southern British Columbia (Canada) and northern Wisconsin (United States) concluded that spacing has only a negligible effect on the productivity of planted poplars [15,24]. In the aforementioned studies, factors such as clone selection, site quality, irrigation, diseases, and weed control had greater effect on yield than stocking level.

Spacing trials also offer an opportunity to evaluate how varying levels of intraspecific competition affect biomass partitioning, C allocation to plantation compartments, soil nutrient status and soil C stocks. In the aboveground compartment, high competition for light in pioneer species tends to enhance the self-pruning process, and trees with proportionally smaller crowns are generally observed in highly stocked stands [25]. Poplars also tend to develop large and deep crowns under high light availability, while small and shallow crowns are typical of crowded stands [19,24,26]. Thus, the ratio of biomass allocated to stems vs. branches is expected to decrease with increasing spacing in planted poplars [19,24,26]. Over the years, root competition (i.e., the reduction in the availability of a soil resource to roots that is caused by other roots [27]) between neighboring trees may also be affected by planting density. More intense root competition in narrow spacings may accelerate soil nutrient depletion, especially in afforested plantations of fast-growing-trees with high nutrient requirements [28,29]. In response to increase competition for soil resources, poplars may enhance C allocation belowground to stimulate root production, root exudation and symbiosis with mycorrhizae [30–33]. Such a change in C allocation may enrich the soil C pool as C inputs derived from roots have been shown to be the major driver of soil carbon sequestration on marginal land afforested with poplars [34]. Over the long-term, increasing planting density in poplar plantations is expected to cause macronutrient depletion in the soil, a decrease in the shoot to root ratio at the stand-level, and an increase in soil C stocks [28–31,33,34]. However, past afforestation studies with fast-growing *Pinus* and *Eucalyptus* species have found little evidence of a planting density effect on soil C [14,35], despite the fact that higher planting density tends to increase total belowground C allocation on the short-term [31].

Additionally, by affecting the time to canopy closure, planting density also has an impact on plantation microclimate. In planted poplars, rapid canopy closure in narrow spacings can lead to a faster decrease of soil and air temperatures during the growing season, which can reduce nitrification and nitrogen mineralization rates [36]. Similarly, narrow spacings may provide favorable conditions for organic C accumulation, in both litter and mineral soil, as the decomposition of unprotected soil organic matter and plant litter is positively affected by temperature [37,38].

This study took place along a site fertility gradient in southern Québec (Eastern Canada). Three replicate stands of hybrid poplars containing six different planting density treatments (ranging 494–1975 trees/ha) and a single genotype (a *P. canadensis* × *P. maximowiczii* hybrid) were sampled. We evaluated to which extent planting density and site affected hybrid poplar productivity and stem dimensions on the mid-term and the longer term (8 vs. 14 years). We also evaluated, after 14 years, the effects of planting density and site on biomass accumulation and C stocks in different plantation compartments, on biomass partitioning at the stand-level, on soil C stocks and on soil nutrient supply rate.

First, we hypothesized that individual stem dimensions (diameter, volume) would be strongly and positively linked to spacing on the mid-term and the longer term. Second, we hypothesized that highest stand productivity would be observed in higher tree density plots in the mid-term only, as planting density effects on productivity tend to decrease as rotation length increases [12]. We also hypothesized that higher stem to branch ratio, but lower shoot to root ratio, would be observed at higher planting densities because intraspecific competition for light and soil resources is more intense. Finally, on the longer term, we expected greater soil C stocks at higher planting densities, but little effect of planting density on total biomass C stocks.

2. Materials and Methods

2.1. Study Sites and Experimental Design

This study was conducted in the Estrie region of southern Québec (Eastern Canada). The three plantation sites (Brompton, Mégantic, and Ogden) were located within a radius of 80 km along a regional elevation and soil fertility gradient. Site elevation above sea level, mean annual temperature (MAT), and mean annual precipitation (MAP) data, obtained from 30 year climatic averages (1981–2010) [39], are provided in Table 1. A continental subhumid moderate climate characterizes the Brompton and Ogden sites, whereas a continental subpolar-subhumid climate characterizes the Mégantic site [40]. Thick glacial till deposits of at least 2 m of depth and gentle slopes (<5%) also characterized the plantation sites [40].

Table 1. Site and soil characteristics of the studied hybrid poplar plantations. Soil variables were measured during the 14th growing season (2013).

Site	Elev. [1] (m)	MAT [1] (°C)	MAP [1] (mm)	Soil Nutrient Supply Rate (μg/10 cm^2/42 days) [2]							pH	Clay (%)	Silt (%)	Sand (%)
				NO_3	NH_4	P	K	Ca	Mg	S				
Brompton	165	5.6	1146	111.8	4.11	5.16	25	1810	246	54.2	5.27	16.2	58.5	25.3
Mégantic	470	4.2	1048	74.1	2.34	6.73	46	1531	287	27.6	5.00	13.1	41.4	45.5
Ogden	265	5.3	1264	2.8	3.83	2.74	138	808	198	37.0	5.17	14.6	43.2	42.2
SE [1]	-	-	-	41.6	0.52	0.64	21	156	33	8.3	0.05	1.6	1.3	1.2
p-value	-	-	-	<0.01	0.08	<0.01	<0.001	<0.01	0.21	0.12	<0.01	0.41	<0.001	<0.001

[1] Abbreviation used in Table: Elev. (Elevation above sea-level), MAT (Mean annual temperature), MAP (Mean annual precipitation), SE (Standard error of the mean). [2] A log transformation and a reciprocal transformation were respectively used in the ANOVA for soil NO_3 and K supply rate.

At each site, the hybrid poplar plantation was established on an abandoned field dominated by herbaceous vegetation. Site preparation included ploughing in the fall of 1999 and disking the following spring. In the spring of 2000, bare-root planting stock (±2 m-long) were planted manually at 30–40 cm depth. Planting stock (1–0) was provided by the Berthierville nursery of the Ministère des Forêts, de la Faune et des Parcs (MFFP) of Québec. For vegetation management, a glyphosate-based herbicide was applied over the entire plantation area in June 2000, and between plantation rows only in June 2001.

The experimental design contained six planting density treatments (494, 741, 988, 1111, 1481 and 1975 trees/ha) per site and three sites for a total of 18 plots. Twenty trees were planted in each plot, for a total of 360 trees. Details of spacing between trees, area per tree and plot size are provided in Table 2 for each planting density. All plots were separated by two buffer rows, with each buffer row having the same planting density as its adjacent plot. The DN×M-915508 clone was used in this study, which is a female hybrid between *Populus canadensis* (DN) and *P. maximowiczii* (M). It was developed in Québec and is recommended for commercial production in the study area [41] because of its high productivity across a wide range of environmental conditions [9,42].

Table 2. Initial planting density, spacing between trees, area occupied per tree and plot area.

Planting Density (trees/ha)	Spacing (m)	Area per Tree (m^2)	Plot Area (m^2/20 trees)
494	4.5×4.5	20.25	405
741	4.5×3	13.5	270
988	4.5×2.25	10.125	203
1111	3×3	9	180
1481	3×2.25	6.75	135
1975	2.25×2.25	5.0625	101

2.2. Soil Nutrient Supply

Soil macronutrient supply rates were determined using Plant Root Simulator (PRS^{TM}-Probes) technology from Western Ag Innovations Inc., Saskatoon, SK, Canada. The PRS-probes consist of an ion exchange membrane encapsulated in a thin plastic probe, which is inserted into the ground with little disturbance of soil structure. Nutrient supplies observed with this method are strongly correlated with nutrients concentrations or stocks obtained with conventional soil extraction methods over a wide range of soil types [43], and in poplar plantations of the study area [42].

On 6 August 2013 (14th growing season), four pairs of probes (an anion and a cation probe in each pair) were buried in the A horizon of each plot for a 42-day period. Probes were always buried in a plot area where there was full stocking. Each pair of probes was placed at the center point between four trees in the plot, in the competition zone between neighboring trees (Figure A1). After probes were removed from the soil, they were washed with distilled water, and returned to Western Ag Labs for analysis (NO$_3$, NH$_4$, P, K, Ca, Mg, S). PRS-probe samples where eluted in a 0.5 M HCl solution and determination of NO$_3$ and NH$_4$ concentrations in the eluate was made colorimetrically using an automated flow injection analysis system [44]. The concentration of other nutrients where determined using inductively-coupled plasma spectrometry [44]. Composite samples were made in each plot by combining the four pairs of probes.

2.3. Mineral Soil Characteristics and Carbon Stocks

In July 2013 (14th growing season), soil samples (0–20 depth) were obtained from a composite sample consisting of four soil cores (diameter of the corer = 5.2 cm) per plot. Prior to coring the mineral soil (A horizon), the O horizon (litter layer) was carefully removed. The soil cores were always extracted from a plot area where there was full stocking (see Figure A1 for the sampling design). Composite soil samples were air dried and sieved (mesh size 2 mm) to remove coarse fragments (stones, roots) and macroinvertebrates. Following sieving, air-dry mass of each soil sample was recorded and a subsample was taken to determine an oven-dry mass (105 °C) to air-dry mass ratio, and also to calculate dry mass of soil samples. Soil bulk density was calculated by dividing the dry mass of the fine earth fraction by the volume of soil cores [45].

Soil C concentration of each sample was determined by high-temperature combustion (960 °C) of the samples, followed by thermo-conductometric detection, on a Vario Macro analyzer (Elementar Analysensysteme, Hanau, Germany). These analyses were done by the Centre d'étude de la forêt (CEF) lab at the University of Sherbrooke (Québec, Canada). Soil C stocks were calculated by multiplying C concentration by the bulk density of soil. In each plot, a subsample from the soil cores was also used to determine soil texture and pH. The analyses were done by the Agridirect Inc. soil analysis lab in Longueuil (Québec, Canada). For particle size analyses, the Bouyoucos [46] method was used. The determination of soil pH was made using a 2:1 ratio of water to soil. Soil pH and texture are provided in Table 1 for each site. At each site, soil textural class was the same across the six spacing treatments (silty loam for the Brompton site, and loam for Mégantic and Ogden sites).

2.4. Fine Root, Herbaceous Vegetation, Leaf Litter and Fine Woody Debris Biomass

Fine root biomass samples (root diameter <2 mm) were collected following sieving of the soil cores (see previous subsection). At Brompton and Mégantic sites, where understory vegetation was almost absent, fine root samples consisted of poplar roots only. At the Ogden site, there was a well-developed herbaceous cover in the understory; poplar and herbaceous fine root biomass were not separated in samples. At all sites, fine root biomass samples included both live and dead roots. Fine root samples were washed and dried at 65 °C to determine dry weight and C concentrations.

Herbaceous vegetation biomass was determined by collecting five samples (50 × 50 cm/sample) in each plot during August 2013 (14th growing season). In each plot, the samples were collected at random locations, where there was full stocking. The five samples from each plot were combined and dried at 65 °C to determine dry weight. An herbaceous vegetation subsample from each plot was used for C concentration determination. Woody vegetation in plantation understory was not sampled as it was only a minor biomass component at each site.

Fine woody debris biomass (large end diameter ranging 1–10 cm) [47] was determined (October 2013, 14th growing season) by collecting a single sample in each plot on an area equivalent to the area occupied by a single tree (see Table 2). In each plot the sample was collected in a randomly selected rectangular or square area delineated by four neighboring trees. Each sample was dried at 65 °C to determine dry weight and a subsample from each plot was used for C concentration determination.

Leaf litter and very fine woody debris (large end diameter <1 cm) biomass was determined by collecting three samples (50 × 50 cm/sample) in each plot in late October 2013 (14th growing season), just after leaf fall. In each plot, the samples were collected directly on the ground at random locations, where there was full stocking. The three samples from each plot were combined and dried at 65 °C to determine dry weight. A litter subsample from each plot was used for C concentration determination.

2.5. Hybrid Poplar Biomass and Stem Volume

In each plot, stem volume per tree and stand volume were estimated after 8 and 14 years, while aboveground woody biomass and coarse root biomass were estimated after 14 years only. Those estimates were obtained using allometric relationships between diameter at breast height (DBH) as the predictor variable of stem volume (outside bark) or compartment biomass (stem with bark, branches and coarse roots) at the tree-level. DBH data (mean of two diameter measurements taken perpendicularly at 1.3 m above ground-level) were collected in late October 2007 and 2013. The DBH of all trees in the experimental design was measured.

For stem volume estimations after 8 years, we used a clone-specific allometric relationship that had been previously developed for clone DN×M-915508 in 8 year-old hybrid poplar plantations of the study region [7]. For estimations of stem volume, stem biomass, branch biomass and coarse root biomass after 14 years, we used site-specific allometric relationships that were developed in the studied experimental design with 14 year-old trees [48]. Site-specific relationships were chosen over generic models because allometry of clone DN×M-915508 was found to be plastic across the studied plantation environments, especially for coarse root biomass, branch biomass and stem volume [48]. During destructive harvest procedures, subsamples from stem, branch and coarse root biomass were collected for C concentration determination in 14-year-old poplars. Given that destructive harvest procedures were only done in the 494, 1111 and 1975 trees/ha density treatments at each site, only three samples of stem, branch and coarse root biomass were collected at each site. For each compartment, each of these subsamples consisted in a composite sample collected from two average-sized trees within the plot (see Fortier et al. [48] for a complete description of stem subsamples collection). At each site, and for a given compartment (stem, branch or coarse root biomass), we used C concentration data obtained in the 494, 1111 and 1975 trees/ha treatment as an estimate of C concentration of 741, 988 and 1481 trees/ha treatments respectively.

2.6. Carbon Concentrations and Stocks in Biomass and in the O Horizon

Biomass and litter subsamples were ground in a mill (Pulverisette 15, Fritsch) to a particle size of <0.5 mm to insure adequate sample homogeneity. For biomass and litter C concentration determination, ground, dried aliquots of samples (approximately 100 mg) were encapsulated in tin prior to analysis. Total C was determined by high-temperature combustion (960 °C), followed by thermo-conductometric detection, on a Vario Macro analyzer (Elementar Analysensysteme, Hanau, Germany). These analyses were done by the CEF lab at the University of Sherbrooke. In each plot, C stocks in the different compartments were obtained by multiplying C concentrations determined in subsamples by the total mass of the compartment. Carbon stocks from litter and fine woody debris were summed to obtain a single C stock value for the O horizon soil layer. All C stocks and biomass data were scaled-up to one-hectare area for comparison purposes with other studies.

2.7. Statistical Analysis

Because each planting density treatment is not replicated within sites, site was used as a blocking factor [49], as in other spacing trials [50]. Site and Planting density effects on studied variables were analyzed using ANOVA in a fixed factorial design [49]. Degrees of freedom were the following: Total, 17; Site, 2; Planting density, 5; Error, 10. All of the ANOVAs were run with the complete set of data (3 Sites × 6 Planting densities = 18 experimental plots).

Following ANOVAs, normality of residual was verified using the Shapiro-Wilk W-test. Variables showing non-normal residual distribution were transformed and ANOVA was repeated on transformed variables. Being proportions, survival data were logit transformed [51], while a log transformation and a reciprocal transformation were respectively used for soil NO_3 and K supply rates. Site and Planting density effects were declared statistically significant for three levels of significance ($p < 0.05$, $p < 0.01$ and $p < 0.001$). *A priori* contrasts were further used to test specific hypotheses between particular sets of means [49,52,53]. Two contrasts were tested on key variables: (1) 494 trees/ha vs. 1975 trees/ha and (2) 741 trees/ha vs. 1481 trees/ha, which corresponds to a four-fold and a two-fold variation in planting density, respectively.

Because the Site (i.e., Block) effect was strong and significant for most of the studied variables, we also tested site-specific correlations between planting density and key variables related to stem dimensions, stand and soil characteristics. The area per tree (m^2/tree) at planting was preferred to planting density in pairwise correlation analysis as it yields stronger correlations, based on the Pearson correlation coefficient (r). Based on results of the pairwise correlations, we then developed regression models (linear and non-linear) between area per tree, as predictor variable, and selected response variables. All statistical analyses were done using JMP 11 from SAS Institute (Cary, NC, USA).

3. Results

3.1. Site and Planting Density Effects on Individual Stem Dimensions

The ANOVA revealed strong and significant Site and Planting density effects ($p < 0.01$ or $p < 0.001$) on all variables related to individual stem DBH and volume (Table 3). Larger trees were produced at the fertile sites of Brompton and Mégantic, while lower planting density treatments produced trees with greater DBH and stem volume. After 8 years, mean tree DBH and stem volume respectively ranged 10.01–15.54 cm and 41.7–132.7 dm^3 across the planting density treatments, while after 14 years DBH and stem volume respectively ranged 14.53–25.18 cm and 128.4–434.9 dm^3 across planting density treatments. Planting density treatments also significantly affected the gain in DBH or in stem volume between the end of the 8th and 14th growing seasons, with greater gains being observed at lower planting density (Table 3). Contrasts analysis further suggests that quadrupling or doubling planting density (from 494 to 1975 or from 741 to 1481 trees/ha) resulted in a significant increase of stem dimensions after 8 and 14 years (Table 4).

Table 3. Site and Planting density effects on survival rate, DBH, stem volume, stand volume and volume yield after 8 and 14 years in hybrid poplar plantations.

Site/Plant. Density	Surv. [1] (%)	DBH (cm) 8 years	DBH (cm) 14 years	DBH (cm) Δ 8–14 years	Stem Volume (dm³/tree) 8 years	Stem Volume (dm³/tree) 14 years	Stem Volume (dm³/tree) Δ 8–14 years	Stand Volume (m³/ha) 8 years	Stand Volume (m³/ha) 14 years	Stand Volume (m³/ha) Δ 8–14 years	Volume Yield (m³/ha/year) 8 years	Volume Yield (m³/ha/year) 14 years
Brompton	100.0	15.43	21.03	+5.60	119.7	307.3	+187.5	115.9	292.7	+176.8	14.48	20.91
Mégantic	86.7	13.24	20.48	+7.24	86.5	285.9	+199.4	69.2	230.3	+161.1	8.65	16.45
Ogden	89.2	9.06	15.01	+5.95	30.8	124.4	+93.6	28.6	108.8	+80.3	3.57	7.77
SE	4.0	0.59	0.69	0.30	10.7	27.3	17.7	6.5	14.2	9.2	0.82	1.02
p-value	<0.05	<0.001	<0.001	<0.01	<0.001	<0.01	<0.01	<0.001	<0.001	<0.001	<0.001	<0.001
494 trees/ha	86.7	15.54	25.18	+9.64	132.7	434.9	+302.2	58.5	189.4	+130.9	7.32	13.53
741 trees/ha	95.0	14.97	22.12	+7.16	113.2	313.1	+199.9	80.9	222.2	+141.4	10.11	15.87
988 trees/ha	93.3	13.07	19.26	+6.20	81.4	238.2	+156.8	78.6	227.5	+148.8	9.83	16.25
1111 trees/ha	83.3	10.01	15.68	+5.67	50.7	160.9	+110.2	51.0	150.4	+99.4	6.37	10.74
1481 trees/ha	93.3	11.58	16.26	+4.68	54.6	159.7	+105.1	76.0	220.6	+144.7	9.50	15.76
1975 trees/ha	100.0	10.29	14.53	+4.24	41.7	128.4	+86.7	82.3	253.6	+171.3	10.29	18.11
SE	5.7	0.83	0.98	0.43	15.2	38.6	25.0	9.2	20.1	13.1	1.15	1.44
p-value	0.34	<0.01	<0.001	<0.001	<0.01	<0.01	<0.01	0.16	0.06	0.05	0.16	0.06

[1] Survival rate (Surv.) was the same after 8 and 14 years.

Table 4. Contrast probabilities (*p*-values) between pairs of planting density treatments for different plantation characteristics in hybrid poplar plantations.

Plantation Characteristics	Response Variable	494 vs. 1975 trees/ha	741 vs. 1481 trees/ha
Stem dimensions	DBH 8 years (cm)	<0.01	<0.05
	DBH 14 years (cm)	<0.001	<0.001
	Stem volume 8 years (dm³)	<0.01	<0.05
	Stem volume 14 years (dm³)	<0.001	<0.05
Stand volume	Stand volume 8 years (m³/ha)	0.10	0.72
	Stand volume 14 years (m³/ha)	<0.05	0.92
Stand biomass (14 years)	Aboveground woody (t/ha)	0.13	0.75
	Belowground (t/ha)	<0.01	0.50
Biomass partitioning (14 years)	Stem to branch ratio	<0.01	<0.05
	Shoot to root ratio	<0.001	<0.05
Biomass C stocks (14 years)	Aboveground (t C/ha)	0.15	0.68
	Belowground (t C/ha)	<0.01	0.53
	Total (t C/ha)	0.09	0.81
Mineral soil C (14 years)	C Concentration (mg C/g soil)	<0.05	<0.05
	C Stocks (t C/ha)	<0.05	<0.05

3.2. Site and Planting Density Effects on Stand Volume, Biomass and Biomass Carbon Stocks

Stand-level variables (stand volume, compartment biomass and biomass C stocks) indicated significant site effects ($p < 0.01$ or $p < 0.001$) on most variables, with the exception of coarse root biomass, total root biomass, and belowground biomass C stocks (Figure 1a, Tables 3 and 5). Overall, C stocks in the belowground and aboveground biomass respectively represented 14% and 86% of the total biomass C stocks at Brompton and Mégantic, while at Ogden those C stocks respectively represented 26% and 74% of the total biomass C stocks (Figure 1a). After 14 years, a near two-fold difference was observed in total (aboveground + belowground) biomass C stocks between sites (60.5 t C/ha at Brompton vs. 32.9 t C/ha at Ogden).

Results also show that at each site, more volume was cumulated between the end of the 8th and of the 14th growing seasons (a 6 year period), than during the first 8 years of the rotation (Table 3). Thus, higher mean annual volume increment was observed after 14 years than after 8 years for a given site (Table 3). After 14 growing seasons the volume yield observed at the higher fertility sites (Brompton and Mégantic) was more than twice the yield at Ogden (20.91 and 16.45 m^3/ha/year vs. 7.77 m^3/ha/year).

For the Planting density effect, the ANOVA showed significant ($p < 0.05$) or non-significant effects on stand-level volume, biomass and biomass C stocks in the different plantation compartments (Figure 1b, Tables 4 and 5). The Planting density effect was also nearly significant ($p = 0.06$) for the total C stocks stored in plantation biomass after 14 years (Figure 1b). The lowest stand volume, aboveground woody biomass and belowground biomass were observed in the 1111 trees/ha treatment, followed by the 494 trees/ha treatment. Biomass C stocks followed the same pattern (Figure 1b). However, in both of these treatments, survival was relatively low (83.3% at 1111 trees/ha and 86.7% at 494 trees/ha) compared to the other treatments, where survival ranged 93.3–100% (Table 3). The survival rate between the 741 and 1481 trees/ha treatments was comparable (93.3% vs. 95%, respectively), and the contrast analysis between this pair of treatments was far from statistical significance for stand volume, aboveground woody biomass, belowground biomass and for biomass C stocks (Table 4). Table A2 also shows the p-value of the Planting density effect when the ANOVA is run without the 1111 trees/ha treatment, where the survival rate was the lowest. In Table A2, all stand-level variables were far from significance, with the exception of belowground biomass and C stocks ($p = 0.05$). Moreover, for the C stocks in total and aboveground biomass, the contrast analysis revealed no significant differences between the 494 and 1975 trees/ha treatments, while a significant difference ($p < 0.01$) was observed for belowground C stocks (Table 4).

Table 5. Site and Planting density effects on main biomass components at the stand level, on the shoot to root biomass ratio and on the stem to branch biomass ratio in 14 year-old hybrid poplar plantations.

| Site/Density | Aboveground Biomass | | | | | Belowground Biomass | | | Total Biomass (t/ha) | Shoot: Root [2] | Stem: Branch |
	Stem (t/ha)	Branch (t/ha)	Woody [1] (t/ha)	Herb.[1] (t/ha)	Total (t/ha)	Coarse Roots (t/ha)	Fine Roots (t/ha)	Total (t/ha)			
Brompton	94.2	21.3	115.4	0.02	115.5	16.1	2.03	18.1	133.6	7.24	4.73
Mégantic	74.2	23.4	97.7	0.01	97.7	14.3	1.31	15.6	113.2	6.84	3.28
Ogden	41.6	11.1	52.7	0.76	53.5	14.7	3.99	18.7	72.2	3.59	3.78
SE	4.6	2.0	6.3	0.07	6.3	1.1	0.30	1.2	7.4	0.11	0.25
p-value	<0.001	<0.01	<0.001	<0.001	<0.001	0.51	<0.001	0.21	<0.001	<0.001	<0.01
494 trees/ha	60.6	22.6	83.2	0.36	83.5	12.1	2.21	14.3	97.8	6.58	2.85
741 trees/ha	73.2	22.7	96.0	0.25	96.2	15.7	2.07	17.8	114.0	6.20	3.29
988 trees/ha	74.8	20.5	95.3	0.16	95.5	15.4	2.45	17.8	113.3	5.96	3.65
1111 trees/ha	50.1	11.1	61.2	0.37	61.6	10.5	2.24	12.7	74.3	5.61	4.35
1481 trees/ha	74.9	16.8	91.8	0.28	92.0	17.0	2.54	19.5	111.6	5.58	4.50
1975 trees/ha	86.4	17.9	104.3	0.16	104.4	19.4	3.16	22.5	127.0	5.43	4.94
SE	6.5	2.9	8.9	0.09	8.9	1.6	0.43	1.7	10.4	0.16	0.35
p-value	<0.05	0.12	0.08	0.46	0.08	<0.05	0.56	<0.05	0.06	<0.01	<0.05

[1] Woody = stem + branch biomass. Herb. = Herbaceous plant biomass. [2] The shoot to root ratio excludes fine root biomass.

On the other hand, stand-level biomass partitioning between shoot and roots, but also between stem and branches was significantly affected by both plantation site and planting density (Table 5). Much higher shoot to root ratio was observed on the more fertile sites (Brompton and Mégantic), while a slight decrease in the shoot to root ratio was observed with increasing planting density (Table 5). Conversely, the stem to branch ratio was found to increase with increasing planting density, from 2.85 at 494 trees/ha, up to 4.94 at 1975 trees/ha. Both contrasts tested were also significant for the shoot to root ratio and the stem to branch ratio (Table 4).

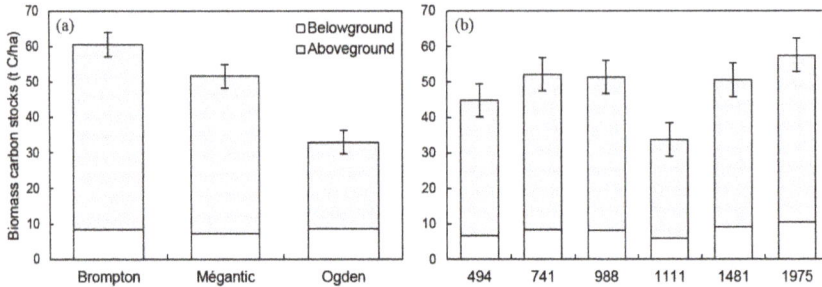

Figure 1. Site effect (**a**) and Planting density effect (**b**) on total biomass C stocks and its distribution between the belowground and aboveground compartments in 14 year-old hybrid poplar plantations. Vertical bars represent standard error of the mean for total C stocks in biomass. *p*-value for the Site effect are the followings: total biomass C ($p < 0.001$), belowground biomass C ($p = 0.21$) and aboveground biomass C ($p < 0.001$). *p*-value for the Planting density effect are the followings: total biomass C ($p = 0.06$), belowground biomass C ($p < 0.05$) and aboveground biomass C ($p = 0.08$).

3.3. Site and Planting Density Effects on Soil Characteristics

The ANOVA showed significant Site effects ($p < 0.01$) for soil NO_3, P, K and Ca supply rates and a marginally significant Planting density effect for soil Mg ($p = 0.05$) supply rate measured during the 14th growing season (Tables 1 and A1). The Ogden site showed the lowest NO_3, P and Ca supply rates, while Mg supply rate was especially high for the 494 and 741 trees/ha treatments (Tables 1 and A1).

For soil C concentration in the fine earth fraction (0–20 cm layer), the Site and Planting density effects were both significant ($p < 0.05$) (Table 6). While the Site effect was highly significant ($p < 0.001$) for soil C stocks (fine earth fraction, O horizon and total soil C), the Planting density effect was nearly significant for C stocks in the fine earth fraction ($p = 0.10$) and for total soil C stocks (mineral soil + O horizon) ($p = 0.07$). There was a trend towards higher soil C concentration, mineral soil C stocks and total soil C stocks with increasing planting density (Table 6). Contrasts analysis showed that C concentrations and stocks in the fine earth fraction (0–20 cm layer) where lower ($p < 0.05$) in the 494 trees/ha vs. the 1975 trees/ha treatment, and in the 741 trees/ha vs. the 1481 trees/ha treatments (Table 4).

3.4. Area per Tree at Planting as a Predictor of Tree-Level, Stand-Level and Soil Characteristics

Because the Site effect was strong on most response variables in this study, we also examined site-specific relationships between area per tree at planting and variables related to individual stem dimensions, stand characteristics and soil characteristics. Several site-specific correlations were strong and significant, mostly for stem dimensions, biomass partitioning and soil characteristics (Table 7).

Table 6. Site and Planting density effects on soil stoniness, bulk density, carbon stocks and concentration in 14 year-old hybrid poplar plantations (0–20 cm layer).

Site/Planting Density	Stoniness (% mass basis)	Bulk Density (g/cm³)	Fine Earth Fraction (Mineral Soil) C Concentration (mg C/g soil)	C Stocks (t C/ha)	O Horizon C Stocks (t C/ha)	Total Soil C Stocks (t C/ha)
Brompton	0.6	1.02	35.1	71.1	2.16	73.2
Mégantic	11.4	0.97	34.2	65.3	2.89	68.1
Ogden	18.8	0.82	27.3	44.7	1.56	46.2
SE	1.6	0.04	1.7	2.4	0.17	2.2
p-value	<0.001	<0.05	<0.05	<0.001	<0.001	<0.001
494 trees/ha	12.9	1.02	25.7	52.0	1.80	53.8
741 trees/ha	7.0	1.01	27.5	56.1	2.28	58.4
988 trees/ha	11.2	0.91	34.0	61.8	2.47	64.3
1111 trees/ha	10.0	0.92	33.8	62.1	2.01	64.2
1481 trees/ha	8.6	0.92	36.3	66.7	2.22	68.9
1975 trees/ha	12.0	0.86	36.0	63.2	2.44	65.6
SE	2.3	0.06	2.4	3.3	0.24	3.2
p-value	0.48	0.34	<0.05	0.10	0.39	0.07

Table 7. Site-specific pairwise correlations (r) between area per tree at planting (m²/tree) and selected variables related to individual stem dimensions, stand-level characteristics and soil characteristics. Correlations with *p* < 0.10 are indicated in bold.

Characteristics	Variables	Brompton r	*p*-value	Mégantic r	*p*-value	Ogden r	*p*-value
Stem dimensions	DBH 8 years (cm)	**0.94**	**<0.01**	**0.88**	**<0.05**	0.51	0.30
	Volume 8 years (dm³/tree)	**0.96**	**<0.01**	**0.94**	**<0.01**	0.58	0.22
	DBH 14 years (cm)	**0.96**	**<0.01**	**0.97**	**<0.01**	**0.81**	**0.05**
	Volume 14 years (dm³/tree)	**0.98**	**<0.001**	**0.98**	**<0.001**	**0.82**	**<0.05**
	Δ DBH 8–14 years (cm)	**0.94**	**<0.01**	**0.97**	**<0.01**	**0.96**	**<0.01**
	Δ Volume 8–14 years (dm³/tree)	**0.96**	**<0.01**	**0.98**	**<0.001**	**0.88**	**<0.05**
Stand volume	Volume 8 years (m³/ha)	**−0.81**	**0.05**	0.13	0.81	−0.54	0.27
	Volume 14 years (m³/ha)	−0.53	0.28	−0.04	0.93	−0.50	0.32
	Δ Volume 8–14years (m³/ha)	−0.28	0.60	−0.15	0.78	−0.47	0.35
Stand biomass (14 years)	Abovegr. woody biomass (t/ha)	−0.23	0.66	0.05	0.92	−0.46	0.36
	Coarse root biomass (t/ha)	**−0.80**	**0.06**	−0.16	0.75	−0.61	0.20
	Fine root biomass (t/ha)	**−0.94**	**<0.01**	0.13	0.81	−0.27	0.61
	Herbaceous biomass (t/ha)	−0.46	0.36	−0.12	0.82	0.53	0.28
Stand-level partitioning (14 years)	Stem to branch biomass ratio	**−0.92**	**<0.05**	**−0.98**	**<0.001**	**−0.75**	**0.09**
	Shoot to root biomass ratio	**0.97**	**<0.01**	**0.99**	**<0.001**	**0.78**	**0.07**
Stand biomass C stocks (14 years)	Aboveground (t C/ha)	−0.20	0.70	0.08	0.87	−0.44	0.38
	Belowground (t C/ha)	**−0.87**	**<0.05**	−0.13	0.81	−0.57	0.24
	Total biomass (t C/ha)	−0.35	0.50	0.06	0.91	−0.48	0.34
Soil characteristics (14 years)	Bulk density (g/cm³)	**0.94**	**<0.01**	**0.82**	**<0.05**	−0.10	0.85
	C conc. mineral soil (mg C/g)	**−0.89**	**<0.05**	**−0.86**	**<0.05**	−0.45	0.37
	Mineral soil C stocks (t C/ha)	**−0.84**	**<0.05**	−0.79	0.06	−0.44	0.39
	O horizon C stocks (t C/ha)	−0.05	0.92	−0.55	0.25	−0.69	0.13
	Total soil C stocks (t C/ha)	**−0.86**	**<0.05**	**−0.83**	**<0.05**	−0.46	0.35
	NO₃ supply (µg/10 cm²/42 days)	**0.87**	**<0.05**	0.77	0.07	**−0.77**	**0.07**
	NH₄ supply (µg/10 cm²/42 days)	−0.32	0.54	−0.56	0.25	0.16	0.76
	P supply (µg/10 cm²/42 days)	**0.80**	**0.06**	0.54	0.26	0.00	1.00
	K supply (µg/10 cm²/42 days)	**−0.91**	**<0.05**	−0.47	0.35	−0.40	0.43
	Ca supply (µg/10 cm²/42 days)	**0.87**	**<0.05**	0.41	0.42	**0.93**	**<0.01**
	Mg supply (µg/10 cm²/42 days)	**0.89**	**<0.05**	0.42	0.40	0.41	0.42
	S supply (µg/10 cm²/42 days)	**0.87**	**<0.05**	0.12	0.82	**0.82**	**<0.05**

For individual stem dimensions, strong and significant positive correlations ($p \leq 0.05$) were observed at Brompton and Mégantic between area per tree and DBH or stem volume (8 years, 14 years and Δ 8–14 years) (Table 7). At the Ogden site, these correlations were only significant ($p \leq 0.05$) for stem DBH and volume after 14 years. Area per tree and stand productivity were only strongly correlated after 8 years at the Brompton site (r = −0.81 and p = 0.05 for stand volume). Significant site-specific relationships ($p \leq 0.05$) between area per tree and tree DBH after 14 years, stem volume per tree after 14 years, and stand volume after 8 years are presented in Figure 2.

For biomass partitioning at the stand-level, negative correlations were observed between area per tree and the stem to branch ratio at Brompton (r = −0.92, p < 0.05), Mégantic (r = −0.98, p < 0.001) and Ogden (r = −0.75, p = 0.09), while positive correlations were observed between area per tree and the shoot to root ratio at Brompton (r = 0.97, $p < 0.01$), Mégantic (r = 0.99, $p < 0.001$) and Ogden (r = 0.78, $p = 0.07$) (Table 7). At the Brompton site, significant or near significant negative correlations between area per tree at planting and fine root biomass (r = −0.94, $p < 0.01$), coarse root biomass (r = −0.80, $p = 0.06$), and belowground C stocks (r = −0.87, $p < 0.05$) were also observed (Table 7). Site-specific relationships between area per tree and the branch to stem ratio, the shoot to root ratio, fine root biomass or coarse root biomass at the stand-level are presented in Figure 3.

For soil nutrients, positive correlations between area per tree and nutrient supply rate were found at Brompton (NO_3, P, Ca, Mg and S, significant at $p = 0.06$ or less), Mégantic (NO_3, $p = 0.07$), and Ogden (Ca and S, significant at $p < 0.05$) (Table 7). However, area per tree and soil K were negatively correlated at Brompton ($p < 0.05$), while area per tree and soil NO_3 were negatively correlated at Ogden ($p = 0.07$). Site-specific relationships between area per tree and soil nutrient supply rates measured during the 14th growing season are presented in Figure 4. For soil C, negative correlations were observed between area per tree and mineral soil C concentration, mineral soil C stocks (0–20 cm layer) and total soil C stocks (mineral soil + O horizon) at Brompton ($p < 0.05$) and Mégantic ($p = 0.06$ or less). At those two sites, mineral soil bulk density (0–20 cm layer) was positively correlated ($p < 0.05$) with area per tree (Table 7). Site-specific relationships between area per tree at planting and soil C or bulk density measured during the 14th growing season are presented in Figure 5.

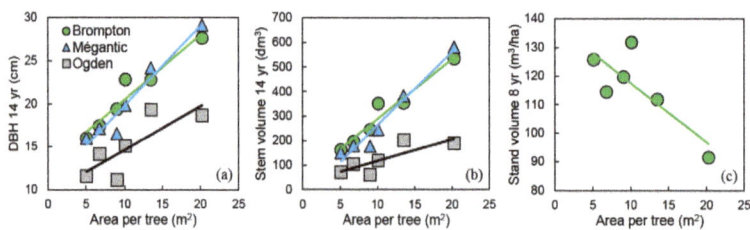

Figure 2. Site-specific linear relationships between area per tree at planting and (**a**) mean tree DBH after 14 years, (**b**) mean stem volume per tree after 14 years and (**c**) stand volume after 8 years in hybrid poplar plantations. Only relationships with $p \leq 0.05$ are presented. Fit and significance level of the relationships are the following for panel (**a**): Brompton site ($R^2 = 0.93$, $p < 0.01$), Mégantic site ($R^2 = 0.95$, $p < 0.01$), Ogden site ($R^2 = 0.65$, $p = 0.05$); for panel (**b**): Brompton site ($R^2 = 0.95$, $p < 0.001$), Mégantic site ($R^2 = 0.96$, $p < 0.001$), Ogden site ($R^2 = 0.67$, $p < 0.05$); for panel (**c**): Brompton site ($R^2 = 0.66$, $p = 0.05$).

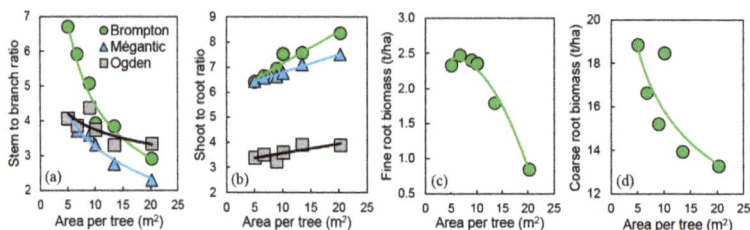

Figure 3. Site-specific relationships between the area per tree at planting and (**a**) the stem to branch biomass ratio at the stand-level, (**b**) the shoot to root biomass ratio at the stand-level, (**c**) fine root biomass and (**d**) coarse root biomass in 14 year-old hybrid plantations. Only relationships with $p < 0.1$ are presented. Model type, fit and significance level of the relationships are the following for panel (**a**): Brompton site (power model, $R^2 = 0.95$, $p < 0.001$), Mégantic site (logarithmic model, $R^2 = 0.97$, $p < 0.001$), Ogden site (power model, $R^2 = 0.56$, $p = 0.09$); for panel (**b**): Brompton site (linear model, $R^2 = 0.94$, $p < 0.01$), Mégantic site (linear model, $R^2 = 0.98$, $p < 0.001$), Ogden site (linear model, $R^2 = 0.61$, $p = 0.07$); for panel (**c**): Brompton site (polynomial model, $R^2 = 0.97$, $p < 0.01$). The shoot to root ratio exclude fine root biomass; for panel (**d**): Brompton site (power model, $R^2 = 0.68$, $p < 0.05$).

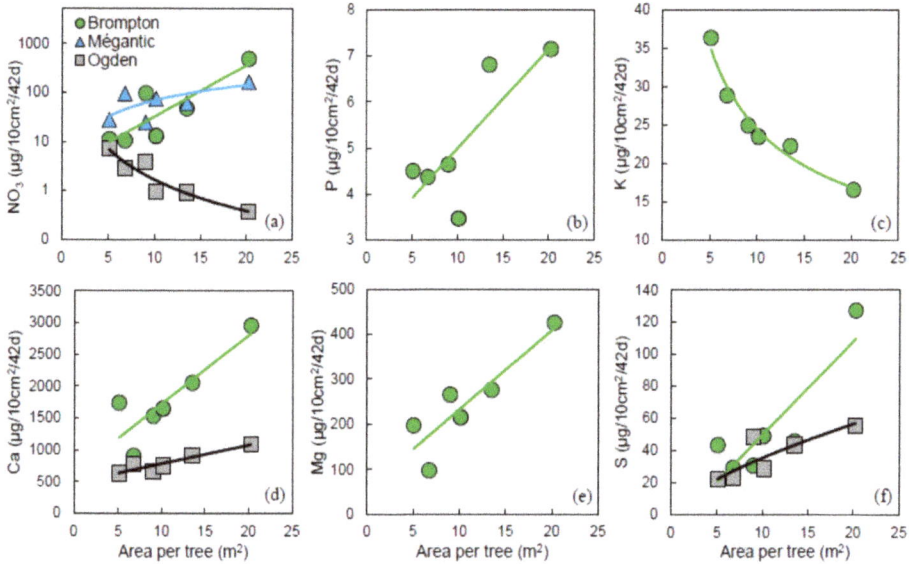

Figure 4. Site-specific relationships between the area per tree at planting and the supply rate of soil (**a**) nitrate, (**b**) phosphorus, (**c**) potassium, (**d**) calcium, (**e**) magnesium and (**f**) sulfur during the 14th growing season in hybrid poplar plantations. Only relationships with $p < 0.1$ are presented. Model type, fit and significance level of the relationships are the following for panel (**a**): Brompton site (exponential model, $R^2 = 0.74$, $p < 0.05$), Mégantic site (linear model, $R^2 = 0.60$, $p = 0.07$), Ogden site (linear model, $R^2 = 0.88$, $p < 0.01$); for panel (**b**): Brompton site (linear model: $R^2 = 0.64$, $p = 0.06$); for panel (**c**): Brompton site (power model, $R^2 = 0.97$, $p < 0.001$); for panel (**d**): Brompton site (linear model, $R^2 = 0.76$, $p < 0.05$), Ogden site (linear model, $R^2 = 0.87$, $p < 0.01$); for panel (**e**): Brompton site (linear model, $R^2 = 0.79$, $p < 0.05$); for panel (**f**): Brompton site (linear model, $R^2 = 0.76$, $p < 0.05$), Ogden site (linear model, $R^2 = 0.71$, $p < 0.05$). Nutrient supply rates were measured in hybrid poplar plantations during 42 days (6 August–17 September 2013).

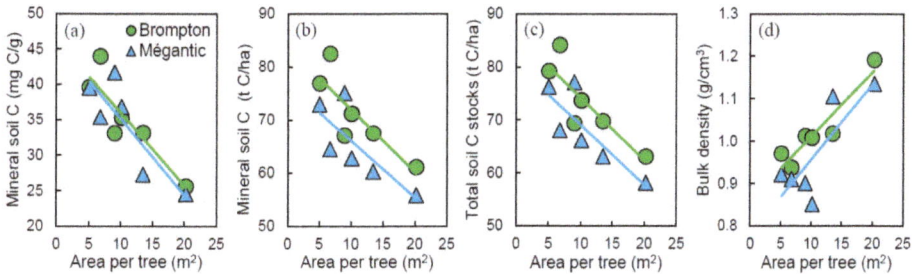

Figure 5. Site-specific linear relationships between the area per tree at planting and (**a**) mineral soil C concentration (0–20 cm layer), (**b**) mineral soil C stocks (0–20 cm layer), (**c**) total soil C stocks (mineral soil + O horizon) and (**d**) mineral soil bulk density (0–20 cm layer). Only relationships with $p < 0.1$ are presented. Fit and significance level of the relationships are the following for panel (**a**): Brompton site ($R^2 = 0.80$, $p < 0.05$), Mégantic site ($R^2 = 0.74$, $p < 0.05$); for panel (**b**): Brompton site ($R^2 = 0.71$, $p < 0.05$), Mégantic site ($R^2 = 0.62$, $p = 0.06$); for panel (**c**): Brompton site ($R^2 = 0.74$, $p < 0.05$), Mégantic site ($R^2 = 0.69$, $p < 0.05$); for panel (**d**): Brompton site ($R^2 = 0.88$, $p < 0.01$), Mégantic site ($R^2 = 0.67$, $p < 0.05$).

4. Discussion

4.1. Planting Density Effects on Stem Dimensions, Stand Productivity and Total Biomass Carbon Stocks

In fast-growing plantations, the selection of the planting density is an important management decision as it can maximize biomass or wood production, and optimize wood log size at harvest, which is important for specific production objectives [12]. As expected, there was a strong effect of planting density on individual stem DBH or volume after 8 and 14 years (Table 3, Figure 2a,b). However, we found little evidence of an important effect of planting density on stand volume after 8 and 14 years, or on woody biomass production after 14 years. Site selection was clearly a more important factor affecting stem volume and woody biomass production at the stand-level (Table 3). Across sites, low stand volume or woody biomass production was mainly observed in planting density treatments were the survival rate was the lowest (Table 3). Yet, when the 1111 trees/ha treatment (lowest survival rate) was excluded from the ANOVA, there was no significant Planting density effect on stand volume or woody biomass production (Table A2). At the Mégantic and Ogden sites, where mortality occurred (Table 3), stand volume and aboveground woody biomass were more strongly correlated with survival than with spacing (Tables 7 and A3). This indicates that if tree mortality occurs early in the rotation, substantial productivity loss will be observed on the long-term.

The strongest evidence of a linkage between planting density and productivity was a significant negative relationship between area per tree and stand volume after 8 years, only observed at Brompton ($R^2 = 0.66$, $p = 0.05$) (Figure 2c). Thus, our second hypothesis suggesting greater mid-term (8 years) stand volume at higher planting density only holds for the most productive site, mainly because of the low productivity of the wider spacing treatment (4.5 × 4.5 m or 484 trees/ha) (Figure 2c). Our results corroborate previous studies showing that productivity differences between planting density treatments in *Populus* plantations are generally minor and tend to decrease as rotation length increases [12,15,16,23,24]. In addition, given that the largest biomass C stock is located in aboveground biomass, planting density only had a marginal effect on the total biomass C storage capacity of 14 year-old poplar plantations (Figure 1b, Tables 7 and A2).

The absence of an aboveground productivity/spacing relationship may be related to the clone selected for this study. Clone DN×M-915508 is known for its high plasticity in biomass allocation to the crown and its high allocation to branch biomass compared to other hybrid types [9,42,48]. Such a biomass allocation strategy allows this clone to rapidly occupy available canopy space, thereby optimizing its light harvesting capacity under both narrow and wide spacings. Clone DN×M-915508 is also known to have a relatively high productivity across a wide range of environmental conditions, even on sites presenting nutrient limitations [7]. This generalist genotype was found to be highly proficient at resorbing growth limiting nutrients (N and P) from senesced leaves, especially when site fertility declined [54]. Such a nutrient cycling strategy may allow clone DN×M-915508 to cumulate aboveground biomass at similar rate in both narrower and wider spacings, despite that the supply of growth limiting nutrients (N and P) in soil tends to decline with increasingly narrow spacing on the more fertile sites (Brompton and Mégantic) (Figure 4a,b).

The small plot size (20 trees/plot at planting) and the presence of only two buffer rows between the planting density treatments were limitations in this study. Such an experimental design was not optimal in completely eliminating the effect of neighboring plots in terms of competition for light. Also, the yield results may not be fully representative of large-scale commercial plantations where mortality rates are rarely 0%, as observed at the Brompton site.

4.2. Planting Density Effects on Biomass Partitioning, Belowground Carbon Allocation, Soil Nutrient Supply and Soil Carbon

This study shows that planting density has important effects on biomass partitioning, on belowground C allocation, on soil nutrient supply rate and on soil C sequestration over the long-term in poplar plantations. As hypothesized, we observed a decrease in the stem to branch biomass ratio with increasing tree spacing (Table 5, Figure 3a), which suggest that poplars responded to high competition for light in narrow spacings by proportionally reducing allocation to branch vs. stem biomass. The negative relationship between spacing and the stem to branch ratio was particularly steep at the productive site of Brompton (Figure 3a), where canopy closure and self-pruning occurred the earliest in the rotation (B. Truax, field observations).

Also, there was evidence of higher root competition for nutrients with decreasing spacing as shown by the positive relationships between spacing and soil NO_3 (Brompton and Mégantic sites), P, Mg (Brompton site), Ca and Mg (Brompton and Ogden sites) (Figure 4). Resource depletion in the narrower spacings likely stimulated biomass allocation belowground (Table 5). As expected, a decrease in the shoot to root ratio was observed at all sites with decreasing spacing (Figure 3b). Significant relationships between spacing and fine or coarse root biomass were also observed at the productive site of Brompton (Figure 3c,d), where signs of nutrient depletion were mainly observed (Figure 4). Thus, at Brompton, higher amounts of soil nutrients may have been taken up by poplars in narrow spacings to enhance biomass production belowground and increase foraging capacity. Furthermore, smaller diameter trees, which were produced in narrower spacings at all sites (Table 3), tend to have greater macronutrient concentrations per unit of woody biomass [55]. This could also have contributed to soil nutrient depletion in narrower spacings, despite the lack of spacing/aboveground productivity relationships after 14 years. Recent studies have suggested that faster canopy closure in narrower spacings can lower nitrification and nitrogen mineralization rates after 5 years in widely spaced (400 trees/ha or less) poplar plantations [36]. However, such a hypothesis is not supported by our results related to soil NO_3 supply at Brompton and Mégantic, as closed canopy characterized all spacing treatments at these sites after 14 years (B. Truax and J. Fortier, field observations).

On the other hand, there was a strong negative relationship between tree spacing and soil K at Brompton (Figure 4c), which suggests little depletion of this soil nutrient with increasing root competition. Weaker negative correlations between spacing and soil K were also observed at Mégantic and Ogden (Table 7). In mature poplar plantations, approximately 90% of total net stand deposition of K originates from canopy leaching, and compared to other base cations, little K is lost by percolation in the soil [56]. Since canopy closure was reached earlier in narrow spacings (B. Truax, field observations), this may have enhanced canopy-derived K input to the soil over the years. A low magnitude negative relationship between soil NO_3 supply and spacing was also observed at the low fertility site of Ogden (Figure 4a), where herbaceous biomass was by far the highest across sites (Table 5). Such a relationship potentially reflects a higher NO_3 uptake by herbaceous vegetation underneath poplars in more widely spaced treatments, since spacing and herbaceous biomass were weakly correlated at Ogden (r = 0.53, $p = 0.28$). Ruderal herbaceous plants that are typically found in the understory of poplar plantations growing on old fields can be strong competitors for soil NO_3 [57,58].

As expected, decreasing spacing between poplars not only led to greater allocation to belowground biomass (Figure 3b–d), it also led to higher soil C concentration and C stocks after 14 years (Tables 4 and 7). Yet, significant relationships between spacing and soil C were only found on the more productive sites (Brompton and Mégantic) (Figure 5a–c). Like many other tree species, poplars growing in less favorable or more competitive soil environments tend to increase root production, and root exudation to support mycorrhizal networks and microorganisms in the rhizosphere, in order to improve the access to soil resources [33,59]. Such a positive feedback of high intraspecific competition on soil C was likely observed in this study. Also, faster canopy closure in narrower spacings may have led to a faster decrease of soil and air temperatures, potentially providing more favorable conditions for soil C accumulation [37].

While mineral soil C concentration decreased with spacing at Brompton and Mégantic (Figure 5a), soil bulk density tended to increase with spacing at those sites (Figure 5d). As a result, soil C concentration and bulk density were negatively correlated at Brompton (r = −0.93, $p < 0.01$) and Mégantic (r = −0.91, $p < 0.05$). Such a trend is consistent with the notion that soil organic matter, which was strongly correlated to soil C concentration across sites (r = 0.92, $p < 0.001$, $n = 18$), has an attractive effect on soil invertebrates that can create pores in the soil [60]. At Brompton, greater fine root biomass in narrower spacings (Figure 3c) could have also contributed to decrease soil bulk density.

4.3. Management Implications

When a particular log size is required for solid wood product applications or for pulpwood, planting density should be carefully selected (Table 3, Figure 2a,b). Particular care should also be given to obtaining a very high survival rate, as higher yields were always achieved in planting density treatments with the highest survival rates (Tables 3 and A3). However, compared to planting density, site selection was of overriding importance for achieving high volume or biomass yields at the stand level. Thus, poplar afforestation on low fertility abandoned farmland sites cannot provide interesting yields (above 10 m^3/ha/year) [9], even if a generalist genotype is planted at a high density (1975 trees/ha) over a relatively long rotation (14 years). On low fertility sites, such as Ogden, hybrid poplars tend to allocate a larger proportion of assimilate C belowground, at the expense of aboveground biomass production [30]. As a result, similar coarse root biomass was observed between sites, while aboveground woody biomass varied by more than two-fold (Table 5).

For the production of bioenergy with single-stem trees, hybrid poplar plantations with planting densities ranging 1000–2000 stems/ha and 5–8 year rotations are recommended in Europe and in North America [2,61]. Yet, this study shows that for planting densities ranging 484–1975 trees/ha, using longer rotations would allow much greater productivity. At all sites and independently of the planting density, more stand volume was cumulated between the end of 8th and 14th growing season (a 6 year period) than during the first 8 years (Table 3). Rotations longer than 10 years were also recommended in Germany for high density (4000–8000 trees/ha) hybrid aspen (*P. tremula* × *P. tremuloides*) plantations on agricultural land [62]. Besides, if nitrogen fertilizers are used, the management of poplar plantations on longer rotations may reduce environmental impacts in terms of NO_3 leaching and soil nitrous oxide emissions [63]. However, longer rotations are not always optimal from an economic perspective [12]. For the specific case of short-rotation coppice bioenergy plantations with high planting densities ($\geq$2500 cuttings/ha), high biomass yields (above 10 t/ha/year) were obtained with hybrid poplars managed on 4-year cutting cycles in southern Québec (Canada) and other temperate regions [21,64]. Coppicing is also known to increase poplar yield in the subsequent rotations [65]. Thus, optimal rotation length may vary greatly between hybrid poplar plantation types (e.g., single-stem vs. coppiced systems) [66].

The use of long rotations also increases the risks of tree exposure to diseases and climatic disturbances [13], such has wind and ice storms, which are common across northeastern America. Substantial stem breakages were observed in the 1975 and 494 trees/ha plots at Brompton during the 15th growing season (B. Truax and J. Fortier, field observations). This suggests that for the studied poplar clone (DN×M-915508), both the smaller diameter trees with narrow crowns and the larger diameter trees with wide crowns may have reached their physical limit to accumulate biomass under the climatic conditions of southern Québec.

The results from this study have important implications for ecosystem services provision in poplar plantations. While planting density had a minor effect on aboveground biomass C stocks, significant increases in belowground biomass C stocks and in mineral soil C stocks were observed with decreasing tree spacing (Table 7, Figures 3 and 5). Soil porosity also tended to increase with decreasing spacing (Figure 5d). Thus, on the longer term, more densely planted poplars may be more effective to sequester atmospheric CO_2 belowground and improve general soil health on marginal agricultural land. Such environmental benefits could be the trade-off compensating for the higher establishing costs of higher density plantations, especially if large planting stocks are used.

More generally, our results have potential implications for the design of multi-functional poplar buffers, which are increasingly used in agricultural watersheds for bioenergy production, streambank stabilization, phytoremediation and on-farm C storage [1,67,68]. Higher soil porosity, reduced soil NO_3 and P, enhanced soil organic C and higher belowground biomass in more densely planted poplars could provide more effective buffering capacity against main agricultural pollutants (sediments, N, P and organic pesticides) reaching streams [69]. Further studies are therefore needed to evaluate the effects of planting density on the provision of various ecosystem services in different types of *Populus* plantation systems.

5. Conclusions

Overall, this long-term study showed that planting density had major effects on stem dimensions, biomass partitioning and soil C, but minor effects on productivity and aboveground biomass C stocks. Site selection was a much more determinant factor in obtaining high yields for the studied hybrid poplar clone. On the higher fertility sites, surface area per tree was a significant negative predictor of fine root biomass, coarse root biomass and soil C, but a significant positive predictor for the supply of several macronutrients. Therefore, in higher density plantations of poplars, the high root competition for soil resources may lead to soil nutrient depletion, which may stimulate C allocation belowground and enhance soil C sequestration potential over the long-term.

Author Contributions: B.T. conceived and planted the experimental design. J.F., B.T. and F.L. were involved in sampling design and field sampling. J.F. and B.T. analyzed the data and J.F. wrote the first draft of the manuscript. B.T., D.G., and F.L. critically revised the manuscript.

Funding: We gratefully acknowledge funding received from Agriculture and Agri-Food Canada (Agricultural Greenhouse Gas Program) and the Ministère des Forêts, de la Faune et des Parcs of Québec (MFFP).

Acknowledgments: We wish to thank A. Déziel of the Berthierville nursery (MFFP) (which provided bare-root planting stock), as well as F. Lemieux and C. Cormier of the MFFP regional office. Thanks also to all the owners of the plantation sites: H. Isbrucker, P. Labrecque, and M. Blais. We greatly appreciated the help of all our tree planters and field assistants (R. Côté, F. Mongeau, J.S. Labrecque, L.P. Gagnon, L. Tétreault-Garneau, D. Adam, L. Godbout, J. Lemelin, M.-A. Pétrin, M. Blais). We would like to thank R. Lamadeleine and M. Poulin for giving us access to the wood drying facilities of Domtar Corp., Windsor, Québec. A special thanks to H. Isbrucker for providing us with a large amount of space for sample storage and preparation. Thanks are also due to R. Bradley and W. Parsons, of the Centre d'étude de la forêt (CEF) laboratory at Université de Sherbrooke, for providing soil C/N analyses and G. Lagacé from Génik Inc. for providing soil corers.

Conflicts of Interest: The authors declare no conflict of interest.

Appendix

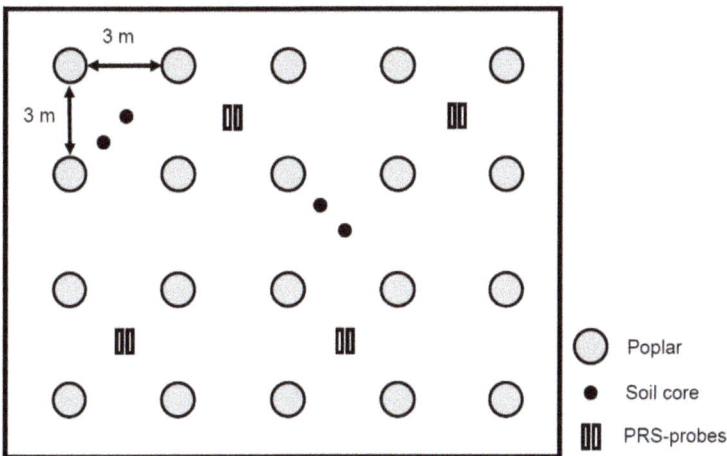

Figure A1. Schematic representation of the sampling design used for the soil core extractions and for the evaluation of soil nutrient supply rate with the PRS-probes. The experimental plot shown represents the 1111 trees/ha (3×3 m spacing) treatment.

Table A1. Planting density effect on soil nutrient supply rate, pH, clay, silt and sand content in 14 year-old hybrid poplar plantations.

Planting Density	Soil Nutrient Supply Rate (μg/10 cm^2/42 days)										
	NO$_3$	NH$_4$	P	K	Ca	Mg	S	pH	Clay	Silt	Sand
494 stems/ha	219.0	3.31	6.02	46	1880	309	70.4	5.20	15.9	45.8	38.3
741 stems/ha	36.5	2.29	4.85	48	1642	387	40.4	5.27	15.9	49.8	34.3
988 stems/ha	29.2	3.38	4.90	55	1324	200	34.5	5.20	14.3	50.1	35.6
1111 stems/ha	41.3	4.91	5.22	63	1123	194	33.3	4.97	13.5	44.4	42.0
1481 stems/ha	36.0	3.25	4.20	149	1136	188	27.6	5.20	13.9	46.7	39.4
1975 stems/ha	15.2	3.42	4.06	56	1194	186	31.7	5.03	14.2	49.4	36.4
SE	58.9	0.73	0.91	29	221	47	11.7	0.06	2.3	1.9	1.7
p-value	0.91	0.33	0.69	0.08	0.17	0.05	0.21	<0.05	0.95	0.24	0.14

Table A2. Probability value (*p*-value) of the Planting density effect on stand-level characteristics following the removal of the 1111 trees/ha treatment in the ANOVA.

Stand-Level Characteristics	*p*-value Planting Density Effect
Volume 8 years (m^3/ha)	0.15
Volume 14 years (m^3/ha)	0.21
Δ Volume 8–14 years (m^3/ha)	0.33
Aboveground woody biomass 14 year (t/ha)	0.49
Belowground biomass 14 years (t/ha)	0.05
Total biomass 14 years (t/ha)	0.37
Aboveground C stocks 14 years (t C/ha)	0.53
Belowground C stocks 14 years (t C/ha)	0.05
Total biomass C stocks 14 years (t C/ha)	0.40

Table A3. Site-specific pairwise correlations (r) between survival and selected stand-level characteristics in hybrid poplar plantations.

Stand-Level Characteristics	Mégantic		Ogden	
	r	*p*-value	r	*p*-value
Volume 8 years (m^3/ha)	0.92	<0.01	0.74	0.09
Volume 14 years (m^3/ha)	0.93	<0.01	0.69	0.13
Δ volume 8–14 years (m^3/ha)	0.86	<0.05	0.65	0.16
Aboveground woody biomass 14 years (t/ha)	0.91	<0.05	0.66	0.15
Aboveground biomass C stocks 14 years (t C/ha)	0.91	<0.05	0.66	0.15
Belowground biomass C stocks 14 years (t C/ha)	0.91	<0.05	0.74	0.09
Total biomass C stocks 14 years (t C/ha)	0.91	<0.05	0.69	0.13

References

1. Isebrands, J.G.; Aronsson, P.; Ceulemans, M.C.; Coleman, M.; Dimitriou, N.D.; Doty, S.; Gardiner, E.; Heinsoo, K.; Johnson, J.D.; Koo, Y.B.; et al. Environmental applications of poplars and willows. In *Poplars and Willows: Trees for Society and the Environment*; Isebrands, J.G., Richardson, J., Eds.; CABI: Wallingford, UK; FAO: Rome, Italy, 2014; pp. 258–336.
2. Stanturf, J.A.; van Oosten, C. Operational poplar and willow culture. In *Poplars and Willows: Trees for Society and the Environment*; Isebrands, J.G., Richardson, J., Eds.; CABI and FAO: Rome, Italy, 2014; pp. 200–257.
3. Volk, T.A.; Berguson, B.; Daly, C.; Halbleib Michael, D.; Miller, R.; Rials Timothy, G.; Abrahamson Lawrence, P.; Buchman, D.; Buford, M.; Cunningham Michael, W.; et al. Poplar and shrub willow energy crops in the United States: Field trial results from the multiyear regional feedstock partnership and yield potential maps based on the PRISM-ELM model. *GCB Bioenergy* **2018**, 1–17. [CrossRef]

4. Fortier, J.; Truax, B.; Gagnon, D.; Lambert, F. Potential for hybrid poplar riparian buffers to provide ecosystem services in three watersheds with contrasting agricultural land use. *Forests* **2016**, *7*, 37. [CrossRef]

5. Zalesny, R.S.; Stanturf, J.A.; Gardiner, E.S.; Perdue, J.H.; Young, T.M.; Coyle, D.R.; Headlee, W.L.; Bañuelos, G.S.; Hass, A. Ecosystem services of woody crop production systems. *BioEnergy Res.* **2016**, *9*, 465–491. [CrossRef]

6. Coleman, M.D.; Isebrands, J.G.; Tolsted, D.N.; Tolbert, V.R. Comparing soil carbon of short rotation poplar plantations with agricultural crops and woodlots in North Central United States. *Environ. Manag.* **2004**, *33*, 299–308. [CrossRef]

7. Truax, B.; Gagnon, D.; Fortier, J.; Lambert, F. Yield in 8 year-old hybrid poplar plantations on abandoned farmland along climatic and soil fertility gradients. *For. Ecol. Manag.* **2012**, *267*, 228–239. [CrossRef]

8. Werner, C.; Haas, E.; Grote, R.; Gauder, M.; Graeff-Hönninger, S.; Claupein, W.; Butterbach-Bahl, K. Biomass production potential from *Populus* short rotation systems in Romania. *GCB Bioenergy* **2012**, *4*, 642–653. [CrossRef]

9. Truax, B.; Gagnon, D.; Fortier, J.; Lambert, F. Biomass and volume yield in mature hybrid poplar plantations on temperate abandoned farmland. *Forests* **2014**, *5*, 3107–3130. [CrossRef]

10. Stanturf, J.A.; van Oosten, C.; Coleman, M.D.; Portwood, C.J. Ecology and silviculture of poplar plantations. In *Poplar Culture in North America*; Dickmann, D.I., Isebrands, J.G., Eckenwalder, J.E., Richardson, J., Eds.; NRC Research Press, National Research Council of Canada: Ottawa, ON, Canada, 2001; pp. 153–206.

11. Powers, R.F. On the sustainable productivity of planted forests. *New For.* **1999**, *17*, 263–306. [CrossRef]

12. Mead, D.J. Opportunities for improving plantation productivity. How much? How quickly? How realistic? *Biomass Bioenergy* **2005**, *28*, 249–266. [CrossRef]

13. Christersson, L. Wood production potential in poplar plantations in Sweden. *Biomass Bioenergy* **2010**, *34*, 1289–1299. [CrossRef]

14. Davis, M.; Nordmeyer, A.; Henley, D.; Watt, M. Ecosystem carbon accretion 10 years after afforestation of depleted subhumid grassland planted with three densities of *Pinus nigra*. *Glob. Chang. Biol.* **2007**, *13*, 1414–1422. [CrossRef]

15. Johnstone, W.D. The effects of initial spacing and rectangularity on the early growth of hybrid poplar. *West. J. Appl. For.* **2008**, *23*, 189–196.

16. DeBell, D.S.; Clendenen, G.W.; Harrington, C.A.; Zasada, J.C. Tree growth and stand development in short-rotation *Populus* plantings: 7-year results for two clones at three spacings. *Biomass Bioenergy* **1996**, *11*, 253–269. [CrossRef]

17. Tun, T.N.; Guo, J.; Fang, S.; Tian, Y. Planting spacing affects canopy structure, biomass production and stem roundness in poplar plantations. *Scand. J. For. Res.* **2018**, 1–11. [CrossRef]

18. Puri, S.; Singh, V.; Bhushan, B.; Singh, S. Biomass production and distribution of roots in three stands of *Populus deltoides*. *For. Ecol. Manag.* **1994**, *65*, 135–147. [CrossRef]

19. Benomar, L.; DesRochers, A.; Larocque, G. The effects of spacing on growth, morphology and biomass production and allocation in two hybrid poplar clones growing in the boreal region of Canada. *Tree Struct. Funct.* **2012**, 1–11. [CrossRef]

20. Fang, S.; Xue, J.; Tang, L. Biomass production and carbon sequestration potential in poplar plantations with different management patterns. *J. Environ. Manag.* **2007**, *85*, 672–679. [CrossRef] [PubMed]

21. Armstrong, A.; Johns, C.; Tubby, I. Effects of spacing and cutting cycle on the yield of poplar grown as an energy crop. *Biomass Bioenergy* **1999**, *17*, 305–314. [CrossRef]

22. Eisenbies, M.H.; Volk, T.A.; Espinoza, J.; Gantz, C.; Himes, A.; Posselius, J.; Shuren, R.; Stanton, B.; Summers, B. Biomass, spacing and planting design influence cut-and-chip harvesting in hybrid poplar. *Biomass Bioenergy* **2017**, *106*, 182–190. [CrossRef]

23. Cañellas, I.; Huelin, P.; Hernández, M.J.; Ciria, P.; Calvo, R.; Gea-Izquierdo, G.; Sixto, H. The effect of density on short rotation *Populus* sp. plantations in the Mediterranean area. *Biomass Bioenergy* **2012**, *46*, 645–652. [CrossRef]

24. Strong, T.; Hansen, E. Hybrid poplar spacing/productivity relations in short rotation intensive culture plantations. *Biomass Bioenergy* **1993**, *4*, 255–261. [CrossRef]

25. Mäkelä, A. A carbon balance model of growth and self-pruning in trees based on structural relationships. *For. Sci.* **1997**, *43*, 7–24.

26. Peterson, E.B.; Peterson, N.M. *Ecology, Management, and Use of Aspen and Balsam Poplar in the Prairie Provinces*; Forestry Canada, Northwest Region, Northern Forestry Centre: Edmonton, AB, Canada, 1992; p. 252.

27. Schenk, H.J. Root competition: Beyond resource depletion. *J. Ecol.* **2006**, *94*, 725–739. [CrossRef]

28. Berthrong, S.T.; Jobbágy, E.G.; Jackson, R.B. A global meta-analysis of soil exchangeable cations, pH, carbon, and nitrogen with afforestation. *Ecol. Appl.* **2009**, *19*, 2228–2241. [CrossRef] [PubMed]

29. Heilman, P.E.; Stettler, R.F. Nutritional concerns in selection of black cottonwood and hybrid clones for short rotation. *Can. J. For. Res.* **1986**, *16*, 860–863. [CrossRef]

30. Fortier, J.; Truax, B.; Gagnon, D.; Lambert, F. Plastic allometry in coarse root biomass of mature hybrid poplar plantations. *BioEnergy Res.* **2015**, *8*, 1691–1704. [CrossRef]

31. Giardina, C.P.; Ryan, M.G. Total belowground carbon allocation in a fast-growing eucalyptus plantation estimated using a carbon balance approach. *Ecosystems* **2002**, *5*, 487–499. [CrossRef]

32. Hjelm, K.; Rytter, L. The influence of soil conditions, with focus on soil acidity, on the establishment of poplar (*Populus* spp.). *New For.* **2016**, *47*, 731–750. [CrossRef]

33. Szuba, A. Ectomycorrhiza of *Populus*. *For. Ecol. Manag.* **2015**, *347*, 156–169. [CrossRef]

34. Hu, Y.-L.; Zeng, D.-H.; Ma, X.-Q.; Chang, S.X. Root rather than leaf litter input drives soil carbon sequestration after afforestation on a marginal cropland. *For. Ecol. Manag.* **2016**, *362*, 38–45. [CrossRef]

35. Hernández, J.; del Pino, A.; Vance, E.D.; Califra, Á.; Del Giorgio, F.; Martínez, L.; González-Barrios, P. *Eucalyptus* and *Pinus* stand density effects on soil carbon sequestration. *For. Ecol. Manag.* **2016**, *368*, 28–38. [CrossRef]

36. Yan, Y.; Fang, S.; Tian, Y.; Deng, S.; Tang, L.; Chuong, N.D. Influence of tree spacing on soil nitrogen mineralization and availability in hybrid poplar plantations. *Forests* **2015**, *6*, 636–649. [CrossRef]

37. Conant, R.T.; Ryan, M.G.; Ågren, G.I.; Birge, H.E.; Davidson, E.A.; Eliasson, P.E.; Evans, S.E.; Frey, S.D.; Giardina, C.P.; Hopkins, F.M.; et al. Temperature and soil organic matter decomposition rates —Synthesis of current knowledge and a way forward. *Glob. Change Biol.* **2011**, *17*, 3392–3404. [CrossRef]

38. Prescott, C.E. Litter decomposition: What controls it and how can we alter it to sequester more carbon in forest soils? *Biogeochemistry* **2010**, *101*, 133–149. [CrossRef]

39. Government of Canada. Station Results—1981–2010 Climate Normals and Averages. Available online: http://climate.weather.gc.ca/climate_normals/station_select_1981_2010_e.html?searchType=stnProv&lstProvince=QC (accessed on 16 February 2017).

40. Robitaille, A.; Saucier, J.-P. *Paysages régionaux du Québec Méridional*; Les publications du Québec: Ste-Foy, QC, Canada, 1998; p. 213.

41. Périnet, P.; Gagnon, H.; Morin, S. *Liste des Clones Recommandés de Peuplier Hybride par Sous-Région Écologique au Québec (mise à jour Octobre 2010)*; Direction de la Recherche Forestière, MRN: Québec, QC, Canada, 2010; p. 1.

42. Fortier, J.; Truax, B.; Gagnon, D.; Lambert, F. Mature hybrid poplar riparian buffers along farm streams produce high yields in response to soil fertility assessed using three methods. *Sustainability* **2013**, *5*, 1893–1916. [CrossRef]

43. Qian, P.; Schoenau, J.J.; Huang, W.Z. Use of ion exchange membranes in routine soil testing. *Commun. Soil Sci. Plant Anal.* **1992**, *23*, 1791–1804. [CrossRef]

44. Western Ag Innovations. Analysis. Available online: https://www.westernag.ca/innovations/technology/analysis_units (accessed on 11 May 2018).

45. Throop, H.L.; Archer, S.R.; Monger, H.C.; Waltman, S. When bulk density methods matter: Implications for estimating soil organic carbon pools in rocky soils. *J. Arid Environ.* **2012**, *77*, 66–71. [CrossRef]

46. Bouyoucos, G.J. Hydrometer method improved for making particle size analysis of soils. *Agron. J.* **1962**, *54*, 464–465. [CrossRef]

47. Harmon, M.E.; Woodall, C.W.; Fasth, B.; Sexton, J. *Woody Detritus Density and Density Reduction Factors for Tree Species in the United States: A Synthesis*; General Technical Report NRS-29; United States Department of Agriculture, Forest Service, Northern Research Station: Newtown Square, PA, USA, 2008; p. 90.

48. Fortier, J.; Truax, B.; Gagnon, D.; Lambert, F. Allometric equations for estimating compartment biomass and stem volume in mature hybrid poplars: General or site-specific? *Forests* **2017**, *8*, 309. [CrossRef]

49. Petersen, R.G. *Design and Analysis of Experiments*; Marcel-Dekker: New York, NY, USA, 1985; p. 429.

50. Burkes, E.C.; Will, R.E.; Barron-Gafford, G.A.; Teskey, R.O.; Shiver, B. Biomass partitioning and growth efficiency of intensively managed *Pinus taeda* and *Pinus elliottii* stands of different planting densities. *For. Sci.* **2003**, *49*, 224–234.

51. Warton, D.I.; Hui, F.K.C. The arcsine is asinine: The analysis of proportions in ecology. *Ecology* **2011**, *92*, 3–10. [CrossRef] [PubMed]

52. Day, R.W.; Quinn, G.P. Comparisons of treatments after an analysis of variance in ecology. *Ecol. Monogr.* **1989**, *59*, 433–463. [CrossRef]

53. Gotelli, N.J.; Ellison, A.M. *A primer of Ecological Statistics*; Sinauer Associated, Inc.: Sunderland, MA, USA, 2004; p. 510.

54. Fortier, J.; Truax, B.; Gagnon, D.; Lambert, F. Linking biomass productivity to genotype-specific nutrient cycling strategies in mature hybrid poplars planted along an environmental gradient. *BioEnergy Res.* **2017**, *10*, 876–890. [CrossRef]

55. Augusto, L.; Meredieu, C.; Bert, D.; Trichet, P.; Porté, A.; Bosc, A.; Lagane, F.; Loustau, D.; Pellerin, S.; Danjon, F.; et al. Improving models of forest nutrient export with equations that predict the nutrient concentration of tree compartments. *Ann. For. Sci.* **2008**, *65*, 808. [CrossRef]

56. Meiresonne, L.; Schrijver, A.D.; Vos, B.D. Nutrient cycling in a poplar plantation (*Populus trichocarpa* × *Populus deltoides* 'Beaupré') on former agricultural land in northern Belgium. *Can. J. For. Res.* **2007**, *37*, 141–155. [CrossRef]

57. Boothroyd-Roberts, K.; Gagnon, D.; Truax, B. Can hybrid poplar plantations accelerate the restoration of forest understory attributes on abandoned fields? *For. Ecol. Manag.* **2013**, *287*, 77–89. [CrossRef]

58. Gebauer, G.; Rehder, H.; Wollenweber, B. Nitrate, nitrate reduction and organic nitrogen in plants from different ecological and taxonomic groups of Central Europe. *Oecologia* **1988**, *75*, 371–385. [CrossRef] [PubMed]

59. Broeckx, L.S.; Verlinden, M.S.; Berhongaray, G.; Zona, D.; Fichot, R.; Ceulemans, R. The effect of a dry spring on seasonal carbon allocation and vegetation dynamics in a poplar bioenergy plantation. *GCB Bioenergy* **2014**, *6*, 473–487. [CrossRef]

60. Adams, W.A. The effect of organic matter on the bulk and true densities of some uncultivated podzolic soils. *J. Soil Sci.* **1973**, *24*, 10–17. [CrossRef]

61. Berthelot, A.; Gavaland, A. *Produire de la Biomasse avec des Taillis de Peupliers*; Institut Technologique Forêt Cellulose Bois-construction Ameublement (FCBA): Champs-sur-Marne, France, 2007; Fiche n° 760; p. 6.

62. Liesebach, M.; von Wuehlisch, G.; Muhs, H.J. Aspen for short-rotation coppice plantations on agricultural sites in Germany: Effects of spacing and rotation time on growth and biomass production of aspen progenies. *For. Ecol. Manag.* **1999**, *121*, 25–39. [CrossRef]

63. Schweier, J.; Molina-Herrera, S.; Ghirardo, A.; Grote, R.; Díaz-Pinés, E.; Kreuzwieser, J.; Haas, E.; Butterbach-Bahl, K.; Rennenberg, H.; Schnitzler, J.P.; et al. Environmental impacts of bioenergy wood production from poplar short-rotation coppice grown at a marginal agricultural site in Germany. *GCB Bioenergy* **2017**, *9*, 1207–1221. [CrossRef]

64. Labrecque, M.; Teodorescu, T.I. Field performance and biomass production of 12 willow and poplar clones in short-rotation coppice in southern Quebec (Canada). *Biomass Bioenergy* **2005**, *29*, 1–9. [CrossRef]

65. Geyer, W.A. Biomass production in the Central Great Plains USA under various coppice regimes. *Biomass Bioenergy* **2006**, *30*, 778–783. [CrossRef]

66. Shooshtarian, A.; Anderson, J.A.; Armstrong, G.W.; Luckert, M.K. Growing hybrid poplar in western Canada for use as a biofuel feedstock: A financial analysis of coppice and single-stem management. *Biomass Bioenergy* **2018**, *113*, 45–54. [CrossRef]

67. Fortier, J.; Truax, B.; Gagnon, D.; Lambert, F. Biomass carbon, nitrogen and phosphorus stocks in hybrid poplar buffers, herbaceous buffers and natural woodlots in the riparian zone on agricultural land. *J. Environ. Manag.* **2015**, *154*, 333–345. [CrossRef] [PubMed]

68. Ferrarini, A.; Fornasier, F.; Serra, P.; Ferrari, F.; Trevisan, M.; Amaducci, S. Impacts of willow and miscanthus bioenergy buffers on biogeochemical N removal processes along the soil–groundwater continuum. *GCB Bioenergy* **2017**, *9*, 246–261. [CrossRef]

69. Dosskey, M.G.; Vidon, P.; Gurwick, N.P.; Allan, C.J.; Duval, T.P.; Lowrance, R. The role of riparian vegetation in protecting and improving chemical water quality in streams. *JAWRA* **2010**, *46*, 261–277.

forests MDPI

Article

Investigating the Effect of a Mixed Mycorrhizal Inoculum on the Productivity of Biomass Plantation Willows Grown on Marginal Farm Land

Thomas Joseph Pray [1], Werther Guidi Nissim [2], Marc St-Arnaud [1] and Michel Labrecque [1,*]

[1] Biodiversity Centre, Institut de Recherche en Biologie Végétale, Université de Montréal and Jardin Botanique de Montréal, Montréal, QC HIX 2B2, Canada; tjpray@gmail.com (T.J.P.); marc.st-arnaud@umontreal.ca (M.S.-A.)

[2] Department of Agrifood Production and Environmental Science, University of Florence, 4,50121 Firenze, Italy; werther.guidinissim@unifi.it

* Correspondence: michel.labrecque@umontreal.ca; Tel.: +1-514-872-1862

Received: 28 February 2018; Accepted: 30 March 2018; Published: 4 April 2018

Abstract: Inoculation with mycorrhizal fungi, proven mediators of soil fertility, has great potential in agricultural and silvicultural systems. This is particularly true in short-rotation coppices (SRCs), where questions of food displacement and fertilization are causes of concern for researchers and policy makers. We set out to thoroughly test if current inoculation methods, coupled with reduced fertilization, can demonstrate a growth benefit in SRC willows on marginal lands. Roughly 21,600 *Salix miyabeana* Seeman ('SX61' and 'SX64') were planted in a hierarchical design with inoculation treatments randomized first, cultivars randomized second, and fertilization treatments randomized third. This process was repeated across three fields of different marginal soil type (which, in our experiment, were given the descriptive names Sandy, Rocky, and Dry). The inoculum species, *Rhizoglomus irregulare* Błaszk., Wubet, Renker & Buscot Sieverd., G.A. Silva & Oehl and *Hebeloma longicaudum* (Pers.) P. Kumm., were chosen as they are most likely to be commercially available, and because they represent both arbuscular and ectomycorrhizal inoculum types. Growth was measured over 2.5 years, or three growing seasons. Fertilization treatment (75 kg/ha Nitrogen), however, was only applied during the second growing season. Our results conclusively showed no benefit from mycorrhizal inoculation across fields that exhibited significantly different growth rates, as well as significant differentiation from fertilization.

Keywords: mycorrhizal fungi; willow; inoculation; agricultural field experiment; rhizospheric soil

---

1. Introduction

Growing short-rotation coppiced (SRC) willows for energy purposes is particularly promising on agriculturally marginal land. Such land would neither come as a sacrifice to pristine, old-growth wilderness, nor would it have a significant impact upon food production, as much of this marginal land has become, or is becoming, unviable within the current competitive agricultural marketplace [1]. However, the main driver of efforts to develop biomass plantations, global warming, can also fuel concerns about fertilizer use [2]. In general, SRC willows can be grown on many types of agricultural land (though wetter land is much better than dry) [3,4]. Due to high biomass yields though, they remove nutrients at a high rate [5]. This means that poor sites are not suitable for SRC cultivation unless fertilizers are supplied. A relevant study showed that fertilizer represents up to 10–20% of the cost of production over several rotations of a willow SRC crop [6]. Estimates based on nutrient off-take measurements vary between 50–130 kg N, 60–83 kg K, and 8–16 kg P per hectare per year for willow SRC, although farmers generally only need to add nitrogen (N) over the first several years, as potassium (K) and phosphorus (P) usually build up excessively in farm soil [7]. This paper addresses whether the central challenge of fertility that arises with SRC farming in a carbon-negative way on marginal land could be addressed or mitigated by inoculation with mycorrhizal fungi.

Arbuscular mycorrhiza (AM), which penetrate the root cells of their host plant, and ectomycorrhiza (EM), which interact just as intimately with their host plant but through a form of root interface that does not penetrate cells, are the two main types of mycorrhizal fungi (e.g., [8]). In exchange for the sugars that the plant produces, both AM and EM provide nutrients harvested from surrounding soil (particularly those trapped in mineral form or difficult-to-degrade organic molecules) with their extensive fungal networks and specialized degradation enzymes [9,10]. There is also extensive evidence of AM and EM providing their hosts with protection from disease and aiding with water stress during droughts [11–13].

Biomass shrub willows—though small and with multiple stems instead of a single trunk—are nevertheless a tree species in the genus *Salix* L., and are typically found with EM symbionts in samples of their roots and rhizosphere (the narrow soil zone under the influence of roots) [14–16]. *Salix* species, however, have been found to be associated with both EM and AM (separately or at the same time) [17–20]. Minor association with AM cannot be ignored, as some evidence suggests that mycorrhiza can provide significant benefits for their host, even at low levels of root colonization [21,22].

Because agricultural land (or marginal, formerly agricultural land) does not typically have a diverse collection of trees growing on it, researchers have conducted experiments inoculating trees with mycorrhiza, reasoning that the specific EM species that would colonize biomass willows could be missing [23]. Similarly, as agricultural land is often exhausted of organic carbon and repeatedly left barren of plant hosts (when crops are harvested in late summer or fall), even AM, that could use crop plants as hosts, might be low in number and/or diversity [24,25]. Some experiments showed a benefit from inoculation, but many others were inconclusive [26–30]. The same is true for experiments specifically using *Salix* (or closely related *Populus* L.) and AM or EM inoculum [31–33].

In most of these studies, the controlled conditions used are significantly different from what willow growers will actually encounter in their fields. Even when young trees were planted into natural settings in some of these experiments, the seedlings were first raised in containers. This means that the control conditions, with completely sterile potting soil, could stunt the trees' early growth, and therefore do not reflect a natural control with a random mix of native mycorrhiza. Experiments with sterile control soils, and sterile soils plus one or more inoculum species throughout the measurement time period, while valuable for basic research, are inherently limited. Agricultural activities almost never deal with sterile soils, so for practical purposes, farmers and agricultural researchers need to know what their treatments do in relation to unsterile controls.

This study addresses the sterile control problem by planting directly into intact farm soil. This is possible in practical terms because willow cuttings do not need to be started in pots, and in theoretical terms because of the sheer number of willow cuttings planted. The myriad of combinations of native mycorrhiza, as well as of plant pathogens, etc., that would threaten a smaller experiment with too many confounding variables, are better dealt with through replication across large fields. If a few willows are negatively affected by small areas of soil with disease pathogens, or conversely if a few willows have increased growth due to a small area of particularly beneficial mycorrhiza, their growth measurements will not significantly shift the mean results from the many more growth measurements we were able to take through a field-scale investigation. Furthermore, the high variability in the growth of plants in fields that are naturally patchy in soil structure and nutrient concentration can be accounted for statistically with enough data points from several different blocks. This "random" variation was part of our ANOVA models, and our field sizes allowed us to do this. Biomass farmers, and those who advise farmers, should take note of this study's results, as they relate closely to real-world conditions. Even the equipment used to set up and implement the experiment was consistent with modern farming realities.

If the use of such energy intensive products as fertilizers can be limited through mycorrhizal inoculation, and likewise marginal agricultural land can be better taken advantage of, then biofuel grown with carbon taken from the atmosphere would inarguably represent a net climate benefit. Knowing whether inoculation is effective or not is a key step for this industry, and more generally, for those working in agriculture and silviculture. Our experiment aimed to take this step: it tested whether or not a mycorrhizal inoculant can have a positive effect on the growth of SRC willows. Furthermore, our experiment was designed extremely robustly, testing the inoculants effect across two different cultivars of willow in three different marginal field-types, and both with and without nitrogen fertilization.

2. Materials and Methods

2.1. Experimental Design

Three similarly designed experimental fields of 108 m × 43.2 m, or ~4670 m^2 each, were established in early June of 2010 at three sites on the Allard family farm (Agro Énergie Inc., Saint-Roch-de-l'Achigan, QC, Canada (45°51′00″ N, 73°36′00″ W). The farm is ~60 m above sea level, flat and open, but sparsely wooded at the edges. It lies within the St. Lawrence River watershed, ~25 km north-northeast of Montreal, QC, Canada. A different soil type defines each experimental field (details are summarized in Table 1). One location, referred to in this paper as the Sandy field, is almost pure sand, with a low pH of 6. Another, named Dry field, is sandy-loam with a close to neutral pH of 7. The last consists of silty-loam, with a high pH of 8, but with ~30% of the surface covered with small and medium sized rocks (~1–5 cm) (therefore named Rocky field). Fields had been cultivated in the past with the standard North American rotation of corn and soybeans (corn most recently), but the Sandy field had also been periodically planted with carrots before that.

Table 1. Characterization of the three field sites, including a soil analysis at two depths.

Depth	pH	Nitrate	Phosphorus	Organic Matter	Clay	Silt	Sand	Soil Type
			Dry Field (45°49′31″ N, 73°37′29″ W) [1]					
(0–20 cm)	7.1	5.77 ppm	130 kg/ha	4.0%	21.0%	40.9%	38.1%	Medium loam
(20–40 cm)	7.3	7.35 ppm	81 kg/ha	3.9%	19.9%	32.7%	47.4%	Edging towards sandy
			Rocky Field (45°49′38″ N, 73°37′36″ W) [2]					
(0–20 cm)	7.9	6.87 ppm	63 kg/ha	3.5%	29.5%	39.2%	31.3%	Medium loam
(20–40 cm)	7.9	5.77 ppm	39 kg/ha	2.8%	24.1%	43.5%	32.4%	Edging towards clay
			Sandy Field (45°49′32″ N, 73°37′04″ W) [3]					
(0–20 cm)	6.0	5.35 ppm	256 kg/ha	2.1%	2.5%	10.4%	87.1%	Loamy sand
(20–40 cm)	6.1	5.91 ppm	192 kg/ha	2.0%	3.4%	6.8%	89.8%	Very close to pure sand

[1] Defining qualitative features: heavily drained with a 0.5 m deep ditch along its west side and a 2 m deep channel along its south side. [2] Defining qualitative features: ~30% of its surface covered with small and medium sized rocks (~1–5 cm). [3] Defining qualitative features: far removed from any drainage, it often had standing water for a week at a time after any rainfall, between the field and the forest bordering its south side and the southern half of its east side.

The experimental design was a modified split-plot design and had twelve full blocks repeated in each field. In this hierarchical design, inoculation treatments were randomized before cultivar, and fertilization treatments were randomized last. Local soil conditions and weather determined our partner farmer's best practices for preparing a weed-free, flat, and loose soil bed (a mix of plowing and disking). Small cuttings (~30 cm long, and ~1–2 cm thick) of two cultivars of willow, *Salix miyabeana* Seeman 'SX64' and 'SX61' were planted. Planting was done using a modified 3-row cabbage planter. Rows were 1.8 m apart, and willows were planted every 36 cm, for a density of 16,103 trees per hectare. Flagged stakes marked every 18 m, showing the edges of each experiment block (6 × 12-row-wide blocks running down each half of the field, for 12 in total in each field—see Supplementary Material, Figure S1 for diagram). The modified aspect of the split-plot design came with the randomization of the cultivar. Instead of being randomized with each block, the random selection from the first block was continued for the entire length of the field, in order to facilitate timely and accurate planting at this large scale (again, see Supplementary Material, Figure S1 for diagram). The three farmers sitting on the planter were informed about which of the two willow types to feed into the rotating planter cylinders at the beginning of each new group of three rows. However, during planting, water delivery of the inoculant was turned off and on by hand as the planter stopped and started each block section, thus allowing a true randomization between each experimental block. Smaller flags on wire stakes subdivided the twelve blocks at 9 m (or six-row) intervals, but these subdivisions directed nitrogen fertilization, which was not applied until the second growing season (when it was applied by hand, due to the relatively small size of each treatment plot—see next paragraph for application rates). Therefore, during the first year each experimental block had four treatment subplots within them, which then became eight treatment subplots within each block from the second year on Similar blocks, albeit with a different randomly assigned treatment pattern, were set up in the three different fields.

Inoculation was done using a mixed inoculum of AM and EM fungi, provided by BioSyneterra Solutions Inc. (L'Assomption, QC, Canada). It was delivered via a water-suspension to each cutting at planting as ~350 propagules of the AM and ~250 of the EM in 50 mL of water. The AM strain used was a *Rhizoglomus irregulare* (Błaszk., Wubet, Renker & Buscot) Sieverd., G.A. Silva & Oehl DAOM197198 [34,35]) from Pont Rouge, QC, Canada. The EM was a proprietary *Hebeloma longicaudum* (Pers.) P. Kumm. Viability of the inoculum was checked by taking a sample of the inoculum suspension on the day of planting, and inoculating several potted willows in autoclaved farm soil. Viability was qualitatively confirmed the following day with root staining and microscopic visualization of AM and EM structures, as well as with macroscopic identification of EM fructifications. Some 7200 cuttings were planted on each site, for a total of 21,600 plants. The experiment ran for two and a half years (through three growing seasons). Only during the second growing season did half of the trees receive nitrogen fertilization (75 kg/ha N, as pelleted chicken feather compost, scattered by hand in May).

2.2. Sampling and Measurements

Before planting had taken place, a baseline soil analysis was conducted. Soil was collected using a 1000 cm^3, screw-boring hand sampler, combining seven samples taken from a diagonal line across each field. Two depths were sampled, 0–20 cm and 20–40 cm. All of the soil from each sample type was mixed thoroughly. A subsample was then taken back to the laboratory and air dried for chemical analysis using recommended methods [36]. Total nitrogen was measured using Kjeldal's method. Total phosphorus was extracted with nitric-perchloric acid or stannous chloride (for the sludge), and measured using plasma emission spectroscopy. Potassium, calcium, and magnesium were extracted with NH$_4$Ac, and analyzed by atomic absorption. Metals were extracted with nitric or hydrochloric acid, and their dosage was done using ICP (Inductively Coupled Plasma) [37].

We measured the height of the longest stem, its diameter at a distance of 10 cm from the ground, as well as the total number of stems, in order to assess shrub growth. Growth measurements were taken on every other tree along the middle row of the three-row treatment groups in early November, 2010. This meant that 16 trees were measured in each treatment group. With four treatment groups during the first year and 12 blocks repeated in each field, that totaled 2304 trees measured (the actual total was slightly lower as one block within the Sandy field was not measured in the first year, due to excessive mortality in that block, presumably as roots had not developed enough to compensate for the higher drainage in an almost 100% sand patch).

At the end of the second growing season (early December 2011), a total of 1152 trees were measured for growth, with 576 also selected randomly to be cut and weighed. A representative 2–3 kg of stem pieces (without leaves) were further subsampled from each field to measure moisture percentage. Although a block in the Sandy field had been dropped from the analysis in the first year due to the aforementioned high mortality, it was kept during the second year. This was because many of the trees that had died were actually concentrated in the middle of the block, and there were enough left to give an accurate representation of all the treatments with the addition of an increased fertilization subdivision.

At the end of the third growing season (late November 2012), two trees per treatment subgroup per block were measured and weighed (a total of 288 trees).

Thirty-six whole root samples were also dug up and collected in late October, 2010, after the first growing season; these consisted of three replicates of the twelve treatment combinations then in place (randomly chosen from the ten trees in the middle of each treatment group, within three randomly chosen blocks in each field). Because nitrogen fertilization was not applied until the second season, the twelve treatment combinations consisted of the two willow cultivar types in each of the three fields, inoculated and not. After shaking off unattached or excess soil, each plant's root system was separately bagged and stored at −30 °C in the laboratory's freezers the same day.

Root systems were later thawed and vigorously rinsed and agitated by hand in distilled water. The dirty rinse water was then allowed to settle, and the sediment was set aside and refrozen as 36 rhizospheric soil samples. After cutting up the roots into 1 cm pieces, they were well mixed and further homogenized (with a washed and sterilized commercial food-beverage blender in milli-Q/0.1TE buffer solution (Black & Decker, Montréal, QC, Canada) and frozen as 36 root samples for the next molecular analysis steps.

2.3. Molecular Fungal Community Analysis

DNA from the first-year's 36 root and 36 rhizosphere samples was extracted using a MoBio Laboratories PowerSoil Extraction kit (MoBio Laboratories Inc., Carlsbad, CA, USA) according to the manufacturer's instructions, except that instead of the standard homogenization and lysis step, an MP Biomedicals FastPrep machine (MP Biomedicals Inc., Solon, OH, USA) at setting 4 was used for 25 s and six repetitions.

For cloning-sequencing, all replicates of both willow cultivars were pooled together by field and, inoculation treatment. The DNA extracted from rhizospheric soil was used to look for AM while DNA from root samples to look for EM since preliminary efforts revealed a tendency of the AM specific primers (AML1/AML2 [38]) to amplify willow DNA, attempts were made to detect both AM in the DNA extracted from rhizospheric soil, and EM in the DNA of the root samples. Conversely, in order to look for EM, universal fungal primers (ITS1F/ITS2 [39]) were used, since no EM fungi-specific primer was available. A search for AM was undertaken in both root extracts from the Sandy field and the rhizospheric soil extract, as soil extracts had not produced any AM sequences. All of this left six pooled samples for AM PCR-cloning and six for EM PCR-cloning.

A nested PCR reaction was run with samples used for AM detection. The initial PCR used the primer pair NS1/NS41 in a 25 μL reaction with reagents from the MoBio PCR CoreKit (MoBio Laboratories Inc., Carlsbad, CA, USA). Initial denaturation was at 94 °C for 3 min, followed by 30 cycles of 94 °C for 45 s, 55 °C for 1 min, and 72 °C for 1 min (modified from [40]). The reaction ended with 10 min at 72 °C One microliter of product from this first PCR was then amplified in a second 25 μL reaction using the AML1/AML2 primer pair. Initial denaturation was at 94 °C for 3 min, followed by 30 cycles of 94 °C for 30 s, 58 °C for 45 s, and 72 °C for 45 s, and ending with 10 min at 72 °C [40]. Eppendorf MasterCycler Pro thermocyclers (Eppendorf Canada Ltd., Mississauga, ON, Canada) were used, and PCR products visualized on 1% agarose gels, according to the directions for Biotium's GelRed dye (Biotium, Inc., Fremont, CA, USA) and BioRad's Molecular Imager Gel Doc XR (Bio-Rad Laboratories, Hercules, CA, USA).

The samples used for EM detection were amplified in a single step using the same MoBio reagents (MoBio Laboratories Inc., Carlsbad, CA, USA) and the primer pair ITS1F/ITS2. Initial denaturation was at 94 °C for 3 min, followed by 30 cycles of 94 °C for 45 s, 60 °C for 1 min, and 72 °C for 1 min (modified from [41]). The reaction ended with 10 min at 72 °C. The same thermocyclers and visualization method were used as described above.

PCR products were cloned using a TOPO TA Cloning Kit (Invitrogen, Carlsbad, CA, USA) for sequencing, from Life Technology, according to the manufacturer's directions. Forty-eight clones for each sample were reamplified, and the products sequenced, at McGill University and Génome Québec Innovation Centre in Montreal, QC, Canada, using the common Sanger-sequencing method.

Sequences were grouped by CD-HIT (http://weizhongli-lab.org/cd-hit/) into operational taxonomic unit (OTUs) of 98% similarity. One representative sequence from each OTU group was then randomly selected for analysis using the BLAST search tool in the National Center for Biotechnology Information (NCBI) database. Only OTUs that were matched with AM organisms were tabulated (the 18S gene is a highly conserved region and the AML1/AML2 primer pair is known to amplify other eukaryotic organisms as well).

2.4. Data Analysis

A full-factorial ANOVA was performed on the measured growth data. Growth is presented in this paper as stem basal area per hectare (SBA m^2/ha) (modified from [42]). We determined SBA by calculating transectional stem areas from diameter measurements, multiplied by mean stem number, and then divided by the average area of land occupied by one willow (in hectares). SBA was calculated for each measured tree before running the ANOVA, thereby normalizing diameter measurements for different mortality rates in each field. The actual average density of willows was quite different between fields, after mortality suffered during dry conditions in the weeks following planting (almost ~25% in Dry field, ~15% in Sandy, and ~5% in Rocky; but scattered enough in each field to be roughly equivalent across treatment combinations and blocks). Finally, because of fertilizer spillover from a nearby field during the third year, an entire row of the Sandy field had to be dropped from the analyses and the blocks in that field redrawn to maintain the correct number of each treatment combination (reducing the number of blocks in that field by four).

Growth was also measured in the second and third years by cutting down and weighing above-ground biomass, and reported as oven dry tons per hectare (ODT/ha). This was calculated from wet mass measurements taken in the field, using a conversion of 0.53 in 2011 and 0.61 in 2012, found from drying samples in the lab. The same density values for each field were then used to calculate per ha, as with SBA/ha.

In the model for our ANOVA, experimental blocks needed to be nested in the field, since block 1 of the Dry field was not the same as block 1 of the Sandy field, and so on. The model was modified accordingly, keeping a full-factorial combination of all variables besides those instances where field and (now nested) blocks were combined. Furthermore, all combinations of blocks had to be designated as random components, since they did not represent a variable we were interested in testing, but were necessary for the ANOVA to incorporate random heterogeneity within the fields. ANOVA residuals did indicate heteroscedasticity, which was corrected by logarithmic transforming our measured data.

3. Results

The experiment found no productivity difference between uninoculated and inoculated plantation willows. As Figure 1 shows, during the first 2 years, when the most trees were measured, no difference was shown between the observed results and inoculation treatment predicted mean SBA/ha values. Similarly, while the third year had fewer measured trees, and did seemingly see a small difference between SBA/ha mean predicted values, this difference was not statistically significant. This result held true whether tree height was measured, or whole trees were cut down and weighed (see Supplementary Material, Tables S1–S10).

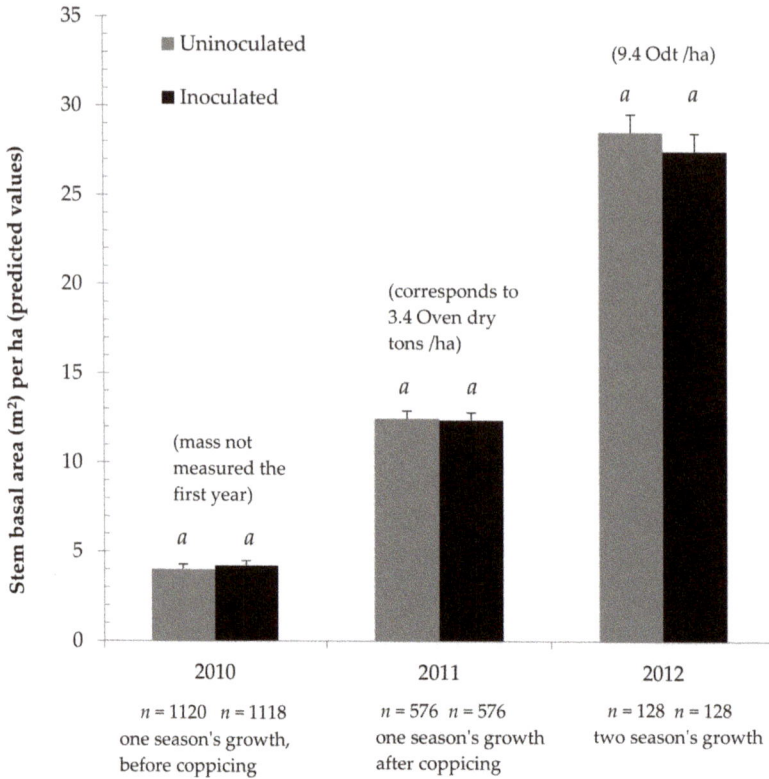

Figure 1. Cumulative growth of *Salix miyabeana* ('SX61' and 'SX64') during establishment on marginal land, treated with mycorrhizal inoculum. Error bars are standard error (SE). Within each year, similar letters (a, b) above the bars indicate that the ANOVA did not find significant differences ($p < 0.05$) between predicted means for inoculation treatment.

Table 2 shows the results of the ANOVA on 2011 SBA/ha data in more depth. This ANOVA confirms that the field sites used were different enough to affect the willows' productivity, and that the nitrogen fertilization treatment was significant as well. Table 3 provides the effect size, showing that fertilization gave a predicted mean increase of 27% SBA/ha in 2011. Even more strikingly, the Sandy field showed a predicted mean increase of 55% SBA/ha over the Rocky field in 2011. Similarly, significant biological and statistical effect sizes were seen in other years, and with the height and weight data, though fertilization was only applied in 2011, and therefore, showed less effect in 2012 (see Supplementary Material, Tables S1–S14). The overall patterns in Figure 1 and Tables 2 and 3 are representative of those seen over the three years, with height and weight as seen in the Supplementary Material (Tables S1–S10). As seen in Figure 2, the different fields did yield different results relative to one another. While the Sandy field showed markedly higher cumulative growth the first two years, cumulative growth in the Dry field caught up and equaled that of the Sandy field by the third year (Figure 2).

Table 2. ANOVA of the effects of soil type, mycorrhizal inoculation and nitrogen fertilization on stem basal area/ha (LOG transformed) of willows after one season's growth in marginal land.

Source	Nparm	DF	DFDen	*F* Ratio	Prob > *F*
Field	2	2	33	21.0017	<0001 *
Inoc	1	1	33	0.0175	0.8955
Field × inoc	2	2	33	1.1145	0.3401
Fert	1	1	33	105.1391	<0001 *
Field × fert	2	2	33	1.8694	0.1702
Inoc × fert	1	1	33	0.0187	0.8920
Field × inoc × fert	2	2	33	0.2658	0.7682
Cultivar	1	1	33	0.0465	0.8305
Field × cultivar	2	2	33	0.4076	0.6686
Inoc × cultivar	1	1	33	0.8624	0.3598
Field × inoc × cultivar	2	2	33	0.2301	0.7957
Fert × cultivar	1	1	33	0.1629	0.6891
Field × fert × cultivar	2	2	33	0.4619	0.6341
Inoc × fert × cultivar	1	1	33	2.0731	0.1593
Field × inoc × fert × cultivar	2	2	33	0.602	0.5536

All combinations of the block treatment, by itself and with the other treatment variables, were part of the model. However, the block was treated differently since it was nested in the field variable, and labeled as a random attribute. Therefore, it does not appear in this table. An asterisk (*) next to the *p*-value denotes a 5% statistical significance.

Table 3. Stem basal area per hectare (m^2/ha) of willows after one season's growth in marginal land. ANOVA predicted values and a posteriori test results according to soil type, mycorrhizal inoculation, nitrogen fertilization or willow cultivar.

Experimental Treatments	Least Squares Mean	Standard Error	Test
	Panel A: Field		Tukey's test
Dry	11.029889	0.66402733	B[1]
Rocky	10.307269	0.66402733	B
Sandy	15.964387	0.66402733	A
	Panel B: Inoculation [1]		Student's *T*-test
Not inoculated	12.478451	0.43045506	A
Inoculated	12.389246	0.43045506	A
	Panel C: Nitrogen fertilization		Student's *T*-test
Fertilized	13.921206	0.4122257	A
Unfertilized	10.946491	0.4122257	B
	Panel D: Cultivar		Student's *T*-test
'SX64'	12.486569	0.44855227	A
'SX61'	12.381127	0.44855227	A

[1] Different letters (A and B) within each panel denote a significant (≥0.05) difference between means.

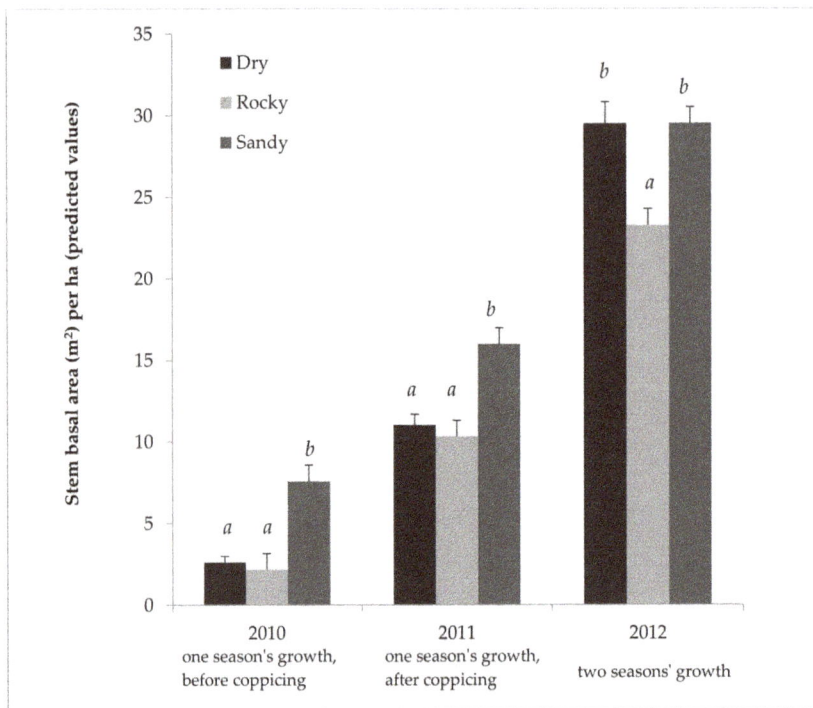

Figure 2. Cumulative growth of *Salix miyabeana* ('SX61' and 'SX64') during establishment on marginal land, in three different fields. Stem basal area predicted values (untransformed) and test results (LOG transformed) by field for all three years. Note: 2012 represents only one cultivar type ('SX61'), because of treatment interaction; however, 2010 and 2011 are the full model. Error bars are standard errors. Letters (a, b) above the bars indicate that the Tukey's test found a significant difference ($p < 0.05$) between predicted means for each field (a separate analysis each year). $n = 768$ each for Normal and Rocky fields in 2010; $n = 704$ for Sandy field in 2010; $n = 384$ in 2011 for each field; $n = 96$ in 2012 for each field.

Figure 3 shows the effect of nitrogen fertilization on growth, using SBA/ha from 2011 and 2012 to illustrate the difference in effect in each year. Figure 4 shows the fertilization effect on mass, using ODT/ha (Oven dry tons per hectare), as opposed to SBA/ha. Figure 4 also focuses on 2011, the year of the main fertilization effect, and taking into account an interaction between field and fertilization as indicated by the ANOVA, breaks down the effect by field. The effect of fertilization is even more marked in this case, as fertilization gave a predicted mean increase of 51% ODT/ha in 2011 in the Sandy field.

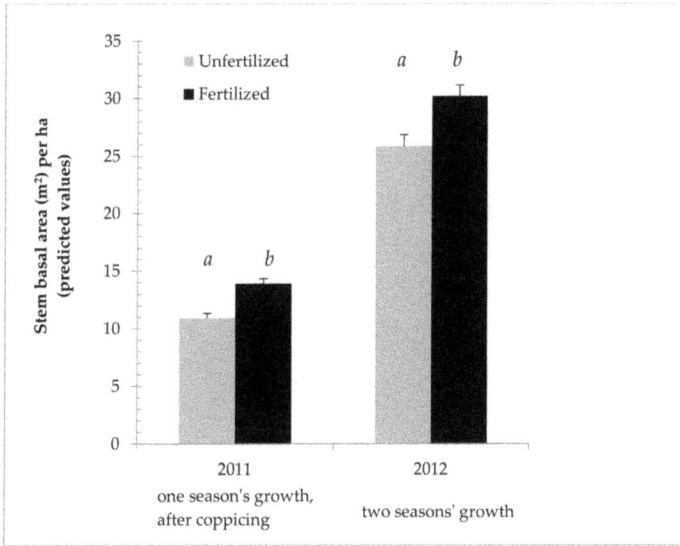

Figure 3. Cumulative growth of *Salix miyabeana* ('SX61' and 'SX64') during establishment on marginal land. Stem basal area predicted values (untransformed) and ANOVA results (LOG transformed) for the year in which fertilization treatment was applied (2011), as well as for the following year (past growth, plus any residual nitrogen in the soil). Note: Error bars are standard errors. Different letters (a, b) above the bars indicate a significant difference ($p < 0.05$). $n = 576$ in 2011 for each fertilization treatment; $n = 144$ in 2012 for each fertilization treatment.

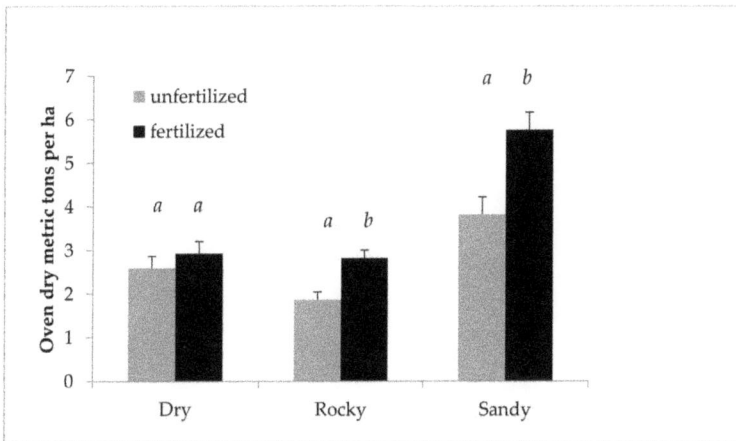

Figure 4. Cumulative growth of *Salix miyabeana* ('SX61' and 'SX64'). Oven dry tons per ha predicted values (untransformed) by field for 2011. Note: Error bars are standard errors. Different letters (a, b) above the bars indicate a significant difference ($p < 0.05$). The ANOVA (performed on LOG transformed values), including all variables in the model, found a likely interaction between field and fertilization, prompting the breakdown by field. $n = 192$ for each field and fertilization treatment.

3.1. Sequence Library Results

Out of a total of 384 sequenced clones from 18S amplicons, 269 were identified as AM sequences and were regrouped into eight OTUs at 98% similarity (Table 4). This represented 70% specificity for primers AML1 and AML2, nested following amplification with primers NS1 and NS41. Forty-three other OTUs were non-specific amplifications, 35 of them being from eukaryotic organisms. Wild AM fungi (in our uninoculated group) included three *Diversispora* OTUs, one OTU of the closely related *Archeospora/Ambispora* Ambispora, and three *Glomeromycota*. OTUs. The inoculum only added a single unique OTU, identified as uncultured *Archaeospora* with one sequence (possibly an artifact), but did appreciably increase the relative *Glomus* sequence numbers, nor did it eliminate the *Diversispora* numbers entirely. Also, somewhat surprisingly, the inoculum increased relative *Ambispora* sequence numbers.

Table 4. Arbuscular mycorrhizal sequences found in rhizosphere samples from biomass plantation willows after one season's growth in marginal land, by mycorrhizal inoculation treatment.

OTU	Not Inoculated	Inoculated	Name and GI [1] of Closest Match in NCBI Database
OTU-10	1	-	*Diversispora celata*: 224586636
OTU-11	32	-	*Diversispora* sp. W4538: 342298391
OTU-12	2	-	Uncultured *Diversispora*: 398649715
OTU-13	19	47	*Glomus* sp. MC27: 334683211
OTU-14	31	39	Uncultured *Glomus*: 401664149
OTU-15	10	78	Uncultured *Ambispora*: 308084344
OTU-16	-	1	Uncultured *Archaeospora*: 308084350
OTU-17	9	-	*Glomeromycota* sp. MIB 8442: 328541374

[1] GI: GenInfo Identifier.

The ITS sequences grouped into 40 OTUs at 98% similarity. Out of them, 16 OTUs represented two or more sequences, and encompassed almost 90% of the 288 total sequenced clones (Table 5). Five OTUs were mycorrhizal, according to NCBI database notes: *Pulvinula* (*P.*) *constellation, Hymenogaster griseus,* uncultured *Sebacinales,* uncultured ectomycorrhizal fungus, and uncultured *Salix* L. associated fungus. The inoculum did not have a clear effect, with *P. constellatio* increasing in number but the uncultured ectomycorrhizal fungus decreasing appreciably. The most numerous sequence, the uncultured *Salix* associated fungus, was unaffected. One OTU identified as *Hebeloma* (Fr.). cf. *crustuliniforme* was not labeled as EM in the NCBI notes, but is known to be ectomycorrhizal [43]. It was only present in the inoculated samples, and is so closely related to the inoculated EM species (*Hebeloma longicaudum*) that the five sequences found could very well be the inoculum detected in low numbers.

Table 5. Ectomycorrhizal sequences (and sequences of other fungi) found in rinsed root samples from biomass plantation willows after one season's growth in marginal land, by mycorrhizal inoculation treatment.

OTU	Not Inoculated	Inoculated	Name and GI [1] of Closest Match in NCBI Database	Phylum (Division)
OTU-2	5	3	*Cladosporium cladosporioides*: 356484684	Ascomycota
OTU-3	2	2	*Epicoccum nigrum*: 404474360	Ascomycota
OTU-6	8	8	*Magnusiomyces* 357934165	Ascomycota
OTU-9	9	24	* *Pulvinula constellatio*: 10178659	Ascomycota
OTU-11	2	-	*Trichurus spiralis*: 237872399	Ascomycota
OTU-17	-	2	Uncultured *Geopora*: 295291451	Ascomycota
OTU-18	2	2	Uncultured *Hyaloscyphaceae*: 193850652	Ascomycota
OTU-21	-	5	*Hebeloma crustuliniforme* 2 UE-2011: 359751813	Basidiomycota

Table 5. *Cont.*

OTU	Not Inoculated	Inoculated	Name and GI [1] of Closest Match in NCBI Database	Phylum (Division)
OTU-22	-	3	* *Hymenogaster griseus*: 387145960	Basidiomycota
OTU-28	2	-	Uncultured 334683052	Basidiomycota
OTU-30	1	3	* Uncultured *Sebacinales*: 264716693	Basidiomycota
OTU-33	16	1	* Uncultured ectomycorrhizal fungus: 404247775	unknown
OTU-34	71	70	* Uncultured fungus (from *Salix* rhiz. [2]): 402535072	unknown
OTU-35	4	3	Uncultured soil fungus: 195964332	unknown
OTU-36	1	5	*Olpidium brassicae*: 87159723	unknown
OTU-40	2	-	*Entrophospora* sp. JJ38: 15809596	Glomeromycota

Names marked with an asterisk (*) are EM species according to NCBI entry notes. [1] GI: GenInfo Identifier; [2] rhizosphere.

4. Discussion

Willow growers with access to marginal land should be interested in our data, as it represents farm-scale, real-world results across different field types, both with and without nitrogen fertilization. The success of the less drained, wetter Sandy field is promising for those with similar land, though the apparent early growth benefit in the Sandy field could simply have resulted from the more clay-rich Dry and Rocky fields being harder for the willow roots to penetrate. At the very least, a farmer with a rock-strewn, clay-heavy field like our Rocky field might decide not to try cultivating willows after seeing our results, especially if it is deficient in phosphorous minerals (as ours was). Likewise, a farmer might try to reduce fertilizer costs (or to minimize the climate impacting effect of such fertilizers), given that our data indicates that a 51% increase in growth is possible in particular fields with a fairly modest nitrogen addition (75 kg/ha). However, although such results are of interest, they must be seen as ancillary to those concerning inoculation. The entire design of the experiment was geared towards studying the effect of inoculation. Field selection and characterization, as well as fertilization treatment, were chosen to test inoculation rather than drainage or fertilizer effect per se. Neither drainage nor fertilization were monitored in a sufficiently systematic manner as to allow us to ascertain their effects across different treatment values, or the upper and lower bounds.

The likeliest conclusion to draw from the experiment is that inoculation with the two mycorrhizal species used, *R. irregulare* and *H. longicaudum*, does not appreciably benefit the growth of biomass willows cultivated in marginal agricultural fields. Neither cultivar of willow tested showed a significant difference across inoculation treatments. Neither did the full factorial ANOVA show any mixed effects between inoculation and field site, or inoculation and fertilization treatment. So, inoculation was not shown to be of benefit within even one of the tested ranges of soil/fertility conditions.

The AM species used, *R. irregulare*, has been shown to benefit willow growth [17] in sterile soil greenhouse experiments. That it did not behave in this manner in this experiment strongly suggests that wild mycorrhizal species mask the treatment. In fact, several wild AM species were found (Table 4), as well as several wild probable EM species (Table 5). This is not that surprising, as corn and soybean crops that would have been present in previous years could host AM species. Annual monocrop systems are not ideal for diverse mycorrhizal communities [44], but many AM species are known to survive [45,46]. Such wild strains could benefit uninoculated control willows just as much as the inoculum might benefit treatment willows, or even outcompete the inoculum strain with the same result. The wild EM species are more unexpected however, since willow specific EM fungi are not likely to be associated with annual agricultural crops [47]. In this case though, wild willows bordered the sandy site, and all three sites were fairly close (within a quarter km) of established willow plantations. Such locally adapted, willow-specific EM strains could easily have outcompeted with, or masked, the introduced strain.

The pooled sequence libraries for the inoculation treatments do indicate that the inoculum was viable and competitive enough to be detected after the first year of growth in the field (the possibility that it was not had to be considered as an alternative hypothesis to explain the data). Furthermore, viability of the inoculum was confirmed with a small greenhouse test, as mentioned in the methods. The AM species (Table 4) were present at roughly twice the numbers of those of the relative *Glomus* species in the inoculated plots. The EM species (Table 5) are more difficult to interpret, but there were some *Hebeloma* sequences in the treated plots that were not present in the untreated plots.

The experimental results suggest that inoculation with AM fungi in general will not benefit willows grown in agricultural settings, as such willows already have the potential to interact with numerous, wild AM strains. A diversity of AM species appears to be ubiquitous to different degrees in agricultural soil [46].

However, we cannot conclude that another strain of EM fungi would not benefit willows grown in such a setting. The *Hebeloma* species selected has been shown to be associated with a very closely related species of tree (*Populus*), but greenhouse or pot experiments to confirm positive interaction with biomass willows had not been conducted prior to this experiment. Not many EM fungi are commercially available in the quantities needed in an agricultural setting, and the EM inoculum was the only one available at the time of our experiment's setup. Because EM fungi are more host specific than AM fungi [48,49], and are less likely to be present in agricultural fields [50,51], we encourage researchers to test other species and strains as they become available for agricultural applications.

For willow growers, our results caution against investing in mycorrhizal inoculation, unless a strain has proven to be beneficial in field tests with willows (or in pot tests using unsterilized farm soil). We would still advise growers to apply mycorrhizal inoculation if planting on semi-sterile soil, such as heavily fungicide-applied sands, mine tailings, etc. However, our results suggest that such an inoculation could simply be a diluted soil slurry from a healthy agricultural field nearby.

Supplementary Materials: The following are available online at http://www.mdpi.com/1999-4907/9/4/185/s1, Table S1: 2010 Height ANOVA results (LOG transformed), Table S2: 2010 Height (cm) ANOVA predicted values and test results, Table S3: 2011 Height ANOVA results (LOG transformed), Table S4: 2011 Height (cm) ANOVA predicted values and test results, Table S5: 2012 Height ANOVA results (LOG transformed), Table S6: 2012 Height cm ANOVA predicted values and test results, Table S7: 2011 Oven dry tons /ha ANOVA results (LOG transformed), Table S8: 2011 Oven dry tons /ha ANOVA predicted values and test results, Table S9: 2012 Oven dry tons /ha ANOVA results (LOG transformed), Table S10: 2012 Oven dry tons /ha ANOVA predicted values and test results, Table S11: 2010 Stem basal area ANOVA results (LOG transformed), Table S12: 2010 Stem basal area per hectare (m^2/ha) ANOVA predicted values and test results, Table S13: 2012 Stem basal area ANOVA results (LOG transformed), Table S14: 2012 Stem basal area per hectare (m^2/ha) ANOVA predicted values and test results, Figure S1: Experimental design (each field randomized according to the schema, rocky field shown as an example). Widths compressed substantially compared to heights to fit on page; actual measurements given on side and top of plan. B refers to block; F to fertilization; M to inoculation. + sign means the subplot was fertilized or inoculated; − sign that it was not. SM stands for cultivar *Salix miyabeana* 'SX61', SS for cultivar 'SX64'.

Acknowledgments: This study was conducted with the financial assistance of Agri-Food Innovation Support Program of the Quebec Ministry of Agriculture, Fisheries and Food. The authors gratefully thank Agro Énergie for their valuable collaboration and Stéphane Daigle for his assistance in the statistical analyses.

Author Contributions: M.S.-A., M.L., and T.J.P. conceived and designed the experiments; T.J.P. performed the experiments; T.J.P. analyzed the data; M.S.-A. contributed reagents, materials, and analysis tools; T.J.P. wrote the paper; W.G.N. provided extensive proof-reading and suggested edits.

Conflicts of Interest: The authors declare no conflict of interest.

References

1. Liu, T.; Ma, Z.; McConkey, B.; Kulshreshtha, S.; Huffman, T.; Green, M.; Liu, J.; Du, Y.; Shang, J. Bioenergy production on marginal land in Canada: Potential, economic feasibility, and greenhouse gas emissions impacts. *Appl. Energy* **2017**, *205*, 477–485. [CrossRef]

2. Don, A.; Osborne, B.; Hastings, A.; Skiba, U.; Carter, M.S.; Drewer, J.; Flessa, H.; Freibauer, A.; Hyvönen, N.; Jones, M.B.; et al. Land-use change to bioenergy production in Europe: Implications for the greenhouse gas balance and soil carbon. *GCB Bioenergy* **2012**, *4*, 372–391. [CrossRef]

3. Labrecque, M.; Teodorescu, T.I. Influence of plantation site and wastewater sludge fertilization on the performance and foliar nutrient status of two willow species grown under SRIC in southern Quebec (Canada). *For. Ecol. Manag.* **2001**, *150*, 223–239. [CrossRef]

4.   Guidi, W.; Pitre, F.; Labrecque, M. Short-rotation coppice of willows for the production of biomass in Eastern Canada. In *Biomass Now-Sustainable Growth and Use*; Matovic, M.D., Ed.; InTech Open Science: Rijeka, Croatia, 2013; pp. 421–448.

5.   Kopp, R.F.; White, E.H.; Abrahamson, L.P.; Nowak, C.A.; Zsuffa, L.; Burns, K.F. Willow biomass trials in Central New York State. *Biomass Bioenergy* **1993**, *5*, 179–187. [CrossRef]

6.   Buchholz, T.; Volk, T. Improving the profitability of willow crops—Identifying opportunities with a crop budget model. *Bioenergy Res.* **2011**, *4*, 85–95. [CrossRef]

7.   Caslin, B.; Finnan, J.; McCracken, A. *Short Rotation Coppice Willow Best Practice Guidelines*; AFBI Agri-Food & Bioscience Institute: Belfast, England, 2010.

8.   Wang, B.; Qiu, Y.L. Phylogenetic distribution and evolution of mycorrhizas in land plants. *Mycorrhiza* **2006**, *16*, 299–363. [CrossRef] [PubMed]

9.   Whiteside, M.D.; Digman, M.A.; Gratton, E.; Treseder, K.K. Organic nitrogen uptake by arbuscular mycorrhizal fungi in a boreal forest. *Soil Biol. Biochem.* **2012**, *22*, 7–13. [CrossRef] [PubMed]

10. Smith, S.E.; Jakobsen, I.; Grønlund, M.; Smith, F.A. Roles of arbuscular mycorrhizas in plant phosphorus nutrition: Interactions between pathways of phosphorus uptake in arbuscular mycorrhizal roots have important implications for understanding and manipulating plant phosphorus acquisition. *Plant Physiol.* **2011**, *156*, 1050–1057. [CrossRef] [PubMed]

11. Liu, J.; Maldonado-Mendoza, I.; Lopez-Meyer, M.; Cheung, F.; Town, C.D.; Harrison, M.J. Arbuscular mycorrhizal symbiosis is accompanied by local and systemic alterations in gene expression and an increase in disease resistance in the shoots. *Plant J.* **2007**, *50*, 529–544. [CrossRef] [PubMed]

12. St-Arnaud, M.; Vujanovic, V. Effect of the arbuscular mycorrhizal symbiosis on plant diseases and pests. In *Mycorrhizae in Crop Production*; Hamel, C., Plenchette, C., Eds.; Haworth Food & Agricultural Products Press: Binghamton, NY, USA, 2007; pp. 67–122.

13. Lekberg, Y.; Koide, R.T. Is plant performance limited by abundance of arbuscular mycorrhizal fungi? A meta-analysis of studies published between 1988 and 2003. *New Phytol.* **2005**, *168*, 189–204. [CrossRef] [PubMed]

14. Puettsepp, U.; Rosling, A.; Taylor, A.F.S. Ectomycorrhizal fungal communities associated with *Salix viminalis* L. and *S. dasyclados* Wimm. clones in a short-rotation forestry plantation. *For. Ecol. Manag.* **2004**, *196*, 413–424. [CrossRef]

15. Paradi, I.; Baar, J. Mycorrhizal fungal diversity in willow forests of different age along the river Waal, The Netherlands. *For. Ecol. Manag.* **2006**, *237*, 366–372. [CrossRef]

16. Ryberg, M.; Andreasen, M.; Bjoerk, R.J. Weak habitat specificity in ectomycorrhizal communities associated with *Salix herbacea* and *Salix polaris* in alpine tundra. *Mycorrhiza* **2011**, *21*, 289–296. [CrossRef] [PubMed]

17. Van der Heijden, E.W. Differential benefits of arbuscular mycorrhizal and ectomycorrhizal infection of *Salix repens*. *Mycorrhiza* **2001**, *10*, 185–193. [CrossRef]

18. Hashimoto, Y.; Higuchi, R. Ectomycorrhizal and arbuscular mycorrhizal colonization of two species of floodplain willows. *Mycoscience* **2003**, *44*, 339–343. [CrossRef]

19. Milne, J.M.; Ennos, R.A.; Holingsworth, R.M. Vegetation influence on ectomycorrhizal inoculum available to sub-arctic willow (*Salix lapponum* L.) planted in an upland site. *Bot. J. Scotl.* **2006**, *58*, 19–34. [CrossRef]

20. Becerra, A.G.; Nouhra, E.R.; Silva, M.P.; McKay, D. Ectomycorrhizae, arbuscular mycorrhizae, and dark-septate fungi on *Salix humboldtiana* in two riparian populations from central Argentina. *Mycoscience* **2009**, *50*, 343–352. [CrossRef]

21. Baxter, J.W.; Dighton, J. Ectomycorrhizal diversity alters growth and nutrient acquisition of grey birch (*Betula populifolia*) seedlings in host–symbiont culture conditions. *New Phytol.* **2001**, *152*, 139–149. [CrossRef]

22. Fransson, P.M.A.; Toljander, Y.K.; Baum, C.; Weih, M. Host plant-ectomycorrhizal fungus combination drives resource allocation in willow: Evidence for complex species interaction from a simple experiment. *Ecoscience* **2013**, *20*, 112–121. [CrossRef]

23. Corredor, A.H.; van Rees, K.; Vujanovic, V. Changes in root-associated fungal assemblages within newly established clonal biomass plantations of *Salix* spp. *For. Ecol. Manag.* **2012**, *282*, 105–114. [CrossRef]

24. Six, J.; Frey, S.D.; Thiet, R.K.; Batten, K.M. Bacterial and fungal contributions to carbon sequestration in agroecosystems. *Soil Sci. Soc. Am.* **2006**, *70*, 556–569. [CrossRef]

25. Douds, D.D.; Nagahashi, G.; Shenk, J.E. Frequent cultivation prior to planting to prevent weed competition results in an opportunity for the use of arbuscular mycorrhizal fungus inoculum. *Renew. Agric. Food Syst.* **2012**, *27*, 251–255. [CrossRef]

26. Garbaye, J.; Churin, J.L. Growth stimulation of young oak plantations inoculated with the ectomycorrhizal fungus *Paxillus involutus* with special reference to summer drought. *For. Ecol. Manag.* **1997**, *98*, 221–228. [CrossRef]

27. Baum, C.; Stetter, U.; Makeschin, F. Growth response of *Populus trichocarpa* to inoculation by the ectomycorrhizal fungus *Laccaria laccata* in a pot and a field experiment. *For. Ecol. Manag.* **2002**, *163*, 1–8. [CrossRef]

28. Duponnois, R.; Plenchette, C.; Prin, Y.; Ducousso, M.; Kisa, M.; Bâ, A.M.; Galiana, A. Use of mycorrhizal inoculation to improve reafforestation process with Australian Acacia in Sahelian ecozones. *Ecol. Eng.* **2007**, *29*, 105–112. [CrossRef]

29. Quoreshi, A.M.; Piché, Y.; Khasa, D.P. Field performance of conifer and hardwood species 5 years after nursery inoculation in the Canadian Prairie Provinces. *New For.* **2008**, *35*, 235–253. [CrossRef]

30. Chapdelaine, A.; Dalpé, Y.; Hamel, C.; St Arnaud, M. Arbuscular mycorrhizal inoculation of ornamental trees in nursery. In *Mycorrhiza Works*; Feldmann, F., Kapulnik, Y., Baar, J., Eds.; Deutsche Phytomedizinische Gesellschaft, Spectrum Phytomedizin: Braunschweig, Germany, 2008; pp. 46–55. ISBN 978-3-941261-01-3.

31. Quoreshi, A.M.; Khasa, D.P. Effectiveness of mycorrhizal inoculation in the nursery on root colonization, growth, and nutrient uptake of aspen and balsam poplar. *Biomass Bioenergy* **2008**, *32*, 381–391. [CrossRef]

32. Fillion, M.; Brisson, J.; Guidi, W.; Labrecque, M. Increasing phosphorus removal in willow and poplar vegetation filters using arbuscular mycorrhizal fungi. *Ecol. Eng.* **2011**, *37*, 199–205. [CrossRef]

33. Bissonnette, L.; St-Arnaud, M.; Labrecque, M. Phytoextraction of heavy metals by two Salicaceae clones in symbiosis with arbuscular mycorrhizal fungi during the second year of a field trial. *Plant Soil* **2010**, *332*, 55–67. [CrossRef]

34. Sieverding, E.; da Silva, G.A.; Berndt, R.; Oehl, F. *Rhizoglomus*, a new genus of the *Glomeraceae*. *Mycotaxon* **2015**, *129*, 373–386. [CrossRef]

35. Stockinger, H.; Walker, C.; Schüßler, A. *Glomus intraradices* DAOM197198, a model fungus in arbuscular mycorrhiza research, is not *Glomus intraradices*. *New Phytol.* **2009**, *183*, 1176–1187. [CrossRef] [PubMed]

36. Conseil de Production Végétale du Québec. *Méthodes D'analyse des Sols, des Fumiers et des Tissus Végétaux*; Ministère de l'Agriculture, des Pêcheries et de l'Alimentation. Gouvernement du Québec: Ste-Foy, QC, Canada, 1998. (In French)

37. U.S. Environmental Protection Agency. Method 200.7. Inductively coupled plasma-atomic emission spectrometric method for trace element analysis of water and waste. In *Methods for Chemical Analysis of Water and Wastes*; U.S. Environmental Protection Agency: Cincmnati, OH, USA, 1983; EPA-600/4-79-020.

38. Lee, J.; Lee, S.; Young, J.P. Improved primers for the detection and identification of arbuscular mycorrhizal fungi. *FEMS Microb. Ecol.* **2008**, *65*, 339–349. [CrossRef] [PubMed]

39. Ghannoum, M.A.; Jurevic, R.J.; Mukherjee, P.K.; Cui, F.; Sikaroodi, M.; Naqvi, A.; Gillevet, P.M. Characterization of the oral fungal microbiome (mycobiome) in healthy individuals. *PLoS Pathog* **2010**, *6*, e1000713. [CrossRef] [PubMed]

40. Hassan, S.E.; Boon, E.; St-Arnaud, M.; Hijri, M. Molecular biodiversity of arbuscular mycorrhizal fungi in heavy metal polluted soils. *Mol. Ecol.* **2011**, *20*, 3469–3483. [CrossRef] [PubMed]

41. Bell, T.H.; Hassan, S.E.; Lauron-Moreau, A.; Al Otaibi, F.; Hijri, M.; Yergeau, E.; St-Arnaud, M. Linkage between bacterial and fungal rhizosphere communities in hydrocarbon-contaminated soils is related to plant phylogeny. *ISME J.* **2014**, *8*, 331–343. [CrossRef] [PubMed]

42. McKnight, J.S. Black Willow (*Salix nigra*) Marsh. Available online: http://vmpincel.ou.edu/oliver/pdf/Salix_nigraBlackWillow.pdf (accessed on 30 March 2018).

43. Aanen, D.K.; Kuyper, T.; Mes, T.H.M.; Hoekstra, R. The evolution of reproductive isolation in the ectomycorrhizal *Hebeloma crustuliniforme* aggregate (basidiomycetes) in northwestern Europe: A phylogenetic approach. *Evolution* **2000**, *54*, 1192–1206. [CrossRef] [PubMed]

44. Verbruggen, E.; Kiers, T.E. Evolutionary ecology of mycorrhizal functional diversity in agricultural systems. *Evolut. Appl.* **2010**, *3*, 547–560. [CrossRef] [PubMed]

45. Beauregard, M.S.; Hamel, C.; Gauthier, M.P.; Zhang, T.; Tan, C.S.; Welacky, T.; St-Arnaud, M. Various forms of organic and inorganic P fertilizers did not negatively affect AM fungi communities and biomass in a maize-soybean rotation system. *Mycorrhiza* **2013**, *23*, 143–154. [CrossRef] [PubMed]

46. Moebius-Clune, D.J.; Moebius-Clune, B.N.; van Es, H.M.; Pawlowska, T.E. Arbuscular mycorrhizal fungi associated with a single agronomic plant host across the landscape: Community differentiation along a soil textural gradient. *Soil Biol. Biochem.* **2013**, *64*, 191–199. [CrossRef]

47. Tedersoo, L.; May, T.W.; Smith, M.E. Ectomycorrhizal lifestyle in fungi: Global diversity, distribution, and evolution of phylogenetic lineages. *Mycorrhiza* **2010**, *20*, 217–263. [CrossRef] [PubMed]

48. Newton, A.C.; Haigh, J.M. Diversity of ectomycorrhizal fungi in Britain: A test of the species–area relationship, and the role of host specificity. *New Phytol.* **1998**, *138*, 619–627. [CrossRef]

49. Kilronomos, J.N. Host-specificity and functional diversity among arbuscular mycorrhizal fungi. *Microb. Biosyst. N. Front.* **2000**, *1*, 845–851.

50. Dickie, I.A.; Reich, P.B. Ectomycorrhizal fungal communities at forest edges. *J. Ecol.* **2005**, *93*, 244–255. [CrossRef]

51. Oehl, F.; Sieverding, E.; Ineichen, K.; Mäder, P.; Boller, T.; Wiemken, A. Impact of land use intensity on the species diversity of arbuscular mycorrhizal fungi in agroecosystems of Central Europe. *Appl. Environ. Microbiol.* **2003**, *69*, 2816–2824. [CrossRef] [PubMed]

forests

MDPI

Article

The Evaluation of Radiation Use Efficiency and Leaf Area Index Development for the Estimation of Biomass Accumulation in Short Rotation Poplar and Annual Field Crops

Abhishek Mani Tripathi [1,*], Eva Pohanková [1,2], Milan Fischer [1,2], Matěj Orság [1,2], Miroslav Trnka [1,2], Karel Klem [1] and Michal V. Marek [1]

[1]   Global Change Research Institute CAS, Bělidla 986/4a, 603 00 Brno, Czech Republic;
    eva.pohankova@seznam.cz (E.P.); fischer.milan@gmail.com (M.F.); orsag.matej@gmail.com (M.O.);
    mirek_trnka@yahoo.com (M.T.); klem.k@czechglobe.cz (K.K.); marek.mv@czechglobe.cz (M.V.M.)
[2]   Department of Agrosystems and Bioclimatology, Faculty of Agronomy, Mendel University in Brno,
    Zemědělská 1, 613 00 Brno, Czech Republic
*   Correspondence: manicfre@gmail.com or tripathi.a@czechglobe.cz; Tel.: +420-511-192-215

Received: 15 February 2018; Accepted: 23 March 2018; Published: 27 March 2018

Abstract: We evaluated the long-term pattern of leaf area index (LAI) dynamics and radiation use efficiency (RUE) in short rotation poplar in uncoppice (single stem) and coppice (multi-stem) plantations, and compared them to annual field crops (AFCs) as an alternative for bioenergy production while being more sensitive to weather fluctuation and climate change. The aim of this study was to evaluate the potential of LAI and RUE as indicators for bioenergy production and indicators of response to changing environmental conditions. For this study, we selected poplar clone J-105 (*Populus nigra* L. × *P. maximowiczii* A. Henry) and AFCs such as barley (*Hordeum vulgare* L.), wheat (*Triticum aestivum* L.), maize (*Zea mays* L.), and oilseed rape (*Brassica napus* L.), and compared their aboveground dry mass (AGDM) production in relation to their LAI development and RUE. The results of the study showed the long-term maximum LAI (LAI_{max}) to be 9.5 in coppice poplar when compared to AFCs, where LAI_{max} did not exceed the value 6. The RUE varied between 1.02 and 1.48 g MJ^{-1} in short rotation poplar and between 0.72 and 2.06 g MJ^{-1} in AFCs. We found both LAI and RUE contributed to AGDM production in short rotation poplar and RUE only contributed in AFCs. The study confirms that RUE may be considered an AGDM predictor of short rotation poplar and AFCs. This may be utilized for empirical estimates of yields and also contribute to improve the models of short rotation poplar and AFCs for the precise prediction of biomass accumulation in different environmental conditions.

Keywords: aboveground biomass; bioenergy; leaf area index; *Populus*

---

1. Introduction

IPCC [1] reported that the energy security and mitigation of greenhouse gas emissions are major challenges to meet global energy demand. Moreover, the entire food system is considered as a source of greenhouse gas emissions and primary production is by far the most important component [2]. To reduce greenhouse gas emissions, mitigate climate change impacts and, at the same time, fulfill the requirements of food and energy demands, it is important to enhance the sources of biomass production in the short term such as short rotation forestry (SRF) and bioenergy feedstocks (woody and non-woody) [3,4].

However, incentives are needed to persuade crop and livestock producers, agro-industries, and ecosystem managers to adopt good practices for mitigating climate change and improving the productivity of crops in different environmental conditions [5]. The cultivation of biomass, mainly poplars (*Populus*) and willows (*Salix*) in SRF, particularly in coppice management, also known as short rotation coppice (SRC), has considerable potential to meet these requirements. Besides being a production source of biomass for bioenergy, SRC can contribute to the improvement of ecosystem services such as increasing biodiversity, reducing soil erosion, improving water use efficiency, phytoremediation, and reducing greenhouse gas emissions [6,7]. Moreover, in some regions, AFCs also represent an alternative solution for biomass and bioenergy production [8,9]. The suitability of the species for bioenergy production can vary with local soil and climatic conditions and may also show differences due to their sensitivity to changing environmental conditions in the context of climate change.

For this reason, indicators that would make easier comparison of species differences and the responses to climate change in aboveground dry mass (AGDM) production are of high interest. The cultivation of biomass under field conditions depends on the productivity and growth of the species. The variation in productivity and growth is closely linked to the amount of intercepted radiation, largely determined by leaf area index (LAI) [10] which is defined as the one sided green leaf area per unit ground surface area [11], the interception of photosynthetically active radiation (PAR; 400–700 nm), and by interactions with numerous environmental factors and physiological characteristics [12,13].

An intercepted PAR depends mainly on the plant's LAI, canopy architecture (affecting how much LAI is effectively exposed to light), and the physiological capacity to intercept radiation. Thus, plant growth and productivity under field conditions are primarily dependent on the potential of the canopy to intercept incoming PAR and to convert it into biomass [14,15].

The amount of biomass produced per unit of intercepted light is called radiation use efficiency (RUE) and largely determines the growth and biomass in both woody (trees) and non-woody (AFCs) plants [16–18]. Monteith [12] defined the concept of RUE as the ratio between the quantity of dry mass (DM) production and the quantity of accumulated intercepted PAR (IPAR).

In numerous studies, a linear relationship between IPAR, RUE, LAI, leaf area duration (LAD), and biomass production has been found in bioenergy feedstocks and AFCs such as poplar, willow, barley, wheat, maize, and oilseed rape [13,19–21]. The actual RUE and LAI are affected by several factors such as environmental conditions (temperature, water, and nutrient availability [22,23]), phenology, and LAD [24,25].

The RUE also varies in different vegetation types for example in woody and herbaceous plants or management type such as coppiced and uncoppiced, as well as plant density and growth rate [26–31]. Usually, RUE values are higher in AFCs compared to fast growing woody plantations. On the other hand, in most studies the maximum LAI (LAI_{max}) is higher in woody plants [32–34].

To evaluate the contribution of RUE and LAI to AGDM accumulation, several studies on bioenergy feedstocks in different regions were conducted. However, the understanding of these relationships is still low and particularly the knowledge of the differences between plant species are needed to improve bioenergy feedstock growth and production analysis [35–40].

In the present study, we evaluated the dynamics of LAI, LAD, RUE, and AGDM for the growth of poplar clone J-105 (*Populus nigra* L. × *P. maximowiczii* A. Henry) in a high-density short rotation (coppiced and uncoppiced management) on former arable land, AFCs-spring barley (SB, *Hordeum vulgare* L.), winter wheat (WW, *Triticum aestivum* L.), silage maize (SM, *Zea mays* L.), and oilseed rape (OR, *Brassica napus* L.).

We hypothesized the role of RUE and LAI development in determining AGDM production. The aims of this study were: (1) to determine the variation of LAI development, LAD, and RUE between SRC poplar and AFCs; (2) to investigate the relationship between LAI, LAD, RUE, and AGDM in SRC poplar and AFCs as this may help to understand differences in AGDM production in different plant types and species, thus, improving the forecasting of the productivity of SRC poplar and AFCs.

2. Materials and Methods

2.1. Site Description and Treatments

The experiment was conducted in a Domanínek research site (the Czech Republic 49°31′ N, 16°14′ E; 530 m a.s.l.). The research site is located on a slight slope of 3–5° with an eastern aspect and characterized by a temperate climate which is typical for this part of Central Europe with mingling continental and maritime influences [41,42].

The long-term (2002–2014) total annual amount of precipitation and mean annual air temperature at the research site was 603 mm and 7.6 °C, respectively. The area is relatively dry due to low annual precipitation; however, the experimental site is appropriate for SRC and AFCs cultivation due to a deep soil profile with optimum water holding capacity [41]. The soil type is luvic cambisol, influenced by gleyic processes [41] and the texture vary from silt loam (0–0.3 m) to loam (0.3–2.0 m). The soil sampling was carried out prior to planting the poplar clone J-105 and sowing SB, WW, SM, and OR, respectively. The basic soil characteristic such as the total nitrogen (N), organic matter, soil type, bulk density, soil pH, and available nutrients are shown in Table 1.

Table 1. The selected soil characteristics of the experimental site taken from Trnka et al. [41].

Soil Characteristics	Units	Depth (cm)			
		0–24	24–66	66–94	94–130+
Silt	wt %	50	46.1	38.7	19.6
Clay	wt %	15.8	26.3	18.6	13.3
Bulk density	g/cm^3	1.55	1.64	1.59	1.64
Organic matter	wt %	2.65	0.28	0.14	0.14
Total nitrogen	wt %	0.16	<0.05	<0.05	<0.05
pH (KCl)		5.9	5.4	4	3.4
Available P	mg/kg	148	1.3	0.9	24
Available K	mg/kg	15	91	62	76
Available Mg	mg/kg	143	230	278	291
Available Ca	mg/kg	1230	1353	748	652

The poplar clone J-105 plantation of 2.85 ha was established in April 2002 on former arable land. Hardwood cuttings were planted in a double-row design with spacing of 0.7 m within the rows and inter-row distances of 2.5 m, resulting in a density 9216 trees ha^{-1}. Irrigation, fertilizers, or pesticides were not applied before or after planting. The measurements were conducted during two consecutive rotations. The first rotation spanned from 2002–2009 (uncoppiced-single stem stand) and the second rotation from 2010–2015 (coppiced-multi stem stand). The single stem stand was cut 10–15 cm above the ground level in 2009 (total age of trees from planting to harvest was eight years) and subsequently in 2010 the multi stem stand coppice culture was established. AFCs were sown in close distance (ca. 500 m to 1 km) to woody poplar plantation. These crops were cultivated between 2011 and 2013 with two cultivars for SB (Tolar-SB-T and Bojos-SB-B) and one cultivar for WW (Etela-WW-E), SM (MON3301-SM-M), and OR (Rohan-OR-R), respectively. The experimental designs for all AFCs were identical for all the years (2011–2013). These AFCs were established on standardized 12 m^2 (1.5 × 8 m) plots and each AFC was cultivated in three replicates (A, B, and C) for measurements [43]. The AFCs were sown with appropriate seeding rates for given conditions and during the sowing, the AFCs nitrogen (N) fertilizer was applied with optimum doses (see Table 2).

Table 2. The sowing of non-woody annual field crops (AFCs) spring barley (SB), winter wheat (WW), silage maize (SM), and oilseed rape (OR), and the fertilizer (Nitrogen—N) rate in different growing seasons.

Annual Field Crops (AFCs)	Cultivars	Sowing Date	Seeding Rate (kg ha^{-1})	N Rate (kg ha^{-1})
Spring barley	Tolar	11 April 2011	220	60
Spring barley	Tolar	17 April 2012	220	70
Spring barley	Bojos	18 April 2013	220	30
Spring barley	Bojos	12 March 2014	220	30
Winter wheat	Etela	4 October 2011	200	15
Winter wheat	Etela	26 September 2012	200	15
Silage maize	MON3301	10 May 2012	35	140
Oilseed rape	Rohan	16 August 2011	4.5	60

2.2. Field Measurements and Data Collection

All measurements (meteorological, LAI, AGDM production, IPAR, and RUE) were carried out for SRC poplar clone J-105 (for uncoppiced in growing seasons 7–8 and for coppiced-growing seasons 1–4), and AFCs. The data sampling dates covered all stages of canopy development, from the beginning of the growing season (leaf flushing) to the end of the growing season (completely fallen leaves) in SRC poplar and crop emergence to ripening (before harvesting) in AFCs.

2.3. Meteorological Measurements

Meteorological data (air temperature, incident global radiation, and precipitation) were recorded continuously by an automatic weather station placed at the turf grass close (~15–20 m) to the SRC poplar plantation [42]. The used instruments consist of combined air temperature/relative humidity sensor EMS 33 (EMS Brno, Brno, the Czech Republic), precipitation tipping bucket rain gauge MetOne 370 (MetOne Instruments, Grants Pass, OR, USA), and global radiation sensors EMS 11 (EMS Brno, Brno, the Czech Republic), respectively. The PAR was calculated as 0.5 of the global radiation [44]. The weather conditions during the measurement period for seven years from 2008–2014 are presented in Figure 1. The mean annual air temperature (in °C) was within the range of normal conditions. The annual precipitation was almost the same (ranging from 511–554 mm) between all the years except 2009 and 2010 when the total precipitation was higher, with the highest reported in 2009. The annual incident global radiation for each growing season (from April–October, for all measurement years) varied between 2971.28–3247.03 MJ m^{-2}, however, a slightly lower value was observed in 2010 (see Figure 1B).

2.4. Leaf Area Development Measurements

In situ LAI measurement methods can be grouped into two main categories such as direct (destructive and non-destructive) and indirect (non-destructive). Indirect non-destructive LAI measurement methods involve several approaches with advantages and disadvantages for different types of vegetation [45,46]. The LAI development was evaluated throughout each of the growing season in SRC poplar and AFCs. The LAI was indirectly measured by a SunScan plant canopy analyzer (Delta-T Devices, Cambridge, UK). For SRC poplar, the LAI was measured at three different places between rows and within rows while for AFCs, the LAI was measured in each individual replicate. The LAI measurements were repeated weekly and biweekly in the same measurement points. This indirect LAI method was validated because first time SunScan was used in woody SRC poplar plantation/referenced ($r^2 = 0.82$, $y = 1.05 - 0.7$, $p < 0.001$) for SRC poplar [47], and for AFCs it followed the calibration from earlier studies in different crops such as soybean, maize, and wheat [48–51]. For daily LAI, the weekly and biweekly LAI data were interpolated using a regression equation with the cumulative mean daily air temperature sum (>5 °C) [52]. The LAI$_{max}$ (during the peak of growing period), LAI$_{mean}$ (the mean of the LAI, during the growing period), $\sum$LAI (the sum of the daily LAI during the growing period), and LAD (the count of the total number of days from bud burst to completely fallen leaves to the ground) was evaluated for each growing season in SRC poplar and AFCs, respectively.

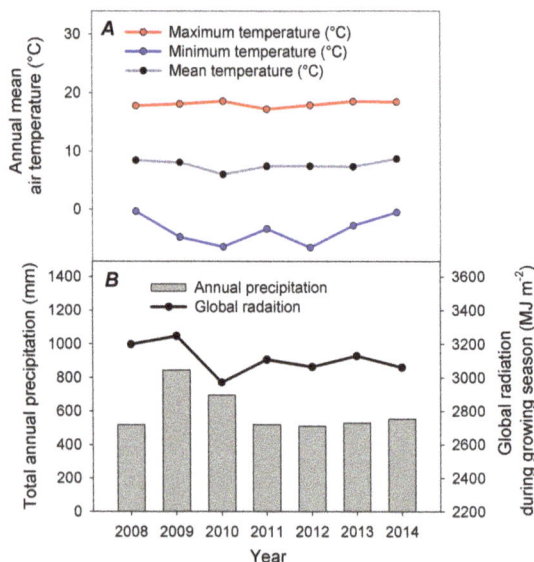

Figure 1. (**A**) The air temperature (°C; maximum, minimum, and mean) and (**B**) the total amount of annual precipitation (mm) and the total sum of incident global radiation (MJ m^{-2}) per growing season (April–October) from 2008–2014 at the research site Domanínek, Czech Republic.

2.5. Aboveground Dry Mass (AGDM) Production Estimation

In uncoppiced poplar, the AGDM was estimated for the seven and eight growing season using an allometric equation which was developed by Fischer et al. [53]. However, for the coppiced plantation, we developed a specific allometric equation using the methodology suggested by Fischer et al. [53]. In winter 2011 (after two years of coppiced growth), 25 trees were randomly selected in the range of 10–40 mm diameter at breast height (DBH, 130 cm above the ground level) and trees were harvested 10–15 cm above the ground level. The DBHs of the tree were measured by using a digital caliper (150 mm digital caliper DC04150, Digital Micrometers Ltd., Sheffield, UK). The whole tree was divided into two parts, stem and branches for chipping, in order to determine the woody dry mass (DM). The chipping samples were stored in aluminum boxes and subsequently dried in an oven at 105 °C until constant weight was achieved [54,55]. The amounts taken on the collected material were used to correlate DBH with the dry weights, and to develop a non-linear (power function) allometric Equation (1) (Figure S1) for the coppiced poplar.

$$DM = a.DBH^b \tag{1}$$

where DM is dry mass, a and b are specific regression coefficients, and DBH is the diameter of the trees at breast height.

The AGDM in SRC poplar was estimated from the DBH inventory of the winter habitus of trees at the end of each growing season and followed by using allometric equation (Figure S1). In AFCs, the AGDM was estimated at least six times during the growing season using a regular harvesting method. For determining the dry matter production, we harvested aboveground biomass of AFCs and dried them until constant weight. For AGDMs, dry matter content was determined per m^2 and scaled up per ha (10,000 m^2). This methodology was followed due to a recommendation for variety testing of the State Institute for Agriculture Supervision and Testing with daily monitoring of weather parameters through automated weather station and daily collection of experimental information through regular oversight by experienced experimenter and technicians [43].

2.6. The Evaluation of Intercepted Photosynthetic Active Radiation (IPAR) and Radiation Use Efficiency (RUE)

In the SRC poplar clone J-105 and AFCs, the RUEs were calculated (see Equation (2)) from the ratio between the produced AGDM (g DM m^{-2}) and the total amount of accumulated IPAR (MJ m^{-2}) during each growing season.

$$RUE = W_s/IPAR \tag{2}$$

where W_s = total AGDM production (g DM m^{-2}) and the IPAR was evaluated using Lambert-Beer's Law function [56]:

$$IPAR = PAR_{above} (1 - e^{-kLAI}) \tag{3}$$

where PAR_{above} represents the cumulated PAR above the canopy and k represents the light extinction coefficient. For fast growing trees k is estimated to vary between 0.4–0.6 [52] while in the case of AFCs such as barley and wheat, k ranges between 0.4 and 0.5, respectively [49,50,57].

In the present study, we calculated k for the poplar clone J-105, SB, WW, SM, and RS, respectively using Equation (4) [58]

$$k = -\ln (PAR_t/PAR_i)/LAI \tag{4}$$

where PAR_t is the transmitted PAR (below the canopy), PAR_i is the incoming PAR (above the canopy), and LAI is the leaf area index.

2.7. Statistical Analysis

The power function of the allometric equation between the AGDM and DBH [59] and the linear regression among the studied variables (LAI$_{max}$, LAD, RUE, and AGDM) of SRC poplar [28] and AFCs were fitted in SigmaPlot® version 11.0 (Systat Software, San Jose, CA, USA). Pearson's correlation coefficients were estimated using the same software for individual relationships.

3. Results

3.1. Leaf Area Development

The time course of LAI for SRC poplar (uncoppiced-PU and coppiced-PC) and for AFCs is shown in Figure 2A–C.

The LAI dynamics changed in SRC poplar (PU and PC) particularly with the development of the canopy and management. In AFCs (SB-T, SB-B, WW-E, SM-M, and OR-R) the LAI dynamics varied mainly due to crop characteristics and also in response to the growing season (Figure 2). The LAI$_{max}$ varied between 3.6–9.5 m^2 m^{-2} in SRC poplar, 3.1–5.6 in SB, 4.1–6.0 m^2 m^{-2} in WW, and reached values of 3.5 in SM and 5.6 m^2 m^{-2} in OR.

In AFCs, the highest measured LAI$_{max}$ was 6.0 m^2 m^{-2} in WW but the LAI$_{max}$ and also the LAI development dynamics varied between the growing seasons (Figure 3).

Besides LAI development, LAD, which is an important parameter to define the plant productivity and is interconnected to LAI development and RUE, was also studied. The LAD for SRC poplar and AFCs is shown in Figure 3. The maximum LAD (186 days) was found in SRC poplar in 2009 (after eight years of plant growth in uncoppiced management) and the minimum LAD (68 days) was observed in SM.

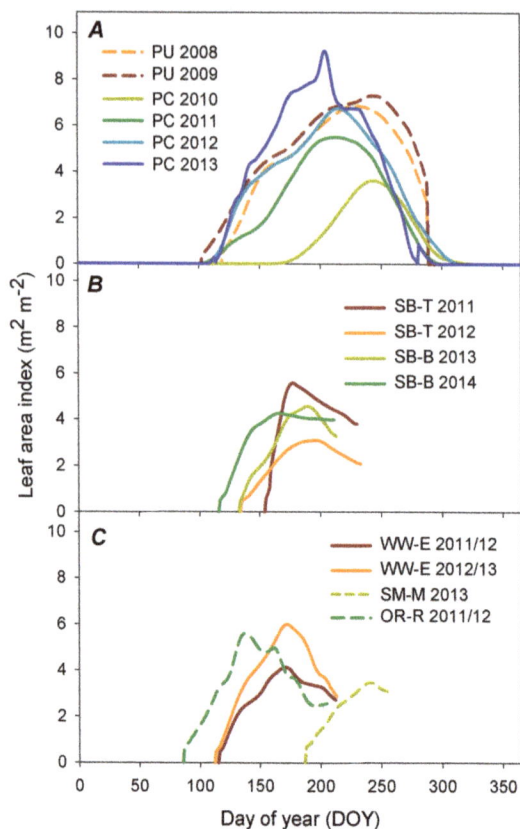

Figure 2. The time course of leaf area index (LAI) in SRC poplar clone J-105 (**A**) uncoppiced (PU) from 2008–2009 and coppiced (PC) from 2010–2013; the time course of LAI in AFCs (**B**) spring barley cultivar Tolar (SB-T) from 2011–2012 and cultivar Bojos (SB-B) from 2013–2014; and (**C**) winter wheat cultivar Etela (WW-E) from 2011/2012–2012/2013, silage maize cultivar MON3301 (SM-M) in 2013, and oilseed rape cultivar Rohan (OR-R) in 2011/12.

3.2. Radiation Use Efficiency (RUE)

RUE is another main determinant of biomass production in SRC plants and AFCs. For the poplar clone J-105, the average RUE was 1.19 g MJ^{-1} over all six years. The maximum RUE was 1.48 g MJ^{-1} in PU 2009 (the last year of uncoppiced poplar cultivation) and the minimum RUE was 1.02 g MJ^{-1} in PC 2010 (the first year of coppiced poplar cultivation). The average RUE for AFCs was 1.57 g MJ^{-1} among all crops, years, and genotypes. The maximum RUE (2.06 g MJ^{-1}) was observed in AFCs SB (cultivar Bojos) in 2013 (SB-B 2013) and the minimum RUE (0.72 g MJ^{-1}) was observed in OR cultivated in 2011/2012 (Figure 3).

3.3. Aboveground Dry Mass (AGDM) Production

The AGDM production in SRC poplar clone J-105 varied from 3.6–16.5 t ha^{-1} $year^{-1}$ in different growing seasons (Figure 3). In the long term six-year observation, the mean AGDM production was estimated to be 10.5 t ha^{-1} $year^{-1}$ including PU and PC. The maximum AGDM production was observed in the last year of PU cultivation (2009) and the minimum AGDM production was observed in the first year of PC cultivation. It is evident that these differences are attributed particularly to the canopy development.

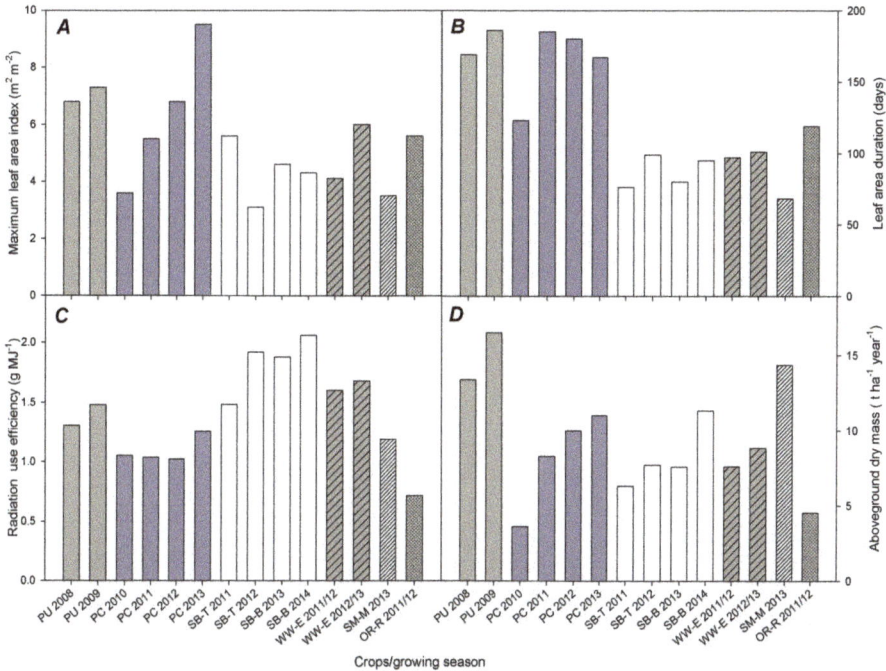

Figure 3. Comparison of (**A**) the maximum leaf area index (LAI$_{max}$, m^2 m^{-2}); (**B**) the leaf area duration (LAD, days); (**C**) the radiation use efficiency (RUE, g MJ^{-1}); and (**D**) the aboveground dry mass (AGDM, t ha^{-1} year^{-1}) in SRC poplar clone J-105 uncoppiced (PU) from 2008–2009 and coppiced (PC) from 2010–2013 and AFCs (spring barley cultivar Tolar (SB-T) from 2011–2012 and cultivar Bojos (SB-B) from 2013–2014, winter wheat cultivar Etela (WW-E) from 2011/2012–2012/2013, silage maize cultivar MON3301 (SM-M) in 2013, and oilseed rape cultivar Rohan (OR-R) in 2011/2012).

The AGDM production in AFCs varied between 4.5 and 14.4 t ha^{-1} year^{-1} and the average AGDM was 8.5 t ha^{-1} year^{-1}. The maximum AGDM was observed in SM-M (14.4 t ha^{-1} year^{-1}) and the minimum was observed in OR-R (4.5 t ha^{-1} year^{-1}) (Figure 3D). These observed differences can be related to RUE, sowing density, crop physiology, and N application.

3.4. Determinants of Aboveground Dry Mass (AGDM) Production in SRC Poplar and AFCs

The AGDM production showed a positive linear regression with LAI$_{max}$, LAI$_{mean}$, $\sum$LAI, LAD, and $\sum$(LAI × IPAR) in only SRC poplar, while a negative linear relationship was observed in AFCs (Figure 4A–D,F, respectively). For SRC poplar, a strong positive (R^2 values varied between 0.86 and 0.94) and linear relationships between AGDM production and LAI (LAI$_{mean}$ and $\sum$LAI) or $\sum$(LAI × IPAR) were found. Using $\sum$LAI (R^2 = 0.91) and $\sum$(LAI × IPAR) (R^2 = 0.94) improved the relationship to AGDM production compared to LAI$_{mean}$ (R^2 = 0.86) and particularly to LAI$_{max}$ (R^2 = 0.46) and thus, provided a better estimator of biomass production.

In SRC poplar and AFCs (excluding SM), the AGDM production was strongly (R^2 ranged from 0.70–0.72) and positively correlated to RUE (Figure 4E). Thus, the use of RUE provided a good estimation of AGDM production for both SRC poplar and AFCs, however, the relationship for AFCs was significantly improved if the C_4 crop SM was excluded from the dataset ($R^2 = 0.7$). The C_4 crop SM AGDM response to RUE fits better in the AGDM versus RUE relationship for SRC poplar.

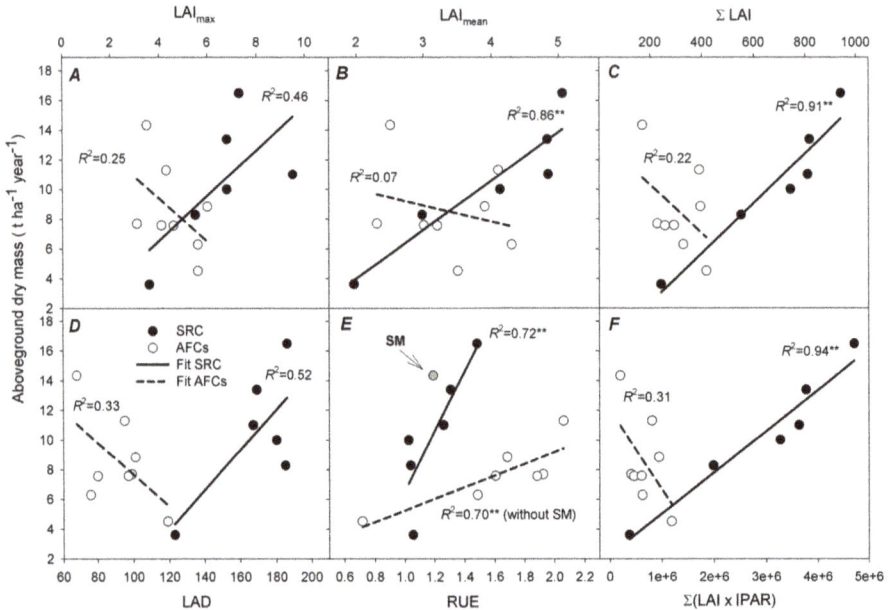

Figure 4. The relationships between aboveground dry mass (AGDM) production and: (**A**) the maximum leaf area index (LAI_{max}); (**B**) the mean LAI (LAI_{mean}); (**C**) the sum of LAI ($\sum LAI$); (**D**) the leaf area duration (LAD); (**E**) the radiation use efficiency (RUE); and (**F**) $\sum (LAI \times IPAR)$ (absorbed photosynthetically active radiation) in SRC poplar clone J-105 and AFCs, respectively. Note: ** significant Pearson correlation coefficients at 0.05.

4. Discussion

Within this study, we analyzed the role of RUE and LAI development in SRC poplar and AFCs in determining AGDM production. We hypothesized the importance of RUE and LAI for AGDM production in SRC poplar and AFCs, and also improved the estimation of AGDM production by using $\sum LAI$ or $\sum (LAI \times IPAR)$ integrating the LAI development over the whole vegetation period.

The results of LAI development in SRC poplar show that the maximum occurs near the end of summer, which closely matches earlier studies for fast growing woody crops [27,47,60,61]. In these earlier studies, the LAI_{max} ranged from 3.5–10 m^2 m^{-2}, which may be attributed to the growing season's weather conditions and development of poplar canopy [27,62]. In our study, an increase of the LAI_{max} from 3.6–9.5 m^2 m^{-2} is evident after transition to coppice management. Likewise, Broeckx et al. [28] reported LAI_{max} values from 0.49–1.68 m^2 m^{-2} for one-year-old (first growing season) and 0.87–4.63 m^2 m^{-2} for two-year-old diverse poplar clones, while in other fast growing trees like eucalyptus and willow the LAI_{max} estimated values varied from 0.8–6.1 m^2 m^{-2} and from 2.4–6.7 m^2 m^{-2}, respectively. In poplar, the LAI development dynamics is also delayed after coppice. The LAI increment rate increases particularly in the first three years after coppice and reaches similar LAI development dynamics as uncoppiced canopy. The data confirm the positive effect of coppice management on the LAI development and canopy closure [28,47,63].

In our study, the LAI_{max} values for AFCs are in close agreement with the LAI_{max} reported in WW, SB, SM, and OR [24,64–67]. Variations in the LAI_{max} values are related to genotype, site (particularly soil conditions), length of the growing season, sowing density, and the availability of water and nutrients [27,68]. The LAI_{max} may be strongly limited by reduced water availability resulting in reduced growth, yield, and yield components [69,70].

Besides LAI development in SRC poplar and AFCs, our LAD results were similar to previous studies where the LAD range varied between 58 and 180 days in AFCs (C_3 and C_4 crops) and SRC [28,47,68,71]. Higher LAI and LAD were achieved in SRC poplar as compared to AFCs because woody crops are taller, multi stemmed, and their developed roots systems may accommodate higher canopy closure as compared to AFCs [25,26,54]. The responses of LAI develoment in SRC poplar and AFCs can be potentially used for the screening of genotypes under stress conditions or for the development of adaptation management strategies to reduce climate change impacts. Some earlier studies have mentioned the importance of LAI measurements in different ecosystems for process-based models such as APEX and ALMANAC, which were used for the prediction of climate change impacts on SRC poplar and crop productivity [72–74]. Thus, our findings may contribute to improving the models as well as to help farmers in better selection of crops for future climates together with fulfilling the growing requirements for food and green-energy production [35].

Plant species show relatively high differences in RUE [16,30], however, the long term average of RUE for fast growing poplar was comparable to data published by Linderson et al. [52] who estimated RUE at 1.40 g MJ^{-1} for willow clones in Scania and also to some studies for poplar plantations where the RUE varied between 1 and 2 g MJ^{-1} [31,36,40]. In contrast, Broeckx et al. [28] reported lower RUE values varying between 0.29 and 0.68 and an average RUE of 0.50 g MJ^{-1} in twelve poplar clones in Flanders, Belgium. Low or high RUE values are characteristic of the biomass accumulation rate, are site specific (water and nutrient availability), depend on the length of growing season, as well as the height and age of trees and genotypes [36]. In our study, the RUE values varied between individual crops and growing seasons, however, the long-term average was higher in AFCs compared to SRC poplar. It may be due to differences in crop physiology [12,75] and nutrient effects [64,76] as the effect of nutrition (particularly N) increases RUE. RUE is increased by higher N doses [16,77] and can be stimulated also by a higher availability of P and K [78–80]. Higher RUE is often reported also for C_4 crops, but the differences in crop physiology depend greatly on environmental conditions, particularly temperature [25,81].

Our RUE values for SB and WW are similar to previous studies reviewed by Kemanian et al. [24]. We observed fairly low RUE for OR (0.79 g MJ^{-1}) and SM (1.19 g MJ^{-1}) as compared to previous studies, where Kuai et al. [67] and Williams et al. [82] reported 1.07–1.09 g MJ^{-1} in OR and 1.71 g MJ^{-1} in maize, respectively. RUE values may be increased by favorable conditions such as higher water and nutrient (particularly N) availability [13,52,68,83]. RUE could also be affected by sowing density and genotype. Higher RUE was found in higher densities [27,52,67].

The growth and biomass production of poplar was in accordance with our earlier data [41]. Compared to other studies, depending on climate and management (plant spacing, irrigation, fertilization, and genotypes), a production of 10–15 t ha^{-1} $year^{-1}$ could be expected [84,85]. Thus, the average productivity observed in our study, despite relatively low precipitation, and soil water recharge, is acceptable. However, in some cases the average AGDM production observed is substantially higher, mainly due to favorable climatic condition, the length of rotation cycle, genotype, higher nutrients, and water availability [86–89]. In our study, AGDM production uniformly increased season by season in SRC poplar (PU and PC) which can be attributed to the development of the root system and the increasing capacity to capture water and nutrients.

In earlier studies, AGDMs of 5.6 t ha^{-1} for barley in Pakistan [64] and 10.3–14.2 t ha^{-1} in the USA [22] were observed, for wheat 5.3–6.7 t ha^{-1} in China [79] and 5.74–7.22 t ha^{-1} in Pakistan [64], for maize 14.5–14.9 t ha^{-1} in different water regimes for many years in China [80], and for oilseed rape 3.6–4.7 t ha^{-1} for many cultivars in the Czech Republic and other European countries [90] which is in accordance with our data.

However, in some studies, the AGDM varied only between 1.82 and 2.9 t ha^{-1} for AFCs [91,92]. Such a low productivity could be related to unfavorable climatic conditions with poor availability of water and nutrients in particular places. Lower AGDM production in OR-R and SM-M within our study could be attributed to lower soil fertility and lower temperatures in a given region, resulting also in a shorter vegetation period. OR requires high nutrient availability, while SM is mainly affected by temperature. Higher temperature requirements are typical for C_4 crops [93,94].

The moderate positive relationships between LAI_{max} and AGDM production in poplar clone J-105 are in accordance with our previous studies [47]. In AFCs, the AGDM production varied between years, as a result of season (weather conditions), fertilizers applied to individual crops, and crop physiology. Our results confirm the earlier observations for a strong and positive correlation between RUE and AGDM production in high density poplar genotypes and AFCs [22,26,29,35].

The relationships between LAI or RUE and AGDM production show shifts between SRC poplar and AFCs indicating changes in the relative contribution to biomass. These results suggest that the biomass production improvement by selecting poplar and AFCs cultivars for specific growth parameters should take crop management, and consequently RUE into account. Our data also demonstrated the role of RUE for AGDM production in C_3 and C_4 crops.

5. Conclusions

Our study shows LAI to be high in SRC poplar compared to AFCs while RUE was shown to be higher in AFCs than SRC poplar. Both high LAI and RUE could be responsible for high biomass accumulation in SRC poplar; while in AFCs, the biomass accumulation is related mainly to RUE. The AGDM estimation by LAI in SRC poplar was improved by using the sum of LAI over the whole vegetation period and the sum of LAI multiplied by IPAR. The RUE can be thus considered as an important determinant of biomass production in both SRC poplar and AFCs while LAI could be a good biomass predictor in only SRC poplar. The evaluation of plant productivity for food and energy security under climate change is unlikely to be achieved without improving the model with LAI and RUE input and the right decisions for crop rotations in the changing environment. This is conditioned by a better understanding of the LAI development, the phenology of the plants, and the role of RUE and LAI in regulating productivity and plant adaption to climate change in different crops such as fast-growing trees and AFCs.

Overall, we can conclude that this long-term study is useful particularly for poplar biomass estimation under different environmental conditions or seasonal changes. The data can serve to improve the modeling of LAI, RUE, and AGDM development for precise forecasting in the climate change era.

Supplementary Materials: The following are available online at http://www.mdpi.com/1999-4907/9/4/168/s1, Figure s1: Development of an allometric relationship between aboveground dry mass (AGDM, kg stump^{-1}) and in diameter at breast height (DBH, mm) of poplar clone J-105 in coppice rotation after two years of plant growth.

Acknowledgments: The study was supported by the Ministry of Education, Youth and Sports (project CzeCOS No. LM2015061 and CzeCOS ProCES No. CZ.02.1.01/0.0/0.0/16_013/0001609). We are also thankful to the internal grant agency (IGA 53/2014) at the Mendel University in Brno, Czech Republic.

Author Contributions: A.M.T., M.F., M.T. and M.V.M. conceived and designed the study. A.M.T., E.P., M.O. and M.F. performed the experiment and also participated in the data collection and cleaning. A.M.T. and K.K. analyzed the data and wrote the manuscript. All authors read and approved the final manuscript.

Conflicts of Interest: The authors declare no conflicts of interest.

Abbreviations

$\sum$LAI	Sum of leaf area index
AFCs	Annual field crops
AGDM	Aboveground dry mass
DBH	Diameter at breast height (mm)
DM	Dry mass

IPAR	Intercepted photosynthetically active radiation (MJ m^{-2})
LAD	Leaf area duration (days)
LAI	Leaf area index (m^2 m^{-2})
LAI_{max}	Maximum leaf area index
LAI_{mean}	Mean leaf area index
OR	Oilseed rape
OR-R	Oilseed rape cultivar Rohan
PAR	Photosynthetically active radiation
PC	Poplar coppiced
PU	Poplar uncoppiced
RUE	Radiation use efficiency (g MJ^{-1})
SB	Spring barley
SB-B	Spring barley cultivar Bojos
SB-T	Spring barley cultivar Tolar
SM	Silage maize
SM-M	Silage maize cultivar MON3301
SRC	Short rotation coppice
WW	Winter wheat
WW-E	Winter wheat cultivar Etela

References

1. IPCC (Intergovernmental Panel on Climate Change). *Climate Change 2014: Synthesis Report. Contribution of Working Groups I, II and III to the Fifth Assessment Report of the Intergovernmental Panel on Climate Change*; IPCC: Geneva, Switzerland, 2014; ISBN 9789291691432.
2. Pinstrup-Andersen, P. Food security: Definition and measurement. *Food Secur.* **2009**, *1*, 5–7. [CrossRef]
3. Popp, J.; Lakner, Z.; Harangi-Rákos, M.; Fári, M. The effect of bioenergy expansion: Food, energy, and environment. *Renew. Sustain. Energy Rev.* **2014**, *32*, 559–578. [CrossRef]
4. Scarlat, N.; Dallemand, J.F. Recent developments of biofuels/bioenergy sustainability certification: A global overview. *Energy Policy* **2011**, *39*, 1630–1646. [CrossRef]
5. Busetto, L.; Casteleyn, S.; Granell, C.; Pepe, M.; Barbieri, M.; Campos-Taberner, M.; Casa, R.; Collivignarelli, F.; Confalonieri, R.; Crema, A.; et al. Downstream services for rice crop monitoring in Europe: From regional to local scale. *IEEE J. Sel. Top. Appl. Earth Obs. Remote Sens.* **2017**, *10*, 5423–5441. [CrossRef]
6. Dimitriou, I.; Baum, C.; Baum, S.; Busch, G.; Schulz, U.; Köhn, J.; Lamersdorf, N.; Leinweber, P.; Aronsson, P.; Weih, M.; et al. *Quantifying Environmental Effects of Short Rotation Coppice (SRC) on Biodiversity, Soil and Water*; IEA Bioenergy Task 43, Report 2011; IEA: Paris, France, 2011; p. 34.
7. Kauter, D.; Lewandowski, I.; Claupein, W. Quantity and quality of harvestable biomass from Populus short rotation coppice for solid fuel use—A review of the physiological basis and management influences. *Biomass Bioenergy* **2003**, *24*, 411–427. [CrossRef]
8. Timmer, C.P. The macro dimensions of food security: Economic growth, equitable distribution, and food price stability. *Food Policy* **2000**, *25*, 283–295. [CrossRef]
9. Wheeler, T.; von Braun, J. Climate change impacts on global food security. *Science* **2013**, *341*, 508–513. [CrossRef] [PubMed]
10. Ewert, F. Modelling plant responses to elevated CO_2: How important is leaf area index? *Ann. Bot.* **2004**, *93*, 619–627. [CrossRef] [PubMed]
11. Chen, J.M.; Black, T.A. Defining leaf area index for non-flat leaves. *Plant Cell Environ.* **1992**, *15*, 421–429. [CrossRef]
12. Monteith, J.L. Reassessment of maximum growth rates for C3 and C4 Crops. *Exp. Agric.* **1978**, *14*, 1–5. [CrossRef]
13. Biscoe, P.V.; Gallagher, J.N. Weather, dry matter production and yield. In *Environmental Effects on Crop Physiology; Proceedings of the a Symposium Held at Long Ashton Research Station, University of Bristol, Bristol, UK, 13–16 April 1975*; Academic Press: New York, NY, USA, 1977.

14. Hangs, R.D.; Van Rees, K.C.J.; Schoenau, J.J.; Guo, X. A simple technique for estimating above-ground biomass in short-rotation willow plantations. *Biomass Bioenergy* **2011**, *35*, 2156–2162. [CrossRef]

15. Hikosaka, K. Leaf canopy as a dynamic system: Ecophysiology and optimality in leaf turnover. *Ann. Bot.* **2005**, *95*, 521–533. [CrossRef] [PubMed]

16. Sinclair, T.R.; Horie, T. Leaf nitrogen, photosynthesis, and crop radiation use efficiency: A review. *Crop Sci.* **1989**, *29*, 90. [CrossRef]

17. Sinclair, T.R.; Shiraiwa, T.; Hammer, G.L. Variation in crop radiation-use efficiency with increased diffuse radiation. *Crop Sci.* **1992**, *32*, 1281–1284. [CrossRef]

18. Medlyn, B.E. Physiological basis of the light use efficiency model. *Tree Physiol.* **1998**, *18*, 167–176. [CrossRef] [PubMed]

19. Liu, W.; Yan, J.; Li, J.; Sang, T. Yield potential of *Miscanthus* energy crops in the Loess Plateau of China. *GCB Bioenergy* **2012**, *4*, 545–554. [CrossRef]

20. Stockle, C.O.; Kiniry, J.R. Variability in crop radiation-use efficiency associated with vapor-pressure deficit. *Field Crop. Res.* **1990**, *25*, 171–181. [CrossRef]

21. Larson, P.R.; Isebrands, J.G. The relation between leaf production and wood weight in first-year root sprouts of two *Populus* clones. *Can. J. For. Res.* **1972**, *2*, 98–104. [CrossRef]

22. Bartelink, H.H. A growth model for mixed forest stands. *For. Ecol. Manag.* **2000**, *134*, 29–43. [CrossRef]

23. Bartelink, H.H.; Kramer, K.; Mohren, G.M.J. Applicability of the radiation-use efficiency concept for simulating growth of forest stands. *Agric. For. Meteorol.* **1997**, *88*, 169–179. [CrossRef]

24. Kemanian, A.R.; Stöckle, C.O.; Huggins, D.R. Variability of barley radiation-use efficiency. *Crop Sci.* **2004**, *44*, 1662–1672. [CrossRef]

25. Jamieson, P.D.; Martin, R.J.; Francis, G.S.; Wilson, D.R. Drought effects on biomass production and radiation-use efficiency in barley. *Field Crop. Res.* **1995**, *43*, 77–86. [CrossRef]

26. Zahoor, A.; Riaz, M.; Ahmad, S.; Ali, H.; Khan, M.B.; Javed, K.; Anjum, M.A.; Zia-Ul-Haq, M.; Khan, M.A. Ontogeny growth and radiation use efficiency of *Helianthus annuus* L., as affected by hybrids, nitrogenous regimes and planting geometry under irrigated arid conditions. *Pak. J. Bot.* **2010**, *42*, 3197–3207.

27. Ceulemans, R.; McDonald, A.J.S.; Pereira, J.S. A comparison among eucalypt, poplar and willow characteristics with particular reference to a coppice, growth-modelling approach. *Biomass Bioenergy* **1996**, *11*, 215–231. [CrossRef]

28. Broeckx, L.; Vanbeveren, S.; Verlinden, M.; Ceulemans, R. First vs. second rotation of a poplar short rotation coppice: Leaf area development, light interception and radiation use efficiency. *iForest-Biogeosci. For.* **2015**, e1–e9. [CrossRef]

29. Kiniry, J.R. Biomass accumulation and radiation use efficiency of honey mesquite and eastern red cedar. *Biomass Bioenergy* **1998**, *15*, 467–473. [CrossRef]

30. Kiniry, J.R.; Jones, C.A.; O'toole, J.C.; Blanchet, R.; Cabelguenne, M.; Spanel, D.A. Radiation-use efficiency in biomass accumulation prior to grain-filling for five grain-crop species. *Field Crop. Res.* **1989**, *20*, 51–64. [CrossRef]

31. Landsberg, J.J.; Wright, L.L. Comparisons among *Populus* clones and intensive culture conditions, using an energy-conservation model. *For. Ecol. Manag.* **1989**, *27*, 129–147. [CrossRef]

32. Iio, A.; Hikosaka, K.; Anten, N.P.R.; Nakagawa, Y.; Ito, A. Global dependence of field-observed leaf area index in woody species on climate: A systematic review. *Glob. Ecol. Biogeogr.* **2014**, *23*, 274–285. [CrossRef]

33. Wright, I.J.; Groom, P.K.; Lamont, B.B.; Poot, P.; Prior, L.D.; Reich, P.B.; Schulze, E.D.; Veneklaas, E.J.; Westoby, M. Leaf trait relationships in Australian plant species. *Funct. Plant Biol.* **2004**, *31*, 551–558. [CrossRef]

34. Asner, G.P.; Scurlock, J.M.O.; Hicke, J.A. Global synthesis of leaf area index observations. *Glob. Ecol. Biogeogr.* **2003**, *12*, 191–205. [CrossRef]

35. Harrington, R.A.; Fownes, J.H.; Meinzer, F.C.; Scowcroft, P.G. Forest growth along a rainfall gradient in Hawaii: *Acacia koa* stand structure, productivity, foliar nutrients, and water- and nutrient-use efficiencies. *Oecologia* **1995**, *102*, 277–284. [CrossRef] [PubMed]

36. Cannell, M.G.R.; Sheppard, L.J.; Milne, R. Light use efficiency and woody biomass production of Poplar and Willow. *Forestry* **1988**, *61*, 125–136. [CrossRef]

37. Haxeltine, A.; Prentice, I.C. BIOME 3: An equilibrium terrestrial biosphere model based on ecophysiological constraints, resource availability and competition among plant functional types. *Glob. Biogeochem. Cycles* **1996**, *10*, 693–709. [CrossRef]

38. Choudhury, B.J. Modeling radiation- and carbon-use efficiencies of maize, sorghum, and rice. *Agric. For. Meteorol.* **2001**, *106*, 317–330. [CrossRef]

39. Wolf, J.; Evans, L.G.; Semenov, M.A.; Eckersten, H.; Iglesias, A. Comparison of wheat simulation models under climate change. I. Model calibration and sensitivity analyses. *Clim. Res.* **1996**, *7*, 253–270. [CrossRef]

40. Green, D.S.; Kruger, E.L.; Stanosz, G.R.; Isebrands, J.G. Light-use efficiency of native and hybrid poplar genotypes at high levels of intracanopy competition. *Can. J. For. Res.* **2001**, *31*, 1030–1037. [CrossRef]

41. Trnka, M.; Trnka, M.; Fialová, J.; Koutecký, V.; Fajman, M.; Žalud, Z.; Hejduk, S. Biomass production and survival rates of selected poplar clones grown under a short-rotation system on arable land. *Plant Soil Environ.* **2008**, *54*, 78–88. [CrossRef]

42. Fischer, M.; Orság, M.; Trnka, M.; Pohanková, E.; Hlavinka, P.; Tripathi, A.M.; Žalud, Z. Annual and intra-annual water balance components of a short rotation poplar coppice based on sap flow and micrometeorological and hydrological approaches. *Acta Hortic.* **2013**. [CrossRef]

43. Pohanková, E.; Hlavinka, P.; Takáč, J.; Žalud, Z.; Trnka, M. Calibration and validation of the crop growth model daisy for spring barley in the Czech Republic. *Acta Univ. Agric. Silvic. Mendel. Brun.* **2015**, *63*. [CrossRef]

44. Bonhomme, R. Beware of comparing RUE values calculated from PAR vs solar radiation or absorbed vs intercepted radiation. *Field Crop. Res.* **2000**, *68*, 247–252. [CrossRef]

45. Bréda, N.J.J. Ground-based measurements of leaf area index: A review of methods, instruments and current controversies. *J. Exp. Bot.* **2003**, *54*, 2403–2417. [CrossRef] [PubMed]

46. Campos-Taberner, M.; García-Haro, F.J.; Confalonieri, R.; Martínez, B.; Moreno, Á.; Sánchez-Ruiz, S.; Gilabert, M.A.; Camacho, F.; Boschetti, M.; Busetto, L. Multitemporal monitoring of plant area index in the Valencia rice district with PocketLAI. *Remote Sens.* **2016**, *8*. [CrossRef]

47. Tripathi, A.M.; Fischer, M.; Orság, M.; Marek, M.V.; Žalud, Z.; Trnka, M. Evaluation of indirect measurement method of seasonal patterns of leaf area index in a high-density short rotation coppice culture of poplar. *Acta Univ. Agric. Silvic. Mendel. Brun.* **2016**, *64*, 549–556. [CrossRef]

48. Liu, G.; Xie, Y.; Gao, X.F.; Duan, X.W. Application of SunScan canopy analysis system in measuring leaf area index of soybean. *Chin. J. Ecol.* **2008**, *27*, 862–866.

49. Oguntunde, P.G.; Olukunle, O.J.; Fasinmirin, J.T.; Abiolu, O.A. Performance of the SunScan canopy analysis system in estimating leaf area index of maize. *Agric. Eng. Int. CIGR J.* **2012**, *14*, 1–7.

50. Wilhelm, W.W.; Ruwe, K.; Schlemmer, M.R. Comparison of three leaf area index meters in a corn canopy. *Crop Sci.* **2000**, *40*, 1179–1183. [CrossRef]

51. Schirrmann, M.; Hamdorf, A.; Giebel, A.; Dammer, K.H.; Garz, A. A mobile sensor for leaf area index estimation from canopy light transmittance in wheat crops. *Biosyst. Eng.* **2015**, *140*, 23–33. [CrossRef]

52. Linderson, M.L.; Iritz, Z.; Lindroth, A. The effect of water availability on stand-level productivity, transpiration, water use efficiency and radiation use efficiency of field-grown willow clones. *Biomass Bioenergy* **2007**, *31*, 460–468. [CrossRef]

53. Fischer, M.; Trnka, M.; Kučera, J.; Fajman, M.; Žalud, Z. Biomass productivity and water use relation in short rotation poplar coppice (*Populus nigra* × *P. maximowiczii*) in the conditions of Czech Moravian Highlands. *Acta Univ. Agric. Silvic. Mendel. Brun.* **2011**, *59*, 141–152. [CrossRef]

54. Ketterings, Q.M.; Coe, R.; Van Noordwijk, M.; Ambagau, Y.; Palm, C.A. Reducing uncertainty in the use of allometric biomass equations for predicting above-ground tree biomass in mixed secondary forests. *For. Ecol. Manag.* **2001**, *146*, 199–209. [CrossRef]

55. Basuki, T.M.; van Laake, P.E.; Skidmore, A.K.; Hussin, Y.A. Allometric equations for estimating the above-ground biomass in tropical lowland Dipterocarp forests. *For. Ecol. Manag.* **2009**, *257*, 1684–1694. [CrossRef]

56. Monsi, M.; Saeki, T. Uber den Lichtfaktor in den Pflanzengesellschaften und seine Bedeutung fur die Stoffproduktion. *Jpn. J. Bot.* **1953**, *14*, 22–52.

57. Kemanian, A.R.; Stöckle, C.O.; Huggins, D.R. Transpiration-use efficiency of barley. *Agric. For. Meteorol.* **2005**, *130*, 1–11. [CrossRef]

58. Heidari, H.; Johansooz, M.R.; Hosseini, S.M.B.; Chaichi, M.R. Alternate furrow irrigation effect on radiation use efficiency and forage quality of foxtail millet (*Setaria italica*). *Ann. Biol. Res.* **2012**, *3*, 2565–2574.

59. Ajit; Das, D.K.; Chaturvedi, O.P.; Jabeen, N.; Dhyani, S.K. Predictive models for dry weight estimation of above and below ground biomass components of *Populus deltoides* in India: Development and comparative diagnosis. *Biomass Bioenergy* **2011**, *35*, 1145–1152. [CrossRef]

60. Howe, G.T.; Saruul, P.; Davis, J.; Chen, T.H.H. Quantitative genetics of bud phenology, frost damage, and winter survival in an F2 family of hybrid poplars. *Theor. Appl. Genet.* **2000**, *101*, 632–642. [CrossRef]

61. Tharakan, P.J.; Volk, T.A.; Nowak, C.A.; Ofezu, G.J. Assessment of canopy structure, light interception, and light-use efficiency of first year regrowth of shrub willow (Salix sp.). *BioEnergy Res.* **2008**, *1*, 229–238. [CrossRef]

62. Ceulemans, R.; Stettler, R.F.; Hinckley, T.M.; Isebrands, J.G.; Heilman, P.E. Crown architecture of *Populus* clones as determined by branch orientation and branch characteristics. *Tree Physiol.* **1990**, *7*, 157–167. [CrossRef] [PubMed]

63. Rae, A.M.; Robinson, K.M.; Street, N.R.; Taylor, G. Morphological and physiological traits influencing biomass productivity in short-rotation coppice poplar. *Can. J. For. Res.* **2004**, *34*, 1488–1498. [CrossRef]

64. Ahmad, S.; Ali, H.; Ismail, M.; Shahzad, M.I.; Nadeem, M.; Anjum, M.A.; Zia-Ul-Haq, M.; Firdous, N.; Khan, M.A. Radiation and nitrogen use efficiencies of C3 winter cereals to nitrogen split application. *Pak. J. Bot.* **2012**, *44*, 139–149.

65. Choudhury, B.J. A sensitivity analysis of the radiation use efficiency for gross photosynthesis and net carbon accumulation by wheat. *Agric. For. Meteorol.* **2000**, *101*, 217–234. [CrossRef]

66. Timlin, D.J.; Fleisher, D.H.; Kemanian, A.R.; Reddy, V.R. Plant density and leaf area index effects on the distribution of light transmittance to the soil surface in maize. *Agron. J.* **2014**, *106*, 1828–1837. [CrossRef]

67. Kuai, J.; Sun, Y.; Zuo, Q.; Huang, H.; Liao, Q.; Wu, C.; Lu, J.; Wu, J.; Zhou, G. The yield of mechanically harvested rapeseed (*Brassica napus* L.) can be increased by optimum plant density and row spacing. *Sci. Rep.* **2015**, *5*. [CrossRef] [PubMed]

68. Watson, D.J. Comparative physiological studies on the growth of field crops. *Ann. Appl. Biol.* **1947**, *11*, 41–76. [CrossRef]

69. García Del Moral, L.F.; Rharrabti, Y.; Villegas, D.; Royo, C. Evaluation of grain yield and its components in durum wheat under Mediterranean conditions: An ontogenic approach. *Agron. J.* **2003**, *95*, 266–274. [CrossRef]

70. Guttieri, M.J.; Stark, J.C.; O'Brien, K.; Souza, E. Relative sensitivity of spring wheat grain yield and quality parameters to moisture deficit. *Crop Sci.* **2001**, *41*, 327–335. [CrossRef]

71. Richards, R.; Townley-Smith, T. Variation in leaf area development and its effect on water use, yield and harvest index of droughted wheat. *Aust. J. Agric. Res.* **1987**, *38*, 983–992. [CrossRef]

72. Johnson, E. Goodbye to carbon neutral: Getting biomass footprints right. *Environ. Impact Assess. Rev.* **2009**, *29*, 165–168. [CrossRef]

73. Guo, L.; Dai, J.; Wang, M.; Xu, J.; Luedeling, E. Responses of spring phenology in temperate zone trees to climate warming: A case study of apricot flowering in China. *Agric. For. Meteorol.* **2015**, *201*, 1–7. [CrossRef]

74. Jiang, R.; Wang, T.; Shao, J.; Guo, S.; Zhu, W.; Yu, Y.J.; Chen, S.L.; Hatano, R. Modeling the biomass of energy crops: Descriptions, strengths and prospective. *J. Integr. Agric.* **2017**, *16*, 1197–1210. [CrossRef]

75. Piedade, M.T.F.; Junk, W.J.; Long, S.P. The productivity of the C4 grass *Echinochloa polystachya* on the Amazon floodplain. *Ecology* **1991**, *72*, 1456–1463. [CrossRef]

76. Muurinen, S.; Peltonen-Sainio, P. Radiation-use efficiency of modern and old spring cereal cultivars and its response to nitrogen in northern growing conditions. *Field Crop. Res.* **2006**, *96*, 363–373. [CrossRef]

77. Cassman, K.G.; Dobermann, A.; Walters, D.T. Agroecosystems, nitrogen-use efficiency, and nitrogen management. *Ambio* **2002**, *31*, 132–140. [CrossRef] [PubMed]

78. Plenet, D.; Mollier, A.; Pellerin, S. Growth analysis of maize field crops under phosphorus deficiency. II. Radiation-use efficiency, biomass accumulation and yield components. *Plant Soil.* **2000**, *224*, 259–272. [CrossRef]

79. Li, Q.; Liu, M.; Zhang, J.; Dong, B.; Bai, Q. Biomass accumulation and radiation use efficiency of winter wheat under deficit irrigation regimes. *Plant Soil Environ.* **2009**, *2009*, 85–91. [CrossRef]

80. Li, Q.; Chen, Y.; Liu, M.; Zhou, X.; Yu, S.; Dong, B. Effects of irrigation and planting patterns on radiation use efficiency and yield of winter wheat in North China. *Agric. Water Manag.* **2008**, *95*, 469–476. [CrossRef]

81. Goyne, P.J.; Milroy, S.P.; Lilley, J.M.; Hare, J.M. Radiation interception, radiation use efficiency and growth of barley cultivars. *Aust. J. Agric. Res.* **1993**, *44*. [CrossRef]

82. Williams, W.; Loomis, R.; Lepley, C. Vegetative growth of corn as affected by population density. I. Productivity in relation to interception of solar radiation. *Crop Sci.* **1965**, *5*, 211–215. [CrossRef]

83. Kiniry, J.R.; Landivar, J.A.; Witt, M.; Gerik, T.J.; Cavero, J.; Wade, L.J. Radiation-use efficiency response to vapor pressure deficit for maize and sorghum. *Field Crop. Res.* **1998**, *56*, 265–270. [CrossRef]

84. Deckmyn, G.; Laureysens, I.; Garcia, J.; Muys, B.; Ceulemans, R. Poplar growth and yield in short rotation coppice: Model simulations using the process model SECRETS. *Biomass Bionergy* **2004**, *26*, 221–227. [CrossRef]

85. Verlinden, M.S.; Broeckx, L.S.; Van den Bulcke, J.; Van Acker, J.; Ceulemans, R. Comparative study of biomass determinants of 12 poplar (*Populus*) genotypes in a high-density short-rotation culture. *For. Ecol. Manag.* **2013**, *307*, 101–111. [CrossRef]

86. Bungart, R.; Hüttl, R.F. Growth dynamics and biomass accumulation of 8-year-old hybrid poplar clones in a short-rotation plantation on a clayey-sandy mining substrate with respect to plant nutrition and water budget. *Eur. J. For. Res.* **2004**, *123*, 105–115. [CrossRef]

87. Scaracia-Mugnozza, G.E.; Ceulemans, R.; Heilman, P.E.; Isebrands, J.G.; Stettler, R.F.; Hinckley, T.M. Production physiology and morphology of *Populus* species and their hybrids grown under short rotation. II. Biomass components and harvest index of hybrid and parental species clones. *Can. J. For. Res.* **1997**, *27*, 285–294. [CrossRef]

88. Dunlap, J.M.; Stettler, R.F. Genetic variation and productivity of *Populus trichocarpa* and its hybrids. X. Trait correlations in young black cottonwood from four river valleys in Washington. *Trees Struct. Funct.* **1998**, *13*, 28–39. [CrossRef]

89. Heilman, P.E.; Stettler, R.F. Genetic variation and productivity of *Populus trichocarpa* and its hybrids. IV. Performance in short-rotation coppice. *Can. J. For. Res.* **1990**, *20*, 1257–1264. [CrossRef]

90. Nedomova, L. *Adriana, Müller 24—Rape Varieties with Excellent Yield from the BOR, Ltd. A Sbornik z Konference Prosperujici Olejniny*; CZU: Prague, Czech Republic, 2012; pp. 120–122.

91. Diepenbrock, W. Yield analysis of winter oilseed rape (*Brassica napus* L.): A review. *Field Crop. Res.* **2000**, *67*, 35–49. [CrossRef]

92. Spiertz, J.H.J.; Ellen, J. Effects of nitrogen on crop development and grain growth of winter in relation to assimilate and utilization of assimilates and nutrients. *Neth. J. Agric. Sci.* **1978**, *26*, 210–231.

93. Massad, R.S.; Tuzet, A.; Bethenod, O. The effect of temperature on C4-type leaf photosynthesis parameters. *Plant Cell Environ.* **2007**, *30*, 1191–1204. [CrossRef] [PubMed]

94. Sage, R.F.; Kubien, D.S. The temperature response of C3 and C4 photosynthesis. *Plant Cell Environ.* **2007**, *30*, 1086–1106. [CrossRef] [PubMed]

forests

MDPI

Article

Tree Willow Root Growth in Sediments Varying in Texture

Ian McIvor [1,*] and Valérie Desrochers [2]

[1]   Sustainable Production Group, The New Zealand Institute for Plant & Food Research Limited,
      Private Bag 11600, Palmerston North 4442, New Zealand
[2]   Institut de recherche en biologie végétale, Montreal, QC H1X 2B2, Canada; valeriedesrochers26@gmail.com
*   Correspondence: ian.mcivor@plantandfood.co.nz; Tel.: +64-6-9537673

Received: 7 May 2019; Accepted: 18 June 2019; Published: 19 June 2019

Abstract: We investigated the early root development of *Salix nigra* L. willow grown from cuttings in the different riverbank sediments; silt, sand and stones. Cuttings were grown for 10 weeks in layered sediment types in five large planter boxes, each box having three separate compartments. The boxes differed in the proportion of silt, sand and stones. At 10 weeks, the roots were extracted and sorted into diameter classes (≥ 2 mm; $1 < 2$ mm; <1 mm) according to sediment type and depth. Root length and dry mass were measured and root length density (RLD) and root mass density (RMD) calculated. Root development of *S. nigra* cuttings varied with the substrate, either silt, sand or stones. Roots initiated from the entire length of the cutting in the substrate but with a concentration of initials located at the bottom and close to the bottom of the cutting. There was substantial root extension into all three substrates and at all depths. Generally, RMD was higher in the stones, influenced by having the bottom of the cuttings in stones for four of the five treatments. RMD was highest for roots <1 mm diameter. RMD of roots <1 mm diameter was least for those roots growing in sand. Whereas RLD for roots >0.5 mm diameter was highest in the sand, RLD of roots with diameter <0.5 mm was lowest in sand. Roots of *S. nigra* cuttings were least effective in binding sand, primarily because of low RLD of roots <0.5 mm diameter. It is surmised that sand lacks water and nutrients sufficient to sustain growth of fine roots compared with silt and even stones. RLD for roots >0.5 mm diameter was lowest in silt likely due to the greater resistance of the substrate to root penetration, or possibly the greater investment into smaller roots with absorption capability.

Keywords: coarse root; fine root; cutting; silt; sand; stone; *Salix nigra* L.

---

1. Introduction

In New Zealand, live willows, primarily male tree willows, are the most important biotechnical tool for frontline protection of riverbanks from water erosion, particularly in periods of high flow. Willows are not native to New Zealand and there is considerable sensitivity to seed spread from planted willows along waterways, hence male willows are preferred. Planting cuttings from known parent trees reduces risk of seed dispersal. Tree willows used for river engineering in New Zealand are established from large cuttings up to 3.5 m long and 70 mm in diameter. Any threat to willow health and effectiveness in binding sediment and stabilizing river banks is also a threat to property, livelihoods and possibly human lives. During the period 1997–2003 willow sawfly (*Nematus oligospilus* Forster) posed such a threat [1–3], with willows along rivers in some parts of the country experiencing severe and repeated defoliations and some mortality from the feeding activity of willow sawfly larvae. Since its arrival in New Zealand in 2014, giant willow aphid (*Tuberolachnus salignus* Gmelin) has posed a significant threat to willow health [4]. The threat of this aphid to biotechnical effectiveness of willows in protecting riverbanks is as yet unclear. This incursion arises at the same time the country is facing possible increased rainstorm frequency and flood risk consequent on climate change [5].

The failure of willows to protect bank sections along the Whakatane River in Bay of Plenty, New Zealand during a severe flood in April 2017 was reported by river engineers, who prompted questions about whether the effectiveness of the willow root system was compromised by the feeding activity of giant willow aphid (GWA). Potentially a reduction in stored carbohydrate in the root system due to feeding activity of GWA could reduce the production of fine roots that bind to sediment and raise biotechnical effectiveness [6–8]. Carbohydrate storage in non-absorptive roots plays an important role in maintaining tree survival after the termination of photosynthate flow from above ground sources [8].

Riverbank sediments get transferred and deposited during floods in layers according to particle size with gravels, stones and boulders being overlaid with sand and with silt dominant in the top layer [9]. In the absence of binding agents, such as tree roots, fine particles are highly mobile in floods, the mean erosion rate increasing with deposit height, deposit width and decreasing grain size [10]. *Salix* spp. are noted as being effective in binding sediment and are not easily washed away with eroded sediment [11].

Published quantitative data on normal root development of tree willows growing along riverbanks is scarce, though other studies have investigated willow root development in different soil types [12–14]. This could be considered of little consequence as willows are seen in practice to be generally effective in binding sediment under flood conditions, with any failings understood as due to the power of the floodwater. However, determining the effect of a pest organism on a tree willow root system cannot be determined without first knowing what a healthy root system is like. Hence, this investigation was designed to first identify how a normal tree willow root system develops.

S. nigra is a colonising floodplain species that produces a massive root system and stabilises streambank sediments [15]. *S. nigra* was used in this experiment as a model tree willow to assess root development from a cutting (the propagating material used for riverbank stabilization in New Zealand) in the different kinds of substrates (stones, sand, silt) found on riverbanks and in varying proportion. We hypothesised that willow root morphology will differ with the sediment type, and contend that knowledge of the differences will contribute to understanding factors that may result in willow failure on riverbanks.

2. Methods

To investigate tree willow root development in river bank sediments a box trial was set up in a research facility at The New Zealand Institute for Plant & Food Research Limited (PFR), Palmerston North. Five plywood boxes (dimensions 1.6 m × 0.8 m × 0.4 m) were constructed, each box having three equal compartments (dimensions 0.5 m × 0.8 m × 0.4 m) separated by plywood bracing. The bracing had a central hole 0.6 m × 0.3 m allowing roots to extend naturally. To observe the roots during the experiment, the plywood on one long side of the boxes was replaced with transparent polycarbonate. Fifteen 0.9 m cuttings taken from two year old shoots of *S. nigra* were cut fresh, pre-soaked in a bucket of water till emerging root primordia were visible, and planted in the boxes in November 2017, three cuttings per box, each cutting in its separate compartment. Cuttings were of similar thickness with top diameter averaging 29.6 mm and basal diameter 46.3 mm. They were then located in an ambient environment. Drainage holes were covered with a fine wire mesh to allow drainage while retaining the substrate. Each box was attributed different proportions of silt, sand and "river run" stones sourced from a local quarry, representative of the river bank variation along a river course (stones phase, mixed silt/sand/stones phase, silt/sand phase). Different substrate proportions were allocated to each box (Table 1). The same proportion of substrate was replicated in each of the three compartments in a single box. Ten percent of volume was marked every 7 cm up the side of the box. A 60 mm diameter polyvinyl chloride pipe was centered into each compartment, extending to the bottom of the compartment, and substrate filled around it to 0.7 m depth, substrate by substrate (Table 1). The 0.9 m *S. nigra* cutting was inserted into the pipe and the pipe removed allowing the substrate to close against the cutting to a depth of 0.7 m without unduly disturbing the substrate layers, leaving 0.2 m of cutting above ground. The media settled as the experiment progressed.

Table 1. Arrangement of the sediment layers in each box.

Layer % (from top)	Depth cm	Box 1	Box 2	Box 3	Box 4	Box 5
0–10	0–7				silt	silt
10–20	7–14			silt		sand
20–30	14–21		silt			
30–40	21–28	silt			sand	
40–50	28–35			sand		
50–60	35–42					stones
60–70	42–49					
70–80	49–56		sand		stones	
80–90	56–63	sand		stones		
90–100	63–70		stones			

The cuttings were provided with natural rainfall, supplemented with water supplied by overhead irrigation twice daily once the foliage emerged. No additional nutrients were added to the mix.

Ten weeks after the cuttings were planted the experiment was terminated. One side of the box was removed and the roots were extracted manually from each compartment in turn. A tarpaulin was employed to collect any substrate that fell from the box and to capture all roots.

Of the substrates, the silt and sand were gently removed from each compartment using a spade, in 7 cm depth intervals (Table 1), to keep the root system as intact as possible and sieved (10 mm mesh) to collect the roots. Where stones were included as a substrate in a compartment (all except those in box 1), they were treated as a single sample, varying in depth (Table 1) because their large size made separation impossible. A total of 100 substrate samples in total were collected from the 5 boxes. The roots in each sample were separated from their substrate and washed, their diameter measured with calipers, then cut and allocated to three diameter classes (≥ 2 mm; $1 < 2$ mm; <1 mm) [16]. Roots in diameter class ≥ 1 mm are subsequently described as coarse roots. For each box, compartment, cutting, substrate type, depth, root length (RL) (measured with a ruler to the nearest cm) and root mass (RM) were recorded. All root diameter classes in each sample were separately oven-dried at 70 °C for 48 h, weighed to 0.01 g and RM recorded. RL (m) of the two root diameter classes ≥ 1 mm was measured. For root diameter class <1 mm, RL of roots with diameter $0.5 < 1$ mm was measured and recorded for a random subset only of the samples from each substrate because of practical constraints. For root diameter class <1 mm, we calculated a relationship between the RL of roots with diameter $0.5 < 1$ mm and RM of the samples for each substrate, assumed this to be a constant relationship for all the samples not measured, and so calculated RL for root diameter class $0.5 < 1$ mm from RM for the remaining unmeasured samples. The mean density of RM and the total RL were calculated for each substrate in the different depth intervals. The root length density (RLD) and the root mass density (RMD) were calculated for the various substrates at each depth interval (7 cm except for stones) and to 0.70 m depth for all the root diameter classes. All shoots emanating from the cutting were counted, regardless of size. Shoot dry mass for each cutting was measured and recorded. The top and bottom diameters of the cuttings were measured at the beginning and the end of the experiment.

Root data were analysed for significant ($p < 0.05$) differences between substrate types and diameter using ANOVA (Genstat 17th Edition; VSNi Limited, Hemel Hempstead, UK, 2014). RLD and RMD data were log transformed to stabilize variance, and a mixed effects model was fitted, with fixed effects for substrate and diameter, and random effects for box and plant within box. Roots 10–20 mm and over 20 mm were combined into one category, as roots over 20 mm were very rare. Raw data are presented in the tables and figures.

3. Results

3.1. Description of the Roots

The live roots observed through the polycarbonate were a mixture of fine and coarse roots. Some roots were white (not lignified) and others golden brown (lignified). New roots were white (Figure 1).

(a)

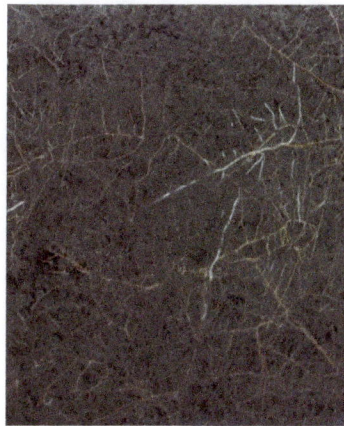

(b)

Figure 1. (a) Contrasting root growth in silt and sand layers. Note the greater number of very fine roots in the silt. (b) Characteristic root development in silt dominated by very fine roots and high production of new roots visible as white roots.

Roots developed from the whole buried length of the cutting, with the greater amount of root development coming from the bottom of the cutting. The finest roots in the stones often adhered to the stones, their growth following the various indentations of the stones where fine sediments were often present. Roots were easiest to recover from the sand, followed by the stones and then silt.

The roots grew in all directions in the substrates. The roots originating in the sand layer tended to grow fine roots upward into the silt. We observed a higher presence of finest roots (roots < 0.5 mm diameter) in the silt than the sand (Figure 1).

3.2. Shoot Production in the Five Boxes

Total shoot dry mass for the cuttings in the five boxes varied (Table 2). Shoot mass of individual trees ranged from 36 g to 128 g, with an overall mean mass of 71.6 g. Except for the trees in box 1, shoot numbers were similar in each box (Table 2). The mean overall was 12.1 shoots per tree for the 15 trees. Differences in shoot number ($p = 0.064$) and shoot dry mass ($p = 0.098$) between plants in the different boxes were not significant. The cuttings in the high silt substrate (box 1) had near-significantly more shoots than the cuttings in the other boxes.

Table 2. Growth in cutting diameter and shoot biomass during the experiment.

Box	Substrate	Cutting Top Diameter (mm)		Cutting Basal Diameter (mm)		Shoots		Cutting Volume (cm³)	Total Root Mass (g)	Coarse Root Length (m)
	Silt:Sand:Stones	Start	End Mean (SE)	Start	End Mean (SE)	Mean No. (SE)	Mean DW (g) (SE)	Mean (SE)	Mean (SE)	Mean (SE)
1	80:20:0	30.7	31.8 (2.5)	47	47.7 (3.1)	16.0 (1.0)	107 (13)	1128 (151)	35 (3)	31 (3)
2	60:30:10	28.3	28.7 (2.9)	44.3	47.2 (4.6)	11.3 (1.5)	68 (22)	1036 (206)	26 (4)	32 (5)
3	30:40:30	30.0	30.7 (2.9)	47	50.2 (4.1)	11.3 (1.9)	73 (13)	1171 (193)	32 (2)	45 (1)
4	20:30:50	29.7	30.3 (1.2)	48.7	49.3 (2.6)	11.7 (0.9)	47 (7)	1127 (106)	24 (3)	11 (3)
5	10:20:70	28.0	30.2 (2.2)	45.3	47.0 (1.2)	10.3 (0.7)	63 (8)	1056 (89)	34 (8)	10 (3)

Differences in total root mass between boxes was not significant ($p = 0.336$). However, differences in coarse RL between boxes was significant ($p < 0.001$; LSD = 10; B3 > B1, B2 > B4, B5).

3.3. Root Distribution within the Five Boxes

Coarse RL was greater in the boxes with the greater proportion of silt (Table 3; F = 20.9 on 4 and 10 df. $p < 0.001$; LSD = 10) and greatest in the box with the highest proportion of sand. The two boxes (4, 5) with a low proportion of silt and a high proportion of stones had the lowest coarse RL. Root mass was greatest in the box with the highest proportion of silt. Root mass was high in the box with the greatest proportion of stones, contributed by a large mass of roots at the bottom of the cuttings. However, with fine roots (<1 mm diameter), the % RM was in similar proportion to the silt present in the box, whereas for sand the finest RM fraction was underrepresented. For example, in box 2, 65% of fine RM was found in silt (60% of the sediment by volume), whereas 18% was found in sand (30% by volume). Fine RM was overrepresented in the stones, contributed by the greater number of roots initiating from the bottom of the cutting.

Table 3. The proportion of root length (RL; roots with diameter ≥1 mm) and root mass (RM) of coarse roots (diameter ≥1 mm) and fine roots (diameter <1 mm) found in the different proportions of sediment in each box.

Box	Sediment	%	% RL ≥ 1 mm Diameter	% RM ≥1 mm Diameter	% RM <1 mm Diameter
1	silt	80	57	58	83
	sand	20	43	42	17
	stones	0	0	0	0
2	silt	60	42	29	59
	sand	30	34	32	28
	stones	10	24	39	13
3	silt	30	16	13	35
	sand	40	44	37	39
	stones	30	39	50	26
4	silt	20	10	1	23
	sand	30	23	37	27
	stones	50	66	62	50
5	silt	10	4	2	8
	sand	20	6	18	8
	stones	70	90	80	84

3.4. Root Density

The mixed effects model indicated significant main effects for substrate and diameter (F = 22.3 on 2 and 185 df. $p < 0.001$, and F = 259.7 on 2 and 185 df. $p < 0.001$, respectively) but no significant interaction with plant or box (F = 0.2 on 4 and 184 df. $p = 0.947$). Mean RLD of coarse roots was significantly lower in silt ($p = 0.03$) than in sand and stones (Figure 2). Coarse root RLD was higher in sand than stones, RLD of fine roots 0.5 < 1 mm diameter was highest in sand (264 m m^{-3}), close to three times higher ($p = 0.04$) than in silt (94 m m^{-3}).

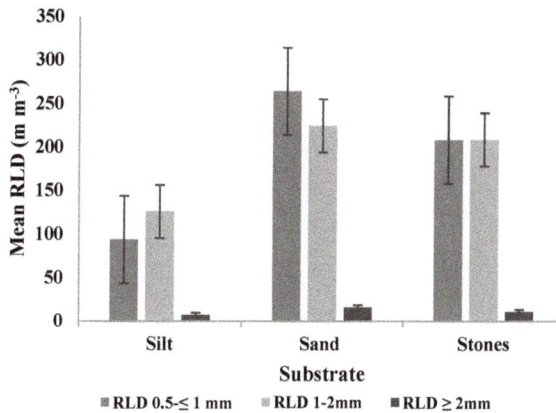

Figure 2. Variation in root length density (RLD) with depth class and substrate. Error bars are 1 SE.

RLD for coarse roots generally increased with depth, whereas for the fine root diameter class (0.5 < 1 mm), RLD was highest in the upper layers in both silt and sand (Figure 3). Differences in RLD with depth were not significant. The effect of depth (but not diameter × depth) was significant, although comparatively small relative to the diameter and substrate effects (F = 3.6, $p < 0.001$).

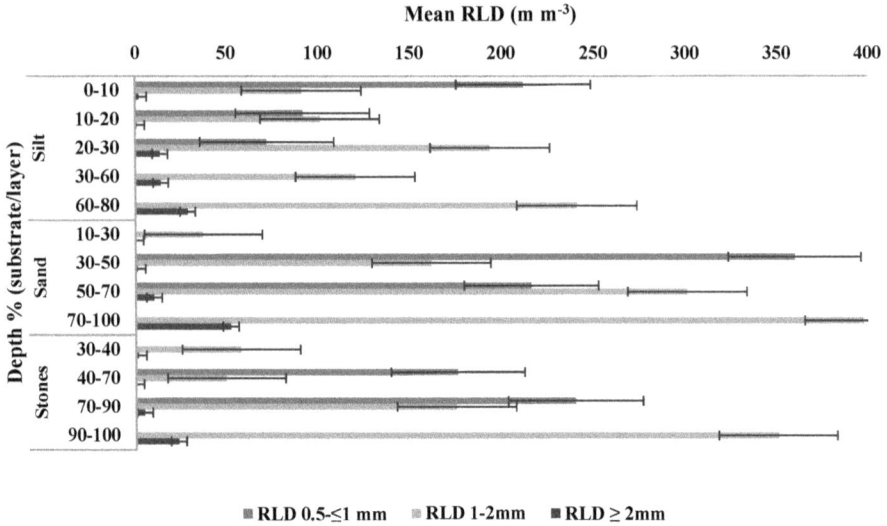

Figure 3. Variation in root length density (RLD) with depth in the three substrates. Error bars are 1 SE. Layers missing a value for the 0.5 < 1 mm root diameter were not sampled.

The mixed effects model indicated significant main effects on RMD for substrate and diameter (F = 36.6 on 2 and 265 df. $p < 0.001$, and F = 506.1 on 2 and 261 df. $p < 0.001$, respectively) but no significant interaction (F = 1.3 on 4 and 261 df. $p = 0.286$). Mean RMD was greater in stones (F = 36.6 on 2 and $p < 0.001$) than in sand or silt. Mean RMD was higher in sand than in silt but not significantly so. Mean RMD was higher for fine roots than for coarse roots in each substrate; and was highest in the stones (254 g m^{-3}) and lowest in the sand (135 g m^{-3}) (Figure 4). Coarse root RMD was highest in stones, intermediate in sand (Figure 4) and significantly lower in silt ($p < 0.001$).

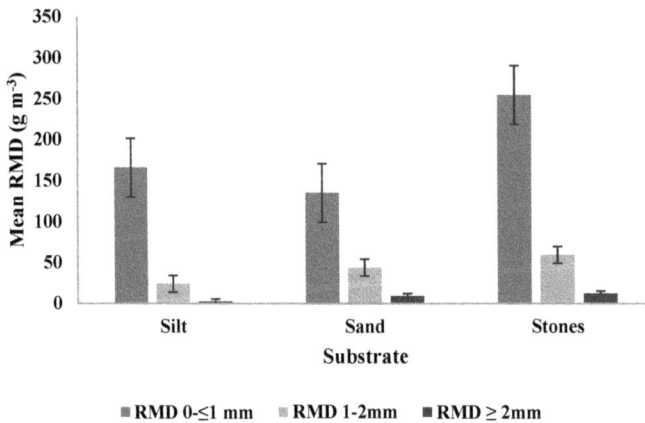

Figure 4. Mean root mass density (RMD) separated by diameter class and substrate. Error bars are 1 SE.

RMD generally increased with depth in each substrate for all root diameter classes (Figure 5), with the highest density in the deepest layer. Fitting the mixed effects model to the log of the RMD, depth (and diameter × depth) were significant, although comparatively small relative to the substrate and diameter effects (F = 6.1, $p < 0.001$ and F = 1.8, $p = 0.019$ for depth and depth × diameter).

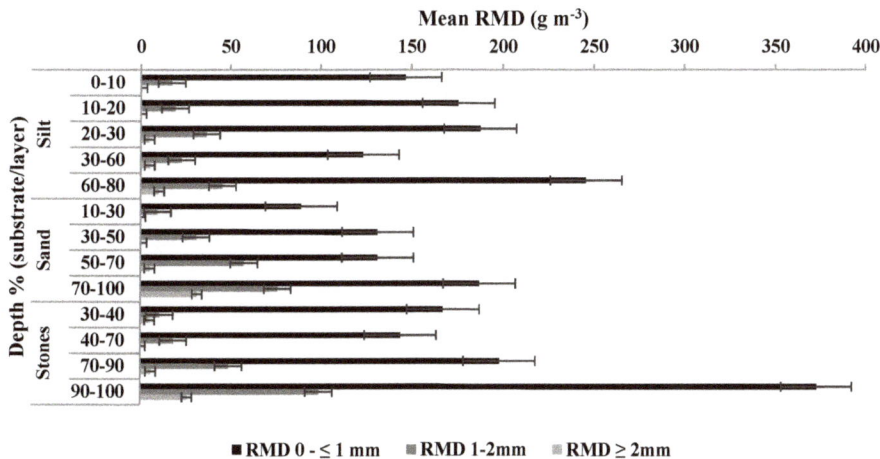

Figure 5. Variation in root mass density (RMD) with depth separated by diameter class and substrate. Depth scale is % (each 10% interval = 7 cm depth from the surface). Error bars are 1 SE.

Mean RMD of roots <1 mm diameter was 246 g m^{-3} in the silt, 187 g m^{-3} in the sand and 374 g m^{-3} in the stones.

3.5. Comparing Root Production in Each Box

The ranking for mean RMD of all root diameter classes was similar in each box; higher RMD of finest roots, followed by the 1 > 2 mm roots, then the ≥2 mm roots. The mean RMD of root classes differed between boxes; the finest (<0.5 mm) root RMD was greatest in box 5 with the highest proportion of stones (261 g m^{-3}), and lowest in box 3 with the highest proportion of sand (139 g m^{-3}), while the 1 < 2 mm diameter were greatest in box 3 (66 g m^{-3}) and lowest in boxes 4 (20 g m^{-3}) and 5 (21 g m^{-3}). However, the RMD of thicker roots was greatest in box 1 (15 g m^{-3}) and lowest in box 4 (2 g m^{-3}).

Total RM was highest in the box with the greatest volume of stones (box 5) followed by the box with the highest silt content (box 1). However, root production varied sufficiently between the trees within each box that the differences in total RM between boxes were not significant.

For the subset of samples for which root length (RL), root mass (RM) and contribution (%) of samples from ≤0.5 mm and between 0.5–1 mm size diameter categories were determined, mean RM contribution of roots 0.5 < 1 mm was 6%, and that of roots ≤0.5 mm was 94%. Mean RL for 0.5–1 mm roots was 1.31 m, and for roots ≤0.5 mm was estimated at 11.8 m following a previous approach [12]. However, these data are more representative of the situation for roots in silt (11 samples) because fewer samples were analysed for sand and stones.

Specific coarse root length (SRL) differed between substrates, being lower ($p < 0.001$) in stones (3.05 m g^{-1}) than in sand (4.94 m g^{-1}) or silt (5.00 m g^{-1}). For fine roots of 0.5 < 1 mm diameter, SRL did not differ between substrates with a range of 9.54–11.52 m g^{-1}. SRL for roots < 0.5 mm diameter was calculated at between 86–104 m g^{-1}.

4. Discussion

Field studies on willow roots in New Zealand focused on larger roots with diameters of ≥2 mm because of the time involved and the challenge of extracting both large structural and fine roots in a single operation [17,18]. It was demonstrated that tree willow with roots growing in recently deposited sandy silt soil grew at a much faster rate than tree willow grown in pastoral soils and that root extension occurred at a much faster rate also [18]. A pot study [12] investigated tree willow fine and coarse root development, but in pastoral soils rather than river sediments. Consequentially, comparisons with these studies will not be made, but there will be comparisons made with aligned studies.

Root length (mass) density, the length (mass) of roots per unit volume of soil, is one of the important parameters required to understand plant performance [19], in this case, ability to bind sediments of various types. In river sediments, unlike other soils, a very high percentage of the roots can be recovered, including the finest roots. RLD for roots >0.5 mm diameter was higher in the sand than the other two substrates. However, the lower RMD for roots <1 mm diameter in sand compared with silt and stones indicates less of the finest roots (<0.5 mm diameter) are present in sand layers. These are the roots that function to absorb water and nutrients. Sand allows easier passage for root extension, however sand does not appear to provide water and nutrients to the same extent as the other two substrates. Roots were able to be separated from the sand relatively easily compared with silt and stones, primarily because of the very low quantities of the finest roots and little adhesion to the sand. As a consequence of this differential root development in sand, we predict that willows will be least effective in binding sand and reducing erosion of sand in floods. RLD for roots >0.5 mm diameter was lowest in silt likely due to the greater resistance of the substrate to root penetration, or possibly the greater investment into smaller roots with absorption capability. Roots were harder to separate from silt than from the other two substrates because of both high fine root densities and silt particles bound to the roots.

A study investigated root reinforcement by fine roots <2 mm diameter for *S. nigra* trees in two creeks in Mississippi, USA, subject to regular flooding and differing particle size distribution (Sardis creek 20% sand, 60% silt and 20% clay; Goodwin creek 61% sand, 26% silt and 13 % clay) [20]. Soil moisture contents for the two soils were 17% (Goodwin creek) and 18% (Sardis). The mean tensile load required to pull *S. nigra* seedlings (mean stem diameter 7.98 mm at Sardis and 7.29 mm at Goodwin creek) was 133.6N at Goodwin and 523.9N at Sardis [20]. Failure at the site with the coarser sediment (Goodwin) was mostly (68%) because the entire root system pulled out of the soil body, whereas at Sardis, 100% of root system failures occurred due to fracture of the larger roots. The finest roots have the capacity to keep soil particles together, and so are effective at soil binding. The nature of substrate, by the predominance of fine particles, will influence the roots–particles association [15]. Likewise, the larger particle sizes of sand did not favour the proliferation of fine roots and did not prevent the removal of particles during flood events [20]. Sand erosion will be best protected when the sand is overlaid or well mixed with silt, which contains a high proportion of those fine roots. Fine roots release exudates that cement particles together, form aggregates, bind the upper layer and protect the sand. The study [20] and this present study strongly suggest that for banks protected by willow trees, the risk of river and stream bank erosion will increase when a higher proportion of sand is present in the soil because of the lower fine root proliferation. While Collins did not excavate roots to discover how much fine root was present in each soil type, our observations confirmed her hypothesis that fine root presence was lower in soils with a high sand content. Fine roots develop the three dimensional nature of root occupancy enabling the root system to better resist wrenching forces arising from different directions. Continuous depth models used to describe tree vertical root distribution for soil based trees [18,21] are not suitable for riverine sediments. The nature of the sediment and moisture availability provide better explanations for riverbank root distribution [15,22]. Schaff and colleagues showed that *S. nigra* pole growth along a riverbank in Mississippi, USA, particularly in the second year of growth was markedly higher (14% increase in biomass) in sandy sediments than in silt/clay sediments, and was lowest where the water table was deepest.

Our findings suggest that proliferation of fine roots is also a characteristic of a stony substrate, however these findings need to be qualified. A large number of roots emerged from the bottom of the cuttings growing in gravel in four of five boxes. This response may be more characteristic of *S. nigra* than other tree willows [12]. However, there was also a large amount of fine root in a substrate offering low resistance to root penetration and a high and constant moisture content. A previous study of root development in *Salix matsudana* × *alba* 'Moutere' tree willow growing along the Hutt river,

New Zealand showed that root development in stones/gravel can be almost nil when the gravel has no water present [23].

We used a mixture of substrate similar to what may occur on a river bank, some having a lot of stones and others composed largely of silt/sand. There were differences in root development between boxes that could be related to different proportions of substrates, such as occur at different reaches of a river. These findings enable some prediction about how the root network will develop as the tree grows, where the greatest pressure on it will come if there is a flood, i.e., which substrate layer is the most mobile and which one is the hardest. The sand had the lowest RMD of fine roots and fell apart so readily when we were extracting roots that we presume the sand will offer low root system anchorage. It will be more easily eroded by forces associated with increased water flow than the other sediments, and will create weak points in the profile amplifying the erosion. Root proliferation was also occurring from the bottom of the cuttings or pole. This strengthens the anchorage of the trees and increases the effectiveness of the shallow roots by reducing mechanical pressure. If the deep roots can absorb the large mechanical forces exerted on the trunk of the tree by high water flows and accompanying debris, then the shallower roots are better positioned to resist the smaller erosive forces. The tree is likely to be better anchored in flood conditions if roots are able to extend into a gravel or boulder layer which will require greater forces to shift than sand or silt. This will be promoted by burying cuttings down to this layer at planting.

While this report has focused on the role of willow roots in stabilising river and stream banks, it can be observed that the root system in sediments is well developed right to the water table and ideally positioned to intercept nutrient runoff from adjoining land before it enters the aquatic system, and that this root system can be developed faster at depth by planting large cuttings to appropriate depths, e.g. water table.

Findings of this study will contribute to understanding possible reasons for tree willow failure during floods, and also the effects of such pests as willow sawfly and giant willow aphid on tree willow small and fine root production.

5. Conclusions

Root development of *S. nigra* grown from cuttings varied with the riverine substrate, either silt, sand or stones. Roots initiated from the entire length of cuttings in the substrate but with a concentration of initials located at and close to the bottom of the cutting. Generally, RMD was higher in the stones. The higher RMD in the stones was influenced by having the bottom of the cuttings located in stones for four of the five treatments. RMD was highest for roots <1 mm diameter. Whereas RLD for roots with >0.5 mm diameter was highest in the sand, RLD of roots with a diameter of <0.5 mm was lowest in sand. RLD for roots >0.5 mm diameter was lowest in silt, likely due to the greater resistance of the substrate to root penetration, or possibly the greater investment into smaller roots with absorption capability in the substrate with higher nutrient and water content. Roots of *S. nigra* were least effective in binding sand, primarily because of low RLD of roots <0.5 mm diameter. It is surmised that sand lacks water and nutrients sufficient to sustain growth of fine roots compared with silt and even stones.

Author Contributions: I.M. was responsible for conceptualization, methodology, validation, resources, formal analysis, writing—review and editing, project administration; V.D. was responsible for methodology, investigation, data curation, formal analysis , writing—original draft preparation.

Funding: This research was funded by Ministry of Business, Innovation & Employment, Project P/442060/10.

Acknowledgments: Technical assistance was provided in setting up and maintaining the experiment and data collection.

Conflicts of Interest: The authors declare no conflict of interest. The funders had no role in the design of the study; in the collection, analyses, or interpretation of data; in the writing of the manuscript, or in the decision to publish the results".

References

1. Charles, J.G.; Allan, D.J. Development of the willow sawfly, Nematus oligospilus, at different temperatures, and an estimation of voltinism throughout New Zealand. *N. Z. J. Zool.* **2000**, *27*, 197–200. [CrossRef]
2. Koch, F.; Smith, D.R. Nematus oligospilus forster (hymenoptera: Tenthredinidae), an introduced willow sawfly in the southern hemisphere. *Proc. Entomol. Soc. Wash.* **2000**, *102*, 292–300.
3. Urban, A.J.; Eardley, C.D. A recently introduced sawfly, nematus oligospilus forster (hymenoptera: Tenthredinidae), that defoliates willows in Southern Africa. *Afr. Entomol.* **1995**, *3*, 23–27.
4. Sopow, S.L.; Jones, T.; McIvor, I.; McLean, J.A.; Pawson, S.M. Potential impacts of tuberolachnus salignus (giant willow aphid) in new zealand and options for control. *Agric. For. Entomol.* **2017**, *19*, 225–234. [CrossRef]
5. Taihoro Nukurangi NIWA. Available online: www.niwa.co.nz/natural-hazards/hazards/climate-change (accessed on 25 March 2019).
6. Ahn, C.; Moser, K.F.; Sparks, R.E.; White, D.C. Developing a dynamic model to predict the recruitment and early survival of black willow (salix nigra) in response to different hydrologic conditions. *Ecol. Model.* **2007**, *204*, 315–325. [CrossRef]
7. Furze, M.E.; Huggett, B.A.; Aubrecht, D.M.; Stolz, C.D.; Carbone, M.S.; Richardson, A.D. Whole-tree nonstructural carbohydrate storage and seasonal dynamics in five temperate species. *New Phytol.* **2019**, *221*, 1466–1477. [CrossRef] [PubMed]
8. Mei, L.; Xiong, Y.M.; Gu, J.C.; Wang, Z.Q.; Guo, D.L. Whole-tree dynamics of non-structural carbohydrate and nitrogen pools across different seasons and in response to girdling in two temperate trees. *Oecologia* **2015**, *177*, 333–344. [CrossRef] [PubMed]
9. Leonard, C.; Legleiter, C.; Overstreet, B. Effects of lateral confinement in natural and leveed reaches of a gravel-bed river: Snake river, wyoming, USA. *Earth Surf. Process. Landf.* **2017**, *42*, 2119–2138. [CrossRef]
10. Gao, C.; Wang, S.J. Evolution of the gravel-bedded anastomosing river within the qihama reach of the first great bend of the yellow river. *J. Geogr. Sci.* **2019**, *29*, 306–320. [CrossRef]
11. Asaeda, T.; Gomes, P.I.A.; Sakamoto, K.; Rashid, M.H. Tree colonization trends on a sediment bar after a major flood. *River Res. Appl.* **2011**, *27*, 976–984. [CrossRef]
12. McIvor, I.R.; Sloan, S.; Pigem, L.R. Genetic and environmental influences on root development in cuttings of selected salix and populus clones—A greenhouse experiment. *Plant Soil* **2014**, *377*, 25–42. [CrossRef]
13. Rabot, E.; Wiesmeier, M.; Schluter, S.; Vogel, H.J. Soil structure as an indicator of soil functions: A review. *Geoderma* **2018**, *314*, 122–137. [CrossRef]
14. Xiao, Q.L.; Huang, M.B. Fine root distributions of shelterbelt trees and their water sources in an oasis of arid northwestern china. *J. Arid. Environ.* **2016**, *130*, 30–39. [CrossRef]
15. Schaff, S.D.; Pezeshki, S.; Shields, F.D. Effects of soil conditions on survival and growth of black willow cuttings. *Environ. Manag.* **2003**, *31*, 748–763. [CrossRef] [PubMed]
16. Phillips, C.J.; Marden, M.; Lambie, S.M. Observations of "coarse" root development in young trees of nine exotic species from a new zealand plot trial. *N. Z. J. For. Sci.* **2015**, *45*. [CrossRef]
17. McIvor, I.R.; Douglas, G.B.; Benavides, R. Coarse root growth of veronese poplar trees varies with position on an erodible slope in New Zealand. *Agrofor. Syst.* **2009**, *76*, 251–264. [CrossRef]
18. Phillips, C.J.; Marden, M.; Suzanne, L.M. Observations of root growth of young poplar and willow planting types. *N. Z. J. For. Sci.* **2014**, *44*. [CrossRef]
19. Moran, C.J.; Pierret, A.; Stevenson, A.W. X-ray absorption and phase contrast imaging to study the interplay between plant roots and soil structure. *Plant Soil* **2000**, *223*, 101–117. [CrossRef]
20. Collins, A. The Role of Willow Root Architecture and Character in Root Reinforcement Potential. Ph.D. Thesis, University of Nottingham, England, UK, 2001.
21. Plante, P.M.; Rivest, D.; Vezina, A.; Vanasse, A. Root distribution of different mature tree species growing on contrasting textured soils in temperate windbreaks. *Plant Soil* **2014**, *380*, 429–439. [CrossRef]

22. Holloway, J.V.; Rillig, M.C.; Gurnell, A.M. Physical environmental controls on riparian root profiles associated with black poplar (populus nigra l.) along the tagliamento river, Italy. *Earth Surf. Process. Landf.* **2017**, *42*, 1262–1273. [CrossRef]
23. Unpublished data.

MDPI

St. Alban-Anlage 66

4052 Basel

Switzerland

Tel. +41 61 683 77 34

Fax +41 61 302 89 18

www.mdpi.com

Forests Editorial Office

E-mail: forests@mdpi.com

www.mdpi.com/journal/forests

www.ingramcontent.com/pod-product-compliance
Lightning Source LLC
Chambersburg PA
CBHW051715210326
41597CB00032B/5493